done
281

Geometry & Symmetry

Geometry & Symmetry

L. Christine Kinsey
Canisius College

Teresa E. Moore
Ithaca College

Efstratios Prassidis
Canisius College

JOHN WILEY & SONS, INC.

VP & Publisher	Laurie Rosatone
Editor:	Jennifer Brady
Marketing Manager:	Sarah Davis
Photo Editor:	Sarah Wilkin
Designer:	RDC Publishing Group Sdn Bhd
Production Manager:	Janis Soo
Senior Production Editor:	Joyce Poh

Cover image © Albright-Knox Art Gallery / Art Resource, NY

This book was set in 10/12 MPS Limited and printed and bound by RR Donnelley. The cover was printed by RR Donnelley.

This book is printed on acid free paper.

Copyright ©2011 John Wiley & Sons, Inc. All rights reserved. No part of this publication may be reproduced, stored in a retrieval system or transmitted in any form or by any means, electronic, mechanical, photocopying, recording, scanning or otherwise, except as permitted under Sections 107 or 108 of the 1976 United States Copyright Act, without either the prior written permission of the Publisher, or authorization through payment of the appropriate per-copy fee to the Copyright Clearance Center, Inc. 222 Rosewood Drive, Danvers, MA 01923, website www.copyright.com. Requests to the Publisher for permission should be addressed to the Permissions Department, John Wiley & Sons, Inc., 111 River Street, Hoboken, NJ 07030-5774, (201)748-6011, fax (201)748-6008, website http://www.wiley.com/go/permissions.

To order books or for customer service please, call 1-800-CALL WILEY (225-5945).

Library of Congress Cataloging-in-Publication Data

Kinsey, L. Christine.
   Geometry and symmetry / L. Christine Kinsey, Teresa E. Moore, and Efstratios Prassidis.
     p. cm.
   Includes bibliographical references and index.
   ISBN 978-0-470-49949-8 (cloth : alk. paper)   1. Mechanics. 2. Geometry. 3. Symmetry (Mathematics)
I. Moore, Teresa E.   II. Prassidis, Stratos, 1962–   III. Title.
   QA807.5.K46 2010
   516—dc22

2010004732

Printed in the United States of America
10 9 8 7 6 5 4 3 2 1

This book is dedicated to
Tappen, William and Chrysoula.

Contents

Preface   xi

I   Euclidean geometry   1

1   A brief history of early geometry   3
- 1.1   Prehellenistic mathematics 3
- 1.2   Greek mathematics before Euclid 5
- 1.3   Euclid 9
- 1.4   *The Elements* 11
- 1.5   Projects 14

2   Book I of Euclid's *The Elements*   15
- 2.1   Preliminaries 15
- 2.2   Propositions I.5–I.26: Triangles 39
- 2.3   Propositions I.27–I.32: Parallel lines 54
- 2.4   Propositions I.33–I.46: Area 59
- 2.5   The Pythagorean Theorem 64
- 2.6   Hilbert's axioms for euclidean geometry 71
- 2.7   Distance and geometry 75
- 2.8   Projects 78

3   More euclidean geometry   80
- 3.1   Circle theorems 80
- 3.2   Similarity 88
- 3.3   More triangle theorems 92
- 3.4   Inversion in a circle 101
- 3.5   Projects 107

4   Constructions   109
- 4.1   Straightedge and compass constructions 109
- 4.2   Trisections 121
- 4.3   Constructions with compass alone 124

	4.4	Theoretical origami	129
	4.5	Knots and star polygons	139
	4.6	Linkages	144
	4.7	Projects	153

II Noneuclidean geometries   155

5 Neutral geometry   157

	5.1	Views on geometry	157
	5.2	Neutral geometry	159
	5.3	Alternate parallel postulates	169
	5.4	Projects	176

6 Hyperbolic geometry   178

	6.1	The history of hyperbolic geometry	178
	6.2	Strange new universe	181
	6.3	Models of the hyperbolic plane	186
	6.4	Consistency of geometries	197
	6.5	Asymptotic parallels	199
	6.6	Biangles	203
	6.7	Divergent parallels	208
	6.8	Triangles in hyperbolic space	210
	6.9	Projects	218

7 Other geometries   220

	7.1	Exploring the geometry of a sphere	220
	7.2	Elliptic geometry	226
	7.3	Comparative geometry	239
	7.4	Area and defect	242
	7.5	Taxicab geometry	254
	7.6	Finite geometries	258
	7.7	Projects	264

III Symmetry   265

8 Isometries   267

	8.1	Transformation geometry	267

	8.2	Rosette groups	289
	8.3	Frieze patterns	294
	8.4	Wallpaper patterns	301
	8.5	Isometries in hyperbolic geometry	314
	8.6	Projects	319

9 Tilings — 320

	9.1	Tilings on the plane	320
	9.2	Tilings by irregular tiles	327
	9.3	Tilings of noneuclidean spaces	341
	9.4	Penrose tilings	345
	9.5	Projects	355

10 Geometry in three dimensions — 357

	10.1	Euclidean geometry in three dimensions	357
	10.2	Polyhedra	369
	10.3	Volume	387
	10.4	Infinite polyhedra	393
	10.5	Isometries in three dimensions	397
	10.6	Symmetries of polyhedra	406
	10.7	Four-dimensional figures	412
	10.8	Projects	417

Appendix A Logic and proofs — 419

	A.1	Mathematical systems	419
	A.2	Logic	420
	A.3	Structuring proofs	424
	A.4	Inventing proofs	427
	A.5	Writing proofs	428
	A.6	Geometric diagrams	429
	A.7	Using geometric software	431
	A.8	Van Hiele levels of geometric thought	431

Appendix B Postulates and theorems — 434

	B.1	Postulates	434
	B.2	Book I of Euclid's *The Elements*	436
	B.3	More euclidean geometry	439

B.4 Constructions 441
B.5 Neutral geometry 442
B.6 Hyperbolic geometry 443
B.7 Other geometries 445
B.8 Isometries 446
B.9 Tilings 448
B.10 Geometry in three dimensions 448

Bibliography 450

Index 455

Preface

Our purpose in writing this book is two-fold. As mathematicians, we know it is important for students to experience and appreciate a rigorous, axiomatic development of a subject. As educators, we realize that many students learn best when they are actively involved in the material. Geometry is an area that lends itself to both rigor and exploration. This text is an introduction to euclidean and noneuclidean geometries. It covers many of the standard results and several topics that are nontraditional. Our hope is that students who use this text will learn the foundations of geometry, rediscover interesting geometric results using the activities, recognize the need to prove all results, and finally write clear and correct proofs.

This text is designed to be used in a junior- and/or senior-level college geometry course. We expect most of the students to be mathematics or math education majors and anticipate that most of them have some experience writing proofs. We have described the historical context for much of the geometry we discuss in the hope that this book will help students appreciate geometry and its importance in the history and development of mathematics. Projects and activities are included to engage the students in discovering some of the geometry for themselves so that they experience the joy of mathematics and see it as a creative endeavor.

In the introductory material, our approach is based on Euclid's axiomatic method, as improved by many generations of succeeding mathematicians. This is not developed in the full rigor that would be appropriate in a more advanced course in geometry. Specifically, our axiom system, though complete, is not meant to be also independent. The reason for this is that the text at hand is intended to introduce students to the basic axiomatic method while keeping the exposition accessible. Some of the more subtle issues are dealt with in the projects following each chapter.

Many of the students who will study from this book are future teachers of mathematics. We have tried to keep their needs in mind when designing the content and sequencing of topics. We believe that these students need a wide variety of experiences, ranging from formal proofs to the activities involving experimentation.

The text is divided into three parts. The first is devoted to euclidean geometry and presents the basic material from *The Elements* while including some more modern material. There is also a chapter on geometric constructions, using the traditional straightedge and compass as well as a variety of other tools.

The second part of the text covers noneuclidean geometry, first extending the results of neutral geometry and then discussing the many attempts to prove the Parallel Postulate. Again, we describe the historical setting for these results. Hyperbolic geometry is developed in some depth. Brief explorations of elliptic, taxicab, and finite geometries are also included. Students are asked to compare the properties of these geometries, discovering the similarities and differences.

The third part of the text is on symmetry. The first chapter discusses the isometries of the euclidean and hyperbolic planes. As applications of these, we discuss rosette groups for finite figures, frieze patterns, and wallpaper patterns. Another chapter is devoted to regular, semiregular, irregular, and nonperiodic tilings of the plane. The last chapter covers three-dimensional geometry. Euclid's geometry of the plane is first extended to three dimensions. Then polyhedra, considered as tilings of the sphere, are investigated, including explorations of infinite polyhedra. These provide examples for the discussion of isometries in three-dimensional space. The symmetries of polyhedra generalize the earlier study of rosette groups. A brief introduction to four-dimensional space and polytopes ends the text.

An appendix gives a brief overview of logic and the art of writing proofs. We do not usually cover this material but assign it as reading for students who are struggling with developing and writing formal proofs. We have tried to keep this material very focused and practical.

Another appendix provides a list of the postulates and theorems for each chapter and is included as a convenience. We include notations indicating which results hold in which geometries.

There are exercises and activities interwoven with the text, rather than collected at the end of each section. We feel that this approach is far more natural. We have tried to encourage students to explore geometry. The instructor should encourage students, even for known results, to experiment, conjecture, and finally prove statements. The activities in the text are opportunities for such experimentation and conjecture. They can be done alone or in small groups, and many are best done using manipulatives or geometric software. Some of the activities and exercises should be done as they occur to make the material that follows more meaningful. We have also included suggestions for projects at the end of each chapter. These may be used for individual presentations and seminar talks. We have tried to include a wide range of topics of varying difficulty.

Designing a course from this text

There is enough material in this text for several different semester-long courses or a year-long course on geometry. The instructor may choose to skip or skim certain sections to concentrate on those sections of particular interest. We have tried, within the natural development of geometry, to allow as much flexibility as possible. We offer some suggestions for possible courses below, but these plans are not meant to be exhaustive.

- A basic semester course on euclidean geometry might include the following:
    Appendix A: Logic and proofs (only if students have very little experience in writing proofs)
    Chapter 1
    Chapter 2
    Chapter 3.1, 3.2, 3.3 (the rest of the chapter is optional)
    Chapter 4.1 (the rest of the chapter is optional)
    Chapter 8.1 (Sections 8.2, 8.3, 8.4 are optional)
- A semester course in noneuclidean geometry for students who have had a strong high school course in euclidean geometry:
    Chapter 2 very briefly as review, emphasizing the choice of axioms
    Chapter 3.1, 3.2, 3.3 briefly, 3.4 if the Poincaré disc model is to be developed in detail
    Chapter 5
    Chapter 6
    Chapter 7
- A semester course on symmetry:
    Chapter 2, briefly
    Chapter 3.1, 3.2, 3.3 briefly
    Chapter 8
    Chapter 9
    Chapter 10

Part I: Euclidean Geometry

   Chapter 1.   This chapter briefly describes the historical context in which Euclid was writing and the achievements of his precursors.

Chapter 2. The second chapter begins with the development of our axiom system, carefully explaining the use and necessity for each axiom and building the foundations for all of the succeeding material. After this structure is in place, we proceed to cover the propositions of Book I of Euclid's *The Elements*, ending with the Pythagorean Theorem. Care is taken to distinguish between the theorems that are true without reference to the parallel postulate and those that depend on this axiom, setting the stage for the later development of noneuclidean geometry.

Chapter 3. This extends the study of euclidean geometry to Euclid's results concerning circles and similar triangles and introduces some more recent results in euclidean geometry, such as concurrence problems and the nine-point circle. The last section of this chapter discusses inversion in a circle. This will be used in the development of the Poincaré disc model for hyperbolic geometry and the Riemann disc model for single elliptic geometry, as well as for the Peaucellier straight-line linkage. However, for those willing to use these models without proof, this section may be omitted.

Chapter 4. The fourth chapter is devoted to the study of geometric constructions, expanding on the traditional straightedge and compass constructions introduced in Chapter 2. We also describe the constructions possible by using a compass alone, by origami, by tying knots in strips of paper, and finally by means of mechanical linkages.

Part II: Noneuclidean Geometries

Chapter 5. In this chapter we extend the study of neutral geometry begun in Chapter 2 and discuss the many attempts to prove the Parallel Postulate. We try to convey the consternation of the mathematical community at the time that noneuclidean geometry was discovered. There follows an extensive study of the development of the parallel postulate and statements that are logically equivalent to it.

Chapter 6. Hyperbolic geometry is developed in depth. Physical models as well as models using geometric software are utilized to aid students in developing some comfort and intuition in this new geometry.

Chapter 7. We begin with an exploration of the geometry of a sphere, then introduce the projective plane, which eliminates the difficulties posed by antipodal points. A modification of Euclid's postulates is developed and the axiomatic approach is used again to explore this geometry. Brief explorations of taxicab and finite geometries are also included. Students are asked to compare the properties of these geometries, discovering the similarities and differences. A section on area in euclidean, hyperbolic, and elliptic geometries allows the students to appreciate the variety of approaches needed.

Part III: Symmetry

Chapter 8. The first chapter of this part defines the isometries of the euclidean and hyperbolic planes, including a proof of the classification of the isometries on the euclidean and hyperbolic planes. As applications of these isometries, we discuss rosette groups for finite figures, frieze patterns, and wallpaper patterns.

Chapter 9. The next chapter is devoted to regular, semiregular, irregular, and nonperiodic tilings of the plane. A discussion of the Penrose tiles is included.

Chapter 10. The final chapter covers three-dimensional geometry. Euclid's geometry of the plane is first extended to three dimensions, paralleling the development of the euclidean geometry in Chapter 2. Then polyhedra, considered as tilings of the sphere, are investigated, including explorations of infinite polyhedra. These provide examples for the discussion of isometries in three-dimensional space. Again, we classify the isometries of euclidean space. Finally, the study of the symmetries of polyhedra extends the earlier discussion of rosette groups. A brief introduction to four-dimensional space and polytopes ends the text.

While all of the chapters contain some opportunities for experimentation, certain sections of the text lend themselves particularly well to extended group exploration. We would suggest that this approach can be used for Sections 4.4, 4.5, 7.1, 7.5, 8.2, 8.3, 8.4, 9.1, 9.2, 10.2, 10.4, 10.6 and 10.7.

A flowchart showing dependencies among the sections follows:

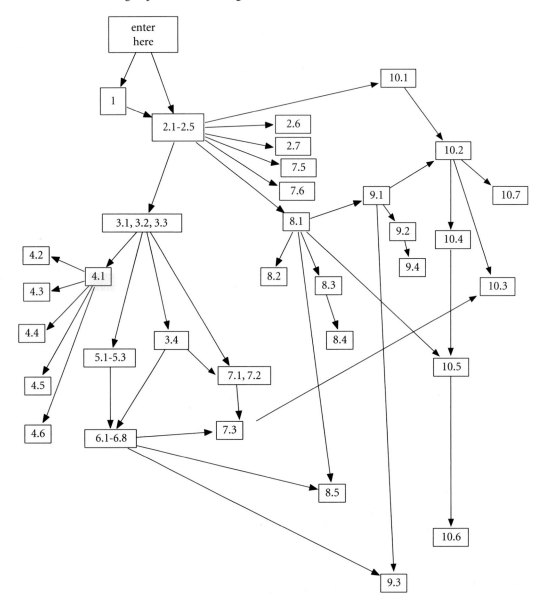

Geometric software

We encourage the use of geometric software, such as *The Geometer's Sketchpad, Cabri, GeoGebra,* or *Cinderella* for general usage and the specialized programs *Non-Euclid* and *Spherical Easel* for the study of noneuclidean geometries. We have found that access to such software is a great help in getting students to experiment and explore mathematics. There are numerous activities that make use of such software. We have also found that using a computer to demonstrate the geometric constructions pertinent to each theorem will give the students an increased familiarity with the software and its capabilities. With a little prompting, this leads them to formulate conjectures of their own, which they can then test. The software also sometimes helps students determine key relationships among geometric objects which can then help them come up with a formal proof of the conjecture. It is particularly useful for showing the students that hyperbolic space is not totally arbitrary and that "curvy lines" have a concrete model. As many high schools have geometric software, it introduces future teachers to a tool that they can use in their classrooms.

Suggestions for the instructor

Some students have always resisted the study of mathematics in general and geometry in particular. Even in 1701, an anonymous writer discussed the reasons for this:

> The aversion of the greatest part of Mankind to serious attention and close arguing; their not comprehending sufficiently the necessity or great usefulness of these in other parts of Learning; An Opinion that this study requires a particular Genius and turn of head, which few are so happy as to be Born with; And the want of . . . able Masters.[1]

It is interesting and a bit discouraging to find oneself in agreement that these are still the primary hurdles to reaching our students. All we can do is offer suggestions on what has worked for us.

Try to get the students actively involved in the development of the material. Many simple theorems and some definitions are left as exercises in the text. Students can discuss their solutions or present them in class. For the more difficult theorems, have students attempt proofs as a class, in smaller groups, or as individuals. The instructor can act as the scribe for the class and let the students make suggestions. Be prepared for frustration, but letting the class work from ideas they have will help the students see that such is the nature of mathematical exploration. Mathematicians know that their initial approach to a problem will not always work. Too many students think that they should be able to solve a problem on the first try.

We support learning through exploration. Students learn material better when they have to figure things out for themselves. The activities in the text are intended to involve the students in mathematics. We often have them work in groups during class or in the computer lab to formulate conjectures that they will prove as homework. Incorporating active learning is the reason that activities and exercises are scattered throughout the text instead of gathered at the end as they were in the books from which we learned. However, the love of student exploration led to unexpected difficulties as we wrote this text.

As anyone who has let students experiment knows, you can't always predict what students will discover. When we ask students to create definitions, we don't always get the ones we expected but the ones we get often work just as well. However, if we want to continue with the class, everyone has to agree to work with one particular definition so we have a class discussion to decide what we agree to use and why.

---

[1] Quoted in Heilbron, *Geometry Civilized*, p. 41.

Unfortunately, we can't have a discussion like that in a textbook. We ask students to explore and conjecture, but we must come to some conclusion if we want to continue in a coherent manner. What this means is that some of the open-ended exploration questions aren't as open as we would like. We have to give an answer in order to proceed.

The same problems occur with many of the activities. Early on, when we ask students to come up with a method for bisecting an angle, we get many different solutions. A wonderful exercise is to have students present their constructions. Then the class as a whole tries to justify the steps – why points of intersection exist, how we know those lengths are equal, why we don't want to use side-side-side yet, etc. However, in the text, we need one construction method that we can prove easily. (Our choice is to follow Euclid.) That doesn't mean we don't value and strongly encourage the students to find other methods. It just means that at some point, we need to agree and move on.

If geometric software is not always available in the classroom, there are many places where you can pull the activities forward and do several at once. When we do Euclid's constructions in the text, we introduce them when we can prove them. If you are in a lab every day, then having the students do the constructions as they arise in the text works easily. If you only have the lab one day a week, then it is perfectly reasonable to have the students do many of the constructions at the same time. Ask them to prove what they can, but explain that the proofs of some may require ideas not yet covered. Similarly, when finding concurrent points for the three medians, altitudes, angle bisectors or perpendicular bisectors of a triangle, one can prove each result after each activity. However, grouping the activities and having the students really explore the triangles gives them more experience making conjectures all at once. They can go back and prove their results when the lab is unavailable or when you want them working by hand instead of on a machine.

As you use these materials, we encourage you to let your students really explore. There are activities that essentially say, "Do this now because reading further will spoil the fun." If you have software available, the beginning of hyperbolic geometry is a great place to turn the students loose and see what they can conjecture or contradict with counter examples. If you only have balls and string or rubber bands, the same type of exploration is possible at the beginning of elliptic geometry. Letting the students figure out when their intuition is right and when it is wrong is a powerful experience for many of them. The results that they find are often not the same and certainly not in the same order that we present them here. In class, we frequently work off what our own students discover, but to make the text usable, we need to include at least one answer or method or proof for the reader. If you are new to the material or uncomfortable letting you students explore, the text follows a path and fills in the gaps. As you become more familiar with the subject, we hope you will see where your students want to go and follow at least a little while in that direction.

Throughout the course, an awareness of the van Hiele levels of geometric thinking (see Appendix A.8) can help both you and your students understand the nature of their difficulties and how they can be overcome.

An instructor's resource manual is available with solutions to the activities and exercises and ideas for classroom demonstrations.

For future teachers

If a class is primarily future teachers, get students involved as educators. Have them give presentations in math classes such as this one rather than only in their education classes. Have them develop demonstrations to enforce a key concept, using technology or manipulatives whenever possible. Ask them to be aware of the learning environment in which they are working and of the different ways they see material presented. Discuss what approaches and techniques they find effective as learners. They will often be surprised that classmates disagree on how they learn best. Solicit their thoughts about making their future students think as part of their students' education. Make it clear that you expect them to use many different approaches to educate their students. Encourage them to involve their students in the experiment-conjecture-prove strategy. But before you have this kind of discussion, make sure you model what you preach.

Try to get pre-service teachers to discuss questions such as: how can they engage students at various grade levels in scientific inquiry? How can they adapt their demonstrations or activities so that primary or secondary students can learn from them? Make sure they understand that justification of an idea can happen at many different levels, from an intuitive explanation to a formal proof. Do not let them push the whole idea of proof off to the experts. Knowledge is evolving. We are part of it. Our students should be part of it. We want our students to teach their students to be part of it, too.

Acknowledgments

Thanks are due to Benita Albert, Collin Bleak, George Boger, Richard Bonacci, Kelly Delp, Caroline DiTullio, Richard Escobales, Steve Ferry, Brad Findell, Sarah Greenwald, Ekaterina Lioutikova, Katie Moran, Joyce Poh, Tricia Profic, Eric Robinson, Brigitte Servatius, Mike Simpson, Alexander Smith, Sarah Wilkin, and all the students who helped test the material and suggest improvements. We would also like to express gratitude to Scott, Preston, and William Moore for their patience and support.

I. Euclidean geometry

1. A brief history of early geometry

♦ 1.1 Prehellenistic mathematics

This brief chapter is not meant as a complete history of geometry or even of Greek geometry prior to Euclid, but contains only enough of the history to put Euclid's *The Elements* in context. As such, we discuss only those civilizations that had a direct impact on the development of Greek mathematics. Thus, we make no mention of the extensive knowledge held by the Chinese, Hindus, and Aztecs, for example.

Pictures from tombs in Egypt show that a rope was the main instrument used by surveyors and engineers. Indeed, the name for this class of professionals translates as "rope pullers" or "rope stretchers." The following group of activities was suggested in Dr. Stephen Luecking's paper, "Introducing Geometry with a Neolithic Tool Kit." You will need a length of string, preferably hemp or cotton since nylon cord tends to stretch; some pushpins; and a nice rock. In doing these activities, pretend you are an Egyptian surveyor, laying out a pyramid. The typical classroom contains entirely too many flat and straight surfaces. A sandbox or beach would be a more realistic setting.

▷ **Activity 1.1.** Take a bit of string about as long as your arm and find its midpoint. Explain your procedure. Tie a knot to mark this midpoint.

▷ **Activity 1.2.** Draw an angle (make this fairly large) and figure out how to bisect it, using only your string with its midpoint knot and some pushpins. Do not use your string as a compass to perform the traditional ruler and compass construction, but come up with another approach. Explain your procedure.

▷ **Activity 1.3.** Draw a straight line (you may use a ruler for this). Using only your string with its midpoint knot and some pushpins, figure out how to erect a perpendicular from a given point on the line. (Do not use your string as a compass to perform the traditional ruler and compass construction, but come up with another approach.) Explain your procedure.

▷ **Activity 1.4.** Draw a straight line (using a ruler). Using only your string with its midpoint knot and some pushpins, figure out how to drop a perpendicular from a given point not on the line. (Do not use your string as a compass to perform the traditional ruler and compass construction, but come up with another approach.) Explain your procedure.

▷ **Activity 1.5.** Now use a marker to make thirteen evenly spaced marks on your string, with the first and last near the ends. Use this to form a 3-4-5 right triangle. Check your angle with a protractor. If you are too far off, then your marks are not evenly spaced, so start over.

▷ **Activity 1.6.** Using your marked string, find at least two ways to construct a line parallel to a given line. Do not use your string as a compass to perform the traditional ruler and compass construction, but come up with another approach. Explain your procedure.

▷ **Activity 1.7.** Now find a rock and tie it to a bit of string. Use this to construct a horizontal line.

With only such simple tools, the Egyptian surveyors were able to carry out the annual survey of the Nile River plots of farmland and build immense structures like the pyramids. From surviving papyri, it was clear that there were precise rules and rigorous checks for the profession. For example, the base of the Great Pyramid at Giza is not quite square, but rather a trapezoid with one side 8 inches shorter than the other. However, the two diagonals are very nearly equal in length, so the requisite check did not find the error.

The Greeks credited the invention of geometry to the Egyptians. Early Egyptian and Babylonian mathematics is fairly well documented as records were kept on papyrus or clay tablets, both of which were sometimes preserved due to the arid climate. These show some evidence of influence from India. The Egyptians and Babylonians were well advanced in arithmetic and simple algebra, including compound interest problems. The annual flooding of the Nile and consequent obliteration of boundary markers and erosion of the river banks show that the Egyptians were accurate and efficient surveyors. Both the Babylonians and the Egyptians were clearly excellent engineers, building the pyramids, large multistory buildings, terraced farms, and extensive irrigation systems. Archaeologists have found tables of multiplication, squares and square roots, cubes and cube roots, and compound interest. Both the Babylonians and the Egyptians knew how to solve the quadratic equation (though only for the positive rational solutions) by completing the square. This and other procedures were evidently taught by rote: a list of steps that lead to the solution, with no explanation.

In geometry, the Egyptians and Babylonians knew of the 3-4-5 right triangle and so at least one special case of the Pythagorean Theorem. The formula $\frac{(AB+CD)(BC+AD)}{4}$ is cited for the area of a quadrilateral. Note that this formula only gives the correct answer for a rectangle.

Here is a problem taken from the Rhind papyrus, a 1550 BC copy of a document written in 1850 BC.[1] This gives a flavor of what their mathematics was like. First, a text description of the problem and its solution are given, then a list of steps.

Find the volume of a cylindrical granary of diameter 9 and height 10: Take away $\frac{1}{9}$ of 9, namely 1; the remainder is 8. Multiply 8 times 8; it makes 64. Multiply 64 times 10; it makes 640 cubed cubits.

1	8
2	16
4	32
\8	64
1	64
\10	640

From this we can see how the mathematics is treated. The procedure is given as a list of steps. We know that the formula for the volume of a cylinder is $V = \pi r^2 h$, or $V = \frac{\pi}{4} d^2 h$, where r is the radius, d the diameter, and h the height. What these Egyptians seem to be doing is starting with the diameter d, then dividing that by 9 and subtracting this from the diameter. Thus the result of the first line is $d - \frac{1}{9}d = \frac{8}{9}d$. They next square this quantity (by doubling it successively to get 8, 16, 32, and finally 64) and then multiply this by $h = 10$. The first four lines of the list of steps show how they approached the problem of squaring a number: begin with eight, two times eight is 16, then two times sixteen (or four times eight) is 32, and finally two times 32 (or eight times eight) is 64. The sixth line indicates the product of the previous computation and the height 10 (the \ marks seems to indicate the result of a computation). Thus the formula implicit in this computation is $V = (\frac{8}{9}d)^2 h = \frac{64}{81}d^2 h$. They are using the approximation $\frac{\pi}{4} \approx \frac{64}{81}$ or $\pi \approx \frac{256}{81} = 3.1604938$, a fairly accurate approximation.

---

[1] From Fauvel and Gray, *The History of Mathematics: A Reader*.

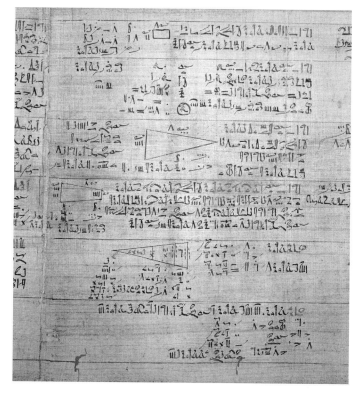

(*Source*: Rhind Mathematical Papyrus, treatise claiming to be a copy of a 12th dynasty work, Thebes, Hyksos period, 15th dynasty, c.1550 BC (papyrus) (see also 116092) by Second Intermediate Period Egyptian (c.1750–c.1650 BC) British Museum, London, UK/The Bridgeman Art Library).

Perhaps one reason for the nature of the mathematics in documents from civilizations such as in Egypt and Babylon is that they seem to have been reasoning by analogy and that their mathematical procedures were based on experimentation and observation. Thus, it appears that they treated mathematical formulae as facts, which would not be subject to discussion or debate.

♦ 1.2 Greek mathematics before Euclid

While Egypt and Babylon had an extensive knowledge of mathematical facts (some correct and some not), it was in Greece that the notion first appeared that formulae and geometric facts should be derived by deductive reasoning rather than by analogy or experiment. It is not clear why this happened, but this was a major change in how people thought. The Greeks have a history of liking to argue and pursuing philosophical inquiries. Public debates were a popular amusement as described, for example, in the plays of Euripides and Aristophanes. For the purposes of these debates and for legal arguments, strict rules of logic were developed by, among others, Socrates, Plato, and Aristotle.

A highly skilled slave class and freemen ran most businesses, farms, and households and provided technical and unskilled labor, while the aristocracy provided the capital. This freed the aristocracy from day-to-day responsibilities. Perhaps this stratification of society led to a corresponding stratification of pursuits: theory versus practice, abstraction and deduction versus experimentation and practical application.

Of early Greek mathematics, very little is known. Few contemporary writings exist, since the writing materials (parchment, usually) and the climate were not as favorable as the desert climates of the Egyptians and Babylonians. Also, Euclid's *The Elements* was so successful that few earlier texts were ever cited afterwards. Of the mathematicians before Euclid, we know a few by name. The first and most influential of these was Thales [ca. 624–546 BC]. He is known to have studied in Egypt. He is credited with being the first to apply the deductive rules of logic to geometry, giving the first mathematical "proofs." He invented words for several major new concepts: cosmos, geometry, and mathematics. Among the theorems which Thales is said to have proved are the following:

> The base angles of an isosceles triangle are equal.
> If two lines intersect, the vertical angles formed are equal.
> Two triangles satisfying the Angle-Side-Angle criterion are congruent.
> An angle inscribed in a semicircle is a right angle.

Most of these facts were already known, some by the Babylonians, but the first evidence of their proofs is attributed to Thales. These proofs were more rudimentary and less sophisticated than those of Euclid but seminal in introducing the idea that geometrical truths must be justified by proof rather than accepted as doctrine or observed by direct experience.

Pythagoras of Samos [ca. 570–500 BC] may have studied under Thales. He traveled to Egypt and may have visited India. He founded a secret society in the Greek colonies in southern Italy which lasted 200 years. This society, really more of a commune, was devoted to mathematics, philosophy, and natural science. All discoveries of the society were attributed to the founder, so it is difficult to figure out who discovered what or even when. The Pythagoreans invested a great deal of mystical significance to numbers and Pythagoras is said to have claimed that "All Things are Number." This gave rise to the study of numerology but also led to solid achievements in mathematics, particularly in the field of number theory. Among their geometric discoveries were the properties of parallel lines, the sum of the angles in a triangle, and, of course, the Pythagorean Theorem.

The most startling discovery of the Pythagoreans was the existence of irrational numbers. Recall that a number is irrational if it cannot be expressed as the quotient of two integers. The proof given below is essentially the same as the one given by Aristotle.[2]

Theorem 1.1. $\sqrt{2}$ is irrational.

Proof: Assume that $\sqrt{2}$ is rational. Then it can be expressed as

$$\sqrt{2} = \frac{p}{q}$$

where p and q are integers, and we additionally assume that the p and q have no common factors, so that the fraction is expressed in lowest terms.

$$2 = \frac{p^2}{q^2},$$
$$2q^2 = p^2.$$

---

[2] Aristotle, *Prior Analectics*, I. 23.

From this, it follows that p^2 must be an even number. By the proof given in Example 2 of Appendix A.3, it follows that p itself must be even. Thus, p can be written in the form $p = 2n$ for some integer n, and so

$$2q^2 = p^2 = (2n)^2 = 4n^2,$$
$$q^2 = 2n^2.$$

Therefore q^2 is also even, and so q must be even. We have thus arrived at the conclusion that both p and q are even, contradicting our assumption that p and q have no common factors. We conclude that $\sqrt{2}$ cannot be rational. □

That $\sqrt{2}$ is irrational was profoundly disturbing to the Pythagoreans, indicating that the precious integers were not adequate to describe all the numbers they could construct. Prior to this discovery, it was thought that a finite line segment is made up of finitely many points. The discovery of the irrational numbers indicated that this could not be true. Two quantities are called *commensurable* if they have a common unit of measure. For example, $\frac{4}{3}$ and $\frac{3}{4}$ can both be expressed in terms of twelfths: $\frac{4}{3} = \frac{16}{12}$ and $\frac{3}{4} = \frac{9}{12}$, so that these numbers are commensurable. Similarly, $\sqrt{2}$ and $\sqrt{18} = 3\sqrt{2}$ are commensurable. The discovery that 1 and $\sqrt{2}$ are incommensurable led to the finding of other irrational numbers.

A vow of secrecy covered this disturbing flaw in the perfection of the logical structure of the universe. It is said that a member of the Pythagorean society divulged the fact to an outsider while traveling by sea, and so a storm blew up out of a clear day and he drowned—clearly an act of divine retribution. (A variant of this legend claims that his colleagues threw him overboard.) The discovery of such nonintuitive facts may have led to the desire to put geometry on a more rational basis. A member of the Pythagorean society during its last years was Hippocrates of Chios [ca. 470–410 BC]. He seems to have been the first to develop a presentation of geometry as a logical chain of propositions based on a few initial definitions.

Plato [428–347 BC] is known to have traveled to Egypt and visited the Pythagoreans in southern Italy before returning to Athens to found his Academy. He was not himself a mathematician but appreciated that geometry was an excellent playground for logical thinking. He is said to have had "Let no one ignorant of geometry enter here" carved over the doorway to the Academy. He wrote that geometry was the finest training for the mind and as such essential for philosophers and statesmen. In describing his idea of a proper education, Plato wrote,

> Geometry will draw the soul towards the truth and create the spirit of philosophy . . . Nothing else will be more likely to have such an effect . . . in all departments of knowledge, as experience proves, any one who has studied geometry is infinitely quicker of apprehension than one who has not.[3]

Plato further describes geometry as "knowledge of what eternally exists."[4] He notes that a perfectly straight line or a perfect circle cannot physically exist. He was thus led to theorize the existence of another universe of ideal forms where our souls lived before birth. The reality we experience daily is a faint echo of these perfect forms, and geometry is the study of these ideal objects that inhabit the human mind. Learning geometry is essentially a process of awakening this inborn knowledge and thus prepares the mind for recovering other knowledge of goodness and justice, the ultimate goal of his philosophy. More than 2000 years later, the role of geometry as an innate product of the human mind formed the cornerstone of Immanuel Kant's philosophy. We will return to the influence of Kant in Chapters 5 and 6.

---

[3] Plato, *The Republic*, VII.
[4] Plato, *The Republic*, VII.

Plato and his vision of perfect geometric forms helped establish mathematics as a purely deductive discipline. He wrote with disdain of experimental science and applied mechanics and firmly designated the (unmarked) straightedge and compass as the proper tools for geometric constructions. All of these ideas can be seen to have influenced the development of Euclid's study of geometry.

The construction of the five regular polyhedra is the culmination of Euclid's *The Elements*, comprising the final chapter, Book XIII. They are known as the Platonic solids for their earliest mention, which occurred in a philosophy text, Plato's *Timaeus*. Timaeus, a fictional Pythagorean, discusses the nature of matter and the matter of nature. Following earlier writers such as Empedocles, Plato described nature as being built of four basic elements: water, earth, air, and fire, in differing combinations and proportions. Earth is represented by the stability of the cube. Fire is associated with the light and pointy tetrahedron. The octahedron represents air and the icosahedron water. Finally, the dodecahedron represents the universe. These associations were later revived by Johannes Kepler [1571–1630].

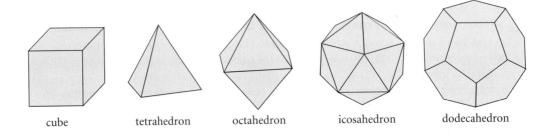

cube   tetrahedron   octahedron   icosahedron   dodecahedron

Plato writes,

> You are aware that students of geometry, arithmetic and the kindred sciences assume the odd and the even and the figures and three kinds of angles and the like in their several branches of science; these are their hypotheses, which they and everybody are supposed to know, and therefore they do not deign to give any account of them either to themselves or others; but they begin with them, and go on until they arrive at last, and in a consistent manner, at their conclusion.[5]

From this, it seems clear that while mathematics was treated as a deductive system, the foundation of definitions and axioms was not yet in place. A mathematician named Theudius wrote the geometry text used at the Academy during this period. This text no longer exists, having been superseded by Euclid.

One of Plato's students who later taught at the Academy was the brilliant mathematician Eudoxus [408–355 BC]. He developed a coherent general theory of proportions (later incorporated as Book V of Euclid's *The Elements*) that successfully dealt with both commensurable and incommensurable quantities by treating these quantities as line segments of appropriate lengths rather than as numbers. Eudoxus also developed the *method of exhaustion* which involves essentially taking successive approximations or limits. This theory was an important precursor to the calculus we use today.

The philosopher Aristotle [384–322 BC] also studied under Plato before founding his own school, the Lyceum. While not a mathematician, he made significant contributions to the foundations of mathematics, providing a systematic treatment of deductive logic and formal recognition of the role of axioms and

---

[5] Plato, *The Republic*, Book VI, sec. 510.

definitions in mathematics. In the works of Aristotle, many examples of logic taken from mathematics are given, and these are probably taken from Theudius's textbook.

Zeno [ca. 490–430 BC] studied with the Pythagoreans for a while. He is best remembered for several paradoxes, raising questions about infinity, which may also be the first examples of proofs by contradiction. These paradoxes, the difficulties they are designed to illustrate, and the attention they received from philosophers and geometers, may explain the wariness that Greek mathematicians used in dealing with infinite objects. Typical of Zeno's paradoxes is the one called the *Dichotomy*, and Aristotle discusses this:

> "Zeno's arguments against motion, which cause so much disquietude to those who try to solve the problems that they present, are four in number. The first asserts the non-existence of motion on the ground that that which is in locomotion must arrive at the halfway stage before it arrives at the goal."[6]

In other words, in order to traverse line segment $\overline{AB}$, Zeno argues that one must first arrive at the halfway point C, but in order to get to C, one must first travel to the point D one quarter of the way from A to B, and in order to get to D, one must first get to the point E one-eighth of the distance, and so on. Thus, in order to traverse the line segment $\overline{AB}$, one must travel through an infinite number of points in a finite amount of time. Therefore, Zeno claimed the motion can never even begin.

All of Zeno's paradoxes attracted a fair amount of attention from mathematicians of the time. The result of these paradoxes was an extremely cautious attitude toward infinity. An immediate consequence of this that we will see reflected in Euclid's *The Elements* is that Euclid assumed that all line segments are finite in extent, though he also assumed that any line segment can be extended to any length required.

Mathematicians can now reply to Zeno's paradox by noting the convergence of the infinite series

$$\sum_{n=1}^{\infty} \frac{1}{2^n} = \frac{1}{2} + \frac{1}{4} + \frac{1}{8} + \frac{1}{16} + \ldots = 1.$$

The rigorous solution to such questions about convergence of infinite sequences and series was thousands of years in the making.

A recently rediscovered text, *The Method* or *The Method of Mechanical Theorems*, by Archimedes [ca. 287–212 BC], shows that in later years (after Euclid's *The Elements* was written), at least some Greek mathematicians handled infinite quantities much as modern mathematicians do. Archimedes shows a clear insight into the importance of the fundamental problems of calculus and applies some of the techniques that were much later refined in the development of that field.

♦ 1.3 Euclid

We know almost nothing of Euclid himself. He flourished (a term used when no one is quite sure when he was born or died, but knew he was alive at one particular time) around 300 BC. He may have studied at the Academy but is known to have lived and taught at the Museum in Alexandria, a Greek colony in Egypt. The only personal information we have consists of two anecdotes. Proclus [410–485], who led the

---

[6]Aristotle, *Physics*, VI, 9.

Academy several centuries later and who wrote a commentary on Euclid's *The Elements*, tells one of these about an exchange between Euclid and King Ptolemy I, though it should be noted that essentially the same story is told of other mathematicians and other kings.

> It is also reported that Ptolemy once asked Euclid if there was not a shorter road to geometry than through the *Elements*, and Euclid replied that there was no royal road to geometry.[7]

Another later commentator, Stobaeus, tells another legend:

> Some one who had begun to read geometry with Euclid, when he had learnt the first theorem, asked Euclid, 'But what shall I get by learning these things?' Euclid called his slave and said 'Give him three obols [coins], since he must needs gain out of what he learns.'[8]

Euclid wrote a number of books, only a few of which have survived, but *The Elements of Geometry* is by far the best known and the most influential. The Greek word for "elements" also means a letter of the alphabet and was commonly used for similar texts, just as there are now a plethora of books entitled *Calculus*. It was written as a textbook, not a research document. It contains a selection of basic theorems of wide and general applicability. Most of these theorems were already known before Euclid's time, but the choice of theorems and their logical arrangement is due to Euclid alone.

The Elements consists of thirteen books or chapters. Roughly, the first of these, with parts of the third and fourth, comprises a standard high school geometry course. The contents are as follows:

I. Plane geometry through the Pythagorean Theorem
II. Geometric algebra, ending with the Law of Cosines
III. Circle geometry: chords and tangents
IV. Construction of the regular polygons of 3, 4, 5, 6, and 15 sides
V. Proportions
VI. Similarity
VII, VIII, and IX. Number theory
X. Irrational numbers
XI. Solid geometry
XII. Volume
XIII. Construction of the regular polyhedra

Euclid's *The Elements* is the most successful (as measured in number of editions printed) and influential mathematics book and has been listed, after the Bible and the Koran, as one of the three most important books of all time. Once considered essential reading for any educated person, it remains a testament to the beauty of mathematics.

The English political philosopher Thomas Hobbes [1588–1679] did not meet the euclidean method until later in life, but then it had a significant impact on his attitude towards mathematics and his approach to philosophy. His biographer describes the encounter:

> He was 40 years old before he looked on geometry; which happened accidentally. Being in a gentleman's library, Euclid's Elements lay open, and it was the 47 El. libri I [i.e., Proposition 47 of Book I of

---

[7] Proclus, *A Commentary on the First Book of Euclid's Elements*.
[8] Stobaeus, *Eclogues*, II.31.

The Elements: the Pythagorean Theorem]. 'By G—, said he, this is impossible!' (He would now and then swear, by way of emphasis.) So he reads the demonstration of it, which referred him back to such a proposition; which proposition he read. That referred him back to another, which he also read. Et sic deinceps [And so back to the beginning], that at last he was demonstratively convinced of that truth. This made him in love with geometry.[9]

The noted mathematician and philosopher Bertrand Russell [1872–1970] describes his first encounter with Euclid:

At the age of eleven, I began Euclid, with my brother as my tutor. This was one of the great events of my life, as dazzling as first love. I had not imagined there was anything so delicious in the world. From that moment until I was thirty-eight, mathematics was my chief interest and my chief source of happiness.[10]

Albert Einstein [1879–1955] had a similar experience, with what he called a "holy geometry booklet":

At the age of twelve I experienced a second wonder of a totally different nature—in a booklet dealing with Euclidean plane geometry, which came into my hands at the beginning of a school year. Here were assertions, as for example the intersection of the three altitudes of a triangle at one point, that—though by no means evident—could nevertheless be proved with such certainty that any doubt appeared to be out of the question. This lucidity and certainty made an indescribable impression on me ... it is marvelous enough that man is capable at all of reaching such a degree of certainty and purity in pure thinking as the Greeks showed us for the first time to be possible in geometry.[11]

◆ 1.4 *The Elements*

Euclid's *The Elements* begins straight off with a list of definitions. There is no prologue or encouraging notes to the the reader or motivational speeches. That the first of these definitions is somewhat impenetrable does not obscure the perfect simplicity and order of the universe we are about to enter.[12]

1. A point is that which has no part.

So what does this mean? Richard Trudeau in *The Non-Euclidean Revolution* gives an argument that this first definition means that points are to be thought of as indivisible: They have no length, width or thickness, unlike the points we draw on the blackboard. Otherwise, if points had physical size and if a line segment were made up of points, then any line segment would consist of a certain number of points. For example, a line segment a inches long might consist of 41,136,978 points, while a line segment b inches long might consist of 32,493,175 points. But then $\frac{a}{b} = \frac{41,136,978}{32,493,175}$, so the ratio of any two lengths would be rational and thus the two line segments are commensurable. Since it was known that incommensurable lengths exist, it must be true that points do not have size. So what Euclid seems to be saying is that

---

[9] John Aubrey, *Brief Lives*.
[10] Bertrand Russell, *The Autobiography of Bertrand Russell*.
[11] Albert Einstein, *Autobiographical Notes*, translated and edited by Paul Schilpp.
[12] All statements attributed to Euclid are from Sir Thomas Heath's 1908 translation of *The Elements*.

a point has no size. That doesn't answer the question of what a point actually is. Nor does it answer the problem of how points, which have no size, can be put together to form a line segment which has positive length. However, in the context of the difficulties raised by the discovery of incommensurable quantities such as $\sqrt{2}$ and the Greeks' attempts to understand the nature of quantity and magnitude, Trudeau's argument makes it clear that this is Euclid's attempt to describe the properties that a point does *not* have.

2. A line is breadthless length.
3. The extremities of a line are points.
4. A straight line is a line which lies evenly with the points on itself.

The second definition is somewhat easier to swallow. Note that what Euclid calls a line might be curved: when he wants a straight line he calls specifically for a straight line. These lines or curves have length but not width, though that begs the question of what length and width mean. The third statement says that the ends of a line segment or curve are points, though there are also curves that do not have endpoints or extremities, such as the circle. The implication of this definition is that all of Euclid's lines and curves are finite in extent. As noted earlier, the Greek mathematicians of this time avoided dealing with infinity because of their inability to resolve the subtleties pointed out in Zeno's paradoxes. What Euclid calls a straight line modern mathematicians call a line segment. The fourth definition makes some sense if one thinks of it as a description: close one eye and hold a pencil up to the open eye so that it is pointed straight away from you. If you sight down the pencil, you should only see a circular cross-section. In this sense, all of the cross-sections are lined up, or "lie evenly with the points on itself." This property is not true for a curve.

In a modern deductive system, undefinable terms, called *primitive terms*, are allowed and are indeed required. This is our practical response to the recognition that one cannot define everything. (As an exercise, look up a random word in a dictionary, then look up the most important descriptive word used in the definition of that word, and so on. You will find that the words you look up will either lead you back in a circle to the original word or give an unending progression of terms needing definition.) While the distinction between primitive undefinable terms and definitions was not recognized in Euclid's time, Euclid does seem to instinctively hold to it. He never cites the more questionable definitions in the propositions that follow. Think of Euclid's early definitions of point and line as descriptions rather than as rigorous definitions, and use them to inform your intuition about the nature of these objects.

Following all of the definitions, there is a list of postulates, followed by some statements called common notions. After these, there are theorems (called propositions) with their proofs. In mathematics, axioms and postulates are things that are to be accepted without proof. As such, both ancient and modern mathematicians insist that the list of such axioms should be kept as short as possible. We should not assume anything we can possibly prove, though exceptions to this principle are occasionally made in the interest of expediency: In an introductory course, authors often adopt as axioms statements that could be proven in a more advanced course. In Euclid's day, it was further stated that axioms ought to be statements that anyone would accept as self-evident: things we all agree are true even without proof. Euclid's five geometric axioms or postulates are as follows.

Euclid's Postulate 1. [It is always possible] to draw a straight line from any point to any point.

Euclid's Postulate 2. [It is always possible] to produce [extend] a finite straight line continuously in a straight line.

Euclid's Postulate 3. [It is always possible] to describe [draw] a circle with any center and distance [radius].

Euclid's Postulate 4. All right angles are equal to one another.

Euclid's Postulate 5. If a straight line falling on two straight lines make the interior angles on the same side less than two right angles, the two straight lines, if produced indefinitely, meet on that side on which are the angles less than the two right angles.

In order to build a rigorous theory of geometry from these, we will need to modify almost all of them. In part this is because we have rejected some of Euclid's definitions and in part this is in response to changes in standards of rigor and in some conventions among mathematicians. In a modern approach, we need to explicate precisely the properties that points and lines, our primitive terms, must have. These properties are implicit in Euclid's definitions, postulates, and usage of the terms, but modern mathematics requires that we spell out the desired properties. In examining precisely how Euclid uses Postulate 1, for example, it is clear that he assumes the line between two points is unique, though he did not feel it necessary to state this. Modern standards demand that this be made explicit. Thus, in the next chapter, we will modify all of Euclid's postulates while trying to retain some historical flavor.

While the Greeks did not distinguish between undefinable and definable terms, they did make another distinction modern mathematicians no longer make. Euclid has two groups of axioms: those he calls postulates as given above and those he calls common notions. This is in agreement with Aristotle's writings: postulates are axioms pertinent to the field of study, while common notions are common to all fields of mathematics. Euclid's postulates all deal with geometric objects—lines, points, and circles—while the common notions deal with common algebraic properties. Euclid's common notions are listed below, with their interpretation in modern algebraic notation.

Common Notion 1. Things that are equal to the same thing are also equal to one another. [If $a = b$ and $c = b$, then $a = c$.]

Common Notion 2. If equals be added to equals, the wholes are equal. [If $a = b$ and $c = d$, then $a + c = b + d$.]

Common Notion 3. If equals be subtracted from equals, the remainders are equal. [If $a = b$ and $c = d$, then $a - c = b - d$.]

Common Notion 4. Things which coincide with one another are equal to one another.

Common Notion 5. The whole is greater than the part. [If $b > 0$, then $a + b > a$.]

However, Euclid does not list all of the common notions that he uses. For example, at some point we will want a statement like $a < b$ and $b < c$ implies $a < c$. This could be proved from the given common notions, but he never does. Also, he uses other simple algebraic properties which cannot be proved from this list, such as $a = b$ implies that $\frac{1}{2}a = \frac{1}{2}b$ (used in Proposition I.37).

Common Notion 4 is a little different. In practice, Euclid uses this, for example, to show that two lines are identical by showing that they have two points in common and then appealing to Postulate 1 and Common Notion 4. He also uses this to show two angles are equal, by showing that their vertex and sides coincide. This usage brings to light another difficulty: Euclid uses the word "equal" (or rather the Greek equivalent) in three senses: when things are identical (or coincide), when they have the same length, angle measure, or area, and when they are what we would call *congruent*.

Some modern axiomatic systems add postulates which associate a line with the real numbers and so guarantee points at every coordinate on the line. One can then use the natural ordering of the real numbers to specify the betweenness relationship among the points of a line. We have chosen not to pursue this approach, because one major issue in Euclid's *The Elements* is constructibility, and assuming all of the real numbers at the start makes this moot.

In the preceding paragraphs, we have indicated some of the problems in reading the text of Euclid's *The Elements* and have also outlined the course we will follow in Chapter 2. In that chapter, we will make the necessary modifications to Euclid's definitions and postulates so that we can recreate the basic foundations of plane geometry with due respect to Euclid's extraordinary accomplishment.

♦ 1.5 Projects

Project 1.1. Read and report on Charles L. Dodgson, "What the Tortoise Said to Achilles."[13] Note that Dodgson also wrote under the pseudonym Lewis Carroll of *Alice's Adventures in Wonderland* fame.

Project 1.2. Research the life of Pythagoras and the history of the Pythagoreans.

Project 1.3. Report on numerology and the Pythagoreans.[14]

Project 1.4. Research and report on the work of Eudoxus.

Project 1.5. The last major classical Greek geometer was Apollonios of Perga [ca. 240–174 BC]. Report on what is known of his work.

Project 1.6. Read *The Clouds* by Aristophanes and comment on his satire on Plato's Academy.

Project 1.7. Build models of the five platonic solids.

Project 1.8. Report on the influence of Euclid's work on modern philosophy.

Project 1.9. Find some other proofs of the irrationality of $\sqrt{2}$ and compare them to the one in this text and to each other.

---

[13] Reprinted in *Mind* 4, pp. 278–280.
[14] A reference is Dudley's *Numerology, or, what Pythagoras wrought*.

2. Book I of Euclid's *The Elements*

◆ 2.1 Preliminaries

We want to build a framework for geometry, a foundation that will allow us to study Euclid's geometry and some modern extensions of the original edifice, as well as alternate geometries such as the geometry of the sphere and less well-known structures. This is much like building an addition to a house. In doing so, we must first carefully examine the foundations of the existing house. This requires excavating the foundation, finding any weak sections and reinforcing them, and adding new piers to support the additional structure we intend to build. A solid understructure, once in place, can be relied on to support the building. Nobody (except perhaps civil engineers) finds foundations exciting, but they are essential and require great attention to detail. Once in place, much of the underpinnings will be covered up and we can proceed to erect the visible and showier parts of the building.

To begin a rigorous study of geometry, we must have objects to study and an axiomatic system to establish the desired properties of these objects. In this text, we have decided to accept the terms "point," "(straight) line," and "plane" as primitive undefinable ideas, rejecting Euclid's attempts to define these objects. These will be our basic building blocks. All properties that we want these objects to possess must be specifically given by the postulates.

We also accept the concepts behind Euclid's original five postulates. However, Euclid has a number of unstated assumptions which we wish to locate and clarify in the form of additional postulates. These were pointed out by the many commentators from the period immediately following the writing of *The Elements* to the definitive axiom set published by David Hilbert [1862–1943] in his *Foundations of Geometry* in 1899. We keep the original numbering of Euclid's postulates and number the required additional postulates consecutively. Appendix B has a list of all of the postulates in order for convenient reference.

◇ 2.1.1 Points and lines

We have accepted "point" and "line" as primitive terms. An additional primitive term "on" describes how a point may be related to a line, so we may say that a point lies *on* a line, or another point might not lie on the line. We can then define another term, "through," by saying if point A lies *on* line ℓ, then ℓ goes *through* A. We will also say sometimes that line ℓ *contains* point A. Since we consider "line," "point," and "on" as undefinable, it is important that we clearly spell out the properties we want these to have in the form of postulates or axioms. Euclid stated in his first postulate that any two points determine a line. If we look at how he uses this postulate, it is clear that he assumed that the line thus specified is unique. We make this explicit in our first postulate:

Postulate 1. Given two distinct points, there exists exactly one line through them.

We use the notation AB for the line through A and B guaranteed by Postulate 1. A point may lie on a line so we can think of a line as a set of points, and instead of saying that point A lies on line ℓ, we can say that A is an element of set ℓ and write $A \in \ell$. This usage of set notation is sometimes convenient. Two lines are defined to *intersect* at a point C if C lies on each of them, i.e., line ℓ intersects line m at C if $C \in \ell \cap m$. Furthermore, we define three or more points to be *collinear* if these points lie on a single line. (Note that by Postulate 1, any two points are collinear.)

▶ **Exercise 2.1.** Using Postulate 1 and the definition of intersection, prove that if two distinct lines intersect, then they have only one point in common. Conclude that if two lines have two points in common, then they must coincide.

One difficulty in reading *The Elements* comes from our rejection of Euclid's first few definitions: do points exist? His definitions of a line and a plane implied that lines were full of points and that the plane was full of lines. It is implicit in most early mathematical writings that anything mentioned had to exist. However, since we have decided to treat lines and planes as undefinable, we must reinstate his assumptions about them. Since we wish to retain the numbering of Euclid's five postulates, our added postulate, which claims the existence of points in various places, will be the sixth. For more on our choice of axioms and the reasons why we made these choices, the interested reader is referred to Section 2.6. (Note that additional clauses considering points in three-dimensional space will be added to this postulate when we come to Chapter 10.)

Postulate 6. [Incidence]

1. There exists at least one plane.
2. Every plane contains at least three noncollinear points.
3. Every line contains at least two points.

Incidence Postulate 6, in combination with Postulate 1, tells us that any two points determine a unique line, and conversely that every line must contain at least two points. The rest of the postulate extends to similar properties of points and planes. The field of finite geometry deals with minimal systems in which Postulates 1 and 6, or even only certain clauses of these postulates, are true. In such geometries, lines are assumed to have only the finite number of points absolutely required to make the postulates true. To explore the consequences of these limitations on our assumptions, see Section 7.6, which can be interpolated at this point if the reader desires.

A further postulate, again implicit in Euclid, concerns the order of points along a line, and will help us not to rely too heavily on diagrams. We introduce a new primitive term, "between": If A, B, and C are collinear points, we talk of a point B lying *between* points A and C. Intuitively, one interprets this as meaning that the three points lie on a single line and occur in that order. If A, B, and C are points on a line ℓ with B lying between A and C, we write $A - B - C$.

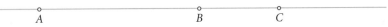

Postulate 7. [Betweenness]

1. If B is between A and C, then A, B, and C lie on a line and B is also between C and A; i.e., if $A - B - C$, then $C - B - A$.
2. Given any two distinct points A and C, there exists at least one other point B on the line that lies between them.

3. Given any two distinct points A and B, there exists at least one other point C so that B lies between A and C.
4. Given three distinct points on a line, only one of them is between the other two.

This postulate gives us additional points on any line, lying between two given points and also on the extension beyond either of the given points. We can use these ideas to define a line segment, ray, triangle, and angle.

Definition 2.1. Given two distinct points A and B on a line, the *line segment* $\overline{AB}$ is the set consisting of the points A and B and all the points lying between them. The points A and B are the *endpoints* of the line segment.

Definition 2.2. Given two distinct points A and B on a line, the *half-line* or *ray* $\overrightarrow{AB}$ is the set consisting of the line segment $\overline{AB}$ and all points C such that B lies between A and C.

▶ **Exercise 2.2.** Let A and B be distinct points. Prove that $AB = \overrightarrow{AB} \cup \overrightarrow{BA}$.

▶ **Exercise 2.3.** Let A and B be distinct points. Prove that $\overline{AB} = \overrightarrow{AB} \cap \overrightarrow{BA}$.

Definition 2.3. Given three noncollinear points A, B, and C, the *triangle* $\triangle ABC$ is the set of points A, B, and C, called the *vertices* of the triangle, and the line segments $\overline{AB}$, $\overline{AC}$, and $\overline{BC}$, called the *edges* or *sides* of the triangle.

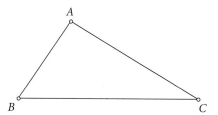

Definition 2.4. Given three noncollinear points, the *angle* $\angle BAC$ is formed by the point A, the *vertex* of the angle, and the rays $\overrightarrow{AB}$ and $\overrightarrow{AC}$, the *sides* of the angle.

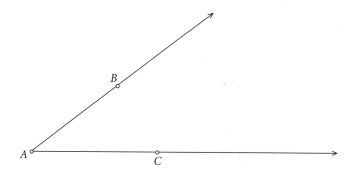

Note that by these definitions, a line segment is a set of points and an angle is just a configuration of rays and points, with no mention of length or angle measure. We will need to postulate exactly what properties length and angle measure must have.

Euclid's Postulate 2 stated that any line segment can be extended. As stated, this can be considered as a consequence of Betweenness Postulate 7(3), but we want more than that. Euclid quite likely did not envision the possibility that one might be able to extend the line segment first by $\frac{1}{2}$ of its length, then by $\frac{1}{4}$, then $\frac{1}{8}$, etc. This would technically satisfy the conditions of his second postulate as well as our betweenness postulate, but would never result in an extension of more than one length. We need to be able to extend lines as far as we like. Thus, we modify his statement:

Postulate 2. One can extend a given line segment from either end to form a ray, or from both ends to form a line.

Postulates 1 and 2 are called the straightedge postulates, since they specify the things that can be done with an unmarked straightedge. We will see later that Postulate 1 and the Incidence and Betweenness Postulates imply that any segment can be extended to form a line or a ray, so Postulate 2 may be considered as a convenient redundancy.

Considering sets of points, we define a property that some sets will have:

Definition 2.5. A set S of points is *convex* if, given any two points A and B in the set S, the line segment $\overline{AB}$ lies inside the set, so $\overline{AB} \subseteq S$.

On the left below is a drawing of a convex polygon, while a nonconvex polygon is shown on the right, with illustrative points A and B:

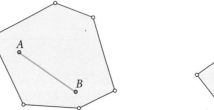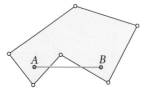

A further postulate will allow us to consider an extension of the concept of betweenness to lines in the plane. Again, this involves formalizing a property that Euclid took for granted and that we make explicit. This is a *continuity axiom*: it asserts that lines are continuous figures that contain no gaps. In other words, it says that one cannot pass from one side of the line to the other without touching the line.

Postulate 8. **[Plane Separation Property]** Given a plane and a line ℓ on that plane, ℓ divides the points on the plane and not on ℓ into two disjoint convex sets, S_1 and S_2, called the *sides* of the line, so that whenever point A is in S_1 and point B is in S_2, the line segment $\overline{AB}$ must intersect ℓ.

Definition 2.6. Given a plane and a line ℓ lying on the plane, let A and B be two points on the plane that do not lie on the line. We define A and B to be on the *same side* of the line if the line segment $\overline{AB}$ does not intersect ℓ. Furthermore, A and B lie on *opposite sides* of ℓ if $\overline{AB}$ intersects ℓ.

Below, A and C lie on the same side of the line. Note that it is important to distinguish between the infinite line AC and the line segment $\overline{AC}$. The line AC would intersect the line ℓ, but the line segment $\overline{AC}$ does not. We can say that C *lies on the A-side of* ℓ. Also, A and B lie on opposite sides of ℓ, and so D lies on the B-side of ℓ.

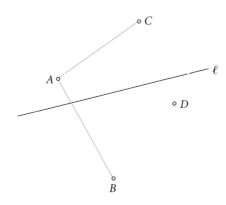

▶ **Exercise 2.4.** Prove that the sides of a given line, S_1 and S_2, are nonempty.

▶ **Exercise 2.5.** Given a line ℓ, we can define a relation on the set of points on the plane and not on line ℓ by $A \sim B$ if A and B lie on the same side of ℓ. Prove that this is an equivalence relation, i.e., show that it is reflexive, symmetric, and transitive.

▶ **Exercise 2.6.** Let ℓ be a line on the plane and A, B, and C distinct points on the plane that do not lie on ℓ. If A lies on the side opposite from B and B lies on the side opposite from C, prove that A and C lie on the same side of ℓ.

The following theorem will be used several times in the succeeding discussion:

Theorem 2.7. Given a line ℓ, let A be a point on line ℓ and B another point that does not lie on ℓ.

1. All of the points on $\overline{AB}$, except A itself, lie on the same side of ℓ as B.
2. All of the points on the ray $\overrightarrow{AB}$, except A itself, lie on the same side of ℓ as B.

▶ **Exercise 2.7.** Prove Theorem 2.7.

▶ **Exercise 2.8.** Given two distinct lines ℓ and m in the plane that do not intersect (so these lines are called parallel), prove that all of the points on ℓ lie on the same side of m.

The concept of plane separation is also precisely what we need to formulate a rigorous definition of what it means for a point to lie inside or outside an angle or a triangle.

▶ **Exercise 2.9.** Use the idea of plane separation to define the interior and exterior of an angle.

We can now extend the idea of betweenness to rays:

Definition 2.8. Ray $\overrightarrow{AD}$ lies *between* rays $\overrightarrow{AB}$ and $\overrightarrow{AC}$ if A, B, and C are not collinear and D lies in the interior of $\angle BAC$.

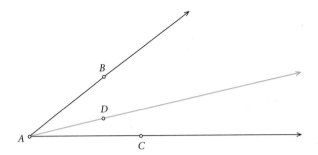

▶ **Exercise 2.10.** Show that if ray $\overrightarrow{AD}$ lies between rays $\overrightarrow{AB}$ and $\overrightarrow{AC}$, then all of the points of $\overrightarrow{AD}$, except A itself, lie in the interior of $\angle BAC$.

▶ **Exercise 2.11.** Use the idea of plane separation to define the interior and exterior of a triangle.

Separation properties are fundamental to our understanding of the relationship between a line and a plane. A similar phenomenon occurs in relation to a point lying on a line, which divides the line into two sections or rays. It seems somewhat counterintuitive not to have proved this first, but in fact the proof makes use of the properties of plane separation.

Definition 2.9. Given a line ℓ and a point B lying on the line, let A and C be two other points on the line. Then A and C are on the *same side* of point B if either $A-C-B$ or $C-A-B$. A and C lie on *opposite sides* of B on ℓ if $A-B-C$.

We need to prove that this definition makes sense and that it implies that a point divides a line into exactly two sides.

Theorem 2.10. [Line Separation] A point P on a line ℓ divides the points on the line distinct from P into two nonempty disjoint sets, called the sides of the point P on line ℓ.

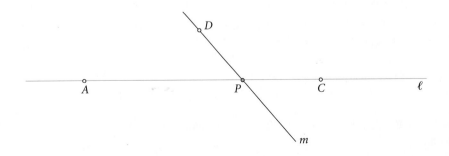

Proof: We are given P on ℓ. By Incidence Postulate 6(2), there is a point D which does not lie on ℓ. Let m be the unique line through P and D. By the Plane Separation Property 8, m divides the plane, and thus ℓ into two disjoint sides. We must show that these sets are nonempty. By Incidence Postulate 6(3), ℓ contains at least two points, P and another point we will denote as A. By Betweenness Postulate 7(3), there is another point C so that $A - P - C$. Thus A and C do not lie on the same side of m, so P separates ℓ into two nonempty disjoint sets. □

▶ **Exercise 2.12.** Let A, B, C, and D be four collinear points. Prove that if $A - B - C$ and $B - C - D$, then $A - B - D$. Furthermore, prove that $A - C - D$.

▶ **Exercise 2.13.** Let A, B, C, and D be four collinear points. Prove that if $A - B - D$ and $B - C - D$, then $A - B - C$. Furthermore, prove that $A - C - D$.

▶ **Exercise 2.14.** Prove that any line contains an infinite number of points.

▶ **Exercise 2.15.** Given three distinct collinear points A, B, and C so that $A - B - C$, prove that $\overline{AC} = \overline{AB} \cup \overline{BC}$ and that $\overline{AB} \cap \overline{BC} = \{B\}$.

In place of the Plane Separation Property 8, we could have used a logically equivalent statement, sometimes called Pasch's postulate, for Moritz Pasch [1843–1930], who originally pointed out the necessity of this statement. Since we have postulated the Plane Separation Property, we prove this as a theorem. Pasch's theorem says that if a line enters a triangle, then it must exit somewhere. The line cannot exit through $\overline{AB}$ since then the line and AB would have two points in common and thus would coincide by Exercise 2.1. The three possibilities are illustrated below, with E as the exit point guaranteed by the theorem.

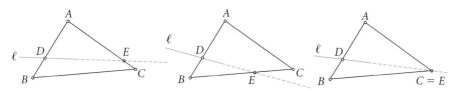

Theorem 2.11. [Pasch's Theorem] If A, B, and C are noncollinear points and a line ℓ intersects $\overline{AB}$ at a single point D between A and B, then ℓ must intersect exactly one of the following: $\overline{AC}$ between A and C, $\overline{BC}$ between B and C, or the vertex C.

Proof: Since $\overline{AB}$ intersects ℓ at D, A and B lie on opposite sides of ℓ.

Case 1: C does not lie on ℓ: If C does not lie on ℓ, then by the Plane Separation Property (Postulate 8), C must lie on one of the two sides of ℓ, either the A-side or the B-side.

Case 1a: C lies on the same side as B: If C lies on the B-side, then C is on the side opposite A, so $\overline{AC}$ must intersect ℓ. Since neither A nor C lie on ℓ, this point of intersection must fall between A and C.
Case 1b: C lies on the same side as A: If C lies on the A-side, then C is on the side opposite B, so $\overline{BC}$ must intersect ℓ. Since neither B nor C lie on ℓ, this point of intersection must fall between B and C.

Case 2: C lies on ℓ: If C lies on ℓ, then we have the last case illustrated above.

Thus the line ℓ intersects exactly one of the given sets. □

Note that there are lines that intersect a triangle exactly once, such as the case shown below. What Pasch's theorem claims is that if a line intersects a side of the triangle, rather than a vertex, then it must have a second point of intersection with the triangle.

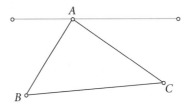

The Plane Separation Property, Pasch's theorem, and the rest of the theorems of this subsection are frequently used to show that lines and line segments intersect where one expects them to. Euclid relied on his drawings to explain when and where things intersected but this reliance on pictures can lead us astray. For example, consider the situation of a triangle $\triangle ABC$ with line AD bisecting $\angle BAC$:

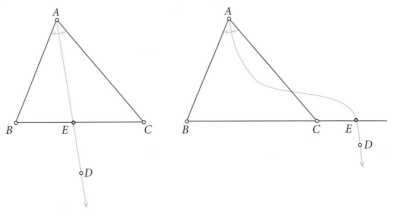

We expect AD to intersect BC as shown on the left, but why could they not intersect differently, for example, as on the right? Remember that we are confined to believing only the things we have spelled out in our postulates and the theorems that follow from these and a mere appeal to common sense has too often led mathematicians (and others) astray. Thus, the axiomatic method demands that we spend time proving some perhaps tedious variations on Pasch's theorem, to cover all possible situations where we will need to justify the existence of points of intersection.

Theorem 2.12. [Cross-bar Theorem] If $\overrightarrow{AD}$ is between $\overrightarrow{AB}$ and $\overrightarrow{AC}$, then $\overrightarrow{AD}$ intersects the segment $\overline{BC}$ at a point between B and C.

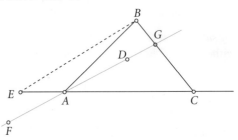

Proof: Using the Betweenness Postulate 7(2) (or Postulate 2), choose a point E on AC so that $E-A-C$, then draw $\overline{EB}$ and consider $\triangle EBC$. By Pasch's Theorem 2.11, AD must intersect $\triangle EBC$ along $\overline{EB}$, along $\overline{BC}$, or at B. We wish to show that AD must intersect the interior of $\overline{BC}$, as shown. If AD were to intersect vertex B, then AD and AB would have two points in common and we would have a contradiction to Postulate 1.

Next we show that $\overrightarrow{AD}$ does not intersect $\overline{BE}$. Note that E was constructed so that E and C lie on opposite sides of AB, and since D and C lie on the same side of AB, it follows that D and E lie on opposite sides of AB. By Theorem 2.7 all of the points of $\overline{BE}$, except for B, lie on the E-side of AB, and all of the points of $\overrightarrow{AD}$, except A itself, lie on the side of AB opposite E. Therefore, $\overrightarrow{AD}$ and $\overline{BE}$ cannot intersect.

Extend AD to F so that $F-A-D$. We now need to show that $\overrightarrow{AF}$ does not intersect $\overline{BE}$ either. Note that F will lie on the side of AC opposite from D and thus on the side of AC opposite from B. By Theorem 2.7 again, all of the points of $\overline{BE}$, except for E, lie on the B side of AC, while all of the points on $\overrightarrow{AF}$, except for A, lie on the opposite side. Therefore, $\overline{BE}$ does not intersect the ray $\overrightarrow{AF}$. Thus, we have proved that $\overline{BE}$ does not intersect the line AD.

Therefore, since line AD must intersect $\triangle EBC$ and cannot intersect at vertex B or along side $\overline{EB}$, AD must intersect the triangle at a point G on the interior of side $\overline{BC}$. It remains to show that G lies on $\overrightarrow{AD}$ rather than on $\overrightarrow{AF}$. Since G is on $\overline{BC}$, by Theorem 2.7, G and B lie on the same side of AC. Since B and D lie on the same side of AC, G and D lie on the same side of AC. Therefore, G and D lie on the same side of A along line AD. Thus, G lies on $\overrightarrow{AD}$. □

The Cross-bar Theorem 2.12 can be used to prove two variations on Pasch's Theorem 2.11. We leave the easier of the two for you to do.

▶ **Exercise 2.16.** Use the Cross-bar Theorem 2.12 to show that if a line contains vertex A of $\triangle ABC$ and an interior point of the triangle, then it must intersect the triangle again along side $\overline{BC}$, strictly between B and C.

Corollary 2.13. If a line segment connects a point outside $\triangle ABC$ to a point inside the triangle, then it must intersect the triangle.

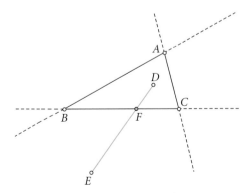

Proof: If E lies outside $\triangle ABC$, then by Exercise 2.11, E must fall into at least one of the following regions: on the side of BC opposite from A, on the side of AC opposite from B, or the side of AB opposite from C. Without loss of generality, we can assume that E lies in the region on the side of BC opposite A. Also, since

D lies inside $\triangle ABC$, we know that D lies on the A-side of BC, on the B-side of AC, and on the C-side of AB. Therefore, D and E lie on opposite sides of BC and so $\overline{DE}$ must intersect line BC at point F. If F lies on $\overline{BC}$ (so either $B-F-C$ or F coincides with either B or C), then we are done.

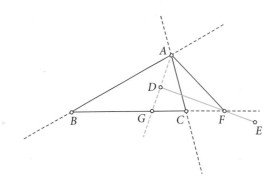

Otherwise, either $B-C-F$ or $F-B-C$ must occur. Without loss of generality, we assume that $B-C-F$. Draw $\overline{AF}$ and consider triangle $\triangle ACF$. If we can show that $\overrightarrow{FD}$ lies between $\overrightarrow{FA}$ and $\overrightarrow{FC}$, then the Cross-bar Theorem 2.12 will give us the desired point of intersection with $\overline{AC}$. Since D lies on the A-side of BF, by Theorem 2.7(2), all of $\overrightarrow{FD}$ except F itself lies on the A-side of CF. Since $\overline{DE}$ intersects AF at F, D and E lie on opposite sides of AF. Now, consider the ray $\overrightarrow{AD}$. Since D lies in the interior of $\triangle ABC$, this ray lies between $\overrightarrow{AB}$ and $\overrightarrow{AC}$. Thus the Cross-bar Theorem 2.12 implies that $\overrightarrow{AD}$ must intersect $\overline{BC}$ at a point G between B and C. By Theorem 2.7, all of $\overrightarrow{AD}$ except A must lie on the same side of AF as D. Thus D and G lie on the same side of AF. Since $G-C-F$, $\overline{GC}$ does not intersect AF, so C and G lie on the same side of AF. Thus C and D lie on the same side of AF. Thus $\overrightarrow{FD}$ lies on the C-side of AF, so $\overrightarrow{FD}$ falls between $\overrightarrow{FA}$ and $\overrightarrow{FC}$. Applying the Cross-bar Theorem 2.12 again to $\angle AFC$ shows that $\overrightarrow{FD}$ must intersect $\overline{AC}$. Since $G-C-F$, F and D lie on opposite sides of the line AC and it follows that $\overline{FD}$ intersects $\overline{AC}$. □

▶ **Exercise 2.17.** Let D lie inside $\triangle ABC$ and consider a ray $\overrightarrow{DE}$ emanating from D. Prove that $\overrightarrow{DE}$ must intersect $\triangle ABC$.

▶ **Exercise 2.18.** Prove that the interior of any triangle is convex.

◇ **2.1.2 Length and congruence**

Thus far, line segments and angles have been treated as physical entities, but we need to develop concepts of length and angle measure. The Greeks used what is called *geometric algebra*; when Euclid seems to be adding two numbers together, he is really joining two line segments of the correct lengths together to create a new longer line segment of length equal to the sum of the lengths of the original two segments. Similarly, when we think of the quantity a^2, we think of multiplying a by itself, but Euclid thinks of a square of side a. The disadvantage to the ancient approach is that if all numbers are represented by physical entities such as lines or rectangles, then it is difficult to deal with zero or negative quantities, which cripples algebraic operations. We would like to adopt a more modern algebraic approach, distinguishing between a line segment and its length, an angle and its measure, and a polygon and its area. The postulates we have so far, Postulates 1 and 2, Incidence, Betweenness, and Plane Separation, describe how points, lines, and line segments behave geometrically.

We need to postulate the algebraic behavior of the related concepts of length, angle measure, and area. Recall the discussion of Euclid's common notions in Chapter 1. Most of these common notions involve simple arithmetic properties and can be applied to the numerical quantities associated to geometric objects. As we mentioned in Chapter 1, Euclid uses Common Notion 4 (Things which coincide with one another are equal to one another.) in several very different settings, depending on the interpretation of the word "equal," which he uses to mean that the objects coincide or to mean that they are equal in length or angle measure or area. We therefore introduce another primitive term, "congruent." This term will be used in two different settings: for line segments and for angles. Intuitively, we think of two line segments or two angles being congruent if they have the same size.

For line segments, if two line segments, $\overline{AB}$ and $\overline{CD}$, are congruent, then we define $\overline{AB}$ and $\overline{CD}$ to *have the same length* and we write $\overline{AB} = \overline{CD}$. Also note that we can define $\overline{BA} = \overline{AB}$, so that the length of the segment does not depend on whether it is measured from A to B or from B to A. Thus, $\overline{AB}$ can indicate either the line segment or its length. The context makes it clear which interpretation is valid. For two congruent angles, $\angle ABC$ and $\angle DEF$, we write $\angle ABC = \angle DEF$ and say that *the two angles have equal angle measures*. Again, $\angle ABC$ can indicate either the angle or its measure, depending on the context. In practice, when speaking informally we will often say that congruent line segments are *equal in length* and congruent angles are *equal in angle measure*, reserving the term congruence for triangles and when we are speaking most formally. Using these concepts of congruence of line segments and angles, we define congruence for triangles:

Definition 2.14. Triangles $\triangle ABC$ and $\triangle DEF$ are *congruent*, written $\triangle ABC \cong \triangle DEF$, if their corresponding parts are equal: $\overline{AB} = \overline{DE}$, $\overline{AC} = \overline{DF}$, $\overline{BC} = \overline{EF}$, $\angle ABC = \angle DEF$, $\angle ACB = \angle DFE$, and $\angle BAC = \angle EDF$.

As for any primitive term, we must specify the properties we wish congruence to have. In the statement of several of the later propositions, Euclid states that the two triangles are equal, meaning that they are equal in area but clearly not congruent, so a last clause is added to this postulate to address the question of the area of congruent triangles.

Postulate 9. [Congruence]

1. Congruence or equality in length is an additive equivalence relation on line segments.
2. Congruence or equality in measure is an additive equivalence relation on angles.
3. Equality in area is an additive equivalence relation on polygons.
4. Congruent triangles have equal areas.

We should clarify what we mean by an *additive equivalence relation*. We define addition for segments first in the case when B lies on a line between A and C. If $A - B - C$, we define $\overline{AC} = \overline{AB} + \overline{BC}$ and say that the length of $\overline{AC}$ is equal to the sum of the lengths of $\overline{AB}$ and $\overline{BC}$. It follows that we can define $\overline{AB} = \overline{AC} - \overline{BC}$ in this situation. Also, if $A - B - C$, we define $\overline{AB} < \overline{AC}$, saying the length of $\overline{AB}$ is less than the length of $\overline{AC}$.

The assumption that congruence of line segments is an additive equivalence relation implies the following properties:

- Reflexivity: If A and B are points in the plane, then $\overline{AB} = \overline{AB}$, i.e., $\overline{AB}$ has the same length as itself.
- Symmetry: If A, B, C, and D are points in the plane and $\overline{AB} = \overline{CD}$, then $\overline{CD} = \overline{AB}$.
- Transitivity: If A, B, C, D, E, and F are points in the plane and $\overline{AB} = \overline{CD}$ and $\overline{CD} = \overline{EF}$, then $\overline{AB} = \overline{EF}$.
- Additivity: If $\overline{A'B'} = \overline{AB}$ and $\overline{C'D'} = \overline{CD}$, then $\overline{A'B'} + \overline{C'D'} = \overline{AB} + \overline{CD}$.

For angles, if $\overrightarrow{AD}$ lies between $\overrightarrow{AB}$ and $\overrightarrow{AC}$, then we define $\angle BAC = \angle BAD + \angle DAC$, saying that the angle measure of $\angle BAC$ equals the sum of the angle measures of $\angle BAD$ and $\angle DAC$. Thus, $\angle BAD = \angle BAC - \angle DAC$. Also, in this case we define $\angle BAC > \angle DAC$. Postulate 9 states that congruence for angles is also an additive equivalence relation.[1]

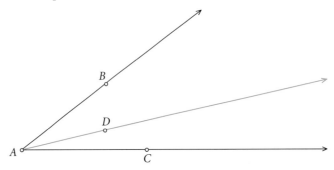

In the next two exercises, do not attempt to prove that midpoints and angle bisectors exist or to give a procedure for constructing these, but merely define what the terms mean in terms of previously defined concepts.

▶ **Exercise 2.19.** Define the *midpoint* or *bisector* of a line segment.

▶ **Exercise 2.20.** Define the *bisector* of an angle.

In practice, we will allow you to cite any simple algebraic fact, whether on Euclid's list or not, and justify it by CN (for Common Notion). Among such basic algebraic properties we are willing to assume, we make an exception for one particularly useful fact, adding a new postulate for this case. This postulate is actually implicit in Euclid's Definition 4 of Book 5 of *The Elements*, which says, "*Magnitudes are said to have a ratio to one another which are capable, when multiplied, of exceeding one another.*" In other words, given two numbers, by adding multiples of the smaller, one can make the sum larger than the other number. Archimedes pointed out that Euclid uses this like an axiom rather than a definition.

Postulate 10. [Archimedes' Axiom] If a and b are positive real numbers with $a < b$, then there exists $n \in \mathbb{N}$ so that $na > b$. In particular,

1. Given line segments $\overline{AB}$ and $\overline{CD}$, there exists $n \in \mathbb{N}$ and a point X on $\overrightarrow{CD}$ so that $\overline{CX} = n \cdot \overline{AB} > \overline{CD}$, and thus $C - D - X$.

---
[1] This can be proved (with some difficulty) from the properties of congruence for line segments: See Project 2.8.

2. Given angles ∠ABC and ∠DEF, there exists $n \in \mathbb{N}$ and a ray $\overrightarrow{EX}$ so that ∠XEF = $n \cdot$ ∠ABC and ∠XEF > ∠DEF.

This simply says that if $\overline{AB}$ and $\overline{CD}$ are two line segments with $\overline{AB} < \overline{CD}$, then we can find a natural number n so that n copies of $\overline{AB}$ will give a line segment longer than $\overline{CD}$. Similarly, given two angles ∠ABC < ∠DEF, there is a natural number n so that n copies of ∠ABC form a new angle ∠ABC' > ∠DEF.[2]

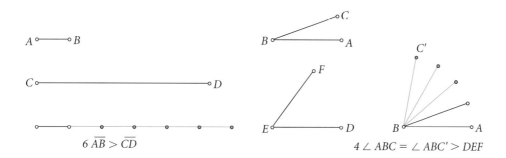

The following variation of Archimedes' Axiom is sometimes useful because in practice it is often easier to double the length of a segment or to bisect it than to create arbitrary multiples.

▶ **Exercise 2.21.** Let a, b, and c be positive real numbers with $a < b < c$.
  a. Prove that there exists $k \in \mathbb{N}$ so that $2^k a > b$.
  b. Prove that there exists $m \in \mathbb{N}$ so that $\frac{c}{2^m} < b$.

In practice, we will frequently need to extend a line segment so that it is longer than a given segment. That follows by applying Postulate 2 in combination with Archimedes' Axiom 10.

◊ 2.1.3 Circles

Now that we have some idea of length, we can consider circles with a given length as radius.

Definition 2.15. Given a point O and another point A, the *circle* centered at O through A is the set of all points X such that $\overline{OA} = \overline{OX}$. Any such segment $\overline{OX}$ is a *radius* of the circle.

Postulate 3 is called the Compass Postulate since it specifies what can be done with a compass. Euclid's original Postulate 3 stated that given two points A and B, one can construct a circle with A as center and passing through B. The Greek compass used to draw circles was what is called a *floppy compass*: essentially two sticks tied together at one end. If either leg was lifted from the paper, the legs would flop together.

---

[2] One can prove the second case of Archimedes' Axiom from the first with some difficulty. We have chosen to include this case in our postulate for convenience but the interested reader is urged to work through Project 2.6.

Euclid shows how to use a floppy compass to transfer a fixed distance from one point to another in his Proposition I.2, as we can easily do with a modern fixable compass. His proof requires the construction of no less than four circles and several lines.

▷ **Activity 2.1.** Make a floppy compass. Some modern compasses can be adapted by loosening a screw, or you can take popsicle sticks and drill small holes and tie them together with a bit of string. Draw a line segment and try to figure out how to transfer its length to another point on your piece of paper using your floppy compass. (Note that this is not easy to do. It's not fair to hold the legs of your compass rigid. Alternatively, figure out how to model such a compass using geometric software and use it to transfer a distance.)

We will modify Euclid's Postulate 3 in two ways: We postulate the convenience of a modern fixable compass and we make explicit his unstated assumption regarding the uniqueness of circles with a given center and radius.

Postulate 3. Given a point and a length, there exists exactly one circle with the given point as center and the given length as radius.

▶ **Exercise 2.22.** Define what it means for a point to lie inside a circle or outside a circle.

▶ **Exercise 2.23.** Define *diameter* and *semicircle*.

Now that we have postulates that give us a (unmarked) straightedge to draw lines and a compass to draw circles, we can begin to investigate Euclid's geometry. All of Euclid's constructions can be done with an actual compass and straightedge, but geometric software such as *The Geometer's Sketchpad*, *Cabri*, and *Cinderella* or the freeware packages *GeoGebra* and *Geometry Playgound* are often more convenient. The dynamic capabilities of these software packages help with exploration and conjecture. Try each construction activity as it arises in the text. It is important that you try to do things for yourself before reading Euclid's solutions. Even if you do not succeed in completing the construction, you will understand and appreciate any difficulties and the proofs will make more sense to you. An *equilateral triangle* is defined to be a triangle with three equal sides.

▷ **Activity 2.2.** Construct a line segment. Then, using only straightedge and compass or the comparable tools in a software package, construct an equilateral triangle whose base is the segment.

It is not enough to merely perform the construction. We must also prove that the construction gives the desired result. We give Euclid's proof of his first theorem or proposition, from Sir Thomas Heath's translation. The proof is written in paragraph style and makes use of the labeling given in the drawing. After some of the statements are citations in brackets of the postulates, definitions and common notions as justification.

Also note that this proof ends with the initials Q.E.F. This is an abbreviation of the Latin phrase "*quod erat faciendum*" which means "that which was to be made." Some of Euclid's propositions give instructions for how to build something, like this equilateral triangle. Others prove some fact about a given triangle or other geometric figure. Those end with Q.E.D., for "*quod erat demonstrandum*" which means "that which was to be demonstrated." The Greeks made many distinctions among types of theorems and their proofs, among which was a distinction between a construction (Proclus, a commentator on Euclid working around 600 AD, calls these "problems") in which a procedure is given to build some specific figure along with a proof that the result was indeed the one desired, and what Proclus calls a "theorem," which demonstrates some property of a figure. The Q.E.D. versus Q.E.F. notation signals this difference. The theorems are numbered as Proposition I.*x*, indicating that they are from Book I of *The Elements*.

Proposition I.1. On a given line segment one can construct an equilateral triangle.

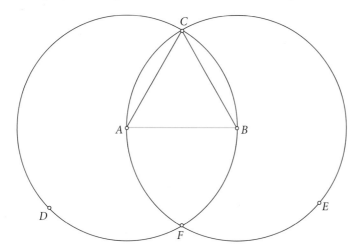

Proof: Let AB be the given finite straight line. It is required to construct an equilateral triangle on the straight line AB. Describe the circle BCD with center A and radius AB. [Post. 3] Again describe the circle ACE with center B and radius BA. Join the straight lines CA and CB from the point C at which the circles cut one another to the points A and B. [Post. 1] Now, since the point A is the center of the circle CDB, therefore AC equals AB. Again, since the point B is the center of the circle CAE, therefore BC equals BA. [Def. 15] But AC was proved equal to AB, therefore each of the straight lines AC and BC equals AB. And things which equal the same thing also equal one another, therefore AC also equals BC. [C.N. 1] Therefore the three straight lines AC, AB, and BC equal one another. Therefore the triangle ABC is equilateral, and it has been constructed on the given finite straight line AB. [Def. 20]   Q.E.F.

For contrast in styles of proofs, the following page is a page from the Byrne *Euclid*.[3] Oliver Byrne [1810–1880] wanted to present Euclid's arguments as much as possible in purely visual terms. Each page was

---

[3] Byrne, *The First Six Books of The Elements of Euclid in which Coloured Diagrams and Symbols are used instead of Letters for the Greater Ease of Learners*, from http://www.sunsite.ubc.ca/DigitalMathArchive/Euclid/byrne.html.

silk-screened in four colors, though it should perhaps be noted that this experiment in pedagogy drove his publisher into bankruptcy. In *Victorian Book Design*, Ruari McLean calls this book "… one of the oddest and most beautiful books of the whole century … a decided complication of Euclid, but a triumph for Charles Whittingham [the printer]." For a full appreciation of this accomplishment, go to the indicated website at the University of British Columbia to see the full-color version.

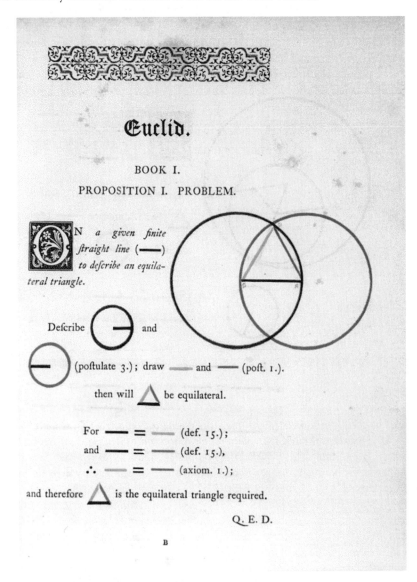

(*Source*: Oliver Byrne, *The First Six Books of the Elements of the Euclid in which Coloured Diagrams and Symbols are used instead of Letters for the Greater Ease of Learners*. Book I, Proposition 1.)

Before analyzing the proof, let us rewrite Euclid's proof in the two-column format that may be familiar from high school geometry. Each claim is written in the left column and the reason justifying this claim is written in the right column.

Proof of Proposition I.1:

1. Let $\overline{AB}$ be the given line segment.
2. Draw a circle centered at A with radius $\overline{AB}$.
3. Draw a circle centered at B with radius $\overline{BA}$.
4. Let C be the intersection of these circles.
5. Draw line segments $\overline{AC}$ and $\overline{BC}$.
6. $\overline{AC} = \overline{AB}$.
7. $\overline{BC} = \overline{AB}$.
8. $\overline{AC} = \overline{BC}$.
9. $\triangle ABC$ is equilateral.

1. Hypothesis
2. Postulate 3
3. Postulate 3
4.
5. Postulate 1
6. Definition "radius"
7. Definition "radius"
8. C.N. 1
9. Definition "equilateral"

□

The two-column approach used to be standard in high school geometry classes, but many students (and teachers) spent all their time trying to fill in the right column with little interest in how to find the actual proof. We will use both approaches in this chapter, gradually moving to mostly paragraph form. However, the two-column approach clearly showed the problem with Euclid's proof of Proposition I.1, which is less blatant in the paragraph-style proof. When we write the proof in this form, it is immediately obvious that there is a problem. Why do the circles intersect? We know nothing about circles except the definition and Postulate 3. Neither of these can be used to show how and when circles intersect.

One very important property of circles that we must make explicit is that circles are continuous objects with no gaps or breaks. This idea is analogous to the valuable Plane Separation Property, Postulate 8. We will need two versions, one for the intersection of lines and circles and another for the intersection of two circles.

Postulate 11. [Circular Continuity Principle]

1. A line segment with one endpoint outside a given circle and the other endpoint inside the circle will intersect the circle exactly once.
2. A circle passing through a point inside a given circle and a point outside that circle will intersect the given circle twice.

Part (1) of the Circular Continuity Principle, sometimes called the Elementary Continuity Principle, guarantees the result illustrated below, in which the line segment $\overline{AB}$ begins at a point A outside the circle and ends at point B inside the circle and so must intersect the circle at C.

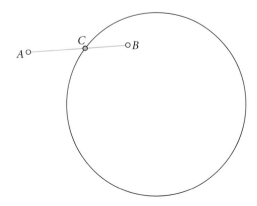

Previous experience with lines and circles would lead us to believe that there are three possibilities: the line and the circle might be disjoint, they could have a single point in common (in which case they are called *tangent*), or they could have two points in common.

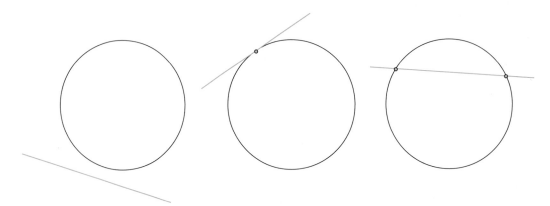

However, our new postulate only guarantees one point of intersection of a line segment and a circle and that only in the situation when one can identify a point inside the circle and another point outside. Therefore, care must be taken to clearly identify the inside point and the outside point when making use of this postulate.

Next, consider the circumstance in which one has a circle and a ray drawn from the center of the circle. It seems obvious that this ray, infinite in extent, cannot wander around inside the circle forever but must eventually come to a point outside the circle and so must intersect the circle by the Circular Continuity Principle. It would at first glance seem that this would follow from Postulate 2, which states that a line segment can be extended indefinitely, but that postulate says nothing about the length of that extension. We need to prove the existence of a point on the ray outside of the circle.

Theorem 2.16. Given a circle with center O, any ray $\overrightarrow{OA}$ emanating from O must intersect the circle.

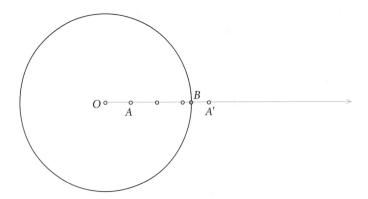

Proof: If A happens to lie on the circle, then A is itself the point of intersection. If A happens to lie outside the given circle, then the Circular Continuity Principle Postulate 11 implies that $\overline{OA}$, and thus $\overrightarrow{OA}$, must intersect the circle. Thus, we assume that $\overline{OA} < r$ where r denotes the radius of the circle. By Archimedes' Axiom Postulate 10, there is a number $n \in \mathbb{N}$ and a point A' on the ray $\overrightarrow{OA}$ so that $\overline{OA'} = n\overline{OA} > r$. Thus, A' lies outside the circle, so $\overline{OA'}$, and thus $\overrightarrow{OA}$, will intersect the circle at some point B. □

Theorem 2.16 is precisely what we need to fill the gap in the proof of Euclid's Proposition I.1. Note that A lies on the circle centered at B. By Betweenness Postulate 7(3) there is a point X so that $A - B - X$. By Theorem 2.16, $\overrightarrow{BX}$ must intersect the circle centered at B at some point Y. Then $A - B - Y$ and $\overline{AB} = \overline{BY}$. Thus, $\overline{AY} > \overline{AB}$, so Y lies outside the circle centered at A with radius $\overline{AB}$.

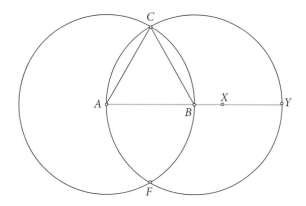

Therefore, the Circular Continuity Principle (2) guarantees that the two circles intersect and completes the proof of Proposition I.1.

▷ **Activity 2.3.** Construct two unequal line segments. Using only straightedge and compass or the software equivalents, cut off from one end of the longer segment a piece equal to the shorter segment.

Proposition I.3. Given two unequal line segments, one can cut off from the longer given segment from either endpoint a line segment equal in length to the shorter given segment.

▶ **Exercise 2.24.** Prove Proposition I.3.

Note that Euclid's Postulate 2 allows us to extend a line segment, but it does not allow us to extend by a given exact amount. Proposition I.3 is frequently used in combination with Postulate 2 to do exactly that: First extend the line longer than you really need, and then cut it off at just the right length.

◇ **2.1.4   Angles**

Now, we turn to the properties and types of angles. Euclid's definition of a right angle states, "When a straight line set up on a straight line makes the adjacent angles equal to one another, each of the equal angles is right."

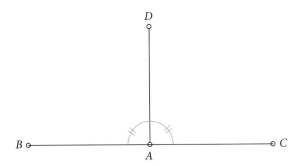

This is a special case of what are now called *supplementary angles*: angles that form a line.

Definition 2.17. Let *B*, *A*, and *C* be collinear points with *A* lying between *B* and *C*, and let *D* be a point that is not collinear with *A*, *B*, and *C*. Then ∡*BAD* and ∡*DAC* are *supplementary angles*.

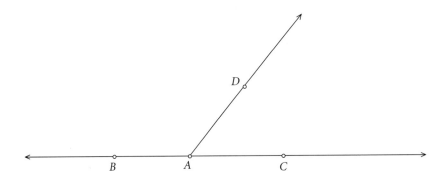

Definition 2.18. ∡*BAD* is a *right angle* if it is congruent to its supplementary angle. We then say that *AD* is *perpendicular* to *BC* and write *AD* ⊥ *BC*.

Euclid's Postulate 4 seems distinctly odd because we are so used to dealing with angles in modern terms of their degree measure. But there is no obvious reason to assume that in moving a right angle from one position to another that it might not warp or droop or change shape somehow. Let us accept that the equality of all right angles cannot be proven and thus that it must indeed be claimed as a postulate.

Postulate 4. All right angles are congruent.

After accepting Postulate 4, we can enjoy the luxury of describing a right angle as having measure 90°. This means that there is a natural standard measure for angles, though there is no equivalent standard for length. We can then proceed to define the measure of any other angle in terms of degrees. In the next exercises, pay attention to developing sound definitions that use precise language applicable in all circumstances.

▶ **Exercise 2.25.** Define *obtuse angle* and *acute angle*.

▶ **Exercise 2.26.** Define what it means for two angles to be *adjacent* and what it means for two angles to be *vertical*.

Consider a triangle △ABC. The angles ∠ABC, ∠BAC, and ∠ACB are the *interior angles*, or simply, the *angles*, of the triangle. We say that ∠BAC is *subtended by* the opposite side $\overline{BC}$ and that side $\overline{BC}$ *subtends* angle ∠BAC. If we extend side $\overline{CA}$ to a point D so that D − A − C, the angle ∠DAB is called an *exterior angle* of the triangle.

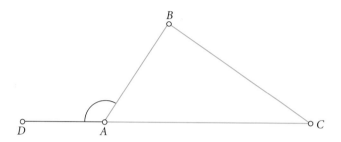

The next of Euclid's propositions is the well-known Side-Angle-Side congruence property, giving the first of a number of conditions by which we can show that two triangles are congruent. Previously, we remarked that when Euclid says, "the triangle will be equal to the triangle," he means that they are equal in area. We added Congruence Postulate 9 (4) to clarify this. We give Euclid's proof from Sir Thomas Heath's translation:

Proposition I.4. If two triangles have the two sides equal to two sides respectively, and have the angles contained by the equal straight lines equal, they will also have the base equal to the base, the triangle will be equal to the triangle, and the remaining angles will be equal to the remaining angles respectively, namely those which the equal sides subtend.

Proof of Proposition I.4: Let ABC, DEF be two triangles having the two sides AB, AC equal to the two sides DE, DF respectively, namely AB to DE and AC to DF, and the angle BAC equal to the angle EDF.

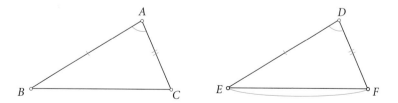

I say that the base BC is also equal to the base EF, the triangle ABC will be equal to the triangle DEF, and the remaining angles will be equal to the remaining angles respectively, namely those which the equal sides subtend, that is, the angle ABC to the angle DEF, and the angle ACB to the angle DFE.

For, if the triangle ABC be applied to the triangle DEF and if the point A be placed on the point D and the straight line AB on DE, then the point B will also coincide with E, because AB is equal to DE.

Again, AB coinciding with DE, the straight line AC will also coincide with DF, because the angle BAC is equal to the angle EDF; hence the point C will also coincide with the point F, because AC is again equal to DF.

But B also coincided with E; hence the base BC will coincide with the base EF. [For if, when B coincides with E and C with F, the base BC does not coincide with the base EF, two straight lines will enclose a space; which is impossible. Therefore, the base BC will coincide with EF] and will be equal to it.

Thus, the whole triangle ABC will coincide with the whole triangle DEF, and will be equal to it. And the remaining angles will also coincide with the remaining angles and will be equal to them, the angle ABC to the angle DEF, and the angle ACB to the angle DFE. Q.E.D.

In the first paragraph, Euclid is setting up and labeling the two triangles. In the second, he is laying out what he wishes to prove. This is always a good idea: to clarify exactly what we intend to do. In the third paragraph, he picks up one triangle and puts it on top of the other. This method is called *superposition*. If point A is placed on top of point D so that $\overline{AB}$ runs along $\overline{DE}$, then point B will land on top of point E and segment $\overline{AC}$ will run along on top of $\overline{DF}$ with point C on top of point F. The remainder of the proof uses Common Notion 4 (*Things which coincide with one another are equal to one another*) repeatedly. The bit in brackets in the next to last paragraph is meant to show that $\overline{BC}$ must coincide with $\overline{EF}$, since otherwise we'd have two different lines connecting point $B = E$ to $C = F$. Euclid notes that if this were true, the two lines would enclose a space. This would contradict his notion of a straight line which "lies evenly with the points on itself." We made this assumption explicit in our version of Postulate 1.

However, to return to the idea of superposition, nothing in the postulates allows such movement of figures. For all we know, when a triangle is picked up and moved, it could shrink, or stretch, or sag, or do something equally weird. If we were to allow such movement, we would need to make up some postulates to clarify exactly when and how we would be allowed to use such an operation. Euclid seems to have viewed superposition with some trepidation. He uses this operation only twice, although there are other places where he could have and where its use would have simplified the proofs. (There is some evidence that the earliest proof of Proposition I.5 by Thales used superposition, but Euclid gives a more complicated proof that avoids its use.) So why did he use it here? Perhaps because he couldn't figure out how to prove this essential theorem without it. Our response is to accept this as a postulate.

Postulate 12. [SAS] If two triangles have two sides equal to two sides respectively, and the angles contained by the equal sides are also equal, then the triangles will be congruent.

Thus, if two triangles satisfy the SAS criterion, then their other corresponding sides and angles are equal. This follows from the Definition 2.14 of congruence for triangles. You may recognize the phrase "corresponding parts of congruent triangles are congruent" (CPCTC) for this.

▶ **Exercise 2.27.** Show that if superposition were allowed, one could prove Postulate 4. Let $AD \perp BC$ and $EH \perp FG$. Pick up the first figure and place on top of the second so that D lies on top of H and B lies on FG (extended if necessary). If A lies on EH, then the right angles $\angle ADB$ and $\angle EHF$ will be equal. Explain why there is a contradiction if A lies inside $\angle EHF$ or $\angle EHG$.

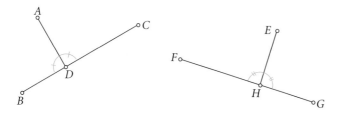

There remains the last of Euclid's postulates, often called the Parallel Postulate, in spite of the fact that it does not mention parallel lines at all. In fact it describes a condition under which lines are *not* parallel.

Definition 2.19. *Parallel lines* are lines that lie in the same plane and do not intersect.

Given two distinct lines, whether parallel or intersecting, a line that crosses both of them is called a *transversal*. Such a transversal forms eight angles with the two lines, shown numbered below:

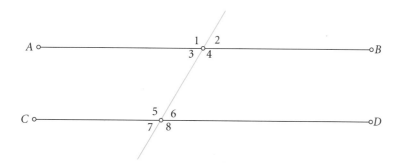

The angles marked #3, #4, #5, and #6 are called *interior angles*. The pair of angles #3 and #6 (or #4 and #5) are called *alternate interior angles*. The pair of angles #3 and #5 (or #4 and #6) are called *adjacent interior angles*.

While Euclid's first four postulates satisfy the criteria that they are simple and acceptable as true without proof, the last of Euclid's postulates stands out:

Euclid's original Postulate 5. If a straight line falling on two straight lines make the interior angles on the same side less than two right angles, the two straight lines, if produced indefinitely, meet on that side on which are the angles less than the two right angles.

Note that this statement is much longer than his other postulates. Once one finally figures out what it says, it doesn't seem that it should be intuitively accepted as true. If we believe it, that is because we are already familiar with geometry. Euclid is describing the situation illustrated below, where two lines are cut by a transversal, and in the case when the two marked angles sum to less than 180° (or, as he says, less than two right angles), he claims that the lines must meet.

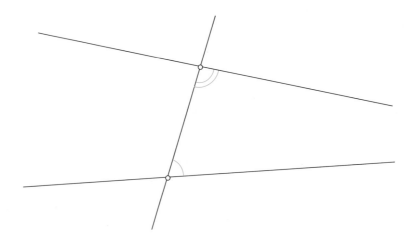

We accept this postulate provisionally, but with some rewriting to take advantage of our comfort with infinite straight lines and degree notation.

Postulate 5. [Parallel Postulate] If two lines are cut by a transversal so that the sum of the interior angles on one side of the transversal is less than 180°, then the two lines meet on that side of the transversal.

In other words, if the interior angles on one side of the transversal sum to less than 180°, then the lines must meet, and so are not parallel. We will treat the Parallel Postulate 5 and its relationship to parallelism in considerable depth later in this chapter and in Chapters 5 and 6.

The remainder of this section is devoted to defining basic terms for use in the succeeding discussion. First we describe different types of triangles. Triangles can be classified in two ways: in terms of their angles or in terms of their sides.

▶ **Exercise 2.28.** Define *right triangle*, *acute triangle*, and *obtuse triangle*.

▶ **Exercise 2.29.** Define *scalene triangle* and *isosceles triangle*.

▶ **Exercise 2.30.** Define *quadrilateral* and *polygon*.

Definition 2.20. A *parallelogram* is a quadrilateral in which each pair of opposite sides are parallel.

Note that it is also true that the opposite sides of a parallelogram are equal in length and that the opposite angles are equal. These facts will be proved later in Proposition I.34. A definition does not need to specify all of the properties of a figure but should spell out exactly what characteristics are required in order to for the figure to be classified properly. As such, the ideal definition should be brief and describe properties which are easy to check. Other properties of the figure will be proved in later theorems. This distinction, between the definition and the set of all properties of a figure, must be kept in mind. Also note that different texts may use different definitions. Many alternate definitions are just different ways of characterizing the same figure but sometimes there are genuine differences. Thus, in using a new term, the first question to ask is, how is the term defined?

Definition 2.21. A *rectangle* is a quadrilateral with four right angles.

Definition 2.22. A *square* is a quadrilateral with four right angles and four equal sides.

Definition 2.23. A *rhombus* is a quadrilateral with four equal sides.

Definition 2.24. A *trapezoid* is a quadrilateral with at least one pair of parallel sides.

▶ **Exercise 2.31.** Figure out the relationships among the terms square, rectangle, parallelogram, rhombus, and trapezoid. For each pair of terms, either state that one implies the other (such as squareness implies rectangularity) or draw a counterexample (as in the following illustration, in which the rectangle is not equilateral, though it is equiangular, and so is not a square).

◆ 2.2 Propositions I.5–I.26: Triangles

Now that we have the foundations in place, we can turn to Euclid and work briskly through the basics of plane geometry. His next two theorems, Propositions I.5 and I.6, are converses of each other. The first says if two sides of a triangle are equal then the angles subtended by those sides are equal, while the second says that if two angles are equal, then the sides subtending those angles are equal. As you will see, this pairing of a proposition and then its converse is common in Euclid. He goes a little further in Proposition I.5 to prove that the angles under the base are also equal. An easier proof would be possible if he did not want this extension, but this result is needed for one of the cases of Proposition I.7.

Proposition I.5. In isosceles triangles, the angles at the base are equal to one another, and, if the equal sides are extended, the angles under the base will be equal to one another.

Proof of Proposition I.5:

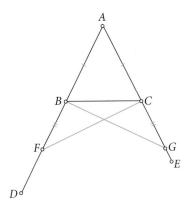

Let $\triangle ABC$ be isosceles, so $\overline{AB} = \overline{AC}$. By Postulate 2 we can extend $\overline{AB}$ to D and $\overline{AC}$ to E. Use Incidence Postulate 6(3) to choose a point F on $\overline{BD}$. Assuming $\overline{AE} > \overline{AF}$, by Proposition I.3 we can cut $\overline{AG} = \overline{AF}$. If $\overline{AE} \leq \overline{AF}$, then use Postulate 2 to extend $\overline{AE}$ first and then cut $\overline{AG} = \overline{AF}$. Postulate 1 allows us to draw $\overline{FC}$ and $\overline{BG}$. Since $\angle A = \angle A$, we have $\triangle FAC \cong \triangle GAB$ by SAS Postulate 12. Thus, $\overline{CF} = \overline{BG}$, $\angle AFC = \angle AGB$, and $\angle ABG = \angle ACF$ by the definition of congruent triangles. Therefore,

$$\overline{BF} = \overline{AF} - \overline{AB} = \overline{AG} - \overline{AC} = \overline{CG}.$$

Thus, $\triangle BFC \cong \triangle CGB$ by SAS, so $\angle FBC = \angle GCB$, and $\angle BCF = \angle CBG$. Therefore,

$$\angle ABC = \angle ABG - \angle CBG = \angle ACF - \angle BCF = \angle ACB.$$

□

▶ **Exercise 2.32.** Explain the logical lapses in the following proof of Proposition I.5: Since $\overline{AB} = \overline{AC}$ and $\angle A = \angle A$, $\triangle ABC \cong \triangle ACB$, so $\angle ABC = \angle ACB$. Thus, $\angle DBC = 180° - \angle ABC = 180° - \angle ACB = \angle ECB$.

The proof of Proposition I.5 is a direct proof, beginning with the hypotheses and proceeding to the desired conclusion. By contrast, the proof of Proposition I.6 will be by contradiction. This provides a useful hint: If the original statement is proved directly, the converse is quite frequently proved by contradiction. We use the two-column style this time to clearly indicate the logic of a proof by contradiction.

Proposition I.6. If a triangle has two equal angles, then the sides which subtend the equal angles will be equal in length.

Proof of Proposition I.6:

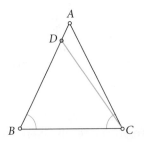

1. $\triangle ABC$ with $\angle ABC = \angle ACB$.	1. Hypothesis
2. Assume $\overline{AB} > \overline{AC}$.	2. Contradiction hypothesis
3. Cut $\overline{DB} = \overline{AC}$.	3. Proposition I.3
4. Draw $\overline{DC}$.	4. Postulate 1
5. $\overline{BC} = \overline{BC}$.	5. C.N.4
6. $\triangle DBC \cong \triangle ACB$.	6. Steps 1, 3, 5, SAS
7. $\angle DCB = \angle ABC$.	7. Definition "congruent"
8. $\angle DCB = \angle ACB$.	8. Steps 1, 7
9. D lies between A and B.	9. Steps 2, 3
10. $\angle DCB < \angle ACB$.	10. Definition "angle measure"
11. $\rightarrow \leftarrow$.	11. Steps 8, 10
12. Therefore, $\overline{AB} \leq \overline{AC}$.	12. Law of contradiction
13. Similarly, if $\overline{AB} < \overline{AC}$, then $\rightarrow \leftarrow$.	13. Repeat Steps 2–11
14. $\overline{AB} \geq \overline{AC}$.	14. Law of contradiction
15. $\overline{AB} = \overline{AC}$.	15. Steps 12, 14

□

Note that the contradiction of Step 11 shows that the contradiction hypothesis (Step 2) must be false, so $\overline{AB} \leq \overline{AC}$. A similar argument (which Euclid omits) shows that $\overline{AC} > \overline{AB}$ cannot be true, therefore, $\overline{AB} = \overline{AC}$.

Before stating Proposition I.7, we encourage you to do the following experiment. This proposition deals with the triangles you can make using different lengths.

▷ **Activity 2.4.** Construct an arbitrary triangle $\triangle ABC$ using three drinking straws cut to different lengths. (Short sections of pipe cleaners slipped into the straws work well as hinges.) We will call the base of this triangle $\overline{BC}$ and the apex A. Now cut some more straws so that you have several of length $\overline{AB}$ (shown as dashed lines below) and several of length $\overline{AC}$ (shown as dotted lines). Construct as many other triangles as you can with base $\overline{BC}$ and using one straw of each of these lengths. What can you say about how these triangles relate to one another?

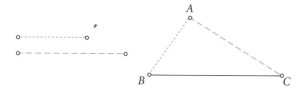

This is a situation where exploration will generally lead to the right conjecture but does not illuminate the proof of that conjecture. Euclid's Proposition I.7 says that a particular configuration cannot occur. To prove this, we again use contradiction and assume that it can.

Proposition I.7. Given $\triangle ABC$ with apex C, we cannot construct another $\triangle ABD$ with D lying on the same side of AB as C and so that $\overline{AD} = \overline{AC}$ and $\overline{BD} = \overline{BC}$.

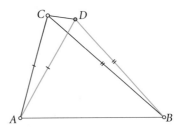

Proof of Proposition I.7: We are given △ABC and must show that no other triangle can be formed with the same lengths on the same side of AB. We will prove this by contradiction, so let us assume that there is a point $D \neq C$ on the same side of AB as C forming △ABD with $\overline{AD} = \overline{AC}$ and $\overline{BD} = \overline{BC}$. We also assume that D lies on the side of BC opposite from A but on the same side of AC as B as pictured above. In this configuration, D is exterior to △ABC and C is exterior to △ABD. Now draw $\overline{DC}$. Since △ACD is isosceles, $\angle ACD = \angle ADC$ by Proposition I.5. If A and D lie on opposite sides of BC, $\angle ACD > \angle BCD$ and thus, by substitution, $\angle ADC > \angle BCD$. Similarly, △BCD is also isosceles and so $\angle BCD = \angle BDC$. Since B and C lie on opposite sides of AD, $\angle BDC > \angle ADC$. Thus we have

$$\angle ADC > \angle BCD = \angle BDC > \angle ADC,$$

which gives us the desired contradiction. Therefore, no such point as D can exist. □

▶ **Exercise 2.33.** Prove Proposition I.7 for the case where *D* lies in the interior of △*ABC*.

Proposition I.7 implies that triangles are *rigid*, which is just another way of saying that they cannot be deformed while retaining the lengths of the edges and the side of the line on which the triangle is constructed. Consider a triangle built from hinged rods. After connecting the sides to the top vertex, the shape of the figure is fixed.

▷ **Activity 2.5.** Construct a quadrilateral out of drinking straws and use this to determine if quadrilaterals are rigid.

Rigidity is generally a desirable property in buildings. Proposition I.7 explains why roofs are peaked in places with a lot of snow or wind. Triangular roofs may collapse entirely, but their rigidity means that the structure will not deflect under load, though the materials it is built from may deform under stress. Flat roofs are usually supported by triangulated trusses to take advantage of the stability of triangles.

Euclid's next proposition is the proof of Side-Side-Side (SSS) congruence and this proof of Proposition I.8 is Euclid's second use of superposition. If we don't mind waiting a bit (until after Proposition I.23) we can give another proof of this that does not make use of superposition. This of course means that we must rewrite the original proofs of all the intervening propositions to use SAS instead of SSS.

The next four of Euclid's propositions are basic geometric constructions which may be familiar to many students. We begin by either remembering or reinventing these constructions, either with actual straightedge and compass or with geometric software. In using geometric software to perform these constructions, you may find that dragging a vertex to a particular position makes the construction disappear. This is usually the result of two objects which you constructed to intersect being dragged to a configuration where they no longer meet. Such failures can sometimes be avoided by replacing line segments with lines or rays. If you are using software, many of these basic constructions will be available as menu options, but try to construct them using only the equivalent of straightedge and compass. Once you can do them in this way from scratch, you can use the menu options.

▷ **Activity 2.6.** Construct an arbitrary angle ∠BAC. Then, using only straightedge and compass or software equivalents, bisect the angle. Check to make sure that your construction works for all possible configurations. Can you prove that your construction gives the desired object?

▷ **Activity 2.7.** Construct an arbitrary line segment $\overline{AB}$. Then, using only straightedge and compass or software equivalents, bisect the segment. Check to make sure that your construction works for all possible configurations. Can you prove that your construction gives the desired object?

▷ **Activity 2.8.** Construct a line segment $\overline{AB}$ and a point C on the segment. (Note that your construction must work even if the point C is an endpoint of the line segment.) Using only straightedge and compass or software equivalents, construct a line CD through the point and perpendicular to the segment. Check to make sure that your construction works for all possible configurations. Can you prove that your construction gives the desired object?

Euclid's proofs of Propositions I.9–I.12 are constructive: He not only explains how to do the construction but also proves that it works. Since he has both purposes in mind, he usually does not choose the most efficient way of doing the construction, but instead chooses the method that is easiest to prove with the few theorems we have in hand.

Proposition I.9. *Given an angle, one can construct the angle bisector.*

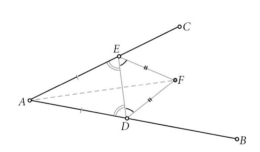

Proof of Proposition I.9: We are given ∠BAC. Choose a point D on $\overline{AB}$ and then construct E on $\overline{AC}$ with $\overline{AD} = \overline{AE}$. Draw $\overline{DE}$ and then construct equilateral △DEF. Now draw $\overline{AF}$. We claim that AF is the desired angle bisector. Note that $\overline{DF} = \overline{EF}$ and so ∠EDF = ∠DEF by Proposition I.5. △ADE is isosceles so ∠ADE = ∠AED. Therefore,

$$\angle ADF = \angle ADE + \angle EDF = \angle AED + \angle DEF = \angle AEF,$$

and so △ADF ≅ △AEF by SAS. Thus, ∠BAF = ∠CAF and $\overrightarrow{AF}$ bisects ∠BAC. □

We next give Euclid's proof of the construction of the bisector of a line segment:

Proposition I.10. *Given a line segment, one can construct the midpoint.*

Proof of Proposition I.10:

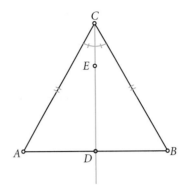

Let AB be the given finite straight line. Thus it is required to bisect the finite straight line AB. Let the equilateral triangle ABC be constructed on it [Prop. I.1], and let the angle ACB be bisected by the straight line CD [Prop. I.9]; I say that the straight line AB has been bisected at the point D. For, since AC is equal to CB, and CD is common, the two sides AC, CD are equal to the two sides BC, CD respectively; and the angle ACD is equal to the angle BCD; therefore the base AD is equal to the base BD [SAS]. Therefore the given finite straight line AB has been bisected at D.   Q.E.F.

Note that Euclid bisects the angle with a line that intersects *AB* at a point *D* that lies between *A* and *B*. While it is clear that the angle bisector must contain an interior point of the triangle, details are missing that justify placing the point *D* where Euclid has it.

▶ **Exercise 2.34.** Let $\overrightarrow{CE}$ be the ray that bisects ∠*ACB* where *E* is a point interior to the triangle △*ABC*. Prove that ray $\overrightarrow{CE}$ intersects △*ABC* at a point *D* on $\overline{AB}$ and that *D* must lie strictly between *A* and *B*.

Proposition I.10 gives us not only the midpoint of the given line segment but constructs the perpendicular bisector of that segment. In his next two propositions, Euclid explains how to draw a line through a point *C* and perpendicular to a given line *AB*. If *C* lies on *AB*, this is called *erecting a perpendicular*, while if *C* does not lie on the line the procedure is called *dropping a perpendicular*.

Proposition I.11. Given a line and a point on the line, one can construct a line through the given point and perpendicular to the given line.

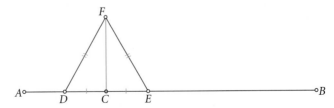

Proof of Proposition I.11: We are given point *C* on line *AB*. Use Incidence Postulate 6(3) to choose another point *D* on *AC*, and then use Proposition I.3 to cut $\overline{CE} = \overline{CD}$ so that $D - C - E$. By Proposition I.1, we can construct equilateral △*DEF* and then, by Postulate 1, draw $\overline{CF}$. Then $\overline{DF} = \overline{EF}$ and so ∠*FDE* = ∠*FED* by

Proposition I.5. Therefore, $\triangle FDC \cong \triangle FEC$ by SAS and so $\angle FCD = \angle FCE$. Thus the line FC makes equal adjacent angles and so $FC \perp AB$ at C by the definition of a right angle (Definition 2.18). □

▷ **Activity 2.9.** Using only straightedge and compass or software equivalents, construct a square. Can you prove that your construction gives the desired object?

Proving that your construction of Activity 2.9 is indeed a square is surprisingly difficult, no matter how you went about the construction. The remaining construction, Euclid's Proposition I.12, which allows us to drop a perpendicular to a line from a point not on that line, we necessarily postpone. The proof requires a variation on the Circular Continuity Principle 11 that we cannot yet prove. Further constructions possible with straightedge and compass are copying angles (Proposition I.23) and constructing parallel lines (Proposition I.31). Section 4.1 discusses straightedge and compass constructions in greater detail, including what lengths, angles, and regular polygons can be constructed. The other sections of Chapter 4 discuss the use of other construction tools.

The next two theorems, Propositions I.13 and 14, are converses of each other. Together, they say that supplementary angles, angles forming a straight line, sum to 180° (or two right angles, as Euclid would say) and, conversely, if a 180° angle is formed by two adjacent angles, then the angles are supplementary, i.e., their sides form a line.

Proposition I.13. The sum of the angle measures of two supplementary angles is 180°.

Proof of Proposition I.13:

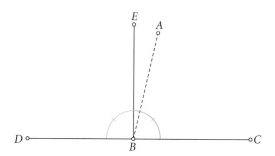

We are given two line segments, $\overline{AB}$ and $\overline{CD}$, that intersect at B.

Case 1: $\angle ABD = \angle ABC$.
By Definition 2.18 and Postulate 4, $\angle ABD = \angle ABC = 90°$, so $\angle ABD + \angle ABC = 180°$.

Case 2: $\angle ABD \neq \angle ABC$.
Use Proposition I.11 to erect $\overline{BE} \perp \overline{CD}$, and without loss of generality assume that $\angle ABC < \angle EBC$ so $\overline{AB}$ lies inside $\angle EBC$. Then $\angle CBE = \angle CBA + \angle ABE$ and $\angle ABD = \angle ABE + \angle DBE$. By the definition of perpendicular lines, $\angle CBE = \angle DBE = 90°$, so by Congruence Postulate 9(2), $\angle CBE + \angle DBE = 180°$. Therefore,

$$180° = \angle CBE + \angle DBE = \angle CBA + \angle ABE + \angle DBE = \angle CBA + \angle ABD.$$

□

Proposition I.14. If the angle measures of two adjacent angles sum to 180°, then the sides of these angles that are not shared form a straight line.

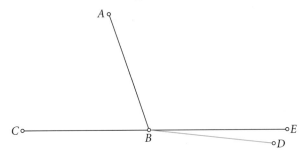

Proof of Proposition I.14: We are given adjacent angles so that $\angle CBA + \angle ABD = 180°$. We want to show that the lines CB and BD form a single line. We will do this by contradiction. Assume that D does not lie on $\overline{CB}$ extended. We can extend $\overline{CB}$ to E, and then since we are assuming that BE and BD are different lines, $\angle ABD \neq \angle ABE$. By Proposition I.13, we know that $\angle CBA + \angle ABE = 180°$. Therefore, $\angle CBA + \angle ABD = 180° = \angle CBA + \angle ABE$, and thus $\angle ABD = \angle ABE$. This contradicts our previous statement, so D must lie on $\overline{CB}$ extended. □

The next proposition, commonly called the Vertical Angle Theorem, is essentially an easy corollary of Proposition I.13.

Proposition I.15. [Vertical Angle Theorem] If two lines intersect, then the vertical angles are equal.

▶ **Exercise 2.35.** Prove Proposition I.15.

One of the basic geometric facts that many people remember long after they have left school is that the interior angles of a triangle sum to 180°. While this is true in euclidean geometry, it is not trivial to prove. Furthermore, there are many geometries where this is not true. The next proposition is a preliminary version of Euclid's Proposition I.32. That theorem is commonly called the Exterior Angle Theorem and will show that the exterior angle of a triangle is equal to the sum of the opposite interior angles. It is also noteworthy that the weakened version given here will remain true in noneuclidean geometry, where the Exterior Angle Theorem will fail. The proof is surprisingly involved.

Proposition I.16. [Weak Exterior Angle Theorem] In any triangle, the exterior angle is greater than either of the nonadjacent interior angles.

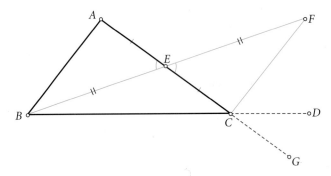

Proof of Proposition I.16: We are given $\triangle ABC$ with $\overline{BC}$ extended to D so $B - C - D$ and $\angle ACD$ is an exterior angle of the triangle. Bisect $\overline{AC}$ at point E and draw $\overline{BE}$. Extend $\overline{BE}$ and then use Proposition I.3 to construct a point F so that $B - E - F$ and $\overline{EF} = \overline{BE}$. Now draw $\overline{FC}$. Extend $\overline{AC}$ to G. Note that $\overline{AE} = \overline{EC}$. By Proposition I.15, $\angle AEB = \angle CEF$ and so $\triangle AEB \cong \triangle CEF$ by SAS. Thus, $\angle BAC = \angle ACF$. Also, note that $\angle ACD > \angle ACF$ and so $\angle ACD > \angle BAC$. Similarly, we can show that $\angle BCG > \angle ABC$, and since $\angle BCG = \angle ACD$, it follows that $\angle ACD > \angle ABC$. □

There is a tiny hole in the proof. Why is $\angle ACF < \angle ACD$? The diagram leads us to believe that this is because F is interior to $\angle ACD$, but how do we know that is true? By Exercise 2.9, we must show that F lies on the A-side of BD and also that F lies on the D-side of AC. By Theorem 2.7, E and F lie on the same side of BD and also A and E lie on the same side of BC. Thus, A and F must lie on the same side of BD. By the construction, B and F lie on opposite sides of AC, as do B and D. Thus, D and F lie on the same side of AC. Therefore, F lies inside $\angle ACD$ as shown.

Proposition I.17 is a preliminary version of the theorem that states that the angles in a triangle sum to 180°. Again, this is of interest since this version will be true in some noneuclidean geometries. This result is essentially a corollary of Proposition I.16.

Proposition I.17. In any triangle, the sum of the angle measures of any two angles is less than 180°.

▶ **Exercise 2.36.** Prove Proposition I.17.

▶ **Exercise 2.37.** From one point, prove that there cannot be drawn to the same line three line segments that are equal in length.

▶ **Exercise 2.38.** Prove: No line can intersect a circle more than twice.

The next two propositions are converses of each other. Again, the first uses a direct proof and so a good strategy for the second is to try a proof by contradiction.

Proposition I.18. In any triangle, the greater side subtends the greater angle.

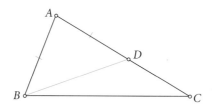

1. $\triangle ABC$ with $\overline{AC} > \overline{AB}$.
2. Cut $\overline{AD} = \overline{AB}$.
3. Draw $\overline{BD}$.
4. $\angle ABC > \angle ABD$.

1. Hypothesis
2. Proposition I.3
3. Postulate 1
4. C.N. 5

▶ **Exercise 2.39.** Finish the proof of Proposition I.18 by showing that ∠ABC > ∠ACB.

Proposition I.19. In any triangle, the greater angle is subtended by the greater side.

▶ **Exercise 2.40.** Prove Proposition I.19. You are given △ABC with ∠ABC > ∠ACB and must show that $\overline{AC} > \overline{AB}$.

Recall that after Proposition I.7 we commented that triangles are rigid and you found that quadrilaterals are not. In other words, if we make a triangle out of three sticks and put hinges at the corners, pulling on the triangle will not change the shape. As you explored this rigidity, you may have chosen some lengths that caused problems, lengths that cannot be used to form a triangle. Your experimentation in the next activity should lead you to form a conjecture about the next proposition (no peeking!).

▷ **Activity 2.10.** Cut drinking straws into lengths of various sizes. Put the straws together to try to form triangles. Can you figure out conditions about which lengths can be used to make triangles?

Proposition I.20. [Triangle Inequality] In any triangle, the sum of the lengths of any two sides is greater than the length of remaining side.

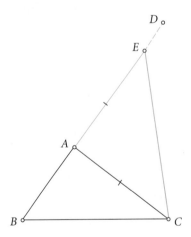

Proof: We are given △ABC. Extend $\overline{BA}$ to D so B − A − D. Use Proposition I.3 to cut $\overline{AE} = \overline{AC}$ so that B − A − E so $\overline{BE} = \overline{BA} + \overline{AE}$. Now draw $\overline{EC}$. By Proposition I.5, ∠AEC = ∠ACE. Since B − A − E, ∠ECB > ∠ECA. Thus ∠ECB > ∠AEC = ∠BEC. By Proposition I.18, $\overline{BE} > \overline{BC}$. Thus, $\overline{BA} + \overline{AC} > \overline{BC}$. □

The Triangle Inequality is just what we need to extend the Circular Continuity Principle 11. In Theorem 2.16, we considered the situation of a ray and a circle and proved that if the ray begins at the center of the circle, then the ray must intersect the circle. We need to prove a similar theorem for the case of a ray or a line emanating from any point inside the circle.

Theorem 2.25. Given a circle and a point A inside the circle, any ray $\overrightarrow{AB}$ emanating from A must intersect the circle.

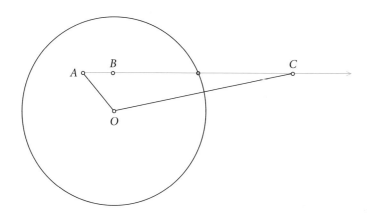

Proof: If A were the center of the circle, this would be Theorem 2.16, so we assume that A is not the center. If B happens to lie on the circle, the result is obvious. If B lies outside the circle, the ray will intersect the circle by the Circular Continuity Principle Postulate 11(1). Thus, we assume that B lies inside the circle. Let r denote the radius of the circle. By Archimedes' Axiom 10, there is a natural number n so that $n\overline{AB} > 2r$. By Proposition I.3, we construct point C on $\overrightarrow{AB}$ with $\overline{AC} = n\overline{AB}$. Draw $\overline{OA}$ and $\overline{OC}$ and consider $\triangle OAC$. By Proposition I.20, $\overline{OC} + \overline{OA} > \overline{AC} > 2r$. Since A lies inside the circle, $\overline{OA} < r$, so $\overline{OC} > 2r - \overline{OA} > 2r - r = r$. Therefore, C lies outside the circle. By the Circular Continuity Principle 11(1), $\overline{AC}$ must intersect the circle. □

▶ **Exercise 2.41.** Given a circle and a point inside the circle, prove that any line through the given point must intersect the circle twice.

▶ **Exercise 2.42.** Prove that a line drawn from a point on a circle to another point inside will intersect the circle exactly one more time.

▶ **Exercise 2.43.** Prove that the interior of any circle is convex.

We will use Theorem 2.25 and Exercises 2.41 and 2.42 frequently to justify the existence of points of intersection, first in the following straightedge and compass construction.

▷ **Activity 2.11.** Construct a line AB and a point C not on the line. Using only straightedge and compass or software equivalents, construct a line CD through the point and perpendicular to AB. Check to make sure that your construction works for all possible configurations.

Proposition I.12. Given a line and a point not on the line, one can construct a line through the given point and perpendicular to the given line.

Proof of Proposition I.12:

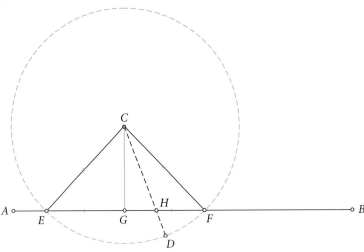

1. Point C is not on line AB.
2. Choose point D on the opposite side of AB to C.
3. Draw a circle with center C and radius $\overline{CD}$.
4. $\overline{CD}$ intersects AB at H.
5. H lies inside the circle.
6. AB intersects this circle at two points E and F.
7. Construct G, the midpoint of $\overline{EF}$.
8. $\overline{EG} = \overline{FG}$.
9. Draw $\overline{CE}, \overline{CF}$, and $\overline{CG}$.
10. $\overline{CE} = \overline{CF}$.
11. $\angle CEG = \angle CFG$.
12. $\triangle CEG \cong \triangle CFG$.
13. $\angle CGE = \angle CGF$.
14. $\overline{CG} \perp AB$.

1. Hypothesis
2. Plane Separation Property 8
3. Postulate 3
4. Plane Separation Property 8
5. Def. "circle", Exer. 2.22
6. Exer. 2.41
7. Proposition I.10
8. Def. "midpoint"
9. Postulate 1
10. Def. "radius"
11. Proposition I.5
13. Steps 8, 10, 11, SAS
13. Def. "congruent"
14. Def. "perpendicular"

□

Proposition I.20 is commonly known as the triangle inequality. Proposition I.22 is the converse. The intervening proposition follows from the conjecture you are asked to make in the following activity.

▷ **Activity 2.12.** Using dynamic geometric software, construct a triangle $\triangle ABC$ on a line segment $\overline{BC}$. Then construct a second triangle $\triangle DBC$ inside the first one using the same base $\overline{BC}$. Move the vertex D of the interior triangle (keeping it inside the original triangle $\triangle ABC$) to conjecture a relationship between the sides of $\triangle ABC$ and $\triangle DBC$ and between $\angle BAC$ and $\angle BDC$.

Proposition I.21. Given $\triangle ABC$, let D be an arbitrary point inside this triangle. Then $\overline{DB} + \overline{DC} < \overline{AB} + \overline{AC}$ and $\angle BDC > \angle BAC$.

▶ **Exercise 2.44.** Prove Proposition I.21.

Proposition I.22. Given three lengths, it is possible to construct a triangle with sides of these lengths if the sum of any two of the given lengths is greater than the third length.

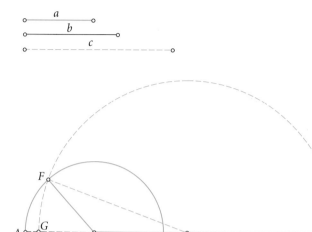

Proof of Proposition I.22: We are given lengths a, b, and c with $a + b > c$, $a + c > b$, and $b + c > a$. Without loss of generality, we may assume that $a \leq b \leq c$. Using the Incidence Postulate 6 and Postulates 1 and 2, we can draw a line segment $\overline{AB}$ and extend it if necessary. By Proposition I.3, we can cut $\overline{AC} = a$, $\overline{CD} = b$, and $\overline{DE} = c$ so that $A - C - D - E$. Draw circles $\mathcal{C}_1$ with center C and radius a and $\mathcal{C}_2$ with center D and radius c. Note that $\overline{AD} = \overline{AC} + \overline{CD} = a + b > c = \overline{DE}$. Thus A lies outside $\mathcal{C}_2$, while D lies inside $\mathcal{C}_2$. By the Circular Continuity Principle Postulate 11, $\overline{AD}$ intersects $\mathcal{C}_2$ at a point G. Since $\overline{DG} = \overline{DE} = c < a + b = \overline{AD}$, $A - G - D$. Let us assume that $c > b$. Then since $\overline{DG} = c > b = \overline{DC}$, we further have $A - G - C - D$. Then $\overline{CG} = \overline{DG} - \overline{DC} = c - b < a$, and so G lies on $\mathcal{C}_2$ and inside $\mathcal{C}_1$. Similarly, $\overline{CE} = \overline{CD} + \overline{DE} = b + c > a$, and so E lies on $\mathcal{C}_2$ and outside $\mathcal{C}_1$. Therefore, by the Circular Continuity Principle 11(2), $\mathcal{C}_2$ intersects $\mathcal{C}_1$ at a point F. Since F lies on $\mathcal{C}_1$, $\overline{CF} = a$ and since F lies on $\mathcal{C}_2$, $\overline{DF} = c$. We have thus constructed the desired $\triangle CDF$. □

▶ **Exercise 2.45.** Modify the proof of Proposition I.22 for the case in which $c = b$.

We must do extensive rewriting to the original proof of Proposition I.23 to avoid the use of SSS congruence, but once that is done, we can use this proposition, which allows us to copy angles, to prove the SSS theorem. We encourage you to try to perform this construction as an activity before reading the solution.

▷ **Activity 2.13.** Construct an angle and a disjoint line segment. Using straightedge and compass or software equivalents, create a copy of the angle at an endpoint of the line segment so that the line segment forms one side of the angle. Try to find a construction that will work for both acute and obtuse angles. Try to prove that your constructed angle is equal to the original.

Proposition I.23. Given an angle and a line segment, one can construct an angle equal to the given angle with the given line segment as one of the sides and one of the endpoints of the segment as the vertex of the angle.

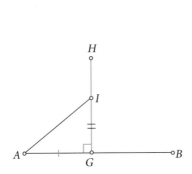
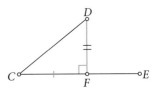

We are given segment $\overline{AB}$ and $\angle DCE$.

Case 1: $\angle DCE < 90°$.
By Proposition I.12, we can drop $DF \perp CE$ (extended if necessary). Cut $\overline{AG} = \overline{CF}$ and use Proposition I.11 to erect $GH \perp AB$. Cut $\overline{GI} = \overline{FD}$ and draw $\overline{AI}$. Then $\angle CFD = \angle AGI$ by Postulate 4. Thus, $\triangle CFD \cong \triangle AGI$ by SAS, so $\angle BAI = \angle ECD$. □

▶ **Exercise 2.46.** Note that the proof of Case 1 above does not seem to use the case hypothesis, though it is implicit in the drawing. Use this hypothesis to explain why point F must lie on the E-side of C rather than on C itself or on the far side of C.

▶ **Exercise 2.47.** Prove the other two cases for Proposition I.23: $\angle DCE = 90°$ and $\angle DCE > 90°$.

Proposition I.8. [SSS] If two triangles have three pairs of equal sides, then the triangles are congruent.

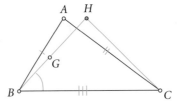
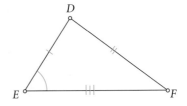

1. $\triangle ABC$, $\triangle DEF$ with $\overline{AB} = \overline{DE}, \overline{AC} = \overline{DF}$, and $\overline{BC} = \overline{EF}$ are given.
2. Assume that $\angle ABC \neq \angle DEF$.

1. Hypothesis

2. Contradiction hypothesis

3. Without loss of generality ∠ABC > ∠DEF.	3. Logic
4. Construct ∠CBG = ∠DEF on the A-side of BC.	4. Proposition I.23
5. ∠CBG < ∠CBA.	5. Steps 3, 4, C.N.
6. Cut $\overline{BH} = \overline{ED}$.	6. Proposition I.3 (Postulate 2)
7. Draw $\overline{CH}$.	7. Postulate 1
8. △HBC ≅ △DEF.	8. Steps 1, 4, 6, SAS
9. ...	
...	
n. → ←	n.
n + 1. ∠ABC = ∠DEF.	n + 1. Law of contradiction
n + 2. △ABC ≅ △DEF.	n + 2. Steps 1, n + 1, SAS

□

▶ **Exercise 2.48.** Fill in the gap in the proof of Proposition I.8.

Propositions I.24 and I.25 are converses of one another. Remember that if you prove one statement directly, then contradiction is a good strategy to consider for proving the converse.

Proposition I.24. [Hinge Theorem] Given △ABC and △DEF with $\overline{AB}=\overline{DE}$ and $\overline{AC}=\overline{DF}$. If ∠BAC > ∠EDF, then $\overline{BC} > \overline{EF}$.

▶ **Exercise 2.49.** Prove Proposition I.24 and explain its nickname.

Proposition I.25. Given △ABC and △DEF with $\overline{AB}=\overline{DE}$ and $\overline{AC}=\overline{DF}$. If $\overline{BC} > \overline{EF}$, then ∠BAC > ∠EDF.

▶ **Exercise 2.50.** Prove Proposition I.25.

We conclude this section with the last of Euclid's congruence theorems. His Proposition I.26 is really two statements, which complete the list of the standard congruence criteria. We consider the two cases separately.

Proposition I.26. [ASA and AAS] If two triangles have two angles of the first triangle equal to two angles of the second triangle respectively, and one side of the first triangle equals to one side of the second, either the side adjoining the equal angles, or the side subtending one of the equal angles, then the triangles are congruent.

Proof of ASA:

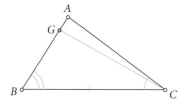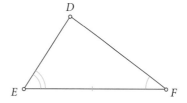

Consider △ABC and △DEF with ∠ABC = ∠DEF, ∠ACB = ∠DFE, and $\overline{BC} = \overline{EF}$. Assume that $\overline{AB} \neq \overline{DE}$. Without loss of generality, we may assume that $\overline{AB} > \overline{DE}$ and then cut $\overline{BG} = \overline{ED}$. Draw $\overline{GC}$ and observe that ∠ACB > ∠GCB. By SAS, △GBC ≅ △DEF, so ∠GCB = ∠DFE. Thus, ∠GCB = ∠ACB, which is a contradiction. Therefore, we must have $\overline{AB} = \overline{DE}$ and thus △ABC ≅ △DEF by SAS. □

▶ **Exercise 2.51.** Prove AAS congruence.

♦ 2.3 Propositions I.27–I.32: Parallel lines

Next we turn our attention to the study of the properties of parallel lines. Recall the basic definition: *Parallel lines are lines that lie in the same plane and do not intersect.*

The next two propositions give criteria to determine if two lines are parallel. Proposition I.27 describes the *alternate interior angle criterion*, while Proposition I.28 describes the *exterior and opposite interior angle criterion* and the *adjacent interior angle criterion*.

Proposition I.27. If two lines are cut by a transversal so that the alternate interior angles are equal, then the given lines are parallel.

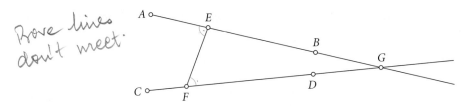

Proof: We are given lines AB and CD with transversal EF intersecting AB at E and CD at F so that ∠AEF = ∠EFD. Assume that AB and CD meet at some point G in the direction of B. Then ∠AEF is exterior to △EFG so by Proposition I.16, ∠AEF > ∠EFD. This contradicts the hypothesis that ∠AEF = ∠EFD. Thus AB cannot meet CD on the B-side. A similar argument shows that AB does not meet CD on the A-side, so it must be true that AB is parallel to CD. □

Proposition I.28 gives other conditions that determine whether the lines are parallel. Let us explore the relationships between all of these angle conditions. A transversal cutting two lines forms eight angles, as shown below:

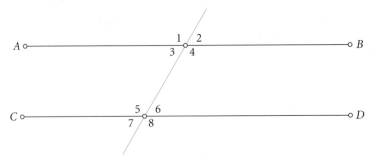

In Proposition I.27, Euclid proved that if ∠3 = ∠6, then the lines are parallel. He neglected to show that equality of the other pair of alternate interior angles, ∠4 = ∠5, also implies parallelism, although this is easy to do. Similarly, in Proposition I.28 he shows that if $\angle 2 = \angle 6$ then the lines are parallel but neglects the other three pairs consisting of an exterior angle and the opposite interior angle. He also proves, in Proposition I.28, that if $\angle 3 + \angle 5 = 180°$, then the lines are parallel. From Propositions I.27 and 28, there are eight angle relationships that imply parallelism. A further note on Lemma 2.26: One could prove 'a $\implies$ b', 'a $\implies$ c', 'a $\implies$ d', ..., 'b $\implies$ a', 'b $\implies$ c', ..., all the way to 'h $\implies$ g', for a total of $8 \cdot 7 = 56$ individual proofs. Each of these is only a few lines, but there are an awful lot of them. A more efficient way of structuring the proof is to construct a circular argument:

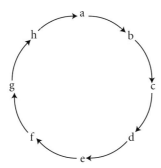

In other words, prove, 'a $\implies$ b', 'b $\implies$ c', 'c $\implies$ d', ..., 'g $\implies$ h', 'h $\implies$ a'. This involves only eight short individual proofs. If, for example, one needed to know that 'b $\implies$ g', that follows from a chain of implications around the circle.

Lemma 2.26. The following are equivalent (i.e., either all true or all false):

a. $\angle 3 = \angle 6$
b. $\angle 4 = \angle 5$
c. $\angle 1 = \angle 5$
d. $\angle 2 = \angle 6$
e. $\angle 3 = \angle 7$
f. $\angle 4 = \angle 8$
g. $\angle 3 + \angle 5 = 180°$
h. $\angle 4 + \angle 6 = 180°$

▶ **Exercise 2.52.** Prove Lemma 2.26.

We can use this lemma in conjunction with Proposition I.27 to immediately conclude that if any one of these angle relationships occurs then the lines are parallel, thus obtaining a proof of Proposition I.28.

Proposition I.28. If two lines are cut by a transversal so that an exterior angle is equal to the opposite interior angle on the same side of the transversal, or so that the sum of the interior angles on one side of the transversal is 180°, then the given lines are parallel.

We next discuss Euclid's Proposition I.31, an easy consequence of Proposition I.27. This is another construction theorem, giving instructions for the construction of parallel lines and also proving that the result has the desired property. While the construction itself could have been performed earlier, the proof requires Proposition I.27. The proof does not require the use of the Parallel Postulate 5, which is why we consider it here, slightly out of order.

Proposition I.31. Given a line and a point not on that line, one can construct a line through the given point and parallel to the given line.

▷ **Activity 2.14.** Given a line and a point not on that line, figure out a method for constructing a line through the given point parallel to the given line. Prove that your construction gives the desired result.

At this point, we have proved that parallel lines exist and you have given instructions on how to construct them. We have also determined conditions that, if fulfilled, guarantee that the lines under consideration are parallel. We now want to determine properties that parallel lines must have. At this point we need to bring the Parallel Postulate 5 into play.

It should be noted that Euclid treats Postulate 5 quite differently from his other postulates. Propositions I.1–28 (and 31) use all of the other postulates freely, but he seems to have put off using the Parallel Postulate 5 as long as he possibly could. From this point on, every proposition in this chapter will require Postulate 5. It is very important to note this, since when we come to study noneuclidean geometry, we wish to deny the Parallel Postulate 5 and so must forswear all of its descendants. This treatment is evidence pointing out the crucial role that the Parallel Postulate plays in euclidean geometry and was cause for the suspicion with which the Parallel Postulate 5 was viewed for millennia. We will discuss this controversy in depth in Chapter 5.

Postulate 5. [Parallel Postulate]. If two lines are cut by a transversal so that the sum of the interior angles on one side of the transversal is less than 180°, then the two lines meet on that side of the transversal.

Thus the Parallel Postulate 5 claims that if, for example, $\angle 3 + \angle 5 < 180°$, then the lines are not parallel. Let us further analyze the logical structure of Propositions I.27 and 28 and the Parallel Postulate 5. By Lemma 2.26, we know if the interior angles on the same side of the transversal sum to less than 180° as in the Parallel Postulate 5, then the other angle relationships will also fail. Given two lines cut by a transversal, consider the statements below as logical entities:

P: One (and therefore all) of the eight angle relationships is false.

Q: The lines are not parallel.

- $P \implies Q$: *If any of the angle relationships is false, then the lines are not parallel.* This statement is essentially the Parallel Postulate 5.
- $\neg Q \implies \neg P$: *If the lines are parallel, then all of the angle relationships are true.* This is the contrapositive and so is logically equivalent to the original statement, the Parallel Postulate 5. This is Proposition I.29, the next theorem and the first whose truth depends on the Parallel Postulate 5.
- $Q \implies P$: *If the lines are not parallel, then all of the angle relationships fail.* This is the converse of the Parallel Postulate 5, and can be rephrased as *If three lines form a triangle, then all of the angle relationships fail.* In particular, the exterior angle will be not equal to the opposite interior angle (Proposition I.16) and the sum of any two interior angles will not be equal to 180° (Proposition I.17).
- $\neg P \implies \neg Q$: *If any of the angle relationships are true, then the lines are parallel.* This is the inverse and is logically equivalent to the converse. The statement is a simple restatement of the combination of Propositions I.27 and 28.

Proposition I.29. If two parallel lines are cut by a transversal, then the alternate interior angles are equal, and consequently, each exterior angle is equal to the opposite interior angle and the sum of the interior angles on the same side of the transversal is 180°.

As noted above, this follows immediately as a logical consequence of the Parallel Postulate 5. Thus, Proposition I.29 implies that if any one of the angle relations of Lemma 2.26 fails, then the lines are not parallel. It is easy to confuse Propositions I.27–28 and I.29. Remember, Propositions I.27 and 28 (which do not depend on the Parallel Postulate 5) say that if any of the angle relationships holds then the lines are parallel. Proposition I.29 (which does depend on the Parallel Postulate 5) says that if the lines are parallel, then all of the angle relationships hold true.

The most commonly used version of the parallel postulate in modern texts was devised by John Playfair [1748–1819] in the eighteenth century:

Playfair's Postulate: *Given a line and a point not on that line, there is exactly one line through the point parallel to the given line.*

Proposition I.31 has shown us that we can always construct one line through the given point parallel to the given line. That construction is true even if the Parallel Postulate 5 is assumed to be false. Thus, the property that this parallel line is unique must be logically dependent on the Parallel Postulate 5.

▶ **Exercise 2.53.** Prove that the Parallel Postulate 5 implies the truth of Playfair's Postulate. You are given a line *AB* and a point *C* not on that line. Using Proposition I.31, construct a line *CD* through *C* and parallel to *AB*. Prove that any other line *CE* through *C* and distinct from *CD* must intersect *AB*.

▶ **Exercise 2.54.** Prove that Playfair's Postulate implies the truth of Proposition I.29. You are given two parallel lines *AB* and *CD* with transversal *EF*. Use Playfair's Postulate (together with Propositions I.1–28 and Proposition I.31) to prove that ∠*AEF* = ∠*DFE* (and thus that the other seven of the angle relationships are also true).

Exercises 2.53 and 2.54 imply that the Parallel Postulate 5 and Playfair's Postulate are logically equivalent. Thus, they may be used interchangeably. The next proposition concerns the transitivity of the parallel relationship: if $\ell_1 \parallel \ell_2$ and $\ell_2 \parallel \ell_3$, then $\ell_1 \parallel \ell_3$. The Parallel Postulate 5 is used by means of its contrapositive, Proposition I.29.

Proposition I.30. Lines parallel to the same line are also parallel to each other.

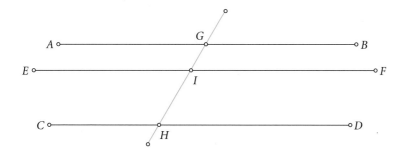

Proof: We are given lines $AB \parallel EF$ and $CD \parallel EF$. Note that A and B (and all of the points on the line AB) must lie on one side of EF by the Plane Separation Property Postulate 8 and Exercise 2.8. Similarly, C and D (and the line CD) must also lie on one side of EF. Then either AB and CD lie on the same side of EF or they lie on opposite sides of EF. Consider the case in which AB and CD lie on different sides of EF. The Incidence Postulate 6(3) lets us choose points G on AB and H on CD. Draw GH and note that by the Plane Separation Property 8, $\overline{GH}$ must intersect EF at I. By Proposition I.29 for the lines $AB \parallel EF$, $\angle AGI = \angle GIF$. The same reasoning applied to lines $EF \parallel CD$ gives us $\angle GIF = \angle IHD$. Therefore, $\angle AGI = \angle IHD$, and so by Proposition I.27, $AB \parallel CD$. □

▶ **Exercise 2.55.** Prove the other case of Proposition I.30, where AB and CD lie on the same side of EF.

Finally, we've gotten to one of the most commonly known geometric facts: that the angles of any triangle sum to 180°. This proposition extends Propositions I.16 and I.17, but is only true in euclidean geometry and depends on the Parallel Postulate 5.

Proposition I.32. [Exterior Angle Theorem] In any triangle, the exterior angle is equal to the sum of the two opposite interior angles. Furthermore, the sum of the three interior angles of the triangle is 180°.

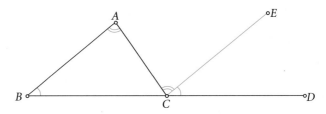

Proof: We are given $\triangle ABC$ with $\overline{BC}$ extended to D. Use Proposition I.31 to draw $CE \parallel AB$. Note that by Proposition I.16, $\angle ACD > \angle ABC$. By Proposition I.29, since CE is parallel to AB, $\angle ECD = \angle ABC$. Thus, $\angle ACD > \angle ECD$ so E lies inside $\angle ACD$ as pictured. By Proposition I.29, we also know that $\angle BAC = \angle ACE$. Note that $\angle ACD = \angle ACE + \angle ECD$. Therefore, $\angle ACD = \angle BAC + \angle ABC$. By Proposition I.13, $\angle BCA + \angle ACD = 180°$, so $\angle BCA + \angle BAC + \angle ABC = 180°$. □

▶ **Exercise 2.56.** Prove that the sum of the interior angles in any quadrilateral is 360°. Make sure that your proof is valid for both of the cases below:

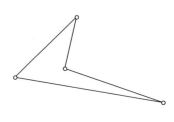

▷ **Activity 2.15.** Can you figure out how to handle angle sums for self-intersecting quadrilaterals like the one pictured below?

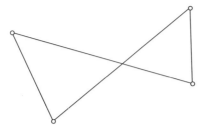

▶ **Exercise 2.57.** Find a formula for the sum of the vertex angles of any *n*-sided convex polygon, and prove that your formula is valid.

▶ **Exercise 2.58.** Find a formula for one of the vertex angles of a regular *n*-sided polygon, in which all of the sides have the same length and all of the vertex angles the same measure.

▶ **Exercise 2.59.** Find a formula for the sum of the exterior angles (formed by extending each side) of any convex *n*-sided polygon, and prove that your formula is valid.

♦ 2.4 Propositions I.33–I.46: Area

We now develop some basic properties of some familiar quadrilaterals such as parallelograms and rectangles and begin a study of the concept of area. The next proposition introduces the parallelogram (defined in Definition 2.20) and provides a link between the previous theorems about parallel lines and the ones that follow concerning parallelograms.

Proposition I.33. *The figure formed by two parallel line segments of equal length, joined by line segments connecting the corresponding endpoints at each side, is a parallelogram. Furthermore, the joining line segments are themselves equal in length.*

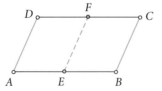

In order to prove this proposition, we must make sense of what it means to be "the endpoints at each side," an idea which is intuitively obvious from the drawing but must be made rigorous. We are given $AB \parallel DC$ with $\overline{AB} = \overline{DC}$. Construct the midpoints E and F of $\overline{AB}$ and $\overline{DC}$, respectively, and draw EF. Then A and B lie on opposite sides of EF, as do C and D. If A and D lie on the same side of EF, then draw $\overline{AD}$ and $\overline{BC}$. Otherwise, A and C must lie on the same side of EF, so draw $\overline{AC}$ and $\overline{BD}$.

▶ **Exercise 2.60.** Finish the proof of Proposition I.33: Assuming that *A* and *D* lie on the same side of *EF*, prove that $\overline{DA} = \overline{CB}$ and that $DA \parallel CB$, and thus that *ABCD* is a parallelogram.

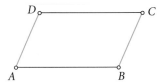

We are used to thinking of area as a formula or function that, given a region in the plane, gives a number designating its area in square units. This notion is completely foreign to *The Elements*. In general, Euclid doesn't seem to think of numbers at all, but rather of line segments and of regions, whose measure we would think of as a number. Some of the modern geometers, especially Hilbert, used the primitive term *content* in place of area. Think of the content of a figure as how much (homogeneous) stuff could be fit inside its boundaries. At this point, we know only two things about content or area, both from Congruence Postulate 9:

(3) *Equality in area is an additive equivalence relation on polygons.*
(4) *Congruent triangles have equal areas.*

We do not know what the area of a figure might mean or how much content might fit in a figure. We can however, in the following activities, exercises, and propositions, determine in certain cases if two figures have the same content (equal areas), and the relationship between the content of some figures.

The next proposition proves some other properties of a parallelogram, besides those guaranteed by the definition, and gives a small start to our study of area, simply claiming that the diagonal divides the parallelogram into two regions of equal area.

Proposition I.34. In a parallelogram, the opposite sides and angles are equal to one another and the diagonal bisects the area.

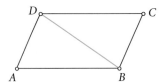

▶ **Exercise 2.61.** Prove Proposition I.34. You are given parallelogram *ABCD*, so you know $AB \parallel DC$ and $AD \parallel BC$. Prove that $\overline{AB} = \overline{DC}$, $\overline{AD} = \overline{BC}$, $\angle DAB = \angle BCD$, and $\angle ABC = \angle CDA$. Finally, prove that $\triangle ABD$ and $\triangle CDB$ have equal areas.

The following group of activities and exercises explore additional properties of quadrilaterals. This is also an ideal use of geometric software to form figures and to make conjectures, followed by the necessity of a formal proof.

▷ **Activity 2.16.** Construct two line segments $\overline{AC}$ and $\overline{BD}$ so that they bisect each other at *E* (thus, *E* is the midpoint of both $\overline{AC}$ and $\overline{BD}$). Formulate a conjecture about the quadrilateral formed by *ABCD*.

▶ **Exercise 2.62.** Prove your conjecture of Activity 2.16.

▷ **Activity 2.17.** Construct a parallelogram *ABCD* and draw the diagonals $\overline{AC}$ and $\overline{BD}$. Formulate a conjecture about these diagonals.

▶ **Exercise 2.63.** Prove your conjecture of Activity 2.17.

▷ **Activity 2.18.** Construct two perpendicular line segments $\overline{AC}$ and $\overline{BD}$ so that they bisect each other at E. Formulate a conjecture about the quadrilateral formed by $ABCD$.

▶ **Exercise 2.64.** Prove your conjecture of Activity 2.18.

▶ **Exercise 2.65.** If $ABCD$ is a rhombus, show that it is a parallelogram.

▶ **Exercise 2.66.** Prove that the diagonals of a rhombus are perpendicular.

▷ **Activity 2.19.** Construct two equal line segments $\overline{AC}$ and $\overline{BD}$ so that they bisect each other at E. Formulate a conjecture about the quadrilateral formed by $ABCD$.

▶ **Exercise 2.67.** Prove your conjecture of Activity 2.19.

▶ **Exercise 2.68.** Prove the converse of the result of Exercise 2.67.

In the next and succeeding propositions, the phrase "in the same parallel" occurs. This means that the top edge of the parallelogram or the top vertex of the triangle in question will lie on a given line parallel to the base. Let us explore variations on this configuration.

▷ **Activity 2.20.** Using geometric software, construct a line segment $\overline{AB}$ and a line CD parallel to the segment. Place another point E on CD and construct a parallelogram $ABEF$, so $\overline{EF}$ lies on line CD. Compute the area of $ABEF$ and drag point E along line CD, making sure not to move the line segment $\overline{AB}$ or the line CD. Make a conjecture about the area of parallelograms described by $ABEF$.

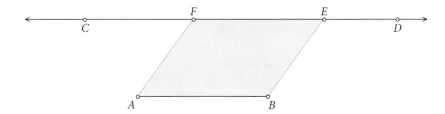

▷ **Activity 2.21.** Using geometric software, construct parallel lines AB and CD. Construct equal line segments $\overline{EF}$ and $\overline{GH}$ on line AB. Construct parallelograms $EFIJ$ and $GHKL$ so that $\overline{IJ}$ and $\overline{KL}$ lie on line CD. Compute the areas of $EFIJ$ and $GHKL$. Without moving lines AB and CD and while keeping $\overline{EF} = \overline{GH}$, drag these parallelograms around. Make a conjecture about their areas.

▷ **Activity 2.22.** Using geometric software, construct parallel lines AB and CD. Construct equal line segments $\overline{EF}$ and $\overline{GH}$ on line AB. Construct triangles $\triangle EFI$ and $\triangle GHJ$ so that I and J lie on line CD. Compute the areas of $\triangle EFI$ and $\triangle GHJ$. Without moving the lines AB and CD, drag these triangles around. Make a conjecture about their areas.

▶ **Exercise 2.69.** Prove your conjecture of Activity 2.20. You are given parallelograms *ABEF* and *ABGH* on the same base and in the same parallel and must prove something about their areas. Note that you may not use the modern formula for the area of a parallelogram. Your experimentation should have shown that you must address several cases, depending on whether and how *AH* and *BE* intersect.

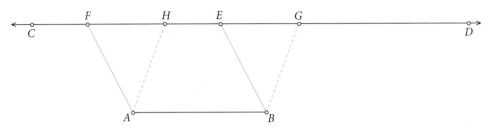

The problem of Exercise 2.69 is Euclid's Proposition I.35. The next proposition should verify your conjecture of Activity 2.21:

Proposition I.36. Parallelograms with equal bases and in the same parallel are equal in area.

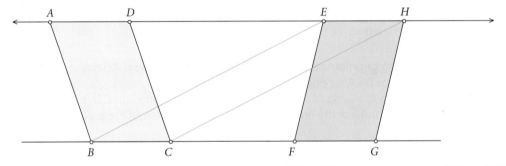

Proof: We are given parallelograms *ABCD* and *EFGH* with $\overline{BC} = \overline{FG}$ and *AH*∥*BG*. Draw $\overline{BE}$ and $\overline{CH}$. By Proposition I.34, $\overline{FG} = \overline{EH}$, so $\overline{BC} = \overline{EH}$. Therefore, by Proposition I.33, *BE*∥*CH* and $\overline{BE} = \overline{CH}$ and so *EBCH* is a parallelogram. Thus by Exercise 2.69, *Area*(*ABCD*) = *Area*(*EBCH*) and *Area*(*EBCH*) = *Area*(*EFGH*). Therefore, *Area*(*ABCD*) = *Area*(*EFGH*). □

Proposition I.36 says that the area of a parallelogram depends only on the base and the parallel. In other words, the area of the parallelogram depends only on the base and the height, given by the perpendicular distance from the base to the opposite parallel line. In modern terms, we are accustomed to the formula $A = bh$ for the area of a parallelogram, since any parallelogram has area equal to the area of the rectangle with the same base and in the same parallel. Except in the congruence propositions, in which Euclid says "the triangle is equal to the triangle," meaning that they have the same area, Propositions I.34, I.35 and I.36 are the first to address the question of area. Technically, these propositions have shown us that the area of a parallelogram is *some* function of the length of the base and the height, but we do not know what this function is. Since area should be measured in square units, we cannot discuss it further until we know more about squares.

▶ **Exercise 2.70.** Prove your conjecture of Activity 2.22. Note that you may not use the modern formula for the area of a triangle.

The next proposition we discuss (after omitting some that we will not need) shows that the area of a triangle is half the area of any parallelogram with the same base and equal height.

Proposition I.41. If a parallelogram has the same base as a given triangle and lies in the same parallel, then the area of the parallelogram is double the area of the triangle.

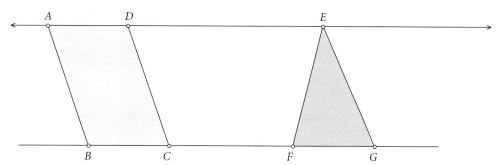

▶ **Exercise 2.71.** Prove Proposition I.41. You are given that $ABCD$ is a parallelogram, $\overline{BC} = \overline{FG}$, and $BG \parallel AE$.

Recall the Definition 2.22 of a square: a quadrilateral with four equal sides and four right angles. In Activity 2.9, you showed that you could construct a square, as Euclid does in the next proposition:

Proposition I.46. Given a line segment, one can construct a square with the given line segment as one of the sides.

▶ **Exercise 2.72.** There are many ways of constructing a square using straightedge and compass. For whichever method you chose in Activity 2.9 above, or what ever method you now prefer, prove that your method does indeed give a square.

▶ **Exercise 2.73.** Prove that if a quadrilateral has three right angles and two adjacent equal sides, then the quadrilateral is a square.

▶ **Exercise 2.74.** Prove that if a quadrilateral has four equal sides and one right angle, then the quadrilateral is a square.

Now that we know that squares do exist of any given side length, let us return to the question of area, which is too valuable a tool to give up. In order to put the theory of area on a firm footing, we must add an additional postulate:

Postulate 13. [Area] The area of a rectangle is the product of the lengths of its base and height.

This gives us the familiar formula

$$A(rectangle) = bh,$$

and thus for a square of side s,

$$A(square) = s^2.$$

▶ **Exercise 2.75.** Give a formula for the area of a parallelogram and justify your formula.

▶ **Exercise 2.76.** Give a formula for the area of a triangle and justify your formula.

▶ **Exercise 2.77.** Give a formula for the area of an arbitrary trapezoid and justify your formula.

Area Postulate 13 only applies to euclidean geometry, i.e., it only makes sense if we have already assumed that the Parallel Postulate 5 is true. In noneuclidean geometry, we will have to devise alternate means of computing area.

♦ 2.5 The Pythagorean Theorem

The Pythagorean Theorem is the culmination of Book I of Euclid's *The Elements*. Almost all of the theorems in Book I contribute to and are necessary for its proof. Euclid's proof of this famous theorem is not the most appealing or intuitive, and probably not the original proof of the Pythagoreans, but we include it for historical interest. Note that sometimes instead of giving the four vertices of a quadrilateral, he will merely describe it as a square or parallelogram and then give two diagonal vertices. This is sufficient to make it clear which figure he is talking about. The side of the triangle subtended by the right angle is called the *hypotenuse*.

Proposition I.47. [Pythagorean Theorem] In a right triangle, the square on the side subtending the right angle is equal to the sum of the squares on the sides containing the right angle.

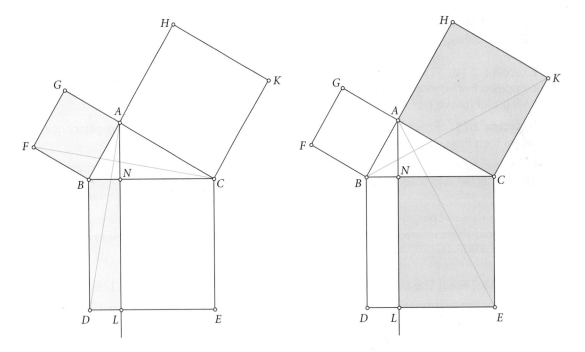

Proof: Let ABC be a right-angled triangle with the angle BAC right; I say that the square on BC is equal to the squares on BA, AC.

For let there be described on BC the square BDEC, and on BA, AC the squares GB, HC [Prop. I.46]; through A let AL be drawn parallel to either BD or CE, and let AD, FC be joined.

Then, since each of the angles BAC, BAG is right, it follows that with a straight line BA, and at the point A on it, the two straight lines AC, AG not lying on the same side make the adjacent angles equal to two right angles; therefore CA is in a straight line with AG. [Prop. I.14]

For the same reason BA is also in a straight line with AH.

And, since the angle DBC is equal to the angle FBA: for each is right: let the angle ABC be added to each; therefore the whole angle DBA is equal to the whole angle FBC. [C.N. 2]

And, since DB is equal to BC, and FB to BA, the two sides AB, BD are equal to the two sides FB, BC respectively, and the angle ABD is equal to the angle FBC; therefore the base AD is equal to the base FC, and the triangle ABD is equal to the triangle FBC. [SAS]

Now the parallelogram BL is double of the triangle ABD, for they have the same base BD and are in the same parallels BD, AL. [Prop. I.41]

And the square GB is double of the triangle FBC, for again they have the same base FB and are in the same parallels FB, GC. [Prop. I.41]

But the doubles of equals are equal to one another. Therefore the parallelogram BL is also equal to the square GB.

Similarly, if AE, BK are joined, the parallelogram CL can also be proved equal to the square HC; therefore the whole square BDEC is equal to the two squares GB, HC. [C.N. 2]

And the square BDEC is described on BC, and the squares GB, HC on BA, AC. Therefore the square on the side BC is equal to the two squares on the sides BA, AC. Q.E.D.

▶ **Exercise 2.78.** Rewrite Euclid's proof in two-column format.

The next proposition, and the last in Euclid's Book I, is the converse of the Pythagorean Theorem.

Proposition I.48. If in a triangle the square on one of the sides is equal to the sum of squares on the remaining two sides, then the angle contained by the two sides of the triangle is right.

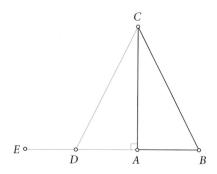

Proof: We are given $\triangle ABC$ with $\overline{BC}^2 = \overline{AB}^2 + \overline{AC}^2$. We must show that $\angle CAB = 90°$. Construct $AE \perp AC$ with E on the side of CA opposite B. Cut $\overline{AD} = \overline{AB}$ and draw $\overline{DC}$. Note that we then have $\overline{AD}^2 + \overline{AC}^2 = \overline{AB}^2 + \overline{AC}^2$. By the Pythagorean Theorem for $\triangle CDA$, we know $\overline{AD}^2 + \overline{AC}^2 = \overline{CD}^2$. Therefore, $\overline{CD}^2 = \overline{BC}^2$ and so $\overline{CD} = \overline{BC}$. Thus, $\triangle CAD \cong \triangle CAB$ by SSS and so $\angle CAD = \angle CAB = 90°$.

□

The Pythagorean Theorem has many useful and easy consequences. We explore some of these in the following set of exercises.

▶ **Exercise 2.79.** Use the Pythagorean Theorem to prove that the shortest distance from a point to a line is measured along the perpendicular from the point to the line.

▶ **Exercise 2.80.** Let $\triangle ABC$ be an equilateral triangle with side 1. Find its height.

▶ **Exercise 2.81.** Give a formula for the area of a regular hexagon and justify your formula.

▶ **Exercise 2.82.** Give a formula for the area of a regular octagon and justify your formula.

▶ **Exercise 2.83.** Consider a cube of side length 1. Find the length of the diagonal from the front left lower corner to the rear right upper corner.

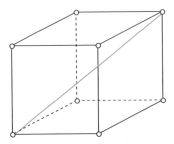

▶ **Exercise 2.84.** Repeat Exercise 2.83 for a rectangular box with width 1, length 2, and height 3.

▶ **Exercise 2.85.** Let $\triangle ABC$ have a right angle at $\angle ACB$. Construct equilateral triangles $\triangle ACD$, $\triangle BCE$, and $\triangle ABF$. Prove that $Area(\triangle ABF) = Area(\triangle ACD) + Area(\triangle BCE)$.

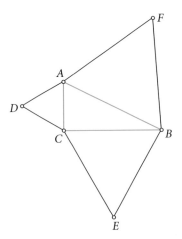

Triples of integers a, b and c so that $a^2 + b^2 = c^2$ are called *Pythagorean triples*, since a triangle with sides of lengths a, b, and c must be a right triangle (with the right angle opposite the side with length c). Examples of such triples are 3-4-5 or 5-12-13. The Egyptians knew of at least the 3-4-5 case. A Babylonian tablet called PLIMPTON 322 from some time between 1900 and 1600 B.C. has a list of 15 Pythagorean triples (including the triple 13500-12709-18541). There is no indication of how these were derived nor of their intended use, though the presence of some very large numbers would tend to support the hypothesis that these were of theoretical rather than practical interest. The first known proof of a general algorithm for generating Pythagorean triples is Lemma 1 to Proposition 29 of Book X in Euclid's *The Elements*. Let m and n be positive integers with $m > n$. Then it is easy to prove that

$$(2mn)^2 + (m^2 - n^2)^2 = (m^2 + n^2)^2.$$

▶ **Exercise 2.86.** Prove the formula above is true for any integers m and n with $m > n$. Then prove that $a = 2mn$, $b = m^2 - n^2$, $c = m^2 + n^2$ is a Pythagorean triple.

Leonhard Euler [1707–1783] proved that the formula above generates all possible Pythagorean triples. Below is a copy of Oliver Byrne's illustrated proof of the Pythagorean Theorem. The colored illustrations make it somewhat easier to envision the relationships among the various angles and among the given triangles and quadrilaterals. Other than the use of colors to indicate dependencies and the corresponding minimal use of text, Byrne gives Euclid's own proof.[4]

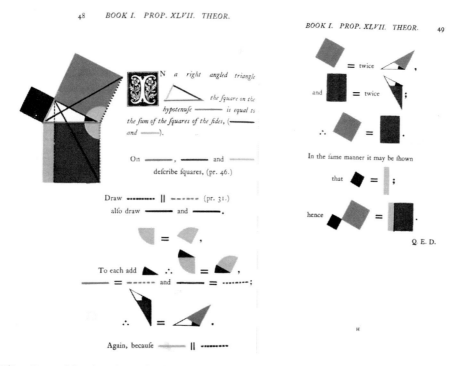

(*Source*: Oliver Byrne, *The First Six Books of the Elements of the Euclid in which Coloured Diagrams and Symbols are used instead of Letters for the Greater Ease of Learners.* Book I, Proposition 47.)

---

[4]Byrne, http://www.sunsite.ubc.ca/DigitalMathArchive/Euclid/byrne.html

There are many different proofs of the Pythagorean Theorem. Elisha Loomis spent his life collecting many of these and published *The Pythagorean Proposition* in 1940 with 371 different proofs. We present sketches of six of these, the first predating Pythagoras himself:

Proof #253: Chinese, ca. 500 B.C.:

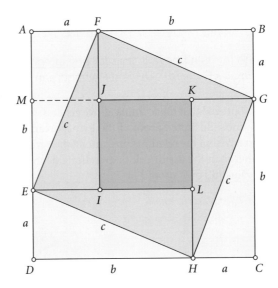

A square of side $a + b$ is divided into eight copies of the triangle together with an internal square. Note that the triangle has $Area(\triangle FBG) = \frac{1}{2}\overline{BF} \cdot \overline{BG} = \frac{1}{2}ab$. Thus the large square can be thought of as formed from four copies of the triangle together with square *FGHE* and so has area

$$Area(ABCD) = Area(FGHE) + 4\left(\frac{1}{2}ab\right) = c^2 + 2ab.$$

Alternatively, the square *ABCD* can be thought of as formed by four rectangles and the square *IJKL*, and so

$$Area(ABCD) = 4Area(AFIE) + Area(IJKL) = 4ab + (b-a)^2 = a^2 + b^2 + 2ab.$$

Therefore, $c^2 = a^2 + b^2$. Note that to make a rigorous proof of this, we would also have to show that all of the lines are straight, using Propositions I.32 and I.14.

Proof #36: Bhaskara, (ca. 1150 AD), India: The square of side c is divided into four copies of the triangle together with a square of sides $b - a$. Thus

$$c^2 = 4\left(\frac{1}{2}ab\right) + (b-a)^2 = 2ab + (b^2 - 2ab + a^2) = a^2 + b^2.$$

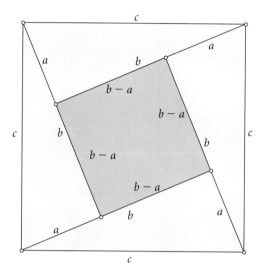

Proof #46: Leonardo da Vinci (1452–1519): Consider the figure below comprised of squares of sides a and b and two copies of the triangle. Cut along the dotted line and reflect the lower portion and rejoin it. The new figure can be redivided into two copies of the triangle and a square of side c. Thus $c^2 = a^2 + b^2$.

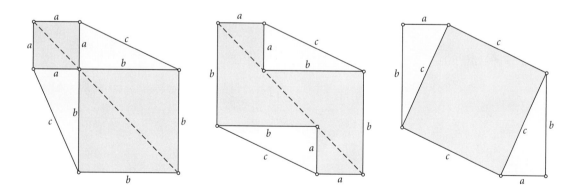

Proof #35: Rev. A. D. Wheeler, 1809: Consider the drawing below, formed from triangles with sides a, b, and c with a right angle between sides a and b. Note that the angles at each interior vertex add up to 180° (using Proposition I.32), and so form a line by Proposition I.14. Thus four copies of the triangle together with a square of side c form a larger square of side $a + b$. In the second picture, four copies of the triangle together with squares of sides a and b form a larger square of side $a + b$. Thus $c^2 = a^2 + b^2$.

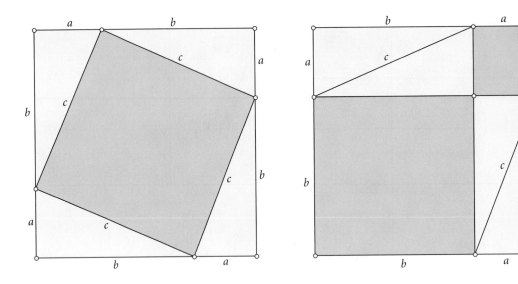

Proof #1: Adrien Legendre, 1858: This is the shortest known proof. Consider right triangle ABC with $\overline{AC}=b$, $\overline{BC}=a$, and $\overline{AB}=c$. Drop $CD \perp AB$, dividing side $\overline{AB}=c$ into lengths $c-y$ and y. Let $x=\overline{CD}$. Note that $\angle A + \angle B + 90° = 180° = \angle A + \angle ACD + 90°$. Thus $\angle ACD = \angle B$. Similarly, $\angle BCD = \angle A$. Thus we have similar triangles $\triangle BCD \sim \triangle CAD \sim \triangle BAC$. From this, nine equations can be formed showing the proportions of the sides. Two of these are:

$$\frac{a}{y} = \frac{c}{a}, \qquad \frac{b}{c-y} = \frac{c}{b},$$

and so

$$a^2 = cy, \qquad b^2 = c^2 - cy.$$

Thus, $b^2 = c^2 - a^2$.

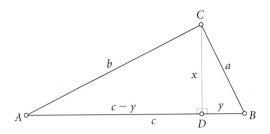

Proof #231: President James Garfield, 1876: Consider right triangle ABC with $\overline{AC}=b$, $\overline{BC}=a$, and $\overline{AB}=c$. Extend $\overline{CB}$ and cut at D so that $\overline{BD}=\overline{AC}=b$. Construct a line through D parallel to AC and cut at E so that $\overline{DE}=\overline{BC}=a$. Draw $\overline{BE}$ and $\overline{AE}$. Then $ACDE$ is a trapezoid. Also note that $\angle ABE = 90°$. The area of this trapezoid is thus

$$\text{Area} = \frac{1}{2}(b_1+b_2)h = \frac{1}{2}(\overline{AC}+\overline{ED})\overline{CD} = \frac{1}{2}(b+a)(a+b) = \frac{1}{2}(a+b)^2.$$

However, the area may also be computed as

$$\text{Area} = \text{Area}(\triangle ABC) + \text{Area}(\triangle ABE) + \text{Area}(\triangle BED),$$

$$\text{Area} = 2\left(\frac{1}{2}ab\right) + \frac{1}{2}c^2.$$

Thus,

$$\frac{1}{2}(a+b)^2 = ab + \frac{1}{2}c^2,$$

and so $c^2 = a^2 + b^2$.

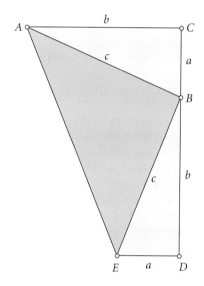

♦ 2.6 Hilbert's axioms for euclidean geometry

Many mathematicians in the late eighteenth century worked to put Euclid on a firm foundation. Some, like Moritz Pasch, pointed out hidden assumptions and suggested amendations to repair these lapses. Others suggested alternate primitive terms and other approaches. For example, in 1898, Mario Pieri succeeded in minimizing the number of primitive terms. He has one undefined term, *point*, and one undefined relation, *point A is equidistant from points B and C*. However, in minimizing the number of undefined terms, it is necessary to increase the number of postulates. Pieri's system for plane geometry requires twenty axioms. His approach is based on the idea of *rigid motions* or *isometries*: one-to-one transformations that preserve equidistance. We will discuss such isometries in Chapter 8.

The definitive set of axioms for euclidean geometry was developed by David Hilbert in 1899 in his *Grundlagen der Geometrie* (*Foundations of Geometry*), culminating 2000 years of piecemeal modifications for Euclid's original set of postulates. Hilbert's axioms have the advantage of being complete and independent. Logical independence means that no axiom depends on the others, so that for each axiom there is a system of geometry (some quite far-fetched) in which all of the axioms are true except the one under consideration, which is false.

We chose not to use Hilbert's set of twenty axioms, since we wanted to stay as close as possible to Euclid's original text. Instead we kept Euclid's five postulates and five common notions, somewhat modified, with additions bearing the content of Hilbert's axioms.

Hilbert's set of axioms is given below:

Primitive terms: point, line, plane, lie, between, congruent

I. Axioms of Incidence

I.1 For every two points A, B there exists a line that contains each of the points A, B.

I.2 For every two points A, B there exists no more than one line that contains each of the points A, B.

I.3 There exist at least two points on a line. There exist at least three points that do not lie on a line.

I.4 For any three points A, B, C that do not lie on the same line there exists a plane α that contains each of the three points A, B, C. For every plane there exists a point that it contains.

II. Axioms of Order (Betweenness)

II.1 If a point B lies between a point A and a point C, then the points A, B, C are three distinct points of a line, and B then also lies between C and A.

II.2 For two points A and C, there always exists at least one point B on the line AC such that C lies between A and B.

II.3 Of any three points on a line there exists no more than one that lies between the other two.

II.4 Let A, B, C be three points that do not lie on a line and let a be a line in the plane ABC which does not meet any of the points A, B, C. If the line a passes through a point of the segment AB, it also passes through a point of the segment AC or through a point of the segment BC.

These assumptions, or their equivalents, are included in our expanded version of Euclid's Postulate 1 and our Incidence Postulate 6 and Betweenness Postulate 7. We have omitted those dealing with planes until we discuss solid geometry in Chapter 10. Note that Axiom II.4 is our Pasch's theorem 2.11. See Project 2.5.

▶ **Exercise 2.87.** Use Hilbert's axioms of incidence and order to prove his Theorem 3: Given any two distinct points A and B, there exists at least one other point C on the line segment between them.

▶ **Exercise 2.88.** Prove Hilbert's Theorem 7: Given any two points, there are infinitely many points on the line segment connecting them.

Hilbert's Theorem 8 is our Plane Separation Property 8.

III. Axioms of Congruence

III.1 If A, B are two points on a line a, and A' is a point on the same or on another line a' then it is always possible to find a point B' on a given side of the line a' through A' such that the segment AB is congruent or equal to the segment $A'B'$.

III.2 If a segment $A'B'$ and a segment $A''B''$ are congruent to the same segment AB, then the segment $A'B'$ is also congruent to the segment $A''B''$, or briefly, if two segments are congruent to a third one they are congruent to each other.

III.3 On the line a let AB and BC be two segments which except for B have no point in common. Furthermore, on the same or on another line a' let $A'B'$ and $B'C'$ be two segments which except for B'

also have no point in common. In that case, if AB is congruent to $A'B'$ and BC is congruent to $B'C'$, then AC is congruent to $A'C'$.

III.4 Let $\angle(h, k)$ [where h and k are rays emanating from point O] be an angle in a plane α and a' a line in a plane α' and let a definite side of a' in α' be given. Let h' be a ray on the line a' that emanates from the point O'. Then there exists in the plane α' one and only one ray k' such that the angle $\angle(h, k)$ is congruent or equal to the angle $\angle(h'k')$ and at the same time all interior points of the angle $\angle(h'k')$ lie on the given side of a'. Every angle is congruent to itself.

III.5 If for two triangles ABC and $A'B'C'$ the congruences $AB = A'B'$, $AC = A'C'$, $\angle BAC = \angle B'A'C'$ hold, then the congruence $\angle ABC = \angle A'B'C'$ is also satisfied.

Hilbert's Axioms III.2 and III.3 are interpretations of Euclid's Common Notions, for the context of congruence (equality in length) of line segments, postulating transitivity and additivity. Axiom III.4 is essentially Euclid's Proposition I.23, which says that angles can be copied. Hilbert finds it necessary to postulate this since he has rejected Postulates 3 and 4.

▶ **Exercise 2.89.** Prove that Hilbert's Axiom III.5 implies the SAS congruence criterion.

▶ **Exercise 2.90.** Prove that if two angles are equal, then their supplementary angles are equal. I.e., $\angle ABC = \angle EFG$, and CB and GF are extended to D and H, respectively, then $\angle ABD = \angle EFH$.

▶ **Exercise 2.91.** Use Hilbert's Axiom III.4 and Exercise 2.90 to prove Euclid's Postulate 4 (Hilbert's Theorem 21).

IV. Axiom of Parallels

IV.1 Let a be any line and A a point not on it. Then there is at most one line in the plane, determined by a and A, that passes through A and does not intersect a.

Axiom IV.1 implies Playfair's Postulate, which we know is logically equivalent to Euclid's Postulate 5.

V. Axioms of Continuity

V.1 (Archimedes' Axiom) If AB and CD are any segments then there exists a number n such that n segments CD constructed contiguously from A, along the ray from A to B, will pass beyond B.

V.2 (Line Completeness) An extension of a set of points on a line with its order and congruence relations that would preserve the relations existing among the original elements as well as the fundamental properties of line order and congruence that follows from Axioms I–III and from V.1 is impossible.

Axiom V.1 is our Postulate 10, Archimedes' Axiom. Axiom V.2 is where we part company with Hilbert entirely. Hilbert is postulating the completeness of the real numbers. In order to use this, one must have some familiarity with the properties and consequences of this completeness, which we are unwilling to assume in our intended audience. Euclid ignores all questions dealing with continuity, as when he assumes the existence of points of intersection of any two lines or certain circles. Instead of adopting Hilbert's Axiom V.2, we have instead given the particular circumstances where this principle is most commonly applied: the Plane Separation Property, Pasch's Theorem, and the Circular Continuity Principle. All of these could be proven from the completeness of the real numbers, but only with a great deal of difficulty. However, note that in our explanation of Theorem 6.12 on the existence of asymptotic parallel lines in hyperbolic geometry, we will need to appeal to the full power of the completeness of the real numbers. Except for that one time, our postulates suffice, and, we feel, are more accessible. We wanted to adopt a set of axioms that are useful and transparent rather than insisting on logical independence.

Hilbert's intention in devising this set of axioms was to include all the suggestions made over the years by various commentators on Euclid, to come up with a consistent and independent set of axioms, and to eliminate the dependence on drawings inherent in Euclid's approach. The order axioms allow one to precisely describe where points fall in relationship with one another, and so the drawings become optional visual aids. He also had the view that we must not be bound by preconceptions about the undefined primitive terms. He once said, "Instead of points, straight lines, and planes, it would be perfectly possible to use the words tables, chairs, and tankards."[5] In this he meant that logically the axioms, and thus, the theorems, should be true for any objects that satisfy his list of axioms.

Raymond Queneau took this idea and reduced it to an absurdity by substituting "word" for point, "sentence" for line, and "paragraph" for plane.[6] We quote some of our favorite parts:

I,1   *A sentence exists containing two given words.*
COMMENT: Obvious. Example: given the two words "a" and "a" there exists a sentence containing these two words—"A violinist gives the vocalist her A."

I,2   *No more than one sentence exists containing two given words.*
COMMENT: This, on the other hand, may occasion surprise. Nevertheless, if one considers the words "years" and "early," once the following sentence containing them has been written, namely "For years I went to bed early," clearly all other sentences such as "For years I went to bed right after supper" or " For years I did not go to bed late" are merely pseudo-sentences that should be rejected by virtue of the above axiom.
SCHOLIUM: Naturally, if "For years I went to bed right after supper" is the sentence written originally, "For years I went to bed early" becomes the sentence to be excluded by virtue of the axiom I,2. In other words, no one can write *A la recherche du temps perdu* twice.

I,3   *There are at least two words in a sentence; at least three words exist that do not all belong to the same sentence.*

...

I,4a   *A paragraph exists including three words that do not all belong to the same sentence.*
COMMENT: A paragraph consequently comprises at least two sentences.

---

[5]Quoted in Gray, *Worlds out of Nothing*, p. 254.
[6]Raymond Queneau, *The Foundations of Literature,* translated and reprinted in *Oulipo Laboratory.*

It is to be noted that the manner in which the axioms I,1 through I,4 are formulated contradicts axiom I,2, since all four require for their articulation the words "words" and "sentences" whereas, according to the said axiom, no more than one sentence containing them should exist.
It is therefore possible to formulate the following metaliterary axiom:
Axioms are not governed by axioms.

...

Theorem 2.7. Between two words of a sentence there exists an infinity of other words.
COMMENT: No doubt a reader surprised by axiom II,2 will deem his surprise justified. To overcome his astonishment and understand these theorems he need only admit the existence of what we shall call, following the example of traditional projective geometry, "imaginary words" and "infinitesimal words." Every sentence contains an infinity of words; only an extremely limited number of them is perceptible; the rest are infinitesimal or imaginary. Many thoughtful minds have had a premonition—but never a clear awareness—of this. No longer will it be possible for students of rhetoric to ignore so crucial a theorem.

There are, of course, many other axiom systems for euclidean geometry. Another system, published in 1932, is due to George David Birkhoff [1884–1944], based on a completely different view. In this system, correspondences are assumed between the points on any line and the real numbers and also between the set of angles through a point and the real numbers. Thus the completeness of the real numbers assumes an even more central role. However, by submerging all geometric difficulties by burying them in the structure of the real numbers, Birkhoff can assume only five axioms.

Primitive terms: point, line, distance, angle

1. (Linear Measure) The points $A, B, \ldots$, of any line m can be placed into a one-to-one correspondence with the real numbers r so that $|r_B - r_A| = d(A, B)$ for all points A and B.
2. (Point-line) One and only one line m contains two given points P and Q ($P \neq Q$).
3. (Angular measure) The half-lines $m, n, \ldots$, through any point O can be placed into a one-to-one correspondence with real numbers $a(mod\ 2\pi)$ so that if $A \neq O$ and $B \neq O$ are points of m and n, respectively, the difference $(a_n - a_m)(mod\ 2\pi)$ is $m\angle AOB$.
4. All straight angles have the same measure.
5. (Similarity) If $\triangle ABC$ and $\triangle A'B'C'$, and for some positive constant, k, $d(A', B') = kd(A, B)$, $d(A', C') = kd(A, C)$, and also $m\angle BAC = \pm m\angle B'A'C'$, then also $d(B', C') = kd(B, C)$ and $m\angle C'B'A' = \pm m\angle CBA$ and $m\angle A'C'B' = \pm m\angle ACB$.

While Birkhoff does not include either Euclid's Postulate 5 nor Hilbert's Axiom V.1, the existence of similar triangles in Birkhoff's Axiom 5 is logically equivalent to the Parallel Postulate 5. Also if the constant k in Axiom 5 is equal to 1, then this axiom implies the SAS property. Note that these axioms are not intended to be independent. The interested reader is referred to *Basic Geometry* by Birkhoff and Ralph Beatley.

♦ 2.7 Distance and geometry

Euclid's approach to distance in geometry has been described as *caliper geometry*. Consider a pair of calipers, like a compass with probes at both arms instead of a pencil at one arm. Using these calipers and given two line segments, one can determine if one line segment is longer or shorter than the other, or if they are the same length. However, there is no concept of length, except in this comparative sense. This concept of comparative length was implicit in our discussion of the length of a line segment in Section 2.1. However, Birkhoff's axiom system of Section 2.6, as well as many modern axiom systems, explicitly relies on a measure of distance. In this section, we discuss the properties of the standard distance measure.

Euclid's geometry made little use of algebra, and what was used was expressed in geometric terms. After the introduction of arabic numerals with their superior adaptation to computation, geometry and algebra tended to mature independently. René Descartes [1596–1650] was the first to bring them back together in a marriage of convenience, giving birth to the field of *analytic geometry*. In this system, a point is considered as an *ordered pair* of real numbers, of the form (x, y). Geometric objects are considered as subsets of the familiar *cartesian coordinate plane*. The distance from point $P(x_1, y_1)$ to $Q(x_2, y_2)$ is measured by the length of the line segment $\overline{PQ}$ and is given by the formula

$$d(P, Q) = \sqrt{(x_2 - x_1)^2 + (y_2 - y_1)^2}.$$

This formula is a simple application of the Pythagorean Theorem, and thus, the standard coordinate system is explicitly founded on euclidean geometry.

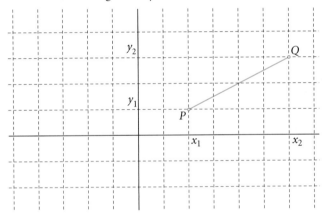

▶ **Exercise 2.92.** Find a formula for the midpoint of the line segment $\overline{PQ}$ where $P(x_1, y_1)$ and $Q(x_2, y_2)$.

The great advantage of the cartesian plane is that geometric objects can be represented as algebraic equations, which are sometimes more amenable. For example, lines are represented by equations of the form $ax + by = c$ for some $a, b, c \in \mathbb{R}$. One can solve such an equation for y to get either $x = a$ (a vertical line), or an equation of the form $y = mx + b$. In this form, m is called the *slope* of the line and measures its vertical inclination. We assume that the standard notation and formulae of coordinate geometry are familiar.

▶ **Exercise 2.93.** Find the equation of the line through the points $(2, 3)$ and $(2, -4)$. Graph the line.

▶ **Exercise 2.94.** Find the equation of the line through the points $(2, 3)$ and $(6, 4)$. Graph the line.

2.7. DISTANCE AND GEOMETRY

▶ **Exercise 2.95.** Prove that two lines are parallel if and only if they have the same slope.

▶ **Exercise 2.96.** Prove that perpendicular lines with slopes m_1 and m_2 will satisfy the equation $m_1 m_2 = -1$.

▶ **Exercise 2.97.** Consider two points $A(x_1, y_1)$ and $B(x_2, y_2)$. Describe algebraic conditions for a point C to lie on AB between A and B.

▶ **Exercise 2.98.** Find the intersection of the lines $3x + 2y = 3$ and $-x + 5y = 16$.

In Definition 2.15, a circle is described as the set of points at a given distance from the center point. In terms of the euclidean distance function, this means that a circle centered at (x_0, y_0) with radius r will have an equation of the form

$$(x - x_0)^2 + (y - y_0)^2 = r^2.$$

We can actually locate the points of intersection guaranteed by the Circular Continuity Principle. Thus, if we assume the existence of a standard metric or distance measure, as Birkhoff did in his axiom system, then the incidence, betweenness, and separation properties, and the Circular Continuity Principle follow as consequences.

▶ **Exercise 2.99.** Find the intersection of the circle $x^2 + y^2 = 1$ and the line $y = x$.

▶ **Exercise 2.100.** Find the intersection of the circle $x^2 + y^2 = 1$ and the circle $(x - 1)^2 + y^2 = 1$.

▶ **Exercise 2.101.** Given a point P and a line ℓ that does not pass through P, define the distance $d(P, \ell)$ from the point to the line.

The techniques of analytic geometry can be used to duplicate many of the results we have already found using purely geometric arguments. For example, consider the following proof of one portion of Proposition I.34:

Proposition: The opposite sides of a parallelogram are equal in length.

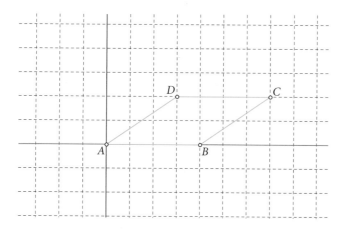

Proof: Consider a parallelogram $ABCD$. Without loss of generality, we may assume that A is placed at the origin and B lies along the x-axis, so $A(0,0)$ and $B(a,0)$. Let C be denoted by $C(b,c)$. These three points suffice to determine the position of $D(x,y)$. Since $ABCD$ is a parallelogram, $AB \parallel DC$, and since AB is given by the line $y=0$, DC must be the line $y=c$. Also, $BC \parallel AD$ and BC has slope $\frac{c}{b-a}$. Thus, the slope of AD is $\frac{y}{x} = \frac{c}{b-a}$, so, substituting $y=c$ into this equation, we find $x = b-a$, so $D(b-a,c)$. We then use the distance formula to compute $d(A,B) = a$, $d(B,C) = \sqrt{(b-a)^2 + c^2}$, $d(C,D) = a$, and $d(A,D) = \sqrt{(b-a)^2 + c^2}$. Therefore, $\overline{AB} = \overline{DC}$ and $\overline{AD} = \overline{BC}$. □

As you can see in the proof above, much of the burden of the proof is placed on doing the relevant computations. In general, some geometric theorems are more easily done by purely geometric means, while others are more amenable to the analytic approach. Determining which approach to apply is a matter of guesswork, informed by previous experimentation. Neither tactic eliminates the occasional need for the other, so a good mathematician is thoroughly familiar with both.

▶ **Exercise 2.102.** Prove that the diagonals of a rectangle bisect each other by finding the midpoints of these diagonals and showing that they coincide.

▶ **Exercise 2.103.** Prove that the diagonals of a square are perpendicular by computing the slopes and then showing that $m_1 m_2 = -1$.

The concept of using a distance measure to define a geometry can be generalized. The following definition is based on the known properties of the distance for the cartesian plane.

Definition 2.27. Let X be a nonempty set of points. A *metric* or distance function on X is a function $d : X \times X \to \mathbb{R}$ such that for all $P, Q, R \in X$,

1. $d(P, Q) \geq 0$,
2. $d(P, Q) = 0$ if and only if $P = Q$,
3. $d(P, Q) = d(Q, P)$,
4. $d(P, Q) \leq d(P, R) + d(R, Q)$.

As we will see in Chapter 7, changing the metric actually changes the geometry of the underlying plane.

▶ **Exercise 2.104.** Prove that the euclidean distance formula defines a metric.

▶ **Exercise 2.105.** Prove that for the euclidean metric, $d(P, Q) = d(P, R) + d(R, Q)$ if and only if P, Q, and R are collinear and R lies between P and Q.

◆ **2.8 Projects**

Project 2.1. Read and report on Charles L. Dodgson's *Euclid and his Modern Rivals*. Note that Dodgson also wrote under the pseudonym Lewis Carroll.

Project 2.2. Investigate some other proofs of the Pythagorean Theorem.

Project 2.3. Consider the set of points defined as ordered pairs (x, y) where both x and y are rational numbers. Define a line in this geometry as a line through any two such points. Prove that Incidence Postulate 6 and Betweenness Postulate 7 are true in this model. Prove that the Plane and Line Separation Properties also hold. Prove that the Circular Continuity Principle fails.

Project 2.4. Consider the set of points defined as ordered pairs (x, y) where both x and y are *dyadic rationals*: numbers of the form $\frac{a}{2^n}$ for some $a, n \in \mathbb{Z}$. Define a line in this geometry as a line through any two such points. Prove that Incidence Postulate 6 and Betweenness Postulate 7 are true in this model. Prove that the Line Separation Property also holds. Prove that the Plane Separation Property fails. [Hint: Consider the intersection of the lines $3x + y = 1$ and $y = 0$.]

Project 2.5. Take Pasch's Theorem 2.11 as a postulate and from that prove the Plane Separation Property. First prove that lying on the same side of a given line defines an equivalence relation and then prove that a line divides the points of the plane not on the given line into exactly two nonempty disjoint sets.

Project 2.6. Prove the second clause of Archimedes' Axiom (Postulate 10) from the first.[7]

Project 2.7. Investigate Dedekind's Axiom: Given a line ℓ and let the points of ℓ be divided into two disjoint sets S and T with the property that no point of either set lies between two points of the other set. Then there exists a unique point X on ℓ so that for any $A \in S$ and $B \in T$, either $A = X$ or $B = X$ or $A - X - B$.

a. Prove that Dedekind's Axiom implies Archimedes' Axiom 10.
b. Prove that Dedekind's Axiom implies our Postulate 11: Circular Continuity Principle.

Project 2.8. Borsuk and Szmielew adopt two additional congruence axioms about line segments and triangles:

1. Given $\triangle ABC$ and line segment $\overline{DE}$ with $\overline{AB} = \overline{DE}$. Then on a given side of DE there is a unique point F so $\overline{AC} = \overline{DF}$ and $\overline{BC} = \overline{EF}$.
2. Given $\triangle ABC$ and $\triangle EFG$ and points D and H so that $A - D - B$ and $E - H - F$, if $\overline{AD} = \overline{EH}$, $\overline{DB} = \overline{HF}$, $\overline{AC} = \overline{EG}$, and $\overline{DC} = \overline{HG}$, then $\overline{BC} = \overline{FG}$.

With these assumptions, we can drop part (2) of the Congruence Postulate 9. Define congruence for angles and then prove Postulate 9(2).[8]

Project 2.9. Given a line segment and a point on the plane, show how to construct a circle with the given point as center and the given line segment as radius using a floppy compass.

Project 2.10. Prove that Hilbert's axioms are true in cartesian coordinate geometry using the standard euclidean distance formula.

---

[7] A reference is Robin Hartshorne's *Geometry: Euclid and Beyond*, Lemma 35.1.
[8] References are K. Borsuk and W. Szmielew, *Foundations of Geometry*, and Martin J. Greenberg, *Euclidean and Non-Euclidean Geometries*.

3. More euclidean geometry

In this chapter, we collect some applications of the euclidean geometry developed in Chapter 2. Much of this material, such as the circle theorems and the theory of similarity, are scattered among several of the earlier books of Euclid's *The Elements*. The last two sections include some more modern geometric theorems and constructions, still in the context of euclidean geometry.

♦ 3.1 Circle theorems

This section concerns some properties of circles and the angles formed inside a circle and how two circles can intersect. We also discuss the construction of tangent lines for circles, an application that does not require calculus. Most of this material is from Book III of Euclid's *The Elements*. A circle is defined to be the set of points at a given distance (the radius) from a given point (the center). From this, we can define other entities: a *diameter* is a straight line segment through the center of a circle which terminates at two points on the circle. Two points on the circle are *opposite* or *antipodal* if they are the endpoints of a diameter. A *chord* $\overline{AB}$ in a circle with center O is a line segment whose endpoints A and B are two points on the circle. Note that two points A and B on a circle will divide that circle into two pieces. The shorter of these is called the *minor arc* and the longer the *major arc*. Of course, if these are equal, they are called *semicircles*. An arc $\widehat{AB}$ of a circle denotes the minor arc between two nonopposite points on the circle. The angle $\angle AOB$ formed by the radii is the *angle subtended by the chord* $\overline{AB}$ or the arc $\widehat{AB}$ and denotes the smaller of the angles formed.

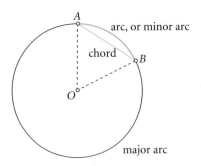

▶ **Exercise 3.1.** Prove that equal chords in a circle subtend equal angles: That is, if $\overline{AB} = \overline{CD}$ in the illustration below, prove that $\angle AOB = \angle COD$. Is the converse of this statement true? Prove or give a counterexample.

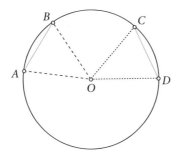

▶ **Exercise 3.2.** Prove that if a chord in a circle is bisected by a diameter, then the diameter and the chord are perpendicular. Is the converse of this statement true? Prove or give a counterexample.

Thales' reputation as the Father of Geometry is based in part on his proof of the following theorem:

Theorem 3.1. [Thales' Theorem] An angle inscribed in a semicircle is a right angle.

▶ **Exercise 3.3.** Prove Thales' Theorem 3.1: consider a triangle $\triangle ABC$ where $\overline{AB}$ is a diameter of a circle and C another point on the circle and then show that $\angle ACB = 90°$.

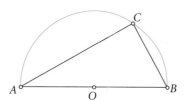

Theorem 3.2. If triangle $\triangle ABC$ is a right triangle with $\angle ACB = 90°$, then there is a circle that passes through C and with diameter $\overline{AB}$.

▶ **Exercise 3.4.** Prove Theorem 3.2, the converse of Thales' Theorem.

We proceed to investigate some generalizations of Thales' Theorem 3.1. Consider a circle with center O and an arc $\widehat{BC}$ on that circle. The angle $\angle BOC$ subtended by the arc $\widehat{BC}$ is also called the *central angle* of the arc. Choose a third point A on the circle but not on the arc $\widehat{BC}$. The angle $\angle BAC$ formed by the endpoints of the arc and the third point is called an *inscribed angle* of the arc.

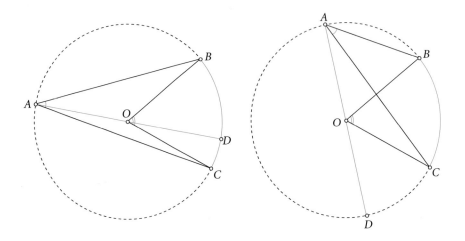

▷ **Activity 3.1.** Given a circle with center O and an arc $\widehat{BC}$ and a point A on the circle but not on the arc, formulate a conjecture regarding the measures of the central and inscribed angles. Note the different cases pictured above.

Theorem 3.3. Given a circle with center O and an arc $\widehat{BC}$ and a point A on the circle but not on the arc, the measure of the central angle is twice the measure of the inscribed angle.

▶ **Exercise 3.5.** Prove Theorem 3.3, considering separately the cases where D, the endpoint of the diameter formed by line AO, falls on either point B or C, falls on the arc $\widehat{BC}$, or falls outside arc $\widehat{BC}$.

Corollary 3.4. The angle formed by the endpoints of a fixed arc or chord of a circle and an arbitrary third point on the circle exterior to the chord is always the same. Thus, in the illustration below, $\angle BAC = \angle BDC$.

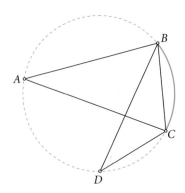

▶ **Exercise 3.6.** Prove Corollary 3.4.

▷ **Activity 3.2.** Consider the case excluded from Corollary 3.4 where D falls on the arc, as shown below. Find a relationship between the angles ∡BAC and ∡BDC.

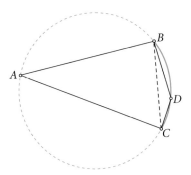

Definition 3.5. A quadrilateral is *cyclic* if it is convex and all its vertices lie on some circle. It is then said that the circle *circumscribes* the quadrilateral.

Cyclic quadrilaterals are easily constructed, so their characterization is very useful and important in geometric constructions.

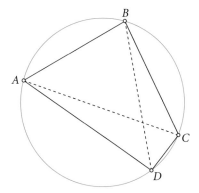

▶ **Exercise 3.7.** Prove that if a quadrilateral ABCD is cyclic, then the pairs of opposite angles sum to 180°, i.e., ∡BAD + ∡BCD = 180° and ∡ABC + ∡ADC = 180°.

▶ **Exercise 3.8.** Prove that if a quadrilateral ABCD is cyclic, then ∡DAC = ∡DBC.

▶ **Exercise 3.9.** Formulate and prove a conjecture about opposite angles for a self-intersecting quadrilateral inscribed in a circle, such as ∡DAB and ∡DCB below.

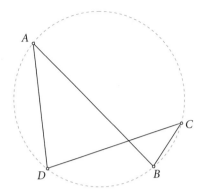

The definition of the tangent line to a circle is the starting point for the modern theory of tangents lines to curves, and more generally, tangent planes for surfaces. The general theory of tangents requires a knowledge of calculus. The Greeks, especially Archimedes, recognized the importance of tangent lines and developed techniques for finding them for certain special curves. Most of their methods were made obsolete by the invention of calculus by Sir Isaac Newton [1643–1727] and Gottfried Leibniz [1646–1716]. However, in the case of the circle, the geometric construction is far simpler than differentiation.

From calculus, we are used to thinking of a tangent line as the best linear approximation to a curve, a concept implicit in the definition of the slope of the tangent line as a limit of slopes of secant lines. In the case of the circle, the tangent line has the property that it touches the circle exactly once:

Definition 3.6. A *secant line* to a circle is a line intersecting a circle at two points. A *tangent line* to a circle is a line which intersects the circle only once.

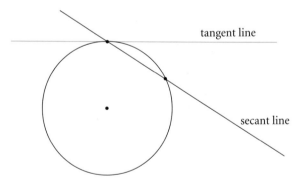

▷ **Activity 3.3.** Given a circle with its center and one marked point on the circumference, use either straightedge and compass or equivalent geometric software to construct a line tangent to the circle through the marked point. Given a circle with its center and one marked point outside the circle, construct a line tangent to the circle through the marked point.

▶ **Exercise 3.10.** Prove that the line perpendicular to a given radius at its endpoint must be tangent to the circle. Is the converse of this statement true? Prove or give a counterexample.

Euclid proves the statement in Exercise 3.10 in Proposition III.16. He then expresses the property of being tangent by saying "into the space between the straight line and the circumference another line cannot be interposed," which is a fairly clear description of our intuitive understanding of tangency from calculus. Euclid proves this statement as follows: We are given a circle centered at *D* with diameter *AB*. Points *C* and *H* lie on the circle. He has constructed *AE* perpendicular to the diameter *AB* and shown that it intersects the circle exactly once. He then considers another line *AF* through *A*, and shows that a certain point *G* on line *AF* cannot lie outside the circle (see his explanation below and the accompanying illustration):

> *I say next that into the space between the straight line AE and the circumference [arc] CHA another straight line cannot be interposed. For, if possible, let another straight line be so interposed, as FA, and let DG be drawn from the point D perpendicular to FA [and intersecting the circle at H]. Then, since the angle AGD is right, and the angle DAG is less than a right angle, AD is greater than DG [Prop. I.19]. But DA is equal to DH; therefore DH is greater than DG: the less than the greater, which is impossible. Therefore another straight line cannot be interposed into the space between the straight line and the circumference.*

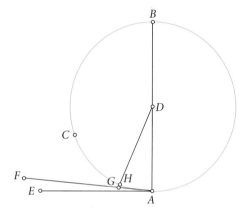

Thus, given a circle and a point on the circle, the procedure of Exercise 3.10 shows us how to construct the tangent to the circle at that point. The next theorem tells us how to construct a tangent line through a point not on the circle:

Theorem 3.7. [Euclid's Proposition III.17] Given a circle and a point outside the circle, one can construct a line through the point tangent to the circle.

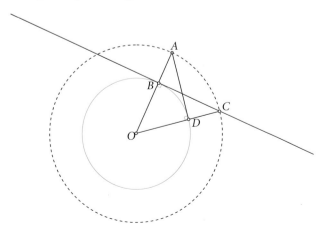

Proof: We are given a circle with center O and point A outside the circle. Draw $\overline{AO}$. By the Circular Continuity Principle 11(1), $\overline{AO}$ intersects the circle at B. Now, draw a circle with center O through A. Construct a line through B perpendicular to AO. Using the variation on Circular Continuity Principle proven in Exercise 2.41, this line must intersect the second circle at a point C. Draw $\overline{OC}$. A third application of Circular Continuity shows that $\overline{OC}$ intersects the first circle at D. Then we know that $\overline{OA} = \overline{OC}$, $\overline{OB} = \overline{OD}$, and of course $\angle AOD = \angle COB$, so by SAS, $\triangle AOD \cong \triangle COB$. Therefore, $\angle ADO = \angle CBO = 90°$. Thus, $AD \perp OC$ and AD is tangent to the original circle by Exercise 3.10. □

▶ **Exercise 3.11.** Given a circle and a point A outside the circle as in Theorem 3.7, show that there are actually two lines through A both tangent to the circle.

We will have occasion to discuss the way two curves or circles intersect. In order to make sense of such an angle formed by two curves, we need the following definition:

Definition 3.8. The *angle formed between two intersecting curves* at their point of intersection is the angle between their tangent lines at that point.

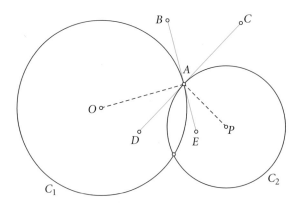

Of course, this makes sense for any two curves for which we can find tangent lines, but we will primarily be concerned with the case of a line and a circle and the case of two circles, as shown above. In this illustration, we have two circles, C_1 centered at O and C_2 centered at P. These intersect at A and another unlabeled point. We have drawn radii $\overline{OA}$ and $\overline{PA}$ and constructed lines BA which is tangent to C_1 and CA which is tangent to C_2. Then the angle formed by the circles C_1 and C_2 is defined to be the angle $\angle BAC$, rather than the angle $\angle BAD$.

▷ **Activity 3.4.** Construct two circles centered at O and P and intersecting at A with tangent lines BA and CA as above. Formulate a conjecture on the relationship between $\angle BAC$ and $\angle OAP$.

▶ **Exercise 3.12.** Prove your conjecture of Activity 3.4.

Two additional theorems, Euclid's Propositions III.36 and III.37, will be needed in Section 3.4. These may be omitted if you decide not to cover that section.

3.1. CIRCLE THEOREMS • 87

Theorem 3.9. [Euclid's Proposition III.36] From a given point outside a given circle, draw a line tangent to the circle and another line through the point intersecting the circle at two points. Then the square of the distance from the given point to the point of tangency is equal to the product of the distances from the point to the endpoints of the chord formed by the second line.

Proof: We are given a circle centered at O and a point A outside the circle. Use Theorem 3.7 to draw $\overline{AB}$ tangent to the circle. Let AC be the second line which intersects the circle at points C and D. We claim that $\overline{AB}^2 = \overline{AC} \cdot \overline{AD}$.

Case 1: $\overline{CD}$ is a diameter.

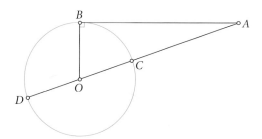

By Exercise 3.10, $\angle ABO = 90°$. By the Pythagorean Theorem,

$$\overline{OB}^2 + \overline{AB}^2 = \overline{AO}^2 = (\overline{AD} - \overline{DO})^2,$$
$$\overline{OB}^2 + \overline{AB}^2 = \overline{AD}^2 - 2\overline{AD} \cdot \overline{DO} + \overline{DO}^2.$$

Since $\overline{DO} = \overline{OB}$ and $\overline{DC} = 2\overline{DO}$,

$$\overline{AB}^2 = \overline{AD}^2 - 2\overline{AD} \cdot \overline{DO} = \overline{AD}(\overline{AD} - 2\overline{DO}) = \overline{AD}(\overline{AD} - \overline{DC}) = \overline{AD} \cdot \overline{AC}.$$

Case 2: $\overline{CD}$ is not a diameter.

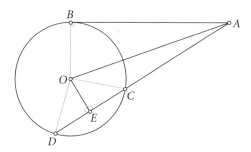

In this case, draw $\overline{OB}, \overline{OD}, \overline{OC},$ and $\overline{AO}$. Drop $OE \perp AD$. Note that by Exercise 3.2, OE is the perpendicular bisector of $\overline{CD}$, so $\overline{CE} = \overline{DE}$. Thus,

$$\overline{AE}^2 = (\overline{AE} + \overline{EC}) \cdot (\overline{AE} - \overline{EC}) + \overline{EC}^2 = \overline{AD} \cdot \overline{AC} + \overline{EC}^2.$$

Therefore, using the Pythagorean Theorem for $\triangle OEC$,

$$\overline{AE}^2 + \overline{OE}^2 = \overline{AD} \cdot \overline{AC} + \overline{EC}^2 + \overline{OE}^2 = \overline{AD} \cdot \overline{AC} + \overline{OC}^2.$$

Using the Pythagorean Theorem again for $\triangle OEA$ and noting that $\overline{OC} = \overline{OB}$,

$$\overline{AE}^2 + \overline{OE}^2 = \overline{AD} \cdot \overline{AC} + \overline{OB}^2 = \overline{AO}^2.$$

Now apply the Pythagorean Theorem to $\triangle ABO$ to get

$$\overline{AB}^2 = \overline{AO}^2 - \overline{OB}^2 = \overline{AD} \cdot \overline{AC}.$$

□

Theorem 3.10. [Euclid's Proposition III.37] From a point outside a given circle, draw a line intersecting the circle at two points and another line touching the circle. If the square of the distance from the given point to the point of intersection of the second line with the circle is equal to the product of the distances from the given point to the endpoints of the chord formed by the first line, then the second line is tangent to the circle.

▶ **Exercise 3.13.** Prove Theorem 3.10 (the converse of Theorem 3.9).

◆ 3.2 Similarity

The definition of similarity is modeled on Definition 2.14 of congruent triangles, substituting proportionality for equality of the lengths of the sides. Our main objective will be to come up with easier ways of proving similarity, just as we earlier proved the various congruence theorems for triangles, SSS, AAS, and ASA. Most of this material is from Book VI of Euclid's *The Elements*. Similarity is a notion that only exists in euclidean geometry. We will see in Chapters 6 and 7 (specifically, in Theorems 6.33 and 7.10) that similar but noncongruent triangles do not exist in some other geometries. Similar triangles also represent a geometric approach to the theory of ratios and proportion.

Definition 3.11. Two triangles are *similar* if their corresponding angles are equal and if their sides are in proportion. That is, triangles $\triangle ABC$ and $\triangle DEF$ are similar, written as $\triangle ABC \sim \triangle DEF$, whenever $\angle A = \angle D$, $\angle B = \angle E$, $\angle C = \angle F$, and $\frac{\overline{AB}}{\overline{DE}} = \frac{\overline{AC}}{\overline{DF}} = \frac{\overline{BC}}{\overline{EF}}$.

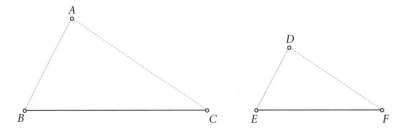

▶ **Exercise 3.14.** Prove that if two triangles are congruent, then they are similar.

▶ **Exercise 3.15.** Prove that similarity defines an equivalence relation for triangles:
   a. Show that for any triangle, $\triangle ABC \sim \triangle ABC$.
   b. Show that if $\triangle ABC \sim \triangle DEF$, then $\triangle DEF \sim \triangle ABC$.
   c. Show that if $\triangle ABC \sim \triangle DEF$ and $\triangle DEF \sim \triangle GHI$, then $\triangle ABC \sim \triangle GHI$.

When Euclid says two triangles are under the same parallel, he means that their bases lie along a line which is parallel to the line connecting their apex vertices. We have rephrased and streamlined his proof of the lemma below in terms of area, letting $A(\triangle ABC)$ denote the area of the triangle.

Lemma 3.12. [Euclid's Proposition VI.1] Triangles which are under the same parallel have areas in the same proportion as their bases.

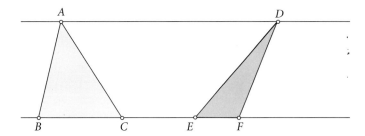

Proof: Consider $\triangle ABC$ and $\triangle DEF$ with $AD \| BC$ and EF lying along BC. These triangles have the same height and so by Exercise 2.76, $A(\triangle ABC) = \frac{1}{2}\overline{BC} \cdot h$ and $A(\triangle DEF) = \frac{1}{2}\overline{EF} \cdot h$. Thus, $\frac{A(\triangle ABC)}{A(\triangle DEF)} = \frac{\overline{BC}}{\overline{EF}}$. □

Theorem 3.13. [Euclid's Proposition VI.2, part 1] If a triangle is cut by a line parallel to one of its sides, then this line cuts the sides of the triangle proportionally.

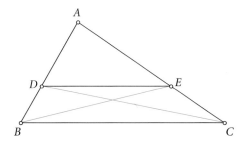

Proof: We are given $\triangle ABC$ cut by a line $DE \| BC$. Draw $\overline{BE}$ and $\overline{CD}$. By Lemma 3.12 above, $\frac{A(\triangle ADE)}{A(\triangle BDE)} = \frac{\overline{AD}}{\overline{BD}}$ and $\frac{A(\triangle ADE)}{A(\triangle CDE)} = \frac{\overline{AE}}{\overline{CE}}$. By Proposition I.37 (which we explored in Activity 2.22 and proved in Exercise 2.70.), $A(\triangle BDE) = A(\triangle CDE)$. Therefore, $\frac{A(\triangle ADE)}{A(\triangle BDE)} = \frac{A(\triangle ADE)}{A(\triangle CDE)}$, and so $\frac{\overline{AD}}{\overline{BD}} = \frac{\overline{AE}}{\overline{CE}}$. □

Corollary 3.14. If a triangle is cut by a line parallel to one of its sides, then the triangle formed is similar to the original triangle.

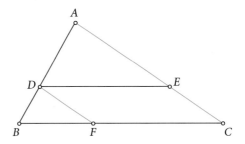

Proof: We are again given $\triangle ABC$ cut by $DE \| BC$. By Proposition I.29, $\angle ADE = \angle ABC$ and $\angle AED = \angle ACB$. By Theorem 3.13, $\frac{\overline{AE}}{\overline{CE}} = \frac{\overline{AD}}{\overline{BD}}$. Taking reciprocals gives us $\frac{\overline{CE}}{\overline{AE}} = \frac{\overline{BD}}{\overline{AD}}$ and so

$$\frac{\overline{CE}}{\overline{AE}} + \frac{\overline{AE}}{\overline{AE}} = \frac{\overline{AC}}{\overline{AE}} = \frac{\overline{BD}}{\overline{AD}} + \frac{\overline{AD}}{\overline{AD}} = \frac{\overline{AB}}{\overline{AD}}.$$

Now, draw $DF \| AC$ and similarly conclude that $\frac{\overline{BF}}{\overline{FC}} = \frac{\overline{BD}}{\overline{DA}}$ and so

$$\frac{\overline{BF}}{\overline{FC}} + \frac{\overline{FC}}{\overline{FC}} = \frac{\overline{BC}}{\overline{FC}} = \frac{\overline{BD}}{\overline{AD}} + \frac{\overline{AD}}{\overline{AD}} = \frac{\overline{AB}}{\overline{AD}}.$$

Note that $DFCE$ is a parallelogram and so $\overline{DE} = \overline{FC}$ by Proposition I.34. Combining the equations above, we get $\frac{\overline{BC}}{\overline{DE}} = \frac{\overline{AB}}{\overline{AD}} = \frac{\overline{AC}}{\overline{AE}}$ and conclude that $\triangle ABC \sim \triangle ADE$ by Definition 3.11. □

Theorem 3.15. [Euclid's Proposition VI.2, part 2] If a triangle is cut by a line so that this line cuts the sides of the triangle proportionally, then the line is parallel to the base.

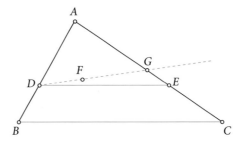

Proof: We are given $\triangle ABC$ cut by $\overline{DE}$ so that $\frac{\overline{AD}}{\overline{DB}} = \frac{\overline{AE}}{\overline{EC}}$. As in Corollary 3.14, a bit of algebra shows that then $\frac{\overline{AD}}{\overline{AB}} = \frac{\overline{AE}}{\overline{AC}}$. Let us assume that DE is *not* parallel to BC. Draw $DF \| BC$. By Pasch's Theorem 2.11, DF intersects $\overline{AC}$ at some point G.

▶ **Exercise 3.16.** Finish the proof of Theorem 3.15, obtaining a contradiction and concluding that *DE* must be parallel to *BC*.

▶ **Exercise 3.17.** If *ABCD* is an arbitrary quadrilateral, let *E* be the midpoint of $\overline{AB}$, *F* the midpoint of $\overline{BC}$, *G* the midpoint of $\overline{CD}$, and *H* the midpoint of $\overline{DA}$. Prove that *EFGH* is a parallelogram.

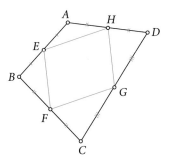

Corollary 3.14 is the foundation for our study of similarity theorems. Lemma 3.12 and Theorem 3.13 are essential steps in proving this corollary, while Theorem 3.15 is its converse. In practice, to show that two triangles are similar, we will usually construct a third triangle congruent to the smaller one and lying inside the larger and then apply Corollary 3.14.

Theorem 3.16 applies this technique to prove that there can be similar triangles that are not congruent. It should be noted that this phenomena only occurs in euclidean geometry. You can happily anticipate proving this in Exercise 5.28. Additionally, Theorem 6.33 will prove that similar triangles must be congruent in hyperbolic geometry, while Theorem 7.10 proves the analogous result in elliptic geometry.

Theorem 3.16. [**Euclid's Proposition VI.4: AAA similarity**] If two triangles have equal angles, then they are similar.

Proof: We are given $\triangle ABC$ and $\triangle DEF$ with $\angle A = \angle D$, $\angle B = \angle E$, and $\angle C = \angle F$. We proceed by cases.

Case 1: $\overline{AB} = \overline{DE}$: In this case, $\triangle ABC \cong \triangle DEF$ by ASA and so by Exercise 3.14, $\triangle ABC \sim \triangle DEF$.

Case 2: $\overline{AB} \neq \overline{DE}$:

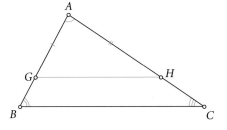
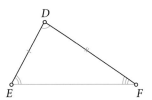

In this case, we assume, without loss of generality, that $\overline{AB} > \overline{DE}$. Cut $\overline{AG} = \overline{DE}$ and $\overline{AH} = \overline{DF}$ and draw $\overline{GH}$. Then $\triangle AGH \cong \triangle DEF$ by SAS and so $\angle AGH = \angle DEF = \angle ABC$. It follows from Proposition I.28 that $GH \| BC$. Thus, by Corollary 3.14, $\triangle ABC \sim \triangle AGH$. By Exercises 3.14 and 3.15, $\triangle ABC \sim \triangle DEF$. □

Corollary 3.17. [**AA similarity**] If two triangles have two equal angles, then they are similar.

▶ **Exercise 3.18.** Prove Corollary 3.17: Given two triangles △ABC and △DEF with ∠ABC = ∠DEF and ∠ACB = ∠DFE, prove that △ABC∼△DEF.

Theorem 3.18. [**Euclid's Proposition VI.5: SSS similarity**] Given two triangles such that the sides are proportional, the triangles are similar.

▶ **Exercise 3.19.** Prove Theorem 3.18: Given △ABC and △DEF with $\frac{DE}{AB} = \frac{DF}{AC} = \frac{EF}{BC}$, prove that △ABC∼△DEF.

Theorem 3.19. [**Euclid's Proposition VI.6: SAS similarity**] Given two triangles such that an angle in the first triangle is equal to an angle of the second triangle and the corresponding sides enclosing these angles are proportional, then the triangles are similar.

▶ **Exercise 3.20.** Prove Theorem 3.19.

▶ **Exercise 3.21.** Devise a definition for similarity between two quadrilaterals.

▶ **Exercise 3.22.** Show that SSSS does not imply similarity for quadrilaterals by exhibiting a counterexample.

▶ **Exercise 3.23.** Formulate and prove a SASAS similarity theorem for quadrilaterals.

▶ **Exercise 3.24.** Formulate and prove an ASASA similarity theorem for quadrilaterals.

♦ 3.3 More triangle theorems

We have four ways of proving triangles are congruent: SAS (Postulate 12), SSS (Proposition I.8), and ASA and AAS (Proposition I.26). In general SSA is not true: for a counterexample, consider the isosceles triangle △ACD shown below. Extend $\overline{DC}$ to B and draw $\overline{AB}$. Then $\overline{AC} = \overline{AD}$, $\overline{AB} = \overline{AB}$, and ∠ABC = ∠ABD, but △ABC and △ABD are clearly not congruent.

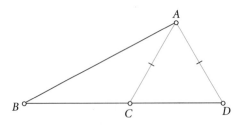

However, with one piece of additional information, SSA is true. We first prove a special case:

Theorem 3.20. [Hypotenuse-Side] If two right triangles have their hypotenuses equal and one other pair of sides equal, then the triangles are congruent.

▶ **Exercise 3.25.** Prove Theorem 3.20: Let $\triangle ABC$ and $\triangle DEF$ be right triangles with $\angle ABC = \angle DEF = 90°$, $\overline{AB} = \overline{DE}$, and $\overline{AC} = \overline{DF}$. Use the Pythagorean Theorem to show that the triangles are congruent.

▶ **Exercise 3.26.** Prove the Hypotenuse-Side Theorem 3.20 again, but without using the Parallel Postulate 5 or any of the propositions after Proposition I.28. [Hint: Try a proof by contradiction.]

The proof of the Hypotenuse-Side Theorem 3.20 in Exercise 3.25 is very easy and quite obvious (with the hint at least). The purpose of reproving it in Exercise 3.26 is to obtain a proof that does not depend on the Parallel Postulate 5. Thus, the theorem will remain true in noneuclidean geometry.

Theorem 3.21. [SSA+] If two triangles have two pairs of equal sides and if the angles subtended by one of these pairs are equal and the angles subtended by the other pair of equal sides are known to be both acute or both obtuse, then the triangles are congruent.

Proof: We are given $\triangle ABC$ and $\triangle DEF$ with $\overline{AC} = \overline{DF}$ and $\overline{AB} = \overline{DE}$. We know that the angles opposite one of the pairs of equal sides are equal, so we assume that $\angle ABC = \angle DEF$. We proceed with cases for the angles opposite the other pair of equal sides, $\angle ACB$ and $\angle DFE$.

Case 1: $\angle ACB$ and $\angle DFE$ are both acute:

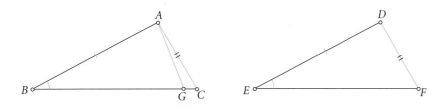

We will prove, by contradiction, that $\overline{BC} = \overline{EF}$ and then congruence follows by SAS. Assume $\overline{BC} \neq \overline{EF}$. Without loss of generality, we can assume that $\overline{BC} > \overline{EF}$. Cut $\overline{BG} = \overline{EF}$ and draw $\overline{AG}$. Then $\triangle ABG \cong \triangle DEF$ by SAS, so $\overline{AG} = \overline{DF}$ and $\angle AGB = \angle DFE$. Therefore, $\overline{AG} = \overline{AC}$ and so $\triangle AGC$ is isosceles. Thus $\angle AGC = \angle ACB$. Note that $\angle AGB = \angle DFE < 90°$. By Proposition I.13, $\angle AGB + \angle AGC = 180°$ and so $\angle AGC = \angle ACB > 90°$. This contradicts Proposition I.17 and so we may conclude that $\overline{BC} = \overline{EF}$ and the triangles are congruent.

Case 2: ∠*ACB* and ∠*DFE* are both obtuse:

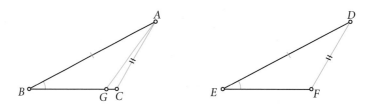

Again, assume $\overline{BC} \neq \overline{EF}$ and so, without loss of generality, $\overline{BC} > \overline{EF}$. Cut $\overline{BG} = \overline{EF}$ and draw $\overline{AG}$. Then $\triangle ABG \cong \triangle DEF$, and so $\overline{AG} = \overline{DF}$ and ∠*AGB* = ∠*DFE*. Therefore, $\overline{AG} = \overline{AC}$, and so ∠*AGC* = ∠*ACB* > 90°. Thus ∠*AGC* + ∠*ACG* > 180°, which is a contradiction to Proposition I.17. We conclude that $\overline{BC} = \overline{EF}$ and therefore $\triangle ABC \cong \triangle DEF$ by SAS. □

▶ **Exercise 3.27.** Let $\triangle ABC$ and $\triangle DEF$ have $\overline{AC} = \overline{DF}$, $\overline{AB} = \overline{DE}$, and ∠*ABC* = ∠*DEF*, and assume that ∠*ACB* = 90°. Prove that $\triangle ABC \cong \triangle DEF$. Do not assume that ∠*DFE* is also right.

The next group of theorems deals with questions of *concurrence*—that is, when three or more lines in the plane all meet at a single point. Two lines in the plane must intersect if they are not parallel. On the other hand, three arbitrary lines on a plane will usually form a triangle. This is called the *generic* configuration, the situation which is most likely to occur.

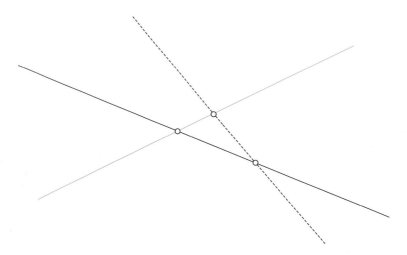

In order to show that three lines are concurrent, first find the point determined by two of the lines, and then try to show that the third line goes through that point. In the activities of this section, you will be asked to make observations: try to formulate your conjectures clearly and precisely and in as much generality as the situation will allow.

▶ **Exercise 3.28.** Euclid's definition states that lines are parallel if they lie in the same plane and do not intersect. Thus, any two lines in the plane must either intersect or be parallel. Describe the possible relationships that can occur with three lines in a plane. Describe the possible relationships between two lines in $\mathbb{R}^3$.

▷ **Activity 3.5.** Using geometric software, draw an arbitrary triangle and label it $\triangle ABC$. Locate the midpoint of each of the sides. Connect each vertex with the midpoint of the opposite side. These three lines (the *medians* of the triangle) meet at a single point, called the *centroid* of the triangle. Measure the distance from each vertex to the centroid and also from the centroid to each of the midpoints of the sides. Move the vertices of the triangle around to consider acute, obtuse, and isosceles triangles. What conjecture can you make?

We will prove the conjecture you should have made in Activity 3.5, but we will first need a lemma:

Lemma 3.22. Given three distinct parallel lines cut by two transversals so that the parallel lines cut the first transversal into equal segments, they also cut the second transversal into equal segments.

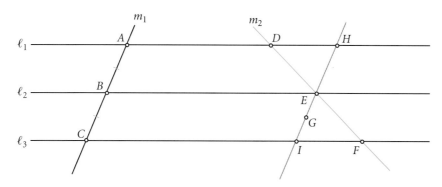

Proof: We are given three parallel lines ℓ_1, ℓ_2, and ℓ_3 in a plane, cut by transversals m_1 and m_2. Let A, B, and C be the points where m_1 intersects ℓ_1, ℓ_2, and ℓ_3, respectively, and let D, E, and F be the points where m_2 intersects ℓ_1, ℓ_2, and ℓ_3, respectively. Note that by the hypothesis, $\overline{AB} = \overline{BC}$. Using Proposition I.31, construct $EG||AB$. By Proposition I.30, EG intersects AD at some point we label H. Similarly, EG intersects CF at I. Then $AHEB$ and $BEIC$ are parallelograms, so $\overline{AB} = \overline{HE}$ and $\overline{BC} = \overline{EI}$ by Proposition I.34. Therefore, $\overline{HE} = \overline{EI}$. Also note that Proposition I.30 implies that $AD||CF$. Thus, $\angle DHE = \angle FIE$ by Proposition I.29. The Vertical Angle Theorem I.15 shows that $\angle DEH = \angle FEI$. Thus, $\triangle DEH \cong \triangle FEI$ by ASA, and so $\overline{DE} = \overline{EF}$. □

Theorem 3.23. The medians of a triangle intersect at a single point, called the centroid, and cut one another so that the distance from a vertex to the centroid is twice the distance from the centroid to the midpoint of the side opposite that vertex.

Proof: We are given $\triangle ABC$. Construct E, the midpoint of $\overline{BC}$ and draw median $\overline{AE}$. Construct G, the midpoint of $\overline{BE}$, and H, the midpoint of $\overline{EC}$, and draw lines through B, G, H, and C parallel to AE, using

Proposition I.31. By Pasch's Theorem 2.11, the parallel through H intersects AC at some point D. Since H is the midpoint of $\overline{EC}$, we know that $\overline{EH} = \overline{HC}$.

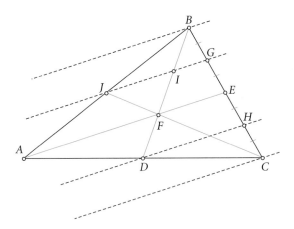

By Lemma 3.22, it follows that $\overline{AD} = \overline{DC}$, so D is the midpoint of $\overline{AC}$. Draw the median $\overline{BD}$. By Pasch's Theorem, $\overline{BD}$ intersects $\overline{AE}$ at F. The parallel through G intersects $\overline{AB}$ at J and $\overline{BD}$ at I. Since $\overline{BG} = \overline{GE}$, by the lemma, $\overline{BJ} = \overline{JA}$, so J is the midpoint of $\overline{AB}$. Draw median $\overline{CJ}$. Let N be the intersection of medians $\overline{CJ}$ and $\overline{BD}$ and let P be the intersection of medians $\overline{AE}$ and $\overline{CJ}$. We must show that $F = N = P$. Applying the lemma again, $\overline{BD}$ is cut by the parallels into equal segments, so $\overline{BI} = \overline{IF} = \overline{FD}$. Therefore, $\overline{BF} = 2\overline{FD}$. Since $\overline{GE} = \overline{EH} = \overline{HC}$, $\overline{CJ}$ is cut into equal segments so $\overline{CP} = 2\overline{PJ}$. Thus, both $\overline{BD}$ and $\overline{CJ}$ are cut in the ratio 2 to 1 at their points of intersection with $\overline{AE}$.

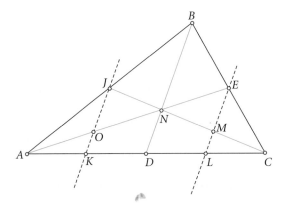

Next, repeat the construction by drawing lines JK and EL parallel to BD. As above, since $\overline{AE}$ intersects $\overline{BD}$ at F, $\overline{AF} = 2\overline{FE}$. Since $\overline{CJ}$ intersects $\overline{BD}$ at N, $\overline{CN} = 2\overline{NJ}$. Previously, we showed that the point P divides $\overline{CJ}$ into thirds, and now we know that N also divides $\overline{CJ}$ into thirds, so P and N must coincide.

Finally, we can repeat the procedure again, drawing lines parallel to CJ. Since $\overline{BD}$ intersects $\overline{CJ}$ at N, $\overline{BN} = 2\overline{ND}$. Previously, we showed that the point F divides $\overline{BD}$ into thirds, so F and N must coincide. Thus, $F = P = N$ and so the three medians meet at a single point and this point cuts each of the medians into thirds. □

The centroid is of interest because, if one imagines the triangle cut from a sheet of metal of uniform thickness, the centroid is the center of gravity. If you placed the triangle on a pin so that the centroid is over the point of the pin, it would balance perfectly. Using calculus, one can show that every line through the centroid divides the triangle into two regions of equal area.

In Activity 3.5, you should have conjectured that the ratio between the distance from the vertex to the centroid and the distance from the centroid to the midpoint of the side opposite that vertex is constant. Such concurrencies and invariant quantities are always important and should be noted. In the following activities, you should look for other points of concurrency and invariant measurements.

▷ **Activity 3.6.** Using geometric software, draw an arbitrary triangle and label it △ABC. Bisect each of the angles. Find a point of concurrency. Drop perpendiculars from this point to each of the sides of the triangle. Find an invariant measure.

▷ **Activity 3.7.** Using geometric software, draw an arbitrary triangle and label it △ABC. Construct the perpendicular bisectors for each of the sides. Find a point of concurrency and then find an invariant measurement. Formulate a conjecture regarding these.

▷ **Activity 3.8.** Using geometric software, draw an arbitrary triangle and label it △ABC. Drop a perpendicular from each vertex to the opposite side. Find a point of concurrency, D. Now consider △ABD and repeat the procedure for this triangle. What can you conjecture?

Theorem 3.24. [Euclid's Proposition IV.4] The angle bisectors of a triangle meet at a single point, the *incenter* of the triangle.

▶ **Exercise 3.29.** Prove Theorem 3.24.

A circle can be drawn centered at the incenter found in Activity 3.6 that intersects each of the sides of the triangle at exactly one point and so is tangent to each of the sides of the triangle. This is called the *inscribed circle* of the triangle and, conversely, the triangle *circumscribes* the circle.

Corollary 3.25. A circle can be inscribed in any triangle.

▶ **Exercise 3.30.** Prove Corollary 3.25.

The result you should have found in Activity 3.7 and will prove in Theorem 3.26 and Exercise 3.31 is another statement that is only true in euclidean geometry. You will meet this again as Alternate Postulate 5.18 and can investigate this property in Project 5.9.

Theorem 3.26. [Euclid's Proposition IV.5] The perpendicular bisectors of the sides of a triangle meet at a single point, the *circumcenter* of the triangle.

▶ **Exercise 3.31.** Prove Theorem 3.26, by showing that two of the perpendicular bisectors of the sides meet and then drawing a line segment connecting this point of intersection with the midpoint of the third side. Now prove that this line segment is the perpendicular bisector of the third side.

Corollary 3.27. Any triangle can be circumscribed by a circle.

▶ **Exercise 3.32.** Prove Corollary 3.27 by showing that a circle can be drawn through the three vertices of the triangle.

▷ **Activity 3.9.** Figure out a procedure to find the center of a given circle and explain why your procedure works.

▶ **Exercise 3.33.** Prove the converse of Exercise 3.7: if a quadrilateral *ABCD* has the property that the pairs of opposite angles sum to 180°, i.e., $\angle BAD + \angle BCD = 180°$ and $\angle ABC + \angle ADC = 180°$, then the quadrilateral is cyclic.

The three lines of Activity 3.8 are called the *altitudes* of the triangle. You should have noticed that they meet at a single point, called the *orthocenter* of the triangle. The proof of this conjecture follows:

Theorem 3.28. The altitudes of a triangle meet at a single point, the orthocenter of the triangle.

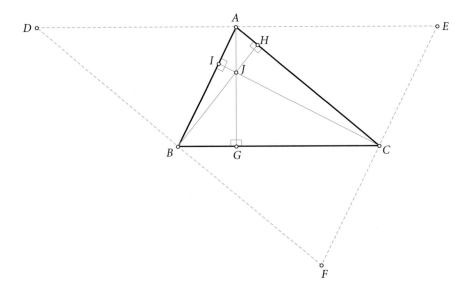

Proof: We are given $\triangle ABC$. Draw lines ℓ_1 through A parallel to BC, ℓ_2 through B parallel to AC, and ℓ_3 through C parallel to AB. As the contradiction hypothesis, assume $\ell_1 \| \ell_2$. Then by Proposition I.30, $\ell_2 \| BC$, which gives a contradiction since ℓ_2 was constructed through B, so ℓ_1 intersects ℓ_2 at some point D. Similarly, ℓ_1 intersects ℓ_3 at point E, and ℓ_2 intersects ℓ_3 at point F. Thus, *ABCE*, *DBCA*, and *ABFC* are parallelograms and so by Proposition I.34, $\overline{DA} = \overline{BC}$ and $\overline{AE} = \overline{BC}$. Therefore, $\overline{DA} = \overline{AE}$. Similarly, $\overline{DB} = \overline{AC} = \overline{BF}$ and $\overline{EC} = \overline{AB} = \overline{CF}$. Drop $AG \perp BC$ and note that AG will be the perpendicular bisector of $\overline{DE}$. Similarly, drop $BH \perp AC$ and $CI \perp AB$. By Theorem 3.26, AG, BH, CI are the perpendicular bisectors of the sides of $\triangle DEF$ and so are concurrent at a point J. □

One interesting note: you may have noticed the use of propositions following from the Parallel Postulate 5 in the proofs of Theorems 3.23 and 3.28. You also should have required the Parallel Postulate or its derivatives in your proof of Theorem 3.26. Surprisingly, Theorem 3.24 can be proved using only the other postulates and Propositions I.1–28 and thus will remain true in noneuclidean geometry. Therefore, in noneuclidean geometry, there exists at least one triangle that has an inscribed circle but cannot be circumscribed.

▷ **Activity 3.10.** Using geometric software, draw an arbitrary triangle $\triangle ABC$ and construct the circumcenter P, the centroid Q, and the orthocenter O. Hide everything except the original triangle and these three points. What can you say about these three points?

▷ **Activity 3.11.** Using geometric software, draw an arbitrary triangle $\triangle ABC$ and construct the orthocenter O. Construct the midpoints M_1, M_2, and M_3 of the three sides of the triangle. Drop the altitudes from each of the three vertices to the opposite side and label their feet as F_1, F_2, and F_3. Construct the midpoints of the segments connecting each of the three vertices to the orthocenter and label these N_1, N_2, and N_3. Hide everything except the original triangle and these nine points. What can you say about these nine points?

The result you should have found in Activity 3.10 was first discovered by Leonhard Euler and bears his name. This and the result of Activity 3.11 are surprising, showing that points generated by quite different constructions align neatly. One can think of this as showing that these distinguished points are, in a sense, natural artifacts of the original triangle.

Theorem 3.29. For any $\triangle ABC$, the circumcenter P, centroid Q, and orthocenter O are collinear, lying on a line called the *Euler line*.

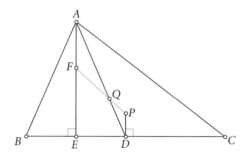

Proof: We are given $\triangle ABC$ with circumcenter P, centroid Q, and orthocenter O. Consider one side $\overline{BC}$ of $\triangle ABC$ and let D be the midpoint of $\overline{BC}$. Then $\overline{PD}$ is the perpendicular bisector of $\overline{BC}$, and $\overline{AD}$ is one of the medians so Q lies on $\overline{AD}$. Drop an altitude from A to BC, intersecting BC at point E. Since AE and PD are parallel, PQ will intersect AE at some point F. We must show that F is the orthocenter O. As PD is parallel to AE, $\angle QAF = \angle QDP$. By the Vertical Angle Theorem, $\angle AQF = \angle DQP$. Therefore, by AA similarity (Corollary 3.17), $\triangle AQF \sim \triangle DQP$. By Theorem 3.23, we know that $\overline{AQ} = 2\overline{QD}$, and it follows that $\overline{QF} = 2\overline{QP}$.

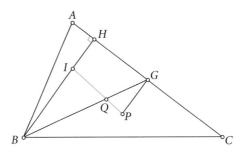

Next, consider side $\overline{AC}$ with midpoint G and altitude BH. The line PQ must intersect $\overline{BH}$ at some point I. Repeating the argument above, we can show that $\overline{QI} = 2\overline{QP}$. Thus $\overline{QF} = \overline{QI}$ and since F and I lie on line PQ, it follows that these points coincide. Therefore, the altitudes meet at this point and so we have $F = I = O$, the orthocenter. □

The next theorem verifies the conjecture of Activity 3.11. The circle we will construct is sometimes called the Feuerbach circle for Karl Feuerbach [1800–1834], a gifted mathematician who published this result in 1822 together with an extension regarding the tangency of this circle to the inscribed circle and to certain circles that can be constructed tangent to the sides and lying outside the triangle. Feuerbach was by all accounts a brilliant mathematician but deeply troubled and finally lost his teaching position after threatening to behead his students with a sword if they failed to solve the problems he posed to them.

Theorem 3.30. Given a triangle, the midpoints of the three sides, the feet of the three altitudes, and the three midpoints of the line segments connecting the orthocenter to the vertices lie on a circle, called the *nine-point circle*.

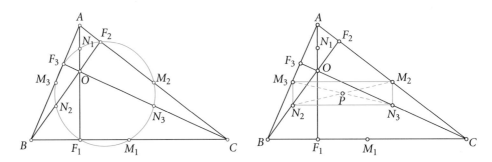

Proof: We are given $\triangle ABC$ with orthocenter O. Let M_1, M_2, and M_3 be the midpoints of the three sides of the triangle. Let F_1, F_2, and F_3 be the feet of the three altitudes. Construct the midpoints of the segments connecting each of the three vertices to the orthocenter and label these N_1, N_2, and N_3.

Considering $\triangle ABC$, note that Theorem 3.15 implies that $M_2M_3 \| BC$. Similarly, from $\triangle BCO$, we have $N_2N_3 \| BC$, from $\triangle ACO$ we have $M_2N_3 \| AO$, and finally from $\triangle ABO$ we have $M_3N_2 \| AO$. Since $AO \perp BC$, it follows that $M_2N_3 \perp M_3M_2$ and $M_2N_3 \perp N_2N_3$. Thus $M_2N_3N_2M_3$ is a rectangle. Draw diagonals M_2N_2 and M_3N_3 and label their intersection as P. Note that a circle can be drawn centered at P that passes through points M_2, M_3, N_2, and N_3. It remains to show that this circle passes through the remaining five points.

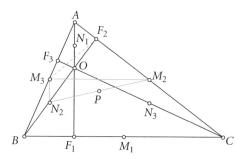

Next, consider the quadrilateral $M_3F_2M_2N_2$. Since $\angle N_2F_2M_2 = 90° = \angle N_2M_3M_2$, by Thales' theorem 3.1, F_2 and M_3 both lie on semicircles with diameter $\overline{N_2M_2}$. Thus, F_2 lies on the circle determined by M_3, M_2, and N_2, which is same circle described above with center P. Similarly, we can show that $M_3F_3M_2N_3$ is cyclic and so F_3 lies on this circle.

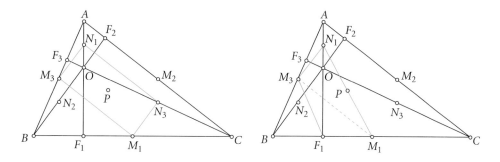

We can also show that $M_3N_1N_3M_1$ is a rectangle and, since it has diagonal M_3N_3 in common with the previous rectangle, this implies that N_1 and M_1 also lie on our circle with center P. Finally, $M_3N_1M_1F_1$ is cyclic, and so F_1 also lies on our circle. □

♦ 3.4 Inversion in a circle

We next investigate a construction called *inversion in a circle*, which turns out to have a surprising number of useful applications, especially in hyperbolic geometry and in number theory. In particular, this construction is used to develop models for the noneuclidean geometries we will study in Chapters 6 and 7. Those who are willing to accept these constructions without proof may want to omit this section.

Definition 3.31. Given a circle C with center O and radius r and a point P distinct from O, the *inverse* of point P in the circle is the point P' which lies on ray $\overrightarrow{OP}$ such that $\overline{OP} \cdot \overline{OP'} = r^2$.

Construction: Given point P inside the circle C, draw the ray $\overrightarrow{OP}$. Construct $PQ \perp OP$, where Q is a point where this perpendicular intersects the circle. Draw the radius $\overline{OQ}$. Construct the line through Q perpendicular to OQ. Extend this line until it meets line OP at the point P'. (Note that these lines must intersect by the Parallel Postulate 5.)

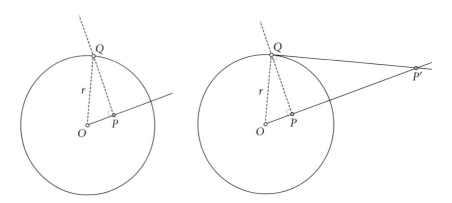

Now we must show that the point P' that we have just found satisfies $\overline{OP} \cdot \overline{OP'} = r^2$. Note that $\triangle OPQ$, $\triangle OQP'$, and $\triangle QPP'$ are right triangles. Also, $\triangle OPQ$ and $\triangle OQP'$ have $\angle POQ$ in common. Thus, by Corollary 3.17, $\triangle OPQ \sim \triangle OQP'$. Therefore, $\frac{\overline{OP}}{r} = \frac{r}{\overline{OP'}}$, and thus $\overline{OP} \cdot \overline{OP'} = r^2$.

▷ **Activity 3.12.** Draw a circle and a point P outside of the circle. Modify the construction above to construct P', the inverse of P in the circle.

▷ **Activity 3.13.** Use geometric software to construct the inversion of a line through a given circle. Make sure your sketch works for lines that intersect the circle of inversion as well as lines that do not. Describe the shape that the inversion of the line forms.

▶ **Exercise 3.34.** Show that if P lies on the circle, then $P' = P$. Show that if P lies inside the circle then P' lies outside, and if P lies outside the circle, then P' lies inside.

▶ **Exercise 3.35.** Let P' be the inverse of point P. What happens if you invert P'?

Lemma 3.32. Given a circle $\mathcal{C}$ and P and R two arbitrary points inside this circle, invert to get points P' and R'. Then $\triangle OPR$ is similar to $\triangle OR'P'$.

▶ **Exercise 3.36.** Prove Lemma 3.32.

Now let us consider the action of this inversion on geometric objects. When we consider the inversion of a line or a circle, we must invert each point of the line or circle and then consider what sort of figure these image points form.

Theorem 3.33. Inversion in a circle centered at point O will take any line not passing through O onto a circle through O, excluding the point O itself. Conversely, inversion takes any circle through O, excluding O itself, to a line that does not pass through O.

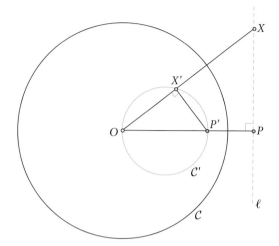

Proof: We are given $\mathcal{C}$, the circle of inversion centered at O and a line ℓ which does not pass through O. Drop a line from O perpendicular to ℓ and label the point where this line intersects ℓ as P. Choose an arbitrary point X on ℓ and draw $\overline{OX}$. Construct the inverse points P' and X'. Then by Lemma 3.32, $\triangle OPX \sim \triangle OX'P'$. Therefore, $\angle OX'P' = \angle OPX = 90°$. By Theorem 3.2, we conclude that X' lies on a circle $\mathcal{C}'$ with diameter $\overline{OP'}$. This shows that all points on the line ℓ land on the circle $\mathcal{C}'$.

Conversely, let X' be a point on $\mathcal{C}'$ with inverse X, and then we must show that X lies on ℓ. Since $\triangle OX'P' \sim \triangle OPX$, $\angle OX'P' = 90° = \angle OPX$, so XP is perpendicular to OP. Thus X lies on ℓ. □

▷ **Activity 3.14.** Theorem 3.33 excludes the case of a line through the center O of the circle of inversion. Explore the action of inversion in this case and describe your findings.

Theorem 3.34. Inversion in a circle centered at O takes any circle that does not pass through O to another circle.

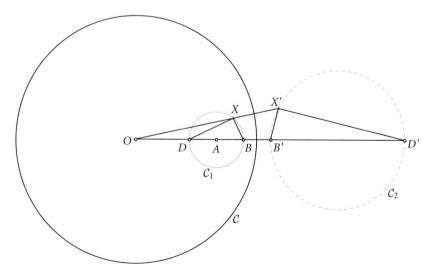

Proof: Let $\mathcal{C}$ be the circle of inversion, and let $\mathcal{C}_1$ be a circle with center A and radius r' which does not pass through O. Note that if $A = O$, then $\mathcal{C}_2$, the image of $\mathcal{C}_1$ under the inversion, will be a circle with the same center O and radius r^2/r', and the result follows. Thus we assume $A \neq O$. Draw the line AO, which will intersect $\mathcal{C}_1$ at points B and D. Let X be any other point on $\mathcal{C}_1$ and take inverses B', D', and X'. By Lemma 3.32, $\triangle OBX \sim \triangle OX'B'$ and $\triangle ODX \sim \triangle OX'D'$. Therefore, $\angle OBX = \angle OX'B'$ and $\angle OXD = \angle OD'X'$. By the Exterior Angle Theorem, $\angle OB'X' = \angle OD'X' + \angle B'X'D'$. Thus, $\angle B'X'D' = \angle OB'X' - \angle OD'X' = \angle OXB - \angle OXD = \angle BXD = 90°$ by Theorem 3.1. Thus, by Theorem 3.2, X' lies on the circle with diameter $\overline{B'D'}$. $\square$

▷ **Activity 3.15.** In the proof of Theorem 3.34 above and in the accompanying illustration, we have assumed that $\mathcal{C}_1$ lies inside $\mathcal{C}$, the circle of inversion. Use geometric software to demonstrate that the proof holds for all cases, i.e., when $\mathcal{C}_1$ lies inside $\mathcal{C}$, when $\mathcal{C}_1$ lies outside $\mathcal{C}$, and when $\mathcal{C}_1$ intersects $\mathcal{C}$.

Recall that in Section 3.1 we defined the angle between two intersecting circles as the angle formed by their tangent lines at the point of intersection. Thus circles are said to be perpendicular if their tangent lines are perpendicular at the point of intersection. We also know that the tangent line at a point on a circle is perpendicular to the radius to that point. Therefore, we can conclude that two circles with centers O and O' that intersect at a point A will be perpendicular if and only if their radii $\overline{OA}$ and $\overline{O'A}$ are perpendicular.

Theorem 3.35. *If a circle $\mathcal{C}'$ is perpendicular to the circle $\mathcal{C}$ of inversion, then inversion takes $\mathcal{C}'$ to itself.*

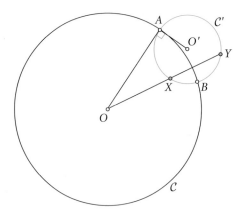

Proof: Let O be the center of $\mathcal{C}$, the circle of inversion, and let $\mathcal{C}'$ have center O'. Label the points where $\mathcal{C}'$ intersects $\mathcal{C}$ as A and B. As we have noted, $OA \perp O'A$, and thus OA is tangent to $\mathcal{C}'$. Choose an arbitrary point $X \in \mathcal{C}'$. Extend OX to point Y on $\mathcal{C}'$, using the Circular Continuity Principle to show that the intersection exists. By Theorem 3.9, $\overline{OA}^2 = r^2 = \overline{OX} \cdot \overline{OY}$. Therefore, $Y = X'$, the inverse of X. This shows that the inverse of every point on $\mathcal{C}'$ lies on $\mathcal{C}'$, and the result follows. $\square$

Theorem 3.36. *If a circle $\mathcal{C}'$ contains a point P and its inverse P' where $P' \neq P$, then $\mathcal{C}'$ is perpendicular to the circle of inversion.*

▶ **Exercise 3.37.** Prove Theorem 3.36.

Note that there are many such circles containing some pair A and A' of inverse points. Theorem 3.36 says that any such circle will be perpendicular to the circle of inversion. Finally, we show that inversion preserves two important measures. The clever proof below is adapted from David Kay's *College Geometry*.

Theorem 3.37. Inversion in a circle preserves angle measure.

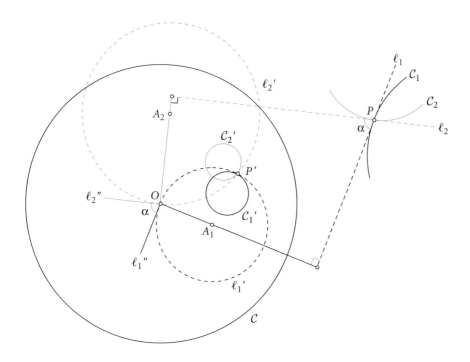

Proof: We will prove the most difficult case of two circles, C_1 and C_2, neither of which contain the center O. The angle of intersection in this case is defined to be the angle formed by their tangent lines. Let P be the point of intersection of C_1 and C_2, and let ℓ_1 and ℓ_2 be the tangent lines to the circles at this point. Let α denote the angle formed at P by ℓ_1 and ℓ_2, which is the angle formed by the two circles. Invert all of these objects to obtain C_1', C_2', P', ℓ_1', and ℓ_2'. Note that by Theorem 3.33, we know that ℓ_1' and ℓ_2' are circles passing through points O and P' with centers we will label A_1 and A_2. By Theorem 3.34, we know that C_1' and C_2' are circles that do not pass through O. We must first show that ℓ_1' and ℓ_2' are tangent to these new circles C_1' and C_2' respectively at point P'. Note that P' must lie on both ℓ_1' and ℓ_2'. Furthermore, for $i = 1, 2$, since C_i and ℓ_i have only the single point P in common, it follows that C_i' and ℓ_i' have only the point P' in common. Thus ℓ_i' must be tangent to C_i' at P' by an extension of Definition 3.6.

We must now show that the angle formed by the circles ℓ_1' and ℓ_2' is the same as α, formed by the original tangent lines. Studying the proof of Theorem 3.33, we see that not only do we know that ℓ_i' is a circle through O, but also ℓ_i is perpendicular to the line OA_i which forms a radius for the circle ℓ_i' for $i = 1, 2$. Since OA_i is also perpendicular to the tangent line to the circle ℓ_i' at O, we know that ℓ_i must be parallel to the tangent line to ℓ_i' at O. We have denoted these tangent lines as ℓ_i'' in the illustration. Thus ℓ_1'' and ℓ_2'' must also meet at angle α. Since the angles formed at the points of intersection O and P' are the same, the tangents to ℓ_1' and ℓ_2' at P' also form the angle α. □

▶ **Exercise 3.38.** Prove Theorem 3.37 for the case of an angle formed by two lines.

▶ **Exercise 3.39.** Prove Theorem 3.37 for the case of an angle formed by a line and a circle.

The fact that inversion preserves the angles formed between lines or circles is part of the reason why this turns out to be useful in certain situations. It is easy to see that inversion does not preserve length, but there is a length-related quantity that is invariant under inversion:

Definition 3.38. Let A, B, C, and D be four distinct points. The *cross-ratio* of these points in that order is the quantity

$$\frac{\overline{AC}/\overline{AD}}{\overline{BC}/\overline{BD}} = \frac{\overline{AC}\,\overline{BD}}{\overline{AD}\,\overline{BC}}.$$

Theorem 3.39. Inversion in a circle preserves the cross-ratio.

Proof: We are given four points A, B, C, and D, none of which coincide with O. Construct their inverses A', B', C', and D'. Then we know that $\overline{OA} \cdot \overline{OA'} = r^2 = \overline{OC} \cdot \overline{OC'}$. Therefore, $\frac{\overline{OA}}{\overline{OC'}} = \frac{\overline{OC}}{\overline{OA'}}$. We must show that

$$\frac{\overline{AC}\,\overline{BD}}{\overline{AD}\,\overline{BC}} = \frac{\overline{A'C'}\,\overline{B'D'}}{\overline{A'D'}\,\overline{B'C'}}.$$

We proceed by cases.

Case 1: O, A, and C are not collinear.

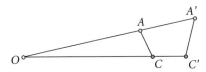

In this case, by Lemma 3.32, we know that $\triangle OAC \sim \triangle OC'A'$, and so $\frac{\overline{AC}}{\overline{A'C'}} = \frac{\overline{OA}}{\overline{OC'}}$.

Case 2: O, A, and C are collinear.

In this case, note that $\overline{AC} = \overline{OC} - \overline{OA}$ and $\overline{A'C'} = \overline{OA'} - \overline{OC'}$. Since $\overline{OA} \cdot \overline{OA'} = \overline{OC'} \cdot \overline{OC}$, it follows that

$$\overline{OA} \cdot \overline{OA'} - \overline{OA} \cdot \overline{OC'} = \overline{OC'} \cdot \overline{OC} - \overline{OA} \cdot \overline{OC'},$$

$$\overline{OA}(\overline{OA'} - \overline{OC'}) = \overline{OC'}(\overline{OC} - \overline{OA}),$$

$$\overline{OA} \cdot \overline{A'C'} = \overline{OC'} \cdot \overline{AC},$$

$$\frac{\overline{AC}}{\overline{A'C'}} = \frac{\overline{OA}}{\overline{OC'}}.$$

Thus, in both cases, we have $\frac{\overline{AC}}{\overline{A'C'}} = \frac{\overline{OA}}{\overline{OC'}}$. Similarly, $\frac{\overline{AD}}{\overline{A'D'}} = \frac{\overline{OA}}{\overline{OD'}}$. Therefore,

$$\frac{\overline{AC}}{\overline{A'C'}} \cdot \frac{\overline{A'D'}}{\overline{AD}} = \frac{\overline{OA}}{\overline{OC'}} \cdot \frac{\overline{OD'}}{\overline{OA}} = \frac{\overline{OD'}}{\overline{OC'}}.$$

Repeat for the points B, C, and D, to find that

$$\frac{\overline{BC}}{\overline{B'C'}} \cdot \frac{\overline{B'D'}}{\overline{BD}} = \frac{\overline{OD'}}{\overline{OC'}},$$

$$\frac{\overline{AC}}{\overline{A'C'}} \cdot \frac{\overline{A'D'}}{\overline{AD}} = \frac{\overline{BC}}{\overline{B'C'}} \cdot \frac{\overline{B'D'}}{\overline{BD}},$$

$$\frac{\overline{AC}}{\overline{BC}} \cdot \frac{\overline{BD}}{\overline{AD}} = \frac{\overline{A'C'}}{\overline{B'C'}} \cdot \frac{\overline{B'D'}}{\overline{A'D'}}.$$

$\square$

What does the cross-ratio represent? It will prove to be useful later (in Chapter 6), but it is difficult to explain why it works. Robin Hartshorne says,

> At this point I can just hear someone asking, "What is the geometrical significance of the cross-ratio?" Although I first encountered cross-ratios as a senior in high school, and have dealt with them many times since then, I must say frankly that I cannot visualize a cross-ratio geometrically. If you like, it is magic. Here is this algebraic quantity whose significance it is impossible to understand, and yet it turns out to do something very useful. It works. You might say it was a triumph of algebra to invent this quantity that turns out to be so valuable and could not be imagined geometrically. Or if you are a geometer at heart, you may say that it is an invention of the devil and hate it all your life.[1]

▷ **Activity 3.16.** Use geometric software to construct the inverses of four points through a given circle. Demonstrate that the cross-ratio is invariant.

♦ 3.5 Projects

Project 3.1. Report, including a proof and applications, on Ptolemy's theorem.[2]

Project 3.2. Report, including a proof and applications, on Varignon's theorem.[3]

Project 3.3. Report, including a proof and applications, on Menelaus' theorem.

---

[1] Robin Hartshorne, *Geometry: Euclid and Beyond*.
[2] References are Tony Crilly and Colin Fletcher, "Ptolemy's Theorem" in *The Changing Shape of Geometry* or Posamentier's *Advanced Euclidean Geometry*.
[3] A reference is Chris Pritchard, "Varignon's Theorem" in *The Changing Shape of Geometry*.

Project 3.4. Report, including a proof and applications, on Ceva's theorem.[4]

Project 3.5. Report, including a proof and applications, on Morley's theorem.[5]

Project 3.6. Report, including a proof and applications, on the Law of Cosines as a generalization of the Pythagorean theorem.[6]

Project 3.7. Report on the relationship between Möbius transformations and the cross-ratio.

Project 3.8. Report on the Pythagoreans and their theory of harmonic ratios.

Project 3.9. Describe the action of inversion in a circle in terms of coordinate geometry. Verify the invariance of angle measure and the cross-ratio using coordinates.

Project 3.10. Report on the relationship of trigonometry and the circumcenter and incenter of a triangle.

---

[4] References are Elmer Rees, "Ceva's Theorem" in *The Changing Shape of Geometry*, Posamentier's *Advanced Euclidean Geometry*, or Bottema's *Topics in Elementary Geometry*.

[5] References are David Burghes, "Morley's Theorem" in *The Changing Shape of Geometry*, or Bottema's *Topics in Elementary Geometry*.

[6] A reference is Neil Bibby and Doug French, "Pythagoras extended: a geometric approach to the cosine rule" in *The Changing Shape of Geometry*.

4. Constructions

◆ 4.1 Straightedge and compass constructions

◇ 4.1.1 Basic operations and constructions

The traditional euclidean tools are a ruler and a compass. To be accurate, the ruler is actually an unmarked straightedge, only one side of which is known to be straight. The first three of Euclid's postulates are sometimes called the Straightedge and Compass Postulates, since they define the things one can do with these tools. In *What is Mathematics?*, Courant and Robbins point out that every straightedge and compass construction is performed by a sequence of steps and each of these steps is one of the following operations:

S & C Operation 1. Given two distinct points, draw the (unique) line connecting them.

S & C Operation 2. Given two nonparallel lines, find their point of intersection.

S & C Operation 3. Given a point and a length, draw a circle with the point as center and with radius equal to the given length.

S & C Operation 4. Given a circle, find the intersection of this circle with a line.

S & C Operation 5. Given a circle, find the intersection of this circle with another circle.

S & C Operation 1 is essentially Postulate 1 and S & C Operation 3 is Postulate 3, so we assume that these operations are always possible. S & C Operations 4 and 5 are possible by the Circular Continuity Principle Postulate 11. In applying S & C Operations 4 and 5, note that we are not assuming that the objects intersect, but rather saying that if they intersect, then we can locate the point of intersection. This list of basic operations are those which can be done easily and which students of any age can figure out how to do with little or no instruction.

Euclid gives instructions for several straightedge and compass constructions. The constructions themselves make use of the five operations above, though of course proving that the construction actually gives the desired result makes use of some of the other postulates and theorems. In this chapter we want to study what lengths, angles, and figures are constructible with a given set of tools.

Here are the basic constructions from Book I of Euclid's *The Elements*, which we studied in Chapter 2:

Proposition I.9. Bisect an Angle: Given any angle, construct a line that bisects that angle.

Proposition I.10. Perpendicular Bisector: Given a line segment, construct the perpendicular bisector.

Proposition I.11. Erect a Perpendicular: Given a line and a point on that line, construct a line perpendicular to the given line through the given point.

Proposition I.12. Drop a Perpendicular: Given a line and a point not on the line, construct a line perpendicular to the given line and through the given point.

Proposition I.23. Copy an Angle: Given an angle and a line segment, construct an angle on the given line segment congruent to the given angle.

Proposition I.31. Construct a Parallel: Given a line and a point not on the line, construct a line through the given point parallel to the given line.

Reexamining the proofs of these propositions shows that each construction can be broken down into a sequence of steps which make use only of the five S & C operations and previous constructions. This section will examine what else one can do with an unmarked straightedge and compass. The other sections of this chapter discuss the same question for a variety of other tools.

◊ 4.1.2 Constructible lengths

We will first discuss what lengths are constructible with straightedge and compass. The straightedge has no markings, so to begin we must be given (or invent) a standard unit of measure. We will assume that someone has given you an inch-long segment. You can extend any line and, using the given line segment and your compass (and Proposition I.3), mark off a line segment of any integral length.

▷ **Activity 4.1.** The line segment below is 1 inch long. Use this to calibrate your compass. (Alternatively, use the measurement and drawing capabilities of a geometric software package to draw an inch-long segment.) Then draw another line and cut off a segment exactly 3 inches long using only the five S & C operations.

Exercise 4.1. Given two line segments of lengths a and b with $a > b$, explain how to construct line segments of length $a + b$ and $a - b$.

Proposition I.10 taught us how to find the midpoint of any line segment and thus allows us to cut any length in half. Therefore, combining this with Exercise 4.1 we can construct any length of the form $\frac{n}{2^k}$ where k and n are nonnegative integers.

▷ **Activity 4.2.** Using the inch-long segment of Activity 4.1 to calibrate your compass, construct a line segment of length $2\frac{3}{8}$, using either straightedge and compass or geometric software.

What other lengths can be constructed? We introduce some terminology to help clarify what we mean by this question.

Definition 4.1. A number a is (straightedge and compass) *constructible* if, given a line segment of length 1, another line segment of length a can be constructed with straightedge and compass.

We noted above that any natural number is constructible by extending the procedure you used in Activity 4.1. You showed in Exercise 4.1 that the sum and difference of two constructible lengths are also constructible by straightedge and compass. The following theorem is basically Euclid's Proposition VI.12:

Theorem 4.2. Given two constructible numbers a and b, their product ab is constructible.

Proof: We are given two line segments of length a and b and assume that there is another line segment of length 1. Let $\overline{BC}$ be a line segment of length 1 and, using Proposition I.11, erect a line through C perpendicular to BC. Using S & C Operation 3, draw a circle centered at C with radius a. By S & C Operation 4 we can locate A, the intersection of the perpendicular line and the circle and note that $\overline{AC} = a$. Draw $\overline{AB}$ (S & C Operation 1). Let $\overline{EF}$ denote the given line segment of length b. Erect a perpendicular to EF at F. Using Proposition I.23, construct an angle along $\overline{EF}$ equal to $\angle ABC$. The side of this angle intersects the perpendicular line at D by S & C Operation 2. Thus we have triangles $\triangle ABC$ and $\triangle DEF$ which are similar by AA similarity. Let $\overline{DF} = x$. By the proportions of the sides of these similar triangles, $\frac{b}{1} = \frac{x}{a}$ and so $x = ab$. □

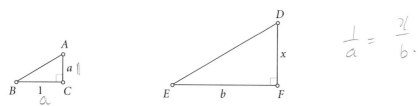

Exercise 4.2. Modify the procedure used in Theorem 4.2 to show how to construct with straightedge and compass a length of $\frac{1}{n}$ for an arbitrary integer n.

Combining the result of Exercise 4.2 with Theorem 4.2 gives us the following result:

Corollary 4.3. Given two constructible numbers a and b, their quotient $\frac{a}{b}$ is constructible.

▷ **Activity 4.3.** Using either straightedge and compass or geometric software and the inch-long segment of Activity 4.1, construct a line segment of length $1\frac{2}{3}$.

Exercise 4.1, Theorem 4.2, and Corollary 4.3 make it clear that we can construct a line segment of any positive rational length. Therefore, the set of all constructible lengths includes the set of all positive rational numbers. The lack of negative numbers is due to Euclid's insistence on considering numbers as actual lengths, rather than as abstract quantities. But there are other lengths which can also be constructed.

Assuming that one has an inch-long segment $\overline{AB}$ to work with, one can erect an inch-long perpendicular $\overline{AC}$ at A. By the Pythagorean Theorem, the hypotenuse of the triangle formed is $\overline{BC} = \sqrt{2}$, as shown below:

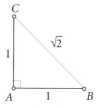

Next, construct another right triangle with one side of length $\sqrt{2}$ as just constructed and the other side of length 1, and the hypotenuse of this triangle will be length c, where

$$c^2 = 1^2 + (\sqrt{2})^2 = 3,$$
$$c = \sqrt{3}.$$

This construction can be continued indefinitely, forming a *Pythagorean spiral*. Note that all of the outside edges are one unit long, and form right angles with the lines radiating from the center.

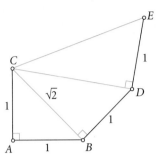

Activity 4.4. Construct a Pythagorean spiral, continuing as far as you can before the triangles start overlapping. Find the lengths (as radicals, not as decimals) of each of the lines radiating from the center C.

Thus, with straightedge and compass, we can construct line segments of all possible lengths which can be expressed as the square root of a positive integer. We can also, by putting segments end on end, construct lengths of $1 + \sqrt{2}$, etc.

▷ **Activity 4.5.** Using either straightedge and compass or geometric software and the inch-long segment below, construct a line segment of length $1 + \sqrt{3}$.

One way to construct a square root of an integer is to build a Pythagorean spiral (or a section of one) up to the desired root. However, that is often cumbersome and doesn't deal directly with roots of rational numbers. Here is another way to construct an arbitrary square root $\sqrt{a}$:

Theorem 4.4. *Given a constructible number a, then $\sqrt{a}$ is also constructible.*

Proof: Let $\overline{AB}$ have length a. Construct a circle with diameter $\overline{AC} = 1 + a$ where $\overline{AB} = a$ and $\overline{BC} = 1$. Erect a perpendicular to the diameter $\overline{AC}$ at the point B, which will meet the circle (by S & C Operation 4) at point D. From Thales' Theorem 3.1 we know that $\angle ADC = 90°$. □

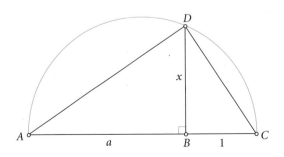

▶ **Exercise 4.3.** Finish the proof of Theorem 4.4 by showing that $\overline{BD} = \sqrt{a}$.

▷ **Activity 4.6.** Using either straightedge and compass or geometric software and an inch-long segment, construct a line segment of length $\sqrt{5/2}$.

▶ **Exercise 4.4.** Given line segments of lengths a and b, explain how to construct a line segment of length $\sqrt{ab}$.

▶ **Exercise 4.5.** Given a line segment of length a, explain how to construct a line segment of length $\sqrt[4]{a}$.

If a right triangle had legs of length 1 and $1 + \sqrt{2}$, then the hypotenuse would have length:

$$c^2 = 1^2 + (1 + \sqrt{2})^2,$$
$$= 1 + [1 + 2\sqrt{2} + (\sqrt{2})^2],$$
$$= 1 + [1 + 2\sqrt{2} + 2],$$
$$= 4 + 2\sqrt{2}.$$
$$c = \sqrt{4 + 2\sqrt{2}}.$$

Similarly, we can construct line segments whose lengths represent positive solutions to any linear or quadratic equation with constructible (for example, rational number) coefficients. For example, given a quadratic equation $ax^2 + bx + c = 0$ with a, b, and c constructible numbers, then the quadratic formula gives us the solutions $x = \frac{-b \pm \sqrt{b^2 - 4ac}}{2a}$. Theorem 4.2 and Exercise 4.1 allow us to construct a segment of length $b^2 - 4ac$ (if this quantity is positive), and Theorem 4.4 allows us to construct a segment of length $\sqrt{b^2 - 4ac}$. Another application of Exercise 4.1 gives us $-b \pm \sqrt{b^2 - 4ac}$ and then Theorem 4.2 and Corollary 4.3 allow us to construct a length equal to $\frac{-b \pm \sqrt{b^2 - 4ac}}{2a}$. Thus, with straightedge and compass, we can construct line segments whose lengths are any finite combination of integers, fractions, and repeated square roots of positive constructible numbers.

Technically, the set of all constructible numbers forms what is called a *field* in abstract algebra. A field is a set of numbers obeying the familiar rules for addition, subtraction, multiplication, and division by any nonzero element so that if one takes two numbers in the field and performs any of these operations, the result is also in the field. For example, $\mathbb{Q}$, the set of rational numbers, is a field, but the set of integers, $\mathbb{Z}$, is not a field since 1 and 2 are integers but their quotient $\frac{1}{2}$ is not an integer. The following theorem may be omitted; it is just a fancy way of stating what we have already observed.

Theorem 4.5. *Given a line segment of unit length, let $\mathcal{F}$ denote the set of lengths constructible by straightedge and compass. Then $\mathcal{F}$ is a field.*

However, in general, cube roots and solutions to irreducible cubic equations (cubic equations that cannot be factored) are not constructible. A famous problem of classical geometry was posed when, after a particularly nasty plague (around 430 B.C.), an oracle instructed the Athenians to double the size of the cubical altar to Apollo at Delos. Theon of Smyrna quotes a now missing work of Eratosthenes:

> In his work entitled *Platonicus*, Eratosthenes says that, when the god announced to the Delians by the oracle that to get rid of a plague they must construct an altar double of the existing one, their craftsmen fell into great perplexity in trying to find how a solid could be made double of another solid, and they went to ask Plato about it. He told them that the god had given this oracle, not because he

wanted an altar of double the size, but because he wished, in setting this task before them, to reproach the Greeks for their neglect of mathematics and their contempt of geometry.[1]

If the original altar was 1 unit on each side, and if they then built a new one in which each side had length 2 units, the new altar would be $2^3 = 8$ cubic units in volume or 8 times as big as the old one. In order to precisely double the volume of the old altar, they would need to construct a cube with sides of length $\sqrt[3]{2}$. Note that $\sqrt[3]{2}$ is the solution to the irreducible cubic equation $x^3 - 2 = 0$. The Greeks, including Eratosthenes himself, gave a number of solutions to this problem, but each of these solutions required the use of additional instruments or curves (typically, one of the conic sections). None succeeded in finding a construction for the duplication of the cube that uses only straightedge and compass.

The proof that cubic roots cannot be constructed in general with straightedge and compass alone requires developments in algebra that were made centuries later and so unavailable to the ancient Greeks. Carl Friedrich Gauss [1777–1855], one of the best mathematicians ever to have lived, stated that neither this problem nor the problem of trisecting an arbitrary angle can be done with straightedge and compass alone but gave no proof, nor was a proof found in his unpublished notebooks. In 1837, Pierre Laurent Wantzel [1814–1848] published a paper "On the means of ascertaining whether a problem in geometry can be solved with ruler and compass" that proved that one cannot duplicate the cube with straightedge and compass alone. He showed that only lengths involving rationals and square roots are constructible by straightedge and compass. For a proof of this result, see Edwin Moise's *Elementary Geometry from an Advanced Standpoint* or George Martin's *Geometric Constructions*.

To summarize, a length is constructible with straightedge and compass if and only if it can be expressed as a finite combination of rational numbers and square roots of rational numbers. Let $\mathbb{Z}$ denote the integers, $\mathbb{Q}$ the rational numbers, $\mathbb{R}$ the real numbers, and $\mathbb{C}$ the complex numbers. The algebraic numbers are all real numbers which represent solutions to polynomial equations, which includes solutions to all cubics and higher roots. Then we have

$$\mathbb{Z} \subseteq \mathbb{Q} \subseteq \{\text{constructible numbers}\} \subseteq \{\text{algebraic numbers}\} \subseteq \mathbb{R} \subseteq \mathbb{C}$$

▶ **Exercise 4.6.** The roots of some particular cubic equations are constructible with straightedge and compass, for example the roots of the reducible equation $x^3 - 7x^2 + 14x - 6 = 0$. Construct line segments whose lengths represent the solutions to this cubic.

The general solution of which quantities are constructible is a deep problem in modern abstract algebra, making use of Galois theory for its complete solution.

◊ 4.1.3 Constructible angles

The next question to ask is: Which angles are constructible? By Proposition I.11 or I.12, we can construct a 90° angle and we can then bisect it by Proposition I.9 to get an angle of 45°.

▷ **Activity 4.7.** Use straightedge and compass or geometric software to construct angles with the following measures:

   a. 22.5°
   b. 11.25°

---

[1] Translated by Ivor Thomas, *Greek Mathematical Works*.

c. 60°
d. 15°
e. 120°
f. 75°

We say that angle α is constructible if and only if the length $\cos \alpha$ is constructible. If we can construct a line segment of length $\cos \alpha$, then we can erect a perpendicular at one end of this line segment and then draw a circle of radius 1 centered at the other end of the line segment. The intersection of these is constructible by S & C Operation 4. Thus we have formed a right triangle as shown below with one leg of length $\cos \alpha$ and hypotenuse of length 1. This triangle then encloses an angle of measure α:

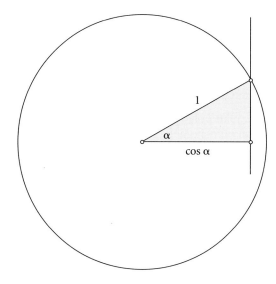

However, we cannot construct with straightedge and compass an angle of 20°, though we can approximate it quite closely using, for example, $\frac{90°}{4} - \frac{90°}{32} = 19.6875°$. To see that a 20° angle is not constructible, recall the trigonometric identities

$$\cos(\alpha + \beta) = \cos(\alpha)\cos(\beta) - \sin(\alpha)\sin(\beta)$$
$$\sin(\alpha + \beta) = \sin(\alpha)\cos(\beta) + \cos(\alpha)\sin(\beta)$$

▶ **Exercise 4.7.** a. Using these identities, show that $\cos(2\alpha) = 2\cos^2(\alpha) - 1$ and $\sin(2\alpha) = 2\sin(\alpha)\cos(\alpha)$.
b. Show that $\cos(3\alpha) = 4\cos^3(\alpha) - 3\cos(\alpha)$.

Let $\alpha = 20°$ in the identity of Exercise 4.7 (b): $\cos(60°) = 4\cos^3(20°) - 3\cos(20°)$. If we let $x = \cos(20°)$, then we have $\frac{1}{2} = 4x^3 - 3x$, or $8x^3 - 6x - 1 = 0$. The only possible rational solutions to this cubic are $\pm 1, \pm \frac{1}{2}, \pm \frac{1}{4}$, and $\pm \frac{1}{8}$. A few simple computations show that none of these are solutions, so this cubic equation is irreducible and its solutions are not constructible. Therefore, the angle 20° cannot be constructed with straightedge and compass.

If we could construct an angle of 20°, we would have trisected a 60° angle. Of course, there are certain angles that can be trisected, for example 90°, just as there are certain cube roots that can be constructed, such as $\sqrt[3]{8}$. The problem of trisecting a given angle is the second of the famous constructibility problems of the Greeks, though no such colorful legend is attached to it as is for the duplication of the cube. Again, many attempts were made on this problem but all of the successful trisections required the use of additional tools beyond the traditional straightedge and compass. Wantzel's paper cited above showed that it is impossible to trisect an arbitrary angle. See also George Martin's *Geometric Constructions* or Benjamin Bold's *Famous Problems of Geometry*. We will discuss some of the trisections possible by nontraditional means in Section 4.2.

◊ 4.1.4 Constructible polygons

Next, we turn our attention to the regular polygons we can construct with straightedge and compass. A regular n-sided polygon will have n equal sides and n equal vertex angles. Euclid assumes that any such regular polygon can be thought of as inscribed in a circle.

▶ **Exercise 4.8.** Given a regular polygon with an even number of sides, explain how to find the center and radius of the circumscribing circle.

▶ **Exercise 4.9.** Given a regular polygon with an odd number of sides, explain how to find the center and radius of the circumscribing circle.

Proposition I.1 tells us how to construct an equilateral triangle, while Proposition I.46 allows us to construct a square. Here is an easier way to construct a square: Construct two perpendicular lines and then draw a circle centered at their intersection:

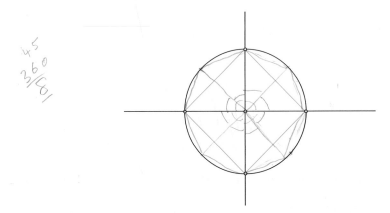

▷ **Activity 4.8.** Using either straightedge and compass or geometric software, construct a regular hexagon.

Once one has a regular polygon, it is (relatively) easy to construct another regular polygon with twice the number of sides, by bisecting the interior angle of each side. More easily, construct the first polygon within a circle so that its vertices are equally spaced around the circle. Then bisect one of the interior angles and locate the point where the angle bisector intersects the circle. Set the compass to the length between one of the neighboring vertices and the point where the angle bisector cut the circle to give the length of the side of the new polygon with twice as many sides. One can then use this compass setting to find the remaining vertices for the new polygon.

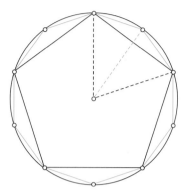

We have constructed regular polygons of three, four, and six sides. Using bisection, we can also construct polygons with 12, 16, 24, 32, etc., sides.

▷ **Activity 4.9.** Using either straightedge and compass or geometric software, construct a regular octagon.

The regular pentagon can be constructed with straightedge and compass with a bit more difficulty, but before we proceed to instructions for this, it is best to meditate for a bit on its structure. Using the formula $\frac{(n-2)180°}{n}$ from Exercise 2.58 with $n=5$, we see that the vertex angles of a regular pentagon each measure $108°$.

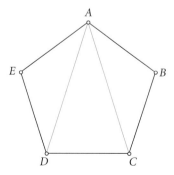

▶ **Exercise 4.10.** The pentagon above is divided into three isosceles triangles. Use this fact to find the measure of each angle formed inside the pentagon.

The isosceles triangle $\triangle ACD$ contained in the pentagon illustrated above is called a *golden triangle*: the base angles are twice the vertex angle of the triangle. The ratio of the lengths of the equal sides of this triangle to the base is called the *golden ratio* and is denoted by the Greek letter ϕ. Take a copy of $\triangle ACD$ and bisect angle $\angle ADC$, with the bisector intersecting side $\overline{AC}$ at point F:

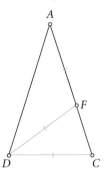

▶ **Exercise 4.11.** Find the measure of each of the angles in the diagram above of $\triangle ACD$.

By the results of Exercise 4.11, we see that the triangles $\triangle CDF$ and $\triangle ADF$ are also isosceles, so $\overline{AF} = \overline{DF} = \overline{DC}$. You should also have found that the triangles $\triangle ACD$ and $\triangle DCF$ are similar by AAA. Therefore, the sides are proportional:

$$\frac{\overline{AC}}{\overline{DC}} = \frac{\overline{DF}}{\overline{FC}}.$$

Since $\overline{AF} = \overline{DC} = \overline{DF}$ in this equation, we have

$$\frac{\overline{AC}}{\overline{AF}} = \frac{\overline{AF}}{\overline{FC}}.$$

Side $\overline{AC}$ is divided at point F so the *the whole is to the larger piece as the larger piece is to the smaller*. This phrase is a verbal description of the golden ratio, ϕ. The golden ratio is also sometimes called the *divine proportion*, or, in the case when a line segment is divided into two pieces by the golden ratio, it is said to be divided in *mean and extreme ratio*.

Definition 4.6. Two lengths x and y with $x > y$ are related by the *golden ratio* if the whole is to the larger piece as the larger piece is to the smaller, i.e., $\frac{x+y}{x} = \frac{x}{y}$.

Thus, in the golden triangle $\triangle ACD$ above,

$$\frac{\overline{AC}}{\overline{AF}} = \frac{\overline{AF}}{\overline{FC}} = \phi.$$

If we assume that the base of the triangle $\overline{CD}$ has length 1, then $\overline{AF}$ is also of length 1, so $\overline{AC} = \phi$ and $\overline{FC} = \frac{1}{\phi}$. Using the quadratic formula, it follows that:

$$\overline{AC} = \overline{AF} + \overline{FC},$$
$$\phi = 1 + \frac{1}{\phi},$$
$$\phi^2 = \phi + 1,$$
$$\phi^2 - \phi - 1 = 0,$$
$$\phi = \frac{1 + \sqrt{5}}{2}.$$

Thus, an alternate form for the golden ratio is given in the next theorem:

Theorem 4.7. Two lengths x and y with $x > y$ are related by the golden ratio if $\frac{x}{y} = \phi = \frac{1+\sqrt{5}}{2}$.

The key to constructing a regular pentagon is building a golden triangle, since any other central triangle will give either a short squat irregular pentagon or a tall thin irregular pentagon. Note that both of these pentagons have five equal sides but do not have five equal angles:

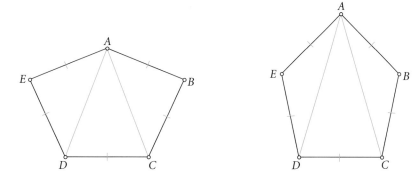

A golden triangle can be constructed if we can find two line segments whose ratio is ϕ. Euclid constructs such line segments in his Proposition VI.30:

Theorem 4.8. [Divide a line segment by the golden ratio] Given a line segment, there is a point dividing it into two parts, so that the whole is to the larger piece as the larger piece is to the smaller.

Proof: Let $\overline{AB}$ be a line segment. We wish to extend $\overline{AB}$ to a point C so that $\frac{\overline{AC}}{\overline{AB}} = \frac{\overline{AB}}{\overline{BC}} = \phi$. Construct a square $ABDE$ with $\overline{AB}$ as one side and then construct the midpoint F of $\overline{AB}$. Draw a circle centered at F with radius $\overline{FD}$. Extend the line $\overline{AB}$ to intersect this circle at C. The proportion of the line segments $\overline{AC}$ and $\overline{AB}$ is ϕ.

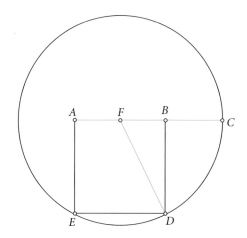

□

▶ **Exercise 4.12.** Verify the conclusion of Theorem 4.8 by showing that if $\overline{AB} = 1$ in the square $ABDE$, then $\overline{AC} = \phi = \frac{1+\sqrt{5}}{2}$.

▷ **Activity 4.10.** Use the construction of Theorem 4.8 to construct a golden triangle with base 1 unit long and sides each of length ϕ, using either a straightedge and a compass or geometric software.

▷ **Activity 4.11.** With a straightedge and a compass or with geometric software and your triangle of Activity 4.10, construct a regular pentagon whose sides are each 1 unit long.

▷ **Activity 4.12.** Construct a regular decagon (10 sides).

▷ **Activity 4.13.** Construct a regular pentakaidecagon (15 sides). [Hint: Begin with your pentagon from Activity 4.11, and construct a circle circumscribing this pentagon. Then construct an equilateral triangle with its vertices on the same circle so that one of the vertices of the triangle coincides with one of the vertices of the pentagon.]

▶ **Exercise 4.13.** Prove that if an n-sided regular polygon is constructible, then one can construct a regular polygon with $2^k n$ sides for any $k \geq 1$.

▶ **Exercise 4.14.** Prove that if an n-sided regular polygon is constructible and $k > 3$ is some integer that divides n, then one can construct a k-sided regular polygon.

▶ **Exercise 4.15.** Prove that if m and n are relatively prime integers (so that there are integers x and y so that $mx + ny = 1$) and it is possible to construct an m-sided regular polygon and an n-sided regular polygon with straightedge and compass, then one can construct a regular polygon with mn sides.

It is possible to construct other regular polygons than the ones we have discussed, but there are some that are not constructible without additional tools. For example, one cannot construct a regular heptagon (7 sides) with straightedge and compass alone. The question of which polygons are constructible was solved in 1796 by Carl Friedrich Gauss (at the age of 19), who proved that if n can be factored as a power of 2 times a product of distinct primes of a certain special form, then a regular n-sided polygon can be constructed. Pierre Wantzel later proved the converse. The proof requires some knowledge of complex numbers and so is omitted.

Theorem 4.9. [**Gauss's Theorem**] A regular n-sided polygon can be constructed with straightedge and compass alone if and only if all the odd prime factors of n are distinct Fermat primes; i.e., if all the odd prime factors are different and if each is of the form $F_k = 2^{2^k} + 1$.

The known Fermat primes are:

$$F_0 = 2^{2^0} + 1 = 2^1 + 1 = 3$$
$$F_1 = 2^{2^1} + 1 = 2^2 + 1 = 4 + 1 = 5$$
$$F_2 = 2^{2^2} + 1 = 2^4 + 1 = 16 + 1 = 17$$
$$F_3 = 2^{2^3} + 1 = 2^8 + 1 = 257$$
$$F_4 = 2^{2^4} + 1 = 2^{16} + 1 = 65537$$

Pierre Fermat [1601–1665] thought that the formula for F_k always gave a prime number, but the next in the series, $F_5 = 2^{2^5} + 1 = 4,294,967,297 = 641 \cdot 6,700,417$ is not prime. It is not known how many Fermat primes there are. Gauss gave instructions for the construction of the regular 17-gon (and had it carved on his gravestone). The regular 257-gon was constructed by Friedrich Julius Richelot [1808–1875] in 1831, while Johann Gustav Hermes [1846–1912] spent ten years on the regular 65,537-gon, completing his work in 1894 and leaving the manuscript in a large box at the University of Göttingen, where it remains a monument to patience.

By Gauss's Theorem, we can construct regular n-sided polygons if the factors of n are the first five Fermat primes listed above and twos (two is, of course, the only even prime). Thus, we can construct polygons with $3, 4 = 2 \cdot 2, 5, 6 = 3 \cdot 2, 8 = 2 \cdot 2 \cdot 2$ sides, etc. The nonagon (nine sides) cannot be constructed with straightedge and compass, although $9 = 3 \cdot 3$ and 3 is a Fermat prime, since the factors are not distinct.

▶ **Exercise 4.16.** List all the regular polygons with up to a hundred sides which can be constructed with straightedge and compass.

◇ 4.1.5 Squaring the circle

The three famous construction problems of the Greeks are doubling the cube and trisecting an angle, both discussed above, and the problem of squaring the circle. This last problem asks if it is possible, with straightedge and compass alone, to construct a square with the same area as a given circle. If we assume the circle has radius 1, then its area will be π and so the square must have edge length $\sqrt{\pi}$. Since we know how to construct square roots, this problem is equivalent to asking if we can construct a line segment of length π. That this is impossible with straightedge and compass alone was proven by Ferdinand von Lindemann in 1882. He showed that π is a *transcendental number*, one that cannot be expressed in terms of rationals and roots. In trying to deal with this problem, some subtle properties of the area function arise that could not be dealt with effectively until the maturing of the calculus.

♦ 4.2 Trisections

Any line segment can be trisected by Corollary 4.3, but one cannot trisect an arbitrary angle with straightedge and compass alone. This famous problem of classical euclidean geometry, solved by Pierre Wantzel in 1837, shows that one cannot, using unmarked straightedge and compass, divide an arbitrary angle into three equal sectors. Of course, one can trisect some special angles, such as $90°$, since we have already shown that we can construct a $30°$ angle. However, with the addition of extra tools, one can do this as well as many other problems unsolvable with the traditional tools. In this section, we will investigate constructions that trisect an arbitrary angle. The remaining sections of this chapter investigate some of these tools in more depth.

◇ 4.2.1 Trisection with marked ruler

If a ruler has markings, Archimedes showed that one can then trisect any angle. Let $\angle ABC$ be an arbitrary angle we wish to divide into three equal angles. Draw a circle centered at B (with any radius you find convenient). Let D and E be the points where the circle intersects the legs of the angle, as shown below. Using the marked ruler, draw a line from D to point F on the extended line CB, intersecting the circle at G, so that $\overline{FG} = \overline{BD}$, the radius of the circle. Then $\angle BFD = \frac{1}{3}\angle ABC$. Note that in this trisection, the ruler only needs two marks, just enough to measure the radius of the circle.

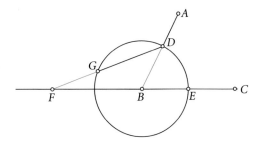

▶ **Exercise 4.17.** Prove that $\angle BFD = \frac{1}{3}\angle ABC$.

◇ 4.2.2 Origami trisection

In 1980, Hisashi Abe showed that it is possible to trisect any acute angle using paper folding techniques:

Trisecting an Angle with origami: *Given an acute angle, construct a line which trisects that angle.*

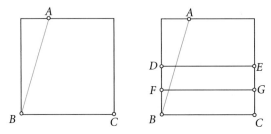

1. Let ABC be the given angle. We assume that the line BC forms the bottom edge of a square piece of origami paper.
2. Fold the origami paper so that BC lies on top of the upper edge of the paper, forming crease DE parallel to BC and cutting the paper in half.
3. Fold so that line BC lies on top of line DE, forming crease FG.
4. Fold so that point D lies on line AB and point B lies on line FG.
5. Without unfolding the crease formed in Step 4, extend the line formed by FG to form line HI.
6. Unfold the crease formed in Step 4, and extend line HI. It will intersect point B.
7. The angle $\angle ABI = \frac{1}{3}\angle ABC$.

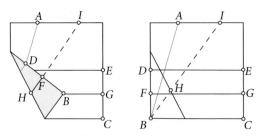

Proof: Let B', D', and F' be the points that B, D, and F lay on top of in Step 4. Drop a perpendicular from point B' to line BC, intersecting BC at point J. Since $\overline{DF} = \overline{BF}$, we have $\overline{DF} = \overline{FB} = \overline{D'F'} = \overline{B'F'}$ and $\overline{B'J} = \overline{BF}$. Since FG is perpendicular to BD, BF' is perpendicular to $B'D'$. Thus $\triangle BD'F'$ is congruent to

△BB'F' by SAS. Thus ∠ABF' = ∠B'BF'. Since ∠BF'B' and ∠BJB' are both right angles and $\overline{B'J} = \overline{B'F'}$, the triangles △BB'F' and △BB'J are congruent by the Hypotenuse-Side Theorem 3.20. Thus ∠JBB' = ∠F'BB', so the angle ∠ABC has been trisected.

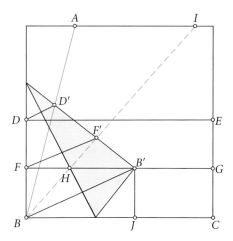

▷ **Activity 4.14.** Trisect an angle as above using origami paper.

There are many other constructions which can be done with origami but not with unmarked straightedge and compass, such as the duplication of the cube. For more on origami constructions, see Section 4.4.

◇ 4.2.3   Three-bar linkage

Archimedes' marked ruler method for trisecting an angle can be used to design a mechanical device called a *linkage* that will trisect an angle. Let ∠ABC be the angle we wish to trisect. The linkage, made of rods and rivets as shown below, is constructed so that $\overline{DE} = \overline{BE} = \overline{BF}$ and point F is constrained to a track running along line DE:

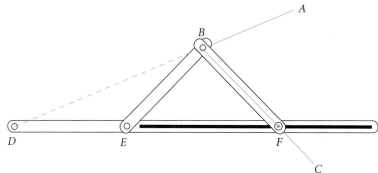

To use the linkage, place B at the vertex of the angle and point F so that BF runs along one side of the angle. Move the linkage so that point D is on line AB, the extension of the other side of the angle.

▶ **Exercise 4.18.** Prove that ∠BDE = $\frac{1}{3}$∠ABC.

◇ 4.2.4 Ceva's pantograph

Another mechanical linkage designed to trisect an arbitrary angle was designed by Giovanni Ceva [1647–1734] in 1695. Again, we are given an angle $\angle ABC$ that we wish to divide into three equal angles. We assume that the angle is drawn in a circle with center B so that $\overline{BA} = \overline{BC}$. We begin by constructing (with straightedge and compass!) the angle bisector BD. The linkage is designed around the rhombus $BEFG$ with $\overline{BE} = \overline{EF} = \overline{FG} = \overline{GB}$, with the legs FE and FG extended beyond their pivot points. Place the rhombus so that pivot point B of the rhombus falls on vertex B of the angle. Manipulate the linkage so that F lies on the extension of the angle bisector DB and so that the extended legs FE and FG fall on the points A and C. Then if we extend lines formed by EB and GB to points H and I, it can be shown that $\angle ABH = \angle HBI = \angle IBC$.

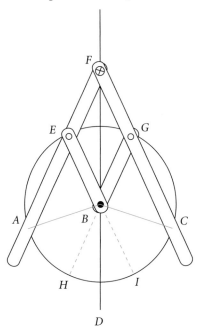

▶ **Exercise 4.19.** Prove that $\angle ABH = \angle HBI = \angle IBC$.

For more linkages and a discussion of constructibility using such devices, see Section 4.6. For a very amusing treatise on the trisection problem and the people who persist in claiming that they have succeeded in finding a straightedge and compass construction in spite of mathematical certainty that it cannot be done, see *The Trisectors* or *Mathematical Cranks* by Underwood Dudley.

♦ 4.3 Constructions with compass alone

We next consider what we could construct if we'd lost our straightedge and had only a compass. Euclid's Proposition I.2 showed us that anything constructible with a fixable compass can be constructed with a floppy compass, though with lots more circles and trouble. Thus we can assume that our compass is a modern fixable compass.

Lorenzo Mascheroni [1750–1800] was an Italian mathematician who investigated constructions with only a compass in 1797 in *The Geometry of Compasses*. In 1928 a mathematician happened on a book in a

Copenhagen bookstore with the rather grandiloquent title *Euclides Danicus* (The Danish Euclid) and was surprised to find that Georg Mohr [1640–1697] had solved the same problem in 1672, though his work had been completely forgotten.

Theorem 4.10. [Mohr-Mascheroni Theorem] Anything that is constructible with a straightedge and a compass can be constructed with a compass alone.

Proof: In studying straightedge and compass constructions we noted that each of these involves a sequence of the five basic S & C Operations, so our first task is to see which of these can be carried out with a compass alone. First, we assume that if we are given two points, we can imagine, but not draw, the line between them, so S & C Operation 1 is possible with reservations. In this context, constructing a line means constructing two points that lie on that line. Obviously, S & C Operation 3, which allowed the construction of a circle with a given center and radius, and S & C Operation 5, finding the intersection of two circles, are possible with compass alone. Thus we need to show that it is possible with compass alone to determine the intersection of two lines (S & C Operation 2) and the intersection of a line and a circle (S & C Operation 4). The first of these is surprisingly difficult, so we'll do the second first. We will use the notation $\mathcal{C}(A, B)$ to denote a circle centered at A that passes through point B and $\mathcal{C}(A, \overline{BC})$ to denote a circle with center A and radius equal to the line segment $\overline{BC}$.

Lemma 4.11. Given two points determining a line and a circle that intersects the line, we can find the points of intersection of the line and the circle using a compass alone.

Proof: We are given two points A and B that determine a line AB and a circle $\mathcal{C}$ intersecting this line. Let the center of the circle be point D and the radius r. We want to find points E and F where $\mathcal{C}(D, r)$ intersects AB.

Case 1: D does not lie on line AB.
Draw circles $\mathcal{C}_1(A, D)$ and $\mathcal{C}_2(B, D)$. These intersect at D and another point G. Now draw circle $\mathcal{C}_3(G, r)$. This will intersect $\mathcal{C}(D, r)$ at two points, E and F.

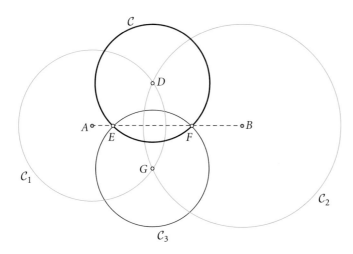

▶ **Exercise 4.20.** Complete the proof of Case 1 of Lemma 4.11 by showing that E and F lie on the line *AB*.

Case 2: D lies on line AB:

Choose a point G on $\mathcal{C}$. Draw circles $\mathcal{C}_1(A, G)$ and $\mathcal{C}_2(B, G)$. These intersect at G and another point H on the circle $\mathcal{C}$. Let L be the intersection of GH and AB, and note that since $\triangle AGH$ and $\triangle BGH$ are isosceles, $\triangle AGL \cong \triangle AHL$, so $GH \perp AB$. Let $x = \overline{GH}$. Draw circles $\mathcal{C}_3(D, x)$ and $\mathcal{C}_4(G, D)$. These will intersect at I on the side of GD opposite from H. Since $\overline{DI} = x = \overline{GH}$ and $\overline{DH} = \overline{DG} = \overline{GI}$, $DHGI$ is a parallelogram. Draw a circle $\mathcal{C}_5(H, D)$. This will intersect $\mathcal{C}_3$ at J on the opposite side of AB from I and, as above, $DGHJ$ is another parallelogram.

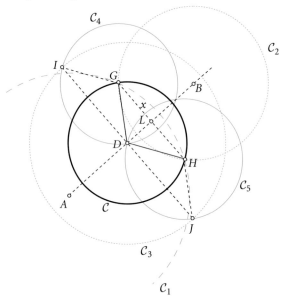

Note that $DHGI \cong DGHJ$, so the diagonals will be equal, $\overline{IH} = \overline{GJ}$. Now draw circles $\mathcal{C}_6(I, H)$ and $\mathcal{C}_7(J, G)$. These will intersect at K and $\overline{IK} = \overline{IH} = \overline{GJ} = \overline{JK}$. Thus, $\triangle IKD \cong \triangle JKD$ so $KD \perp IJ$. Since IJ is parallel to GH, $KD \perp GH$, so K lies on line AB. Let $y = \overline{DK}$. Last, draw circle $\mathcal{C}_8(I, y)$. This will intersect $\mathcal{C}$ at E and F, which we claim are the intersection of AB and $\mathcal{C}$.

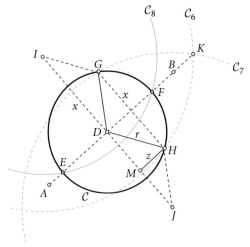

Consider the right triangle $\triangle DKI$. From this we get $y^2 = \overline{IK}^2 - \overline{ID}^2 = \overline{IK}^2 - x^2 = \overline{IH}^2 - x^2$. Drop $HM \perp IJ$ and let $z = \overline{HM}$. Note that $x = \overline{GH} = \overline{ID} = \overline{JD}$ and $\overline{DM} = \frac{1}{2}x$. From $\triangle IMH$, $\overline{IH}^2 = z^2 + (\frac{3}{2}x)^2$. Thus, $y^2 = z^2 + \frac{9}{4}x^2 - x^2 = z^2 + \frac{5}{4}x^2$. From $\triangle DHM$, we see that $z^2 = r^2 - (\frac{1}{2}x)^2$, so $y^2 = r^2 + \frac{5}{4}x^2 - \frac{1}{4}x^2 = r^2 + x^2$. Thus since $y = \overline{DK} = \overline{IE} = \overline{IF}$, E and F lie at the intersection of the circle $\mathcal{C}$ and the line AB. $\square$

The most difficult construction is the one that shows how to find the intersection of two lines:

Lemma 4.12. Given four points determining two nonparallel lines, we can find the point of intersection of the two lines using compass alone.

Proof: Given points A, B, C, and D that determine lines AB and CD, we want to find point X where AB intersects CD. Specifically, we must find two circles that intersect at X.

Draw circles $\mathcal{C}_1(A, C)$, and $\mathcal{C}_2(B, C)$. These circles will intersect at C and another point C'. Now draw $\mathcal{C}_3(A, D)$ and $\mathcal{C}_4(B, D)$ which will intersect at D and another point D'. The following exercise shows that CD and $C'D'$ also intersect at X, the desired intersection of AB and CD.

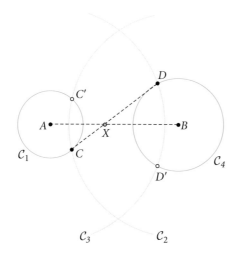

▶ **Exercise 4.21.** Prove first that $\triangle ACB \cong \triangle AC'B$, and then that $\triangle ACX \cong \triangle AC'X$. Similarly, $\triangle BDA \cong \triangle BD'A$ and $\triangle BDX \cong \triangle BD'X$. Prove that $C'XD'$ is a straight line.

▶ **Exercise 4.22.** Prove that $AB \perp CC'$ and $AB \perp DD'$ and so $CC' \| DD'$.

Now construct circles $\mathcal{C}_5(C, \overline{DD'})$ and $\mathcal{C}_6(D', \overline{CD})$ and label the intersection shown on the C-side of AB as E.

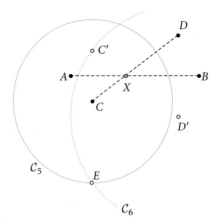

▶ **Exercise 4.23.** Prove that $CDD'E$ is a parallelogram. Then show that $\triangle C'D'E \sim \triangle C'XC$, that $\frac{\overline{CX}}{\overline{CD}} = \frac{\overline{CC'}}{\overline{C'E}}$ and finally, that $\overline{C'E} = \overline{CC'} + \overline{DD'}$.

Construct circles $\mathcal{C}_7(C', E)$ and $\mathcal{C}_8(E, \overline{CD})$ and label their intersection as F. Construct circles $\mathcal{C}_9(C', C)$ and $\mathcal{C}_{10}(E, \overline{CF})$ and label their intersection as G.

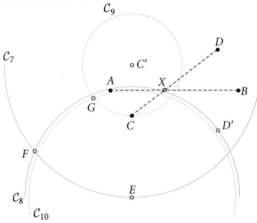

▶ **Exercise 4.24.** Prove (by contradiction) that that G lies on line FC' and then show $\triangle EFG \cong \triangle FEC$. Prove that $\triangle C'EF$ and $\triangle C'CG$ are similar isosceles triangles and so $\frac{\overline{CG}}{\overline{EF}} = \frac{\overline{C'C}}{\overline{C'E}}$.

From the exercises above, it follows that $\overline{CG} = \overline{CX} = \overline{C'X}$, and thus X is determined by the intersection of $\mathcal{C}_{11}(C, G)$ and $\mathcal{C}_{12}(C', \overline{CG})$.

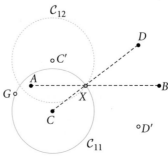

Thus with the assistance of these two lemmata we have a proof of the Mohr-Mascheroni Theorem 4.10. □

A straightedge alone does not allow for the construction of all of Euclid's constructions. However, Jacob Steiner [1796–1863] published a work in 1833 showing that all of the traditional constructions can be carried out with a straightedge if a single circle with its center is also given. Jean-Victor Poncelet [1788–1867] had previously announced this result but had not published a proof, so this is usually called the Poncelet-Steiner Theorem. For details of this, see George Martin's *Geometric Constructions*.

♦ 4.4 Theoretical origami

Although the art of origami or paper folding has been pursued in Japan for millennia, the idea of studying geometry by means of paper folding seems to have originated with an otherwise unknown Indian mathematician and teacher, Sundara Row (or Rao) [1853–?] in a book first published in 1893. More recently, interest has grown in the field of mathematical origami and its application to the classroom. For this approach one needs only thin paper and a pencil (to mark and label points) and nimble fingers. Of course, it is difficult to make the folds as precise as they need to be, but it is also difficult in practice to be precise with straightedge and compass. Origami has the definite advantage in that it avoids the common hazard of students stabbing each other with the pointy ends of their compasses. Paper cuts remain a danger.

◊ 4.4.1 Basic operations and constructions

In discussing mathematical origami, we try to parallel the development of straightedge and compass constructions in Chapter 4.1. The basic origami operations we will use are due to Humiaki Huzita in 1989, with an additional operation introduced by Koshiro Hatori in 2002. These are now commonly called the Huzita-Hatori axioms. We adapt the approach of the Austrian mathematician Robert Geretschläger in his "Euclidean Constructions and the Geometry of Origami."

Origami Operation 1. Given two distinct points, fold the (unique) line connecting them.

▷ **Activity 4.15.** Mark two points at random (but not too close together) on a piece of tracing paper and fold, forming a straight line passing through both points.

This operation is the precise counterpart of the first straightedge and compass operation. We also assume that we can find the point of intersection of two nonparallel folded lines, the analog of S & C Operation 2. S & C Operations 3, 4, and 5 refer to the drawing of circles. This doesn't make sense in origami terms, so the rest of the operations differ from those of straightedge and compass, though we will eventually see that origami allows us to perform the traditional constructions.

Origami Operation 2. Given any two distinct points, one can fold so that one point lands exactly on top of the other.

▷ **Activity 4.16.** Mark two points at random on a piece of tracing paper and fold so that one point lies over the other. Discuss the geometric relationship between the fold line and the line connecting the two points.

Origami Operation 3. Given two lines, one can fold so that the first line falls exactly over the second line.

Note that for intersecting lines there are two ways of carrying out Origami Operation 3.

▷ **Activity 4.17.** Draw (or fold) two intersecting lines at random on a piece of tracing paper and fold so that one line lies over the other. Discuss the geometric relationship between the fold line and the two original lines.

▷ **Activity 4.18.** Draw (or fold) two parallel lines on a piece of paper and fold so that one line lies over the other. Discuss the geometric relationship between the fold line and the two original lines.

Origami Operation 4. Given a point and a line, one can fold so that the creased line goes through the given point and is perpendicular to the given line.

▷ **Activity 4.19.** Draw (or fold) a line on a piece of tracing paper. Mark a point on the paper. Fold to form a line through the marked point and perpendicular to the given line. Consider separately the cases where the given point lies on the given line and where the point does not lie on the line.

Origami Operation 5. Given two points A and B and a line ℓ, one can fold the paper forming a single crease so that point B lies on the crease while point A lands on ℓ.

Note that in certain configurations there may be zero, one, or two possible solutions for the fold of Origami Operation 5.

▷ **Activity 4.20.** Mark two points A and B and draw (or fold) a line on a piece of paper. Fold so that point A lies on the line while point B lies on the crease formed. Do this separately for the cases illustrated below:

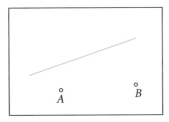 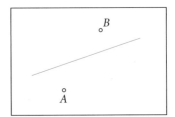

▶ **Exercise 4.25.** Give a condition under which Origami Operation 5 will have at least one solution and explain why your condition is necessary.

▶ **Exercise 4.26.** Discuss at least one case where there are two different solutions for the fold line of Origami Operation 5.

Using just these five Origami Operations, let us try to see which of Euclid's straightedge and compass constructions can be performed. For example, let $\angle ABC$ be an arbitrary angle. We can use Origami Operation 3 to fold BA onto BC, forming the angle bisector. Alternatively, we could use Origami Operation 5 to fold so that point B lies on the fold while A falls on the line BC:

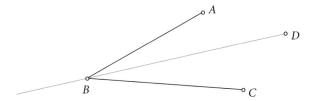

▷ **Activity 4.21.** Figure out how to construct the midpoint of a given line segment. Explain which of the Origami Operations are used in your construction.

Origami Operation 4 shows that we can drop and erect perpendiculars from any point to any line. Thus with origami we can do the constructions from Euclid's Propositions I.9–12. The question remains of how to manage without circles. Certainly we cannot fold a circle. In Section 4.3 when we performed constructions with only a compass, we noted that it sufficed to merely imagine the relevant lines as long as we could locate precisely any points that we needed, such as the points of intersection of these lines with other lines and circles. We take a similar approach to the existence of circles. We need to show that given a point O to use as the center of a circle and a radius $r = \overline{AB}$, we can locate any point on the circle we can imagine. First use Origami Operation 2 to fold so that A lands on O. In doing this, point B lands on point P and, since the fold line is the perpendicular bisector of $\overline{OA}$, we can show that $\overline{OP} = \overline{AB} = r$. Thus we have constructed one point P on the desired circle centered at O with radius r:

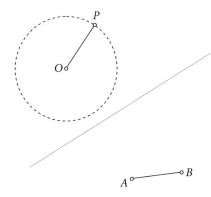

To construct other points on the circle, consider any line ℓ through O. We can find the point of intersection of the circle and line ℓ by using Origami Operation 5 to fold so that the crease passes through O and P lands on ℓ. Then P lands on point Q on ℓ, which also lies on the circle.

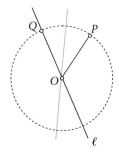

Thus given a center and a radius, we can construct as many points as we like on the circle, including those in any given direction. We can effectively construct with origami all of the points that S & C Operation 3 gave with straightedge and compass. It remains to consider the points that S & C Operations 4 and 5 give.

If we are given a center O and a radius r, as above we can construct a point P lying on the circle. Let ℓ be any line intersecting this circle. We need to show that we can construct the point of intersection of the line and the circle. Use Origami Operation 5 to fold so that the crease goes through point O and P lands on line ℓ. Label the point on ℓ where P falls as Q. Note that the fold line is the perpendicular bisector of $\overline{PQ}$. Thus, we have shown that we can construct with origami the intersection of any line with any circle as in S & C Operation 4.

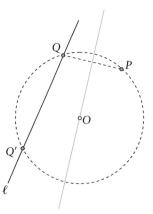

▶ **Exercise 4.27.** Prove that Q lies on the circle. Explain how to construct Q', the other point of intersection of the line and the circle.

▷ **Activity 4.22.** One of the main uses of circles in geometry is to cut a length off a line equal to a given line segment. Given a line segment $\overline{AB}$ and a point C on a line ℓ, construct a point D on ℓ so that $\overline{CD} = \overline{AB}$. Justify each of the steps of your construction by one of the Origami Operations.

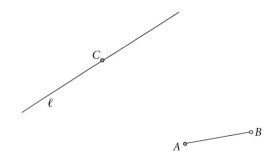

▷ **Activity 4.23.** If we are given two lengths, a and b, with $a > b$, by Activity 4.22 we can assume that we have a single line ℓ and points A, B, C, and D on ℓ such that $\overline{AB} = a$ and $\overline{CD} = b$. Show how to construct line segments of length $a + b$ and $a - b$. Consider separately the cases shown below where $\overline{AB}$ and $\overline{CD}$ are disjoint and where they overlap. Justify each of the steps of your constructions by one of the Origami Operations.

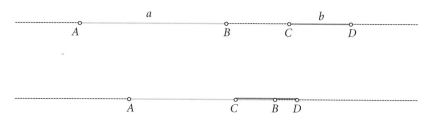

▷ **Activity 4.24.** Draw a line ℓ and a point P not on this line on a piece of tracing paper. Figure out how to fold a line through P parallel to ℓ. Justify each of the steps of your construction by one of the Origami Operations.

S & C Operation 5 claimed that one can find the points of intersection of two circles and we must consider how to do this using origami. Consider the intersection of two circles, one with center O and radius r and the other with center P and radius s, assuming that $\overline{OP} < r + s$ so the circles intersect. We want to find the points A and B. We will do this by showing that we can construct the chord $\overline{AB}$ and appealing to the previous construction to find the intersection of this line with either circle. Let C denote the intersection of AB with OP and note that AB is perpendicular to OP at C. Thus, it suffices to locate the point C.

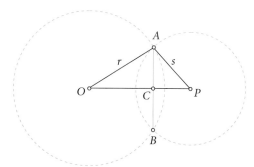

Let $\overline{OC} = x$, $\overline{AC} = \overline{BC} = y$, and $\overline{OP} = z$. Note that $\overline{CP} = \overline{OP} - \overline{OC} = z - x$. Applying the Pythagorean Theorem to $\triangle OAC$ and to $\triangle PAC$, we have

$$x^2 + y^2 = r^2,$$
$$(z-x)^2 + y^2 = z^2 - 2xz + x^2 + y^2 = s^2.$$

Thus,

$$x = \frac{z^2 + r^2 - s^2}{2z}.$$

We need to show that we can construct a length equal to x.

If we are given a length a, note that we can construct a length of a^2 by constructing a length of 1 along a line, erecting a perpendicular at one end (Origami Operation 4) and measuring off a length of a (Activity 4.22), thus forming a right triangle with base of length 1 and height a. Construct a similar triangle with base of length a. By similar triangles, the height of this triangle will have length a^2.

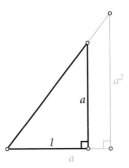

With this construction and the results of Activity 4.23 we can construct a line segment of length $z^2 + r^2 - s^2$. The following exercise shows that we can construct a line segment of length $x = \frac{z^2 + r^2 - s^2}{2z}$.

▶ **Exercise 4.28.** Given two lengths a and b, explain how to construct a line segment of length $\frac{a}{b}$. Justify each of the steps by one of the Origami Operations.

Now, using the length x we can construct point C on $\overline{OP}$ and then fold the line through C perpendicular to OP to form the chord $\overline{AB}$. We previously showed that we can locate the intersection of a line with a given circle, so we can find points A and B where the perpendicular line meets either one of the circles. Thus we can achieve the result of S & C Operation 5 and so have the following theorem:

Theorem 4.13. Anything that is constructible with a straightedge and a compass can be constructed with origami.

We have shown that anything that can be constructed using an unmarked straightedge and compass can also (theoretically) be constructed using origami folds. We therefore should be able to construct using origami any regular polygon that is constructible with straightedge and compass.

▷ **Activity 4.25.** Tear an irregular piece of paper and fold a square.

If you can do Activity 4.25, then you can make squares anywhere, so you can use standard origami paper for the rest of the activities.

▷ **Activity 4.26.** Fold an equilateral triangle using square origami paper. Hint: Consider the diagram below, and think about how the triangle and its altitude will fit inside the square paper if the base of the triangle runs along the lower edge of the paper.

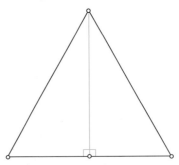

▷ **Activity 4.27.** Fold a regular hexagon using square origami paper. Explain why your procedure results in a regular figure.

▷ **Activity 4.28.** Fold a regular octagon using square origami paper. Explain why your procedure results in a regular figure.

We will also consider how to divide a line segment into equal portions. By Activity 4.21, we can fold the midpoint of any line segment. We can thus fold a line segment into fourths, eighths, etc. Thus we need only consider how to divide a line segment into an odd number of pieces.

▷ **Activity 4.29.** Using a piece of square paper or geometric software, let $ABCD$ be a square. Let E and F be the midpoints of $\overline{AD}$ and $\overline{BC}$ and fold $\overline{EF}$. Fold $\overline{BD}$ and $\overline{EC}$. These fold lines intersect at a point G. Make a conjecture about G.

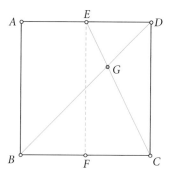

▶ **Exercise 4.29.** Prove your conjecture of Activity 4.29.

▶ **Exercise 4.30.** Generalize the procedure of Activity 4.29 to fold a line segment into fifths.

▶ **Exercise 4.31.** Generalize the procedure of Activity 4.29 to fold a line segment into $n$ths for an odd number n.

The procedure of Exercise 4.31, together with bisection, allows the construction of any rational length. As for straightedge and compass constructions, we can fold square roots by utilizing the Pythagorean Theorem. In particular, we should be able to construct the irrational number ϕ. The key to folding a pentagon is, of course, finding two lengths whose ratio is the golden ratio or ϕ. The following construction is from Sundara Row's *Geometric Exercises in Paper Folding*.

Theorem 4.14. *Given a line segment, we can fold so as to divide this line segment by the golden ratio, so that the whole is to the larger piece as the larger piece is to the smaller.*

Proof: Let $\overline{AB}$ be the given line segment and construct a square $ABCD$. Fold the square in half, laying $\overline{AB}$ on top of $\overline{DC}$ and marking point E, the midpoint of $\overline{AD}$. Fold through E and B. Fold so that line $\overline{AE}$ lies on top of $\overline{EB}$ and mark point F where this fold intersects $\overline{AB}$ and point G on $\overline{EB}$ so that $\overline{EG} = \overline{EA}$. Fold line $\overline{BA}$ on top of line $\overline{BE}$ and mark point H so that $\overline{BH} = \overline{BG}$. Then H divides $\overline{AB}$ by the golden ratio.

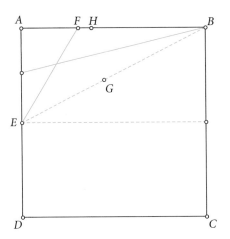

▶ **Exercise 4.32.** In Theorem 4.14, assume that $\overline{AB} = 1$. Show that $\frac{\overline{AB}}{\overline{BH}} = \phi = \frac{1+\sqrt{5}}{2}$.

To construct a regular pentagon, first mark the point H on side $\overline{AB}$ as in Theorem 4.14.

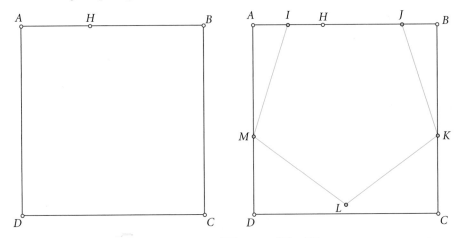

Let I be the midpoint of $\overline{AH}$, and mark point J on $\overline{AB}$ so that $\overline{AI} = \overline{BJ}$. Fold so that point J lies on line $\overline{AD}$ while point I lies on the crease and mark point M so that $\overline{IJ} = \overline{IM}$. Fold so that point I lies on line $\overline{BC}$ while point J lies on the crease and mark point K so that $\overline{JK} = \overline{JI}$. Fold so that $\overline{JK}$ is doubled on itself, so that J lies over K and M lies on the crease. Mark point L where I lands. Note that L will not lie on edge $\overline{CD}$. Then IJKLM is a regular pentagon.

▷ **Activity 4.30.** Construct a pentagon using the directions above.

◊ **4.4.2 Additional operations and their consequences**

There are two additional operations that are allowable in origami and these have no counterparts for straightedge and compass. Origami Operation 6 is one of Huzita's original axioms and opened the door to many new constructions using origami techniques. Additionally, in certain configurations there may be up to three possible solutions. Operation 7 is a recent discovery of Koshiro Hatori in 2002. It should be noted that all seven origami operations were noted by Jacques Justin in 1989, the same year that Huzita published his mathematical treatment of origami. Unfortunately, his findings were not widely publicized until after Hatori independently rediscovered the last of the operations.

Origami Operation 6. Given two points, A and B, and two lines, ℓ_1 and ℓ_2, one can fold forming a single crease so that point A lies on top of line ℓ_1 while point B lies on top of line ℓ_2.

▷ **Activity 4.31.** Lay out two points and two lines on a piece of tracing paper as shown in each of the four diagrams below. For each, find one fold so that point A lies on top of the line ℓ_1 while point B lies on top of line ℓ_2.

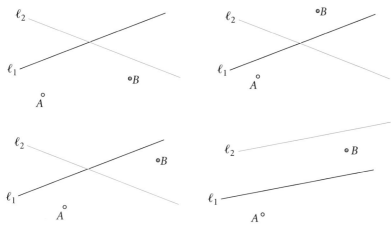

Note that Origami Operation 6 may have zero, one, two, or three possible solutions.

Origami Operation 7. Given a point A and two lines, ℓ_1 and ℓ_2, one can fold forming a single crease so that point A lies on top of line ℓ_1 and the fold is perpendicular to ℓ_2.

▷ **Activity 4.32.** Lay out a point and two lines on a piece of tracing paper as shown in the diagram below. Fold a crease so that point A lies on top of the line ℓ_1 and the fold is perpendicular to ℓ_2.

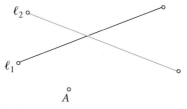

Robert Lang has proved that this list of operations or origami axioms is complete, so every origami construction can be expressed as some combination of these eight. Origami Operations 6 and 7 are the ones that distinguish the straightedge and compass universe from the origami universe. In fact, with the use of these operations (in particular, Origami Operation 6), we can trisect an arbitrary angle as in Abe's construction of Chapter 4.2. We can also construct cube roots of given lengths, impossible with an unmarked straightedge and compass and thus solve the problem of the duplication of the cube. The following solution appeared as Problem #1054 in *Crux Mathematicorum* in 1986 posed by Peter Messer, with the solution given by Stanley Rabinowitz.

Theorem 4.15. It is possible to duplicate the cube using origami.

Proof: We wish to construct a line segment of length $\sqrt[3]{2}$. Let $ABCD$ be a square. Fold $ABCD$ in thirds as in Activity 4.29, by creases $\overline{EF}$ and $\overline{GH}$. Fold so that corner C falls on point C' along the edge $\overline{AB}$ while H falls on point H' on $\overline{EF}$. Then we claim that $\frac{AC'}{C'B} = \sqrt[3]{2}$.

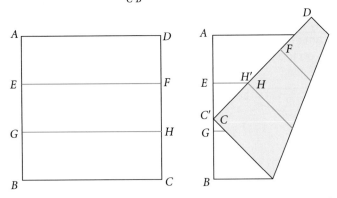

To prove that this works, consider the drawing below. We label the fold point along $\overline{BC}$ as I. Without loss of generality, we may assume that $\overline{C'B} = 1$. Let $\overline{AC'} = x$ and $\overline{BI} = y$. Thus the square $ABCD$ has sides of length $1+x$.

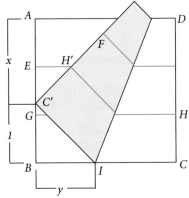

▶ **Exercise 4.33.** Complete the proof of Theorem 4.15 by performing the following computations.

a. Use the Pythagorean Theorem for $\triangle BC'I$ to find an equation relating x and y. Solve this equation for y.
b. Find the lengths of $\overline{EC'}$ and $\overline{C'H'}$ in terms of x.
c. Show that $\triangle EC'H'$ and $\triangle BIC'$ are similar.
d. Use parts (b) and (c) to derive another equation relating x and y. Solve this equation for y.
e. Set the equations for y found in (a) and (d) equal to each other and solve for x.

In Section 4.1, we showed that a length is constructible by straightedge and compass if it involves only rational numbers and square roots. We have now shown that it is possible to construct at least one cube root with origami. In fact the set of origami constructible numbers consists precisely of all numbers that are finite combinations of rational numbers and square and cube roots.

Paralleling Gauss's Theorem 4.9, in 1988 Andrew Gleason proved that a regular n-sided polygon can be constructed with origami (or any other tool that allows the trisection of an arbitrary angle) if and only if $n = 2^n 3^m (p_1 p_2 \ldots p_i)$ where the p_j's are distinct primes of the form $p_j = 2^k 3^\ell + 1$. These are called *Pierpont primes*, generalizing the definition of the Fermat primes of Chapter 4.1. This allows the construction of regular polygons with seven, nine, and thirteen sides, among others. Robert Geretschläger has written up the details of these and many other origami constructions.

▶ **Exercise 4.34.** List all the regular polygons with up to a hundred sides that can be constructed with origami.

◆ 4.5 Knots and star polygons

There are many ways of constructing regular polygons other than with straightedge and compass or origami. One easy way of forming a regular pentagon is by knotting a strip of paper in a simple overhand knot. We show both a drawing of the knot as it would appear if done with a bit of string and a picture of how the knotted strip of paper will look.

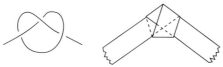

If the long sides of the strip of paper are parallel and the knot is snug but not too tight, this will form a regular pentagon. Cash register tape or wide paper ribbon has parallel edges and is handy to use.

▷ **Activity 4.33.** Knot two strips of paper to form a regular hexagon as below:

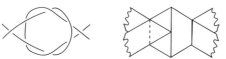

▷ **Activity 4.34.** Knot a strip of paper to form a regular heptagon. Below is a picture of the completed knot.

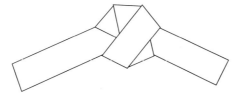

▷ **Activity 4.35.** Tie an octagonal knot from a single strip of paper. The following instructions are taken from the *Ashley Book of Knots*, knot number 2590. First tie a modified overhand knot exactly as shown below on the left:

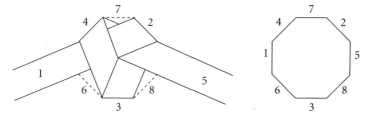

We wish to end up with an octagon as on the right. The folds of the knot you have made form three sides, labeled 2, 3, and 4, and the loose ends form two other sides, labeled 1 and 5. Fold the loose end down at position 5 and thread the end through the fold at 3 to position 6. Next fold down at position 6 and go over the loose end at position 1 and through the fold at position 4 to position 7. Fold down at position 7 and go through the folds at positions 2 and 5 to position 8. Fold the loose end down at position 8 and go through the folds at 3 and 6 to end at position 1. After tightening you will have an octagonal knot with both loose ends at position 1.

To see that these knotted strips of paper form regular polygons is an exercise in classical euclidean geometry. The following proof is adapted from Alexander Bogomolny's *Cut-the-Knot* website.

Theorem 4.16. Knotting a strip of paper with parallel edges with an overhand knot forms a regular pentagon.

Proof: Consider the first fold. We are given that $XA \parallel YB$, $ZA \parallel WB$. Drop $AH \perp WB$ and $BG \perp XA$.

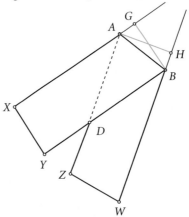

Since the width of the strips is constant, $\overline{AH} = \overline{BG}$. Then $\triangle AHB \cong \triangle BGA$ by the Hypotenuse-Side Theorem 3.20. Therefore, $\angle GAB = \angle HBA$, so $\angle XAB = \angle WBA$. Since $XA \parallel YB$, $\angle XAD = \angle ADB$. Similarly, $ZA \parallel WB$ implies that $\angle ADB = \angle WBD$. Thus, $\angle DAB = \angle DBA$, so $\overline{AD} = \overline{BD}$.

Now consider the second fold: By the argument above, $\angle EAD = \angle ADB = \angle CBD$. Similarly, $\angle EDA = \angle DAC = \angle ACB$, $\angle ADC = \angle ACD$, and $\overline{AD} = \overline{AC}$.

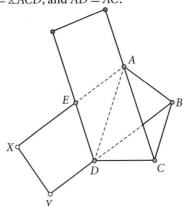

Similarly, in the picture of the final fold (below), $\angle DEC = \angle ECA = \angle CAB$, $\angle CEA = \angle CAE$, and $\overline{CE} = \overline{CA}$. Therefore, $\overline{AD} = \overline{BD} = \overline{AC} = \overline{CE}$. Drop $EI \perp BA$ and $BJ \perp EA$. Since $\overline{EI} = \overline{BJ}$, $\triangle EIA \cong \triangle BJA$ by AAS, so $\overline{EA} = \overline{BA}$. Dropping perpendiculars and using congruent right triangles shows that $\overline{DE} = \overline{AE}$ and $\overline{CB} = \overline{AB}$. Thus $\overline{DE} = \overline{EA} = \overline{AB} = \overline{BC}$.

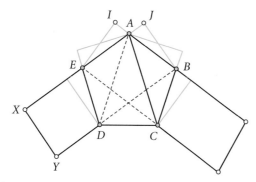

Therefore, $\triangle DEA$ and $\triangle CBA$ are isosceles, so $\angle EDA = \angle EAD$ and $\angle BAC = \angle BCA$. Thus all of the angles marked with dots in the picture below are equal. Thus $\angle DAB = \angle EAC$. Since we know that $\angle DAB = \angle DBA$ and $\angle CEA = \angle CAE$, it follows that $\angle DBA = \angle CAE$ and so $\angle EAB = \angle ABC$. Therefore, the trapezoids $EABC$ and $ABCD$ are congruent, so $\overline{CD} = \overline{BC}$ and $\angle ABC = \angle BCD$. Thus $\overline{EA} = \overline{AB} = \overline{BC} = \overline{CD} = \overline{DE}$ and $\angle DEA = \angle EAB = \angle ABC = \angle BCD = \angle CDE$, and so $ABCDE$ is a regular pentagon.

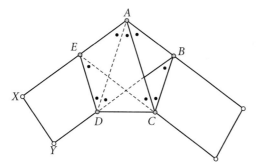

□

We now turn our attention to a seemingly unrelated way of drawing polygons. Assume that an n-sided polygon has been constructed by some means. Such a polygon and its circumscribing circle form a ring of n equally spaced points around the circle. Other figures can be drawn using these n regularly spaced points. Below is a circle with eight such points:

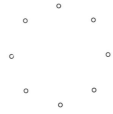

If we connect each point in order, we get a regular octagon. If we connect every other point, we get two squares. Thus, connecting every second point is said to form two *cycles* of length four.

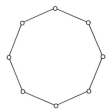 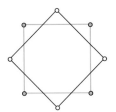

Connecting every kth dot around a circle of n equally spaced dots forms a *star polygon*. Note that sometimes the resulting figure connects every point in one path. It is then called a *regular star polygon*. Sometimes the figure consists of several separate cycles, such as the one above in which we connected every other point of eight equally spaced points or in the familiar Star of David. These are *compound star polygons*. The notation we shall use is $\{{}^n_k\}$ for the star polygon in which we connect every kth point of n equally spaced points around a circle. Note that $\{{}^n_1\}$ is always the regular n-sided polygon. Also note that some of these star polygons are identical with others, the only difference being that the points are connected in a different order.

You can either find dot paper to use for the following exercises or make your own using geometric software.

▷ **Activity 4.36.** Find or make some rings of 8 dots and then draw the star polygon $\{{}^8_3\}$ formed by connecting every third dot. Draw $\{{}^8_4\}$, $\{{}^8_5\}$, $\{{}^8_6\}$, and $\{{}^8_7\}$.

▷ **Activity 4.37.** Draw some rings of 9 dots and then draw $\{{}^9_2\}$, $\{{}^9_3\}$, $\{{}^9_4\}$, etc.

▶ **Exercise 4.35.** State a rule for when two star polygons are identical.

▷ **Activity 4.38.** Draw some rings of 12 dots and draw all the different $\{{}^{12}_k\}$ star polygons. For each star polygon, note whether it is regular and or compound. You need not duplicate identical figures.

▷ **Activity 4.39.** Draw some rings of 15 dots and draw all the different $\{{}^{15}_k\}$ star polygons. For each star polygon, note whether it is regular and or compound. You need not duplicate identical figures.

▷ **Activity 4.40.** Draw a ring of 18 dots and draw the star polygon $\{{}^{18}_4\}$.

▷ **Activity 4.41.** It was noted above that $\{{}^8_2\}$ consists of two cycles of length four. Which $\{{}^9_k\}$ star polygons are compound? How many cycles do these have and how long are the cycles? Which $\{{}^{12}_k\}$ star polygons are compound? How many cycles do these have and how long are the cycles? Repeat for $\{{}^{15}_k\}$ and $\{{}^{18}_4\}$.

▶ **Exercise 4.36.** State a rule for when a star polygon is regular. State a rule for when a star polygon is compound, how many cycles it will have, and how long the cycles will be.

▶ **Exercise 4.37.** How many cycles does the compound star polygon $\{{}^{24}_9\}$ have? How long are those cycles?

▶ **Exercise 4.38.** List all the regular star polygons of the form $\{{}^{21}_k\}$.

If one were to walk around the polygon $\{{}^8_1\}$, one would make one trip around the center point. However, traveling around $\{{}^8_2\}$ requires two clockwise revolutions around the center. What we here refer to as the number of revolutions is related to the concept of the *winding number* of a closed curve. Note that to count the number of revolutions correctly, each cycle of $\{^8_4\}$ is considered to have length 2: go from each point to the point directly opposite and then back again.

▷ **Activity 4.42.** For each of the star polygons drawn in Activities 4.36, 4.37, 4.38, and 4.39 above, note how many revolutions around the center are made.

▶ **Exercise 4.39.** State a rule for how many revolutions around the center are made in a star polygon.

▶ **Exercise 4.40.** What do the knots of Activities 4.33, 4.34, and 4.35 have to do with star polygons? Explain the connection carefully.

▷ **Activity 4.43.** Can you make a regular nine-sided knot? Make one or explain why you can't.

Next we compute the angle measure for the vertex angle of any star polygon, generalizing the formula $\frac{(n-2)180°}{n}$ for the vertex angle of a regular n-sided polygon. Consider the star polygon $\{^5_2\}$, also called the *pentagram*. Notice that the edges of any star polygon form a smaller regular polygon in the center of the circle. This star polygon can be considered as a central regular pentagon with isosceles triangles attached to each side. Since the vertex angle for a regular pentagon measures 108°, the base angles of the isosceles triangles must be 72°. Since the sum of the angles in a triangle must be 180°, the vertex angle of the isosceles triangle, and thus the vertex angle of the star polygon, must measure 36°.

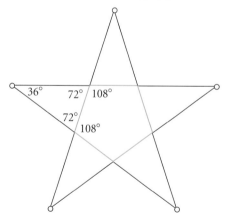

▶ **Exercise 4.41.** Fill in the following table. In the last column, write the difference between the vertex angles of $\{^n_k\}$ and $\{^n_{k+1}\}$.

$n\downarrow$ $k\rightarrow$	1	2	3	4	Difference
5	$\angle\{^5_1\}=$	$\angle\{^5_2\}=$	•	•	
6	$\angle\{^6_1\}=$	$\angle\{^6_2\}=$	$\angle\{^6_3\}=$	•	
8	$\angle\{^8_1\}=$	$\angle\{^8_2\}=$	$\angle\{^8_3\}=$	$\angle\{^8_4\}=$	
9	$\angle\{^9_1\}=$	$\angle\{^9_2\}=$	$\angle\{^9_3\}=$	$\angle\{^9_4\}=$	

▶ **Exercise 4.42.** Figure out a formula for the difference between the vertex angle of $\{{n \atop k}\}$ and $\{{n \atop k+1}\}$ in terms of n. Then find a formula for the vertex angle of the star polygon $\{{n \atop k}\}$ in terms of n and k.

▶ **Exercise 4.43.** Prove that your formula of Exercise 4.42 is valid. [Hint: Consider the diagram below.]

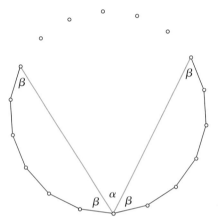

◆ 4.6 Linkages

We have seen in Section 4.2 that simple mechanisms can be used to perform some constructions impossible with only a straightedge and compass. The Greeks were well aware of this and gave mechanical solutions for each of the classical problems as well as many other constructions, though it should be noted that Plutarch [ca. 46–120] twice mentions Plato's disdain for such solutions. But for many problems for which there was no traditional straightedge and compass solution, these provided the only practicable solutions. The system of arithmetic and algebra used by the Greeks did not allow the methods of solving equations we are familiar with, so these geometric constructions were their only means of addressing complicated curves and problems concerning them. Linkages and similar mechanisms were given to construct figures such as ellipses and the other conic sections and other special curves.

In 1637 René Descartes published *La Géométrie*, a fundamental text in the evolution of geometry that initiated the idea of expressing geometry in terms of coordinates and equations. For each curve he discusses, he first gives a means of construction, sometimes by straightedge and compass but often by mechanical means. In this section, we will discuss some constructions by linkages.

A mechanical linkage is an assembly of straight rods with pins at the joints. In this section we will play with some of the simpler ones, with special attention to their usage in drawing lines and curves. The set of points traced out by the movement of a point is called a *locus of points*. The conventions we use are:

- ⊗ is used to denote a fixed point: a point that is nailed to the table, but with enough free play that any bars through the point can rotate freely around the fixed point.
- ○ is used to denote a pivot that is not fixed to the table and can move freely.
- • denotes a point whose movement is constrained in some way, usually by tracing a given path or curve.
- ⊙ denotes a point where a pencil is inserted.

To actually build these linkages, one can use Lego's or an Erector set. Alternatively, stiff cardboard, a hole punch, and prong fasteners make a less durable but more inexpensive equivalent, or one can use popsicle sticks or coffee stirrers and pins. It is also an interesting exercise to model such linkages using geometric software.

It is easy to draw a circle with a single link: just nail down one end of the rod and place a pencil at the other end.

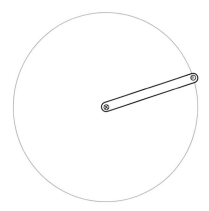

It is much harder to draw a straight line. James Watt [1736–1819] renewed interest in this problem when he designed a steam engine that required some sort of mechanism to move the pistons in a straight line within their cylinders. His solution was approximately straight, but close enough for his purpose of reducing the friction generated. Here is a schematic drawing of Watt's linkage. Note that $\overline{EF} = \overline{GH}$ and M is the midpoint of $\overline{FG}$, which traces out an approximately straight line:

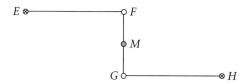

▷ **Activity 4.44.** Given lengths $\overline{AB}$ and $\overline{CD}$, using either straight rods and some sort of pins or geometric software, model Watt's linkage, building it so that $\overline{AB} = \overline{EF} = \overline{GH}$ and $\overline{CD} = \overline{FG}$. Describe the locus traced out by the midpoint M.

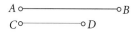

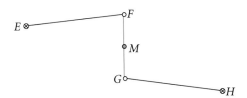

Numerous other mathematicians and engineers offered improved but still approximate solutions to the problem of straight line motion. In 1864 a French army engineer, Charles-Nicholas Peaucellier [1832–1913] produced a linkage that he called a *compas composé*, which produces a genuinely straight line. Now called *Peaucellier's linkage*, this is illustrated below. Note that there are two fixed points at points X and Y in the line drawing. The linkage must be assembled so that $\overline{XY} = \overline{YD}$, $\overline{XA} = \overline{XC}$, and $\overline{AB} = \overline{BC} = \overline{CD} = \overline{DA}$.

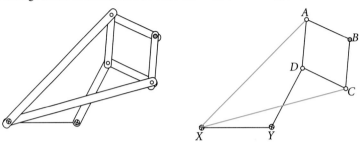

▷ **Activity 4.45.** Using either straight rods and some sort of pins or geometric software, build a model of the Peaucellier linkage and place a pen at point B. Describe the locus of points traced by the movement of B.

To understand the action of Peaucellier's linkage, consider the schematic drawing below. Drop perpendiculars from A to P on XB and from B to R on XY. Erect a perpendicular at D to XB, intersecting XY at Q.

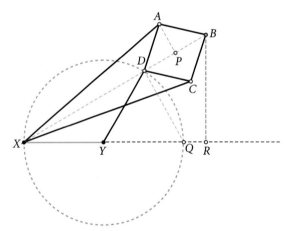

Recall that $\overline{XY} = \overline{YD}$, $\overline{XA} = \overline{XC}$, and $\overline{AB} = \overline{BC} = \overline{CD} = \overline{DA}$. By Theorem 3.2, since $\angle XDQ = 90°$, Q must lie on the circle with center Y and radius $\overline{XY}$. Therefore, the distance $\overline{XQ}$ is constant. Also note that $\triangle XDQ \sim \triangle XRB$ by AA similarity (Corollary 3.17), so $\frac{\overline{XD}}{\overline{XQ}} = \frac{\overline{XR}}{\overline{XB}}$, and thus $\overline{XD} \cdot \overline{XB} = \overline{XR} \cdot \overline{XQ}$. The Pythagorean Theorem for $\triangle XAP$ gives us $\overline{XA}^2 = \overline{XP}^2 + \overline{AP}^2$. Similarly, for $\triangle ABP$, we have $\overline{AB}^2 = \overline{AP}^2 + \overline{BP}^2$. Therefore,

$$\overline{XA}^2 - \overline{AB}^2 = \overline{XP}^2 - \overline{BP}^2 = (\overline{XP} - \overline{BP})(\overline{XP} + \overline{BP}) = \overline{XD} \cdot \overline{XB}.$$

Since the quantities $\overline{XA}$ and $\overline{AB}$ are fixed, it follows that the product $\overline{XD} \cdot \overline{XB}$ is constant. This is precisely the defining equation for inversion in a circle centered at X. Thus, D and B are inverses (as are

R and Q). Since D is constrained to move in a circle that passes through X, by Theorem 3.33 its inverse B must lie on the line that forms the inversion of that circle.

Peaucellier's discovery of this straight-line mechanism at first attracted little notice, until the independent rediscovery almost ten years later by a Russian mathematician, Lippman Lipkin. James Joseph Sylvester [1814–1897], a famous English mathematician, learned of this from Lipkin's advisor, Pafnuty Lvovich Chebyshev [1821–1894]. Sylvester gave an influential lecture on linkages in 1874 and told of the occasion when he showed a model of Peaucellier's linkage to William Thomson (later Lord Kelvin) [1824–1907] and said that Kelvin "nursed it as if it had been his own child, and when a motion was made to relieve him of it, replied 'No! I have not had nearly enough of it—it is the most beautiful thing I have ever seen in my life.'"[2] From this point on, many other mathematicians came up with variations on the Peaucellier linkage which produce the same straight-line action. In 1877, a London barrister and amateur mathematician, Alfred Bray Kempe [1849–1922], published a delightful small book, *How to Draw a Straight Line: a Collection of Linkages*, citing many of these. In it, he describes a variation of Peaucellier's linkage and states, "In this very compact form, the mechanism has been successfully applied to the machines used to ventilate [London's] Houses of Parliament. The smoothness of movement due to the absence of noise and friction is really remarkable. The machines were built on the basis of Peaucellier's mechanism by Mr. Prim, engineer of the Houses of Parliament, to whose courtesy I owe the chance of having seen them: I assure you they are worth a visit."

Peaucellier's linkage can also be used to answer the chicken-and-egg question of how does one draw a straight line without first making a straightedge, and how can you determine if a straightedge is straight without a straight line to compare it to? Since a circle can be draw with a piece of string or a stick, no matter how crooked, by fixing one end and attaching a pencil to the other end, we can use this linkage to create a straight line, which can then be used as the standard for a straightedge.

Now that we have a perfectly straight line, we investigate the *variable-based triangle linkage*: Build a linkage as pictured below, with $\overline{AB} = \overline{BC} = \overline{BD}$. The point C must stay on the straight line AC as shown.

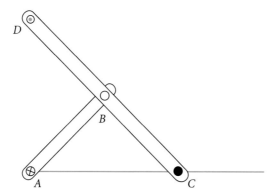

▷ **Activity 4.46.** Using either straight rods and some sort of pins or geometric software, build the mechanical linkage illustrated above and draw the line *AC*.

    a. Place a pencil at point *B*. As *C* moves back and forth along the line, describe the locus of point *B*. Explain why this particular shape results.

    b. Move the pencil to point *D*. Describe the locus traced by the movement of *D*.

---

[2]Sylvester, *Collected Works*, Vol. 3, Paper 2.

▶ **Exercise 4.44.** Prove your conjecture of Activity 4.46 about the locus traced by D.

▷ **Activity 4.47.** Describe how an ironing board folds up.

A *pantograph* is a drawing device used to duplicate and enlarge a drawing, in common use by draftsmen, sign makers, and craftsmen in the days before most illustrations were done on computers. It usually consists of 4 sticks of equal length, with a bolt at point A, a pointy thing at B used to trace over a given figure, a pencil-holding device at point C, and movable pins at points D, E, and F, put together so that $BDEF$ is a parallelogram and A, B, and C are collinear. Thus $\overline{AD} = \overline{BD} = \overline{EF}$ and $\overline{DE} = \overline{BF} = \overline{FC}$.

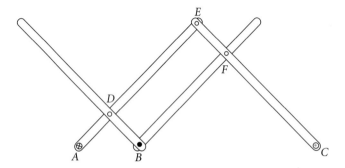

To begin analyzing the action of a pantograph, let us consider first the simple configuration below, where D is placed at the midpoint of $\overline{AE}$ and F at the midpoint of $\overline{EC}$:

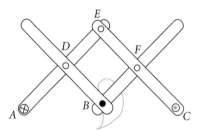

▶ **Exercise 4.45.** Prove that in the configuration above in which D and F are the midpoints of $\overline{AE}$ and $\overline{EC}$ then it follows that $\overline{AC} = 2\overline{AB}$.

▷ **Activity 4.48.** Build or borrow a pantograph. Draw a scribble and trace it with the tracer at point B. What does the pencil at point C draw?

▷ **Activity 4.49.** Using the pantograph in the configuration of Exercise 4.45, switch the positions of the pencil and tracer, so that the pencil is at point B and the tracer at point C. Trace your scribble. What does the pencil at B draw?

The pantograph should give an enlargement or reduction in scale of the figure traced. Trace a circle with your pantograph, moving in a clockwise direction. If the image drawn by the pantograph is also drawn in the clockwise direction, the scaling factor is positive. A negative scaling factor indicates a reflection combined with an enlargement or reduction. With a negative scaling factor, the pencil will move counterclockwise as the tracer moves clockwise.

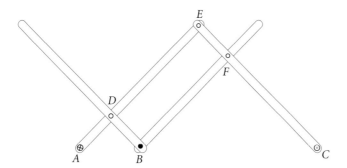

▷ **Activity 4.50.** Build a pantograph in the configuration shown above, so $\overline{DE}=3\overline{AD}$ and also $\overline{AD}=\overline{DB}=\overline{EF}$ and $\overline{DE}=\overline{BF}=\overline{CF}$. Draw a scribble. For each configuration below, trace your scribble and determine the scaling factor:

   a. Place the pencil at point C and the tracer at B.
   b. Place the pencil at point B and the tracer at C.
   c. Fix point B, place the tracer at A, and the pencil at C.
   d. Fix point B, place the tracer at C, and the pencil at A.

▷ **Activity 4.51.** Use geometric software to model a pantograph as above, so that $\overline{DE}=3\overline{AD}$, $\overline{AD}=\overline{DB}=\overline{EF}$, and $\overline{DE}=\overline{BF}=\overline{CF}$. Describe the locus formed by C as B traces a figure.

▶ **Exercise 4.46.** Below is another parallelogram linkage. Point A is fixed, point B is used to trace a figure, and there is a pencil at point C. Using similar triangles, show that this linkage also acts like the pantograph and find the scaling factor.

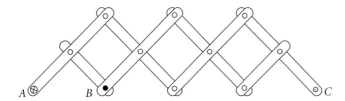

▶ **Exercise 4.47.** Figure out how to to move the fixed point, tracing point, and pencil around for the parallelogram linkage of Exercise 4.46 so that it acts as a pantograph with scaling factor:

   a. $+\frac{1}{4}$
   b. $+3$
   c. $+2$
   d. -2
   e. -3
   f. $\frac{4}{3}$
   g. $\frac{3}{4}$
   h. $\frac{2}{3}$
   i. $-\frac{1}{2}$

We already know that there is a linkage with which we can trisect any angle. We now have linkages capable of drawing circles and straight lines, constructing perpendiculars, and dividing a segment into any number of pieces. It is natural to ask precisely what else is constructible with such linkages.

▷ **Activity 4.52.** Invent a linkage that constructs the bisector of any acute angle.

The question of constructibility was answered (somewhat sketchily) by Kempe in 1876. He claimed that one can theoretically form a chain of linkages that can trace out any algebraic curve, provided that there are linkages able to do the following operations:

- draw a straight line,
- draw a circle,
- construct a perpendicular from a point to a given line,
- construct a line through a given point and parallel to a given line,
- add a constant to a arbitrary length,
- multiply an arbitrary length by a constant,
- add two arbitrary lengths,
- construct the reciprocal of a given length, and
- multiply two arbitrary lengths.

We already have many of these in hand: the Peaucellier linkage draws a straight line, circles are easy, and the variable-based triangular linkage allows us to construct perpendiculars. To construct a parallel line, which Kempe called a *translator*, build two quadrilaterals such that $\overline{AB} = \overline{CD} = \overline{EF}$, $\overline{AC} = \overline{BD}$, and $\overline{CE} = \overline{DF}$:

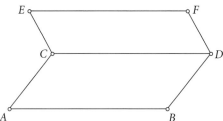

▷ **Activity 4.53.** Using geometric software or rods and pins, build a model of the translator as above, constructing it so that $\overline{EC} = \overline{FD}$, $\overline{AC} = \overline{BD}$, and $\overline{AB} = \overline{CD} = \overline{EF}$ but not explicitly requiring that the lines be parallel. Drag vertex *E* around. Does the mechanism ever fail to keep *EF* parallel to *AB*? Describe any such failures.

One flaw in Kempe's argument is that the translator can fail, as you should have found in Activity 4.53. Such failures can be prevented by adding additional bars $\overline{GH} = \overline{AC} = \overline{BD}$ and $\overline{HI} = \overline{CE} = \overline{DF}$ to *rigidify* the parallelograms:

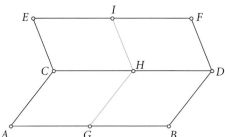

▷ **Activity 4.54.** Using geometric software, build a model of the rigidified translator as above. Drag vertex E around. Does this mechanism ever fail to keep EF parallel to AB?

▶ **Exercise 4.48.** Given a fixed length $a = \overline{AB}$ and a parallel line segment $x = \overline{OX}$, explain how to use the translator to construct points Y and Z on OX so that $\overline{OY} = x + a$ and $\overline{OZ} = x - a$.

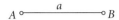

We have seen that the pantograph acts as a *scalar multiplicator* in some configurations. Of course, the pantographs must also be rigidified. Given a with $a > 1$, construct a pantograph so that $\overline{AE} = \overline{EC} = a$ and $\overline{AD} = \overline{BD} = \overline{EF} = 1$. Take the pantograph and set it so $\overline{AB} = x$. By the similarity of $\triangle AEC$ and $\triangle ADB$, $\frac{AB}{AD} = \frac{AC}{AE}$, so $\frac{x}{1} = \frac{AC}{a}$ and thus $\overline{AC} = ax$.

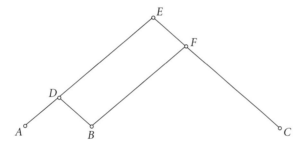

▶ **Exercise 4.49.** Explain how to modify the pantograph to construct ax where $0 < a < 1$.

▶ **Exercise 4.50.** Explain how to use linkages to construct a line segment representing $-x$ where x is a variable length represented by a line segment $\overline{AB}$.

We could also use the variable-based triangular linkage and the translator to construct similar triangles such as those that allowed the multiplication of constructible numbers in Chapter 4.1.

We can now use the pantograph as an *additor* for two variable lengths. We are given two lengths x and y and a line with marked points so $\overline{GH} = x$ and $\overline{GI} = y$. Configure the pantograph so that D is the midpoint of $\overline{AE}$. Set point A of the pantograph on H and point C on I. Then point B of the pantograph will lie on J where $\overline{HJ} = \frac{y-x}{2}$. Thus $\overline{GJ} = x + \frac{y-x}{2} = \frac{x+y}{2}$. Use the pantograph again to double this length to obtain a point K so that $\overline{GK} = x + y$.

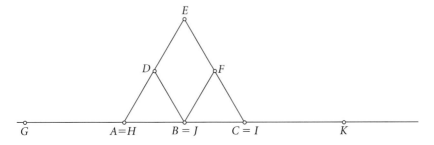

As we saw at the beginning of this section, Peaucellier's linkage has the property that $\overline{XD} \cdot \overline{XB} = \overline{XA}^2 - \overline{AB}^2$. If we construct such a linkage with rods of lengths so that $\overline{XA}^2 - \overline{AB}^2 = 1$, then by using D as the input for the linkage and setting $\overline{XD} = x$, then $\overline{XB} = \frac{1}{x}$. Thus Peaucellier's linkage can be used as an *inversor*.

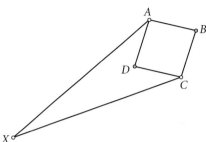

We can then use our previous linkage constructions to construct first the square of a variable length and then the product of two arbitrary lengths. Note that

$$\frac{1}{x - \frac{1}{2}} - \frac{1}{x + \frac{1}{2}} = \frac{1}{x^2 - \frac{1}{4}}.$$

Thus, we can use the additor and the inversor to construct x^2. Given two variable lengths x and y, we can use the identity

$$xy = \frac{(x+y)^2 - (x-y)^2}{4}$$

to construct the product xy. Thus we have constructed linkages capable of performing all of the operations we listed above.

Assuming the rigidification of all of our linkages, then if we are given any algebraic equation of the form

$$f(x, y) = \sum_{n,m} a_{n,m} x^n y^m = 0$$

we can, by adding and multiplying terms, construct a linkage that would plot the points on any section of this curve.

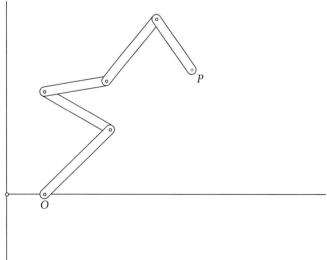

Thus we have a rather cursory outline of the proof of the following theorem. The full proof involves many extremely subtle issues that are well beyond the scope of this text. Kempe's proof works well most of the time, but it breaks down in certain degenerate configurations. These difficulties were resolved recently by Henry King and by Michael Kapovich and John Millson. This theorem has sometimes been restated as saying that it is always possible to design a linkage that will sign your name.

Theorem 4.17. [Kempe's Universality Theorem] If $f(x,y)$ is any algebraic curve, then there is a linkage that will trace out a section of this curve.

♦ 4.7   Projects

Project 4.1. Read the play *The Birds* by Aristophanes and comment on his satire on squaring the circle.

Project 4.2. Research the golden ratio and the golden rectangle.[3]

Project 4.3. Research some of the arithmetic properties of the Fibonacci numbers. Explain the connection between the Fibonacci numbers and the golden ratio.

Project 4.4. Investigate the connection between phyllotaxis and the Fibonacci numbers.[4]

Project 4.5. Report on the nonconstructability of the transcendental numbers e and π.

Project 4.6. Carry out the construction of a 17-gon using geometric software.[5]

Project 4.7. Report on a proof of Gauss's theorem.

Project 4.8. Investigate several methods for duplicating the cube.

Project 4.9. Read Underwood Dudley, *Mathematical Cranks* or *The Trisectors* and report.

Project 4.10. Use geometric software to figure out how to carry out the following constructions using compass alone: bisect an angle, construct the midpoint of a line segment, erect and drop a perpendicular, and copy an angle.

Project 4.11. Read and report on several of the construction tools described in Robert C. Yates, *Geometrical Tools* or George Martin, *Geometric Constructions*.

Project 4.12. Read Humiaki Huzita, "Drawing the regular heptagon and the regular nonagon by origami" and fold a heptagon.

Project 4.13. Read B. Carter Edwards and Jerry Shurman, "Folding quartic roots" and demonstrate how to fold $\sqrt[4]{2}$.

Project 4.14. Investigate Peaucellier's linkage with $XY \neq YD$.

Project 4.15. Investigate and build models of some of the other straight line linkages described in Kempe's *How to Draw a Straight Line*.

---

[3] One reference is George Markowsky, "Misconceptions about the Golden Ratio," *College Mathematics Journal* 23(1), 1992, pp. 2–19.
[4] Some references are Ian Stewart, "Daisy, Daisy, Give Me Your Answer, Do" or Roger V. Jean, *Phyllotaxis*.
[5] Some references are George Martin, *Geometric Constructions*, or Howard Eves, *Fundamentals of Modern Elementary Geometry*.

II. Noneuclidean geometries

5. Neutral geometry

♦ 5.1 Views on geometry

To the classical Greeks, mathematical objects were ideal, purely imaginary objects. Plato describes geometry as "knowledge of what eternally exists." Geometry described perfect objects, and real circles were poor approximations of the divine ideal circles that exist in Euclid's *The Elements*. This view of mathematics and its objects of study held for many centuries.

In the 17th and 18th centuries, astounding progress was made in the application of mathematics to natural phenomena, as witnessed by the successes of mathematicians and physicists such as Galileo Galilei [1564–1642] and Isaac Newton in tackling problems of gravitation, optics, electromagnetism, chemistry, and astronomy. Newton's laws of gravity and motion can be viewed as axioms for physics. Their use in predicting the future, such as the return of comets and eclipses, led to a search for perfect mathematical laws to describe all natural processes. This led to the view of mathematics as the language in which one describes reality. Mathematics gave control and understanding of nature. And Euclid's *The Elements* held a special position in this view of mathematics, as a perfect example of the power of logic and clarity.

René Descartes advocated applying the methods of geometry to all studies, stating in his *Discours de la Methode* of 1637,

> These long chains of perfectly simple and easy reasonings by means of which geometers are accustomed to carry out their most difficult demonstrations had led me to fancy that everything that can fall under human knowledge forms a similar sequence; and that so long as we avoid accepting as true what is not so, and always preserve the right order of deduction of one thing from another, there can be nothing too remote to be reached in the end, or to well hidden to be discovered.

The hopes for mathematics as a universal language did not stop with the sciences. Gottfried Wilhelm Liebniz (the coinventor of calculus) hoped to invent a calculus of thought. This led him to the development of predicate and symbolic logic, which he hoped would lead to a bloodless solution to all of mankind's problems. Philosophers and theologians discussed the nature of truth. One of the two main schools of thought of the time was naturalism, espoused by Thomas Hobbes [1588–1679] and John Locke [1632–1704]. They viewed mathematics as perfect knowledge. A common metaphor for the world was to see it as a machine in which physical forces were sensed, not known: One can only know ideas produced by the action of external sensory input. In opposition were George Berkeley [1685–1753] and David Hume [1711–1776], who claimed that only ideas exist and that the external world was only a delusion. Simultaneously, Jean-Jacques Rousseau [1712–1778] was calling on people to abandon reason and live with the senses only.

Later, philosophy was revolutionized by Immanuel Kant [1724–1804] and his theory of determinism. In his *Critique of Pure Reason*, he classified knowledge into types: empirical (accepted as true based on one's experience) or *a priori* (true without reference to experience), and analytic (true based on one's understanding) or synthetic (true but requiring some additional information from the outside world). Using these distinctions, he had four categories: analytic/empirical, synthetic/empirical, analytic/*a priori*, and synthetic/*a priori*.

An analytic statement is true by virtue of the terms used. Examples of analytic statements are "All bachelors are unmarried" or "All triangles have three sides." These statements are true because of the definitions of the terms used. In opposition to analytic statements are synthetic statements, such as "All bachelors are happy" or "All triangles have angle sum 180°." To determine whether these synthetic statements are true, we must gain further information about the nature of bachelors or triangles.

Empirical or *a posteriori* knowledge is subject to revision as one gains additional experience. For example, "Bachelors don't live as long as married men" is an empirical judgment, based on data collected by the census. *A priori* knowledge merely says what we, in a sense, know intuitively, such as "All bachelors die."

Prior to Kant, most philosophers thought that all synthetic knowledge was gained through experience and thus was empirical, but Kant claimed that some knowledge is both synthetic and *a priori*. Such synthetic *a priori* knowledge is ideal, since it says something nontrivial yet is based on innate human understanding rather than cultural conditioning. However, the only example Kant gives of synthetic *a priori* knowledge is mathematics. He claimed that geometry is clearly *a priori* since it remains true regardless of experience, and he thus classified Euclid's postulates as synthetic knowledge. He claims that "the concept of [euclidean] space is by no means of empirical origin, but is an inevitable necessity of thought." In other words, euclidean geometry is universally hard-wired in our brains. As Henri Poincaré says, "we must first of all ask ourselves, what is the nature of geometric axioms? Are they synthetic *á priori* intuitions, as Kant affirmed? They would then be imposed upon us with such a force that we could not conceive of the contrary proposition."[1] Thus euclidean geometry was the linchpin of Kant's theory about knowledge and truth. Albert Einstein describes his childhood acquaintance with euclidean geometry as:

> The objects with which geometry is concerned seemed to be of no different type from the objects of sensory perception, 'which can be seen and touched.' This primitive conception, which probably also lies at the bottom of the well-known Kantian inquiry concerning the possibility of 'synthetic judgments *a priori*,' rests obviously upon the fact that the relation of geometric concepts to objects of direct experience (rigid rod, finite interval, etc.) was unconsciously present. If thus it appeared that it was possible to achieve certain knowledge of the objects of experience by means of pure thinking, this 'wonder' rested upon an error.[2]

We will return to this concept and its consequences later.

The perfection of geometry and mathematics was marred, in the view of many mathematicians, by the Parallel Postulate 5. Euclid's Parallel Postulate 5 did not seem like a self-evident truth, unlike the other postulates. Its converse (Proposition I.17), contrapositive (Proposition I.29), and inverse (Proposition I.28) are all propositions, and the feeling was that the Parallel Postulate 5 should be provable. This view on the Parallel Postulate 5 was prevalent almost from the minute Euclid put his pen down. Poseidonios (100 B.C.) offers a proof, unfortunately flawed by a hidden assumption which turns out to be logically equivalent to Euclid's own Postulate 5. This assumption was that parallel lines are everywhere the same distance apart. Proclus (450 A.D.) points out this error and offered another proof of the Parallel Postulate 5, but this also contains a hidden assumption, that the distance between two parallel lines must be bounded. This statement also turns out to be logically equivalent to the Parallel Postulate 5.

There follows 2000 years of mathematicians, including almost every one of prominence, trying to come up with a proof of the Parallel Postulate 5, each new generation finding a fault in the previous

---

[1] H. Poincaré, *Science and Hypothesis*, p. 48.
[2] Albert Einstein, *Autobiographical Notes*, translated and edited by Paul Schilpp.

proof. Some attempts focused on revising the definition of line or of parallelism, but none were successful. Jean d'Alembert [1717–1783] wrote "La definition et les propriétés de la ligne droite, ainsi que des lignes parallèles sont l'écueil et pour ainsi dire le scandale des éléments de Géométrie." (The definition and properties of a straight line and also of parallel lines are the difficulty and, so to speak, the scandal of the elements of geometry.)

In the process, geometers extended Euclid's geometry, especially those parts that did not depend on the Parallel Postulate 5. Thus we have two tracks to follow: one extending the portion of Euclid's geometry that does not depend on the Parallel Postulate 5, and the other creating a list of alternative postulates, all logically equivalent to the Parallel Postulate 5.

♦ 5.2 Neutral geometry

Neutral (or absolute) geometry is what you get by assuming Euclid's Postulates 1–4 and our new Postulates 6–12: everything except the Parallel Postulate 5 and the Area Postulate 13. Thus, Euclid's Propositions I.1–28 and Proposition I.31 are true in neutral geometry, as well as Theorems 3.20, 3.21, 3.24, and Corollary 3.25. Note that the Area Postulate 13 also fails to hold, since we will see that the existence of a rectangle is logically equivalent to the Parallel Postulate 5. The purpose of this chapter is to see what remains without the Parallel Postulate 5. Neutral geometry was developed, in context of various faulty proofs of the Parallel Postulate 5, by, among others:

- Gerbert (later Pope Sylvester II) [ca. 950–1103],
- the poet and mathematician Omar Khayyam [1048–1131], who proved Theorem 5.4 below,
- John Wallis [1616–1703], the most prominent English mathematician before Newton,
- Gottfried Liebniz [1646–1716],
- Fr. Giovanni Girolamo Saccheri [1667–1733], a Jesuit priest,
- Johann Heinrich Lambert [1728–1777],
- Adrien Marie Legendre [1752–1833].

The most exhaustive of these attempts was that of Saccheri in his book *Euclides ab Omni Naevo Vindicatus* (Euclid Freed of all Blemish), published in 1733. The blemish referred to is Euclid's Parallel Postulate 5, and the entire work can be viewed as a lengthy proof by contradiction, intended to show that the Parallel Postulate 5 can be proven from the other postulates. He finally thinks he has a contradiction in two straight lines meeting perpendicularly at infinity, announces that this is "contrary to the nature of a straight line," and therefore states that the Parallel Postulate 5 is true, though the behavior of lines at infinity has never been postulated. He does not recognize that lines may behave entirely differently at infinity than they do elsewhere. In Part 2 of his book, he developed many theorems of noneuclidean geometry, claiming his intent was to show how ridiculous the negation of the Parallel Postulate 5 would be.

At first, working in neutral geometry feels like trying something with one hand tied behind your back. Be careful. It may help to revert to the two-column style of proof, since this will help you avoid Euclid's later propositions. One common mistake is to use Proposition I.29. You are allowed to use Propositions I.27 and I.28, which say that if any of the eight angle conditions for parallelism given in Lemma 2.26 are true, then the lines are parallel. But you cannot use Proposition I.29, which says if the lines are parallel, then all eight of the angle relationships are true.

▶ **Exercise 5.1.** Reexamine the proof of Proposition I.32 and explain where the Parallel Postulate 5 is used.

▶ **Exercise 5.2.** Reexamine your proof of Proposition I.34 (Exercise 2.61) and explain where the Parallel Postulate 5 is used.

▶ **Exercise 5.3.** Reexamine the proof of Proposition I.47 (the Pythagorean Theorem) and explain where the Parallel Postulate 5 is used.

Recall Euclid's Proposition I.17, which stated that in any triangle, the sum of any two of the angles must be less than 180°. Proposition I.32 extended this and said that the sum of the three angles of a triangle must equal 180°, but we cannot use that in neutral geometry. Instead, we can extend Proposition I.17 with the following theorem, first proved by Saccheri and later independently by Legendre, whose proof we give.

Theorem 5.1. [Saccheri-Legendre Theorem] The angles in any triangle sum to less than or equal to 180°.

In order to prove this, we need to recall the second part of Archimedes' Axiom:

Postulate 10. Archimedes' Axiom: If a and b are positive real numbers with $a < b$, then there exists $n \in \mathbb{N}$ so that $na > b$. In particular,

1. Given line segments $\overline{AB}$ and $\overline{CD}$, there exists $n \in \mathbb{N}$ and a point X on $\overrightarrow{CD}$ so that $\overline{CX} = n \cdot \overline{AB} > \overline{CD}$ and thus $C - D - X$.
2. Given angles $\angle ABC$ and $\angle DEF$, there exists $n \in \mathbb{N}$ and a ray $\overrightarrow{EX}$ so that $\angle XEF = n \cdot \angle ABC$ and $\angle XEF > \angle DEF$.

We will also prove a lemma, which will make the proof of the main theorem easier.

Lemma 5.2. Given $\triangle ABC$, there exists $\triangle A_1 B_1 C_1$ so that the sum of the angles in $\triangle A_1 B_1 C_1$ is equal to the sum of the angles in $\triangle ABC$ and $\angle A_1 \leq \frac{1}{2} \angle A$.

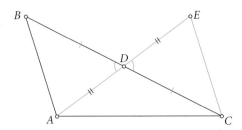

Proof: We are given $\triangle ABC$. Construct midpoint D of $\overline{BC}$ and draw AD. Cut $\overline{DE} = \overline{AD}$ and draw $\overline{CE}$. By Proposition I.15, $\angle BDA = \angle CDE$. Thus $\triangle BDA \cong \triangle CDE$ by SAS. By this congruence, we have $\angle BAD = \angle CED$ and $\angle ABC = \angle BCE$, and so

$$\angle ABC + \angle ACB + \angle BAC = \angle ABD + \angle BAD + \angle DAC + \angle ACD$$
$$= \angle DCE + \angle CED + \angle DAC + \angle ACD = \angle EAC + \angle ACE + \angle CEA$$

Thus, $\triangle ABC$ has the same angle sum as $\triangle ACE$. This new triangle is the one we want for the conclusion of the lemma, but we may have to relabel the vertices to ensure that $\angle A_1 \leq \frac{1}{2} \angle A$. Also, we have $\angle BAC = \angle BAE + \angle EAC$. Thus, one of $\angle BAE$ and $\angle EAC$ is less than or equal to $\frac{1}{2} \angle BAC$.

Case 1: $\angle EAC \leq \frac{1}{2}\angle BAC$

Let $A_1 = A$, $B_1 = E$, and $C_1 = C$, and then $\angle B_1 A_1 C_1 = \angle EAC \leq \frac{1}{2}\angle A$ and the sum of the angles in $\triangle A_1 B_1 C_1$ is the same as the sum of the angles in $\triangle ABC$.

Case 2: $\angle EAC > \frac{1}{2}\angle BAC$

Since $\angle BAC = \angle BAE + \angle EAC$, in this case, $\angle BAE \leq \frac{1}{2}\angle BAC$. Note the $\angle BAE = \angle AEC$. In this case, let $A_1 = E, B_1 = A$, and $C_1 = C$, so again $\angle B_1 A_1 C_1 = \angle AEC \leq \frac{1}{2}\angle A$, and the sum of the angles in $\triangle A_1 B_1 C_1$ is the same as the sum of the angles in $\triangle ABC$. □

Proof of Saccheri-Legendre Theorem 5.1: We assume the theorem is false, so that there exists a $\triangle ABC$ with $\angle ABC + \angle ACB + \angle BAC = 180° + \epsilon > 180°$ for some positive ϵ (which measures the amount by which the angle sum exceeds 180°). By Archimedes' Axiom Postulate 10, there is a number $n \in \mathbb{N}$ so that $\frac{\angle BAC}{2^n} < \epsilon$. Use Lemma 5.2 to construct a $\triangle A_1 B_1 C_1$ with $\angle B_1 A_1 C_1 \leq \frac{1}{2}\angle BAC$ and with angle sum equal to $180° + \epsilon$. Repeat this construction n times to get $\triangle A_n B_n C_n$ with $\angle B_n A_n C_n \leq \frac{1}{2^n}\angle BAC$ and with angle sum equal to $180° + \epsilon$. By the condition of Archimedes' Axiom, we know that $\angle B_n A_n C_n < \epsilon$. We have $\angle A_n B_n C_n + \angle A_n C_n B_n + \angle B_n A_n C_n = 180° + \epsilon$, and so $\angle A_n B_n C_n + \angle A_n C_n B_n > 180°$. This contradicts Proposition I.17, which states that the sum of any two angles in a triangle must be less than 180°. Therefore, we must have $\angle ABC + \angle ACB + \angle BAC \leq 180°$. □

Corollary 5.3. The sum of the angles in any convex quadrilateral is less than or equal to 360°.

▶ **Exercise 5.4.** Prove Corollary 5.3.

▶ **Exercise 5.5.** Prove that if all triangles have the same angle sum, then the sum of the angles in any triangle must be 180°.

In this section, we will prove some of Saccheri's more notable theorems. First, we prove a preliminary result:

Theorem 5.4. [Omar Khayyam's Theorem] If $ABCD$ is a quadrilateral with $\angle B = \angle C = 90°$, then

$\overline{AB} = \overline{DC} \iff \angle A = \angle D$,
$\overline{AB} > \overline{DC} \iff \angle A < \angle D$,
$\overline{AB} < \overline{DC} \iff \angle A > \angle D$.

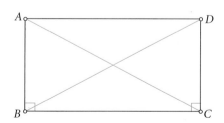

Proof: We are given a quadrilateral $ABCD$ with $\angle DCB = \angle ABC = 90°$.

Case 1: $\overline{AB} = \overline{DC}$.
Draw $\overline{AC}$ and $\overline{DB}$ and note that $\triangle ABC \cong \triangle DCB$ by SAS. Thus $\overline{AC} = \overline{BD}$ and so $\triangle ABD \cong \triangle DCA$ by SSS. Therefore, $\angle BAD = \angle CDA$.

Case 2: $\overline{AB} > \overline{DC}$.

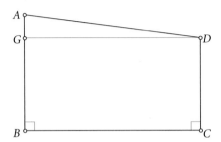

Cut $\overline{BG} = \overline{DC}$. Apply Case 1 above to quadrilateral $GBCD$ to find that $\angle BGD = \angle CDG$. Note that $\angle CDA > \angle CDG$ since G lies between A and B. By Proposition I.16, $\angle BGD > \angle BAD$. Therefore, $\angle CDA > \angle BAD$.

Case 3: $\overline{AB} < \overline{DC}$: The proof is similar to that of Case 2. □

▶ **Exercise 5.6.** Prove the other three cases of Theorem 5.4: $\angle A = \angle D \Longrightarrow \overline{AB} = \overline{DC}$, $\angle A > \angle D \Longrightarrow \overline{AB} < \overline{DC}$, and $\angle A < \angle D \Longrightarrow \overline{AB} > \overline{DC}$.

Saccheri's main weapon in his study of parallel lines was what are now called *saccheri quadrilaterals*. These are potential rectangles.

Definition 5.5. A *saccheri quadrilateral* is a quadrilateral $ABCD$ with base $\overline{BC}$, $\angle ABC = \angle DCB = 90°$, and equal legs $\overline{AB} = \overline{DC}$. The side $\overline{AD}$ is called the *summit* of the quadrilateral and $\angle BAD$ and $\angle CDA$ are the *summit angles*.

Saccheri then theorized that there were three possibilities:

- *The Hypothesis of the Obtuse Angle:* The summit angles of a saccheri quadrilateral are obtuse. This case is eliminated by the Corollary 5.3 to the Saccheri-Legendre Theorem 5.1.
- *The Hypothesis of the Right Angle:* The summit angles are right angles, so that the saccheri quadrilateral is a rectangle. This case gives us classical euclidean geometry.
- *The Hypothesis of the Acute Angle:* The summit angles of a saccheri quadrilateral are acute.

Most of his book is devoted to studying the acute case and trying to find a contradiction that eliminates this possibility.

In euclidean geometry, Theorem 5.4 would imply that the summit angles of a saccheri quadrilateral were right angles, since the sum of the angles must equal 360° by Exercise 2.56. However, in throwing out the Parallel Postulate 5, we have also thrown out this proposition. Therefore, we no longer know what the summit angles might be, only that they are equal and either right or acute. Three corollaries to Theorem 5.4 follow, giving further properties of saccheri quadrilaterals.

Corollary 5.6. The summit angles of a saccheri quadrilateral are equal and the *midline* connecting the midpoints of base and summit is perpendicular to both the base and the summit.

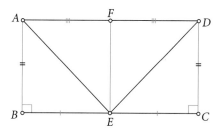

Proof: We are given a saccheri quadrilateral $ABCD$, and thus know that $\angle ABC = \angle DCB = 90°$ and $\overline{AB} = \overline{DC}$. By Theorem 5.4, $\angle BAD = \angle CDA$. Construct the midpoints E and F of $\overline{BC}$ and $\overline{AD}$, and draw $\overline{EF}, \overline{AE}$, and $\overline{DE}$. By the definition of a midpoint, we have $\overline{BE} = \overline{CE}$ and $\overline{AF} = \overline{DF}$. Then $\triangle ABE \cong \triangle DCE$ by SAS. Thus $\overline{AE} = \overline{DE}$ and $\angle AEB = \angle DEC$. Therefore, $\triangle AFE \cong \triangle DFE$ by SSS. By this congruence, $\angle AFE = \angle DFE$ and $\angle AEF = \angle DEF$. We have equal adjacent supplementary angles $\angle AFE = \angle DFE$, so $EF \perp AD$. Also $\angle BEF = \angle BEA + \angle AEF = \angle CED + \angle DEF = \angle CEF$, so we also have $FE \perp BC$. □

Corollary 5.7. The summit and base of a saccheri quadrilateral are parallel.

▶ **Exercise 5.7.** Prove Corollary 5.7.

Corollary 5.8. If $ABCD$ is a quadrilateral with $\angle ABC = \angle DCB = 90°$ and $\angle BAD = \angle CDA$, then $ABCD$ is a saccheri quadrilateral.

▶ **Exercise 5.8.** Prove Corollary 5.8.

Thus, a saccheri quadrilateral is a parallelogram with right angles at the base. The next construction provides a way of associating a saccheri quadrilateral with any triangle, and is used to translate questions about the sum of the angles of a triangle to the corresponding questions about the summit angles of the saccheri quadrilateral.

Theorem 5.9. Given $\triangle ABC$, let D and E be the midpoints of $\overline{AB}$ and $\overline{AC}$. Drop perpendiculars $BF \perp DE$ and $CG \perp DE$. Then $BFGC$ is the *saccheri quadrilateral associated with* $\triangle ABC$ with base $\overline{FG}$. Furthermore, $\overline{FG} = 2\overline{DE}$, and $\angle ABC + \angle ACB + \angle BAC = \angle FBC + \angle GCB$.

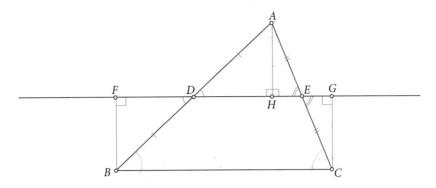

Proof: We are given $\triangle ABC$. Construct midpoints D and E of $\overline{AB}$ and $\overline{AC}$ and draw DE. Drop $BF \perp DE$, $CG \perp DE$, and $AH \perp DE$. By Proposition I.15, $\angle BDF = \angle ADH$ and $\angle AEH = \angle CEG$. Thus by AAS, $\triangle BDF \cong \triangle ADH$ and $\triangle AEH \cong \triangle CEG$. Therefore, $\overline{BF} = \overline{AH}$, $\overline{FD} = \overline{DH}$, $\overline{AH} = \overline{CG}$, $\overline{HE} = \overline{EG}$, $\angle FBD = \angle HAD$ and $\angle HAE = \angle GCE$. Thus $\overline{BF} = \overline{CG}$ and $BFGC$ is a saccheri quadrilateral with base $\overline{FG}$.

Also note if the points F, H, and G lie as shown above, $\overline{FG} = \overline{FD} + \overline{DH} + \overline{HE} + \overline{EG} = 2\overline{DH} + 2\overline{HE} = 2\overline{DE}$. Considering the angle sum, we have

$$\angle ABC + \angle BAC + \angle ACB = \angle ABC + \angle BAH + \angle HAC + \angle ACB$$
$$= \angle ABC + \angle FBD + \angle GCE + \angle ACB = \angle FBC + \angle GCB.$$

$\square$

▶ **Exercise 5.9.** The proof above of Theorem 5.9 assumes that H falls between D and E. Prove the other two cases shown below: H falls outside $\overline{DE}$ (assume that H falls on the far side of D from E) and H falls on D or E (assume H falls on D). In the latter case, you will need to show that F also falls on D.

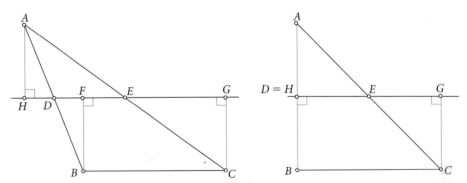

Corollary 5.10. If the angles in a triangle sum to $180°$, then the associated saccheri quadrilateral is a rectangle.

▶ **Exercise 5.10.** Prove Corollary 5.10.

The corollary above indicates the intimate relation between the existence of rectangles and the sum of the angles in a triangle.

Corollary 5.11. The summit of a saccheri quadrilateral is greater than or equal to the base, and the midline is less than or equal to the legs.

▶ **Exercise 5.11.** Prove Corollary 5.11.

▶ **Exercise 5.12.** Theorem 5.9 gave a way of associating a saccheri quadrilateral with any triangle. Reverse the process: Given a saccheri quadrilateral, explain how to construct an associated isosceles triangle with the properties of Theorem 5.9.

Next follows a series of congruence theorems for saccheri quadrilaterals:

Theorem 5.12. [Base-Leg Congruence] If the base of one saccheri quadrilateral is equal to the base of a second saccheri quadrilateral, and if the legs of the first are equal to the legs of the second, then the saccheri quadrilaterals are congruent (i.e., their summits and summit angles are equal).

Theorem 5.13. [Base-Midline Congruence] If the base of one saccheri quadrilateral is equal to the base of a second saccheri quadrilateral, and if the midline of the first is equal to the midline of the second, then the saccheri quadrilaterals are congruent (i.e., their legs, summits, and summit angles are equal).

▶ **Exercise 5.13.** Prove Theorem 5.12.
▶ **Exercise 5.14.** Prove Theorem 5.13.

The next theorem, also proved by Saccheri, is astounding: it says that if one saccheri quadrilateral has acute angles, then all other possible saccheri quadrilaterals must also have acute summit angles. The proof is also rather complex and requires two lemmata. After proving this fundamental theorem, we get a number of easy corollaries, so the time spent understanding the proof is well spent.

Lemma 5.14. Let $ABCD$ be a saccheri quadrilateral and let X be a point on the summit $\overline{AD}$ so that $A - X - D$. Drop $XY \perp BC$. Then

1. $\overline{XY} < \overline{AB}$ if and only if $\angle DAB = \angle ADC < 90°$,
2. $\overline{XY} = \overline{AB}$ if and only if $\angle DAB = \angle ADC = 90°$,
3. $\overline{XY} > \overline{AB}$ if and only if $\angle DAB = \angle ADC > 90°$.

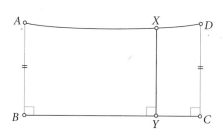

Proof: If $\overline{XY} < \overline{AB} = \overline{DC}$, then applying Theorem 5.4 to the quadrilateral $ABYX$, we see that $\angle DAB < \angle AXY$. Similarly from quadrilateral $XYCD$ we have $\angle ADC < \angle DXY$. Thus,

$$2\angle DAB = \angle DAB + \angle ADC < \angle AXY + \angle DXY = 180°,$$

so $\angle DAB < 90°$. Similarly, if $\overline{XY} = \overline{AB}$ then $\angle DAB = 90°$ and if $\overline{XY} > \overline{AB}$, then $\angle DAB > 90°$.

▶ **Exercise 5.15.** Finish the proof of Lemma 5.14 by showing that

    a. if $\angle DAB = \angle ADC < 90°$, then $\overline{XY} < \overline{AB}$,
    b. if $\angle DAB = \angle ADC = 90°$, then $\overline{XY} = \overline{AB}$,
    c. if $\angle DAB = \angle ADC > 90°$, then $\overline{XY} > \overline{AB}$.

Lemma 5.15. Let $ABCD$ be a saccheri quadrilateral and let X be a point on the extended summit $\overline{AD}$ so that $A - D - X$. Drop $XY \perp BC$. Then

1. $\overline{XY} > \overline{AB}$ if and only if $\angle DAB = \angle ADC < 90°$,
2. $\overline{XY} = \overline{AB}$ if and only if $\angle DAB = \angle ADC = 90°$,
3. $\overline{XY} < \overline{AB}$ if and only if $\angle DAB = \angle ADC > 90°$.

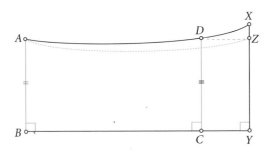

Proof: If $\overline{XY} > \overline{AB} = \overline{DC}$, then cut $\overline{YZ} = \overline{BA} = \overline{CD}$ and draw $\overline{DZ}$ and $\overline{AZ}$. Then $ABYZ$ and $DCYZ$ are both saccheri quadrilaterals, so by Corollary 5.6, $\angle ZAB = \angle AZY$ and $\angle ZDC = \angle DZY$. By Proposition I.16, $\angle XDZ > \angle DAZ = \angle DAB - \angle ZAB$. Since $\overrightarrow{ZA}$ lies between $\overrightarrow{ZD}$ and $\overrightarrow{ZY}$, $\angle AZY < \angle DZY = \angle ZDC$. Thus

$$180° = \angle ADC + \angle ZDC + \angle XDZ > \angle DAB + \angle AZY + \angle DAB - \angle ZAB = 2\angle DAB.$$

Therefore, $\angle DAB < 90°$.

If $\overline{XY} < \overline{AB}$, a similar argument shows that $\angle DAB > 90°$. Finally, if $\overline{XY} = \overline{AB} = \overline{DC}$, then $ABYX$ is a saccheri quadrilateral. Applying Lemma 5.14, it must be true that $\angle DAB = 90°$.

▶ **Exercise 5.16.** Finish the proof of Lemma 5.15 by showing that

    a. if $\angle DAB = \angle ADC < 90°$, then $\overline{XY} > \overline{AB}$,
    b. if $\angle DAB = \angle ADC = 90°$, then $\overline{XY} = \overline{AB}$,
    c. if $\angle DAB = \angle ADC > 90°$, then $\overline{XY} < \overline{AB}$.

Theorem 5.16. If there exists one saccheri quadrilateral with acute summit angles, then all saccheri quadrilaterals have acute summit angles.

Proof: We are given a saccheri quadrilateral *ABCD* with acute summit angles $\angle BAD = \angle CDA$. Let *GHIJ* be another arbitrary saccheri quadrilateral. Construct $\overline{EF}$ and $\overline{KL}$, the midlines of *ABCD* and *GHIJ*.

Case 1: $\overline{KL} = \overline{EF}$: We illustrate the case in which $\overline{BC} < \overline{HI}$. The other cases are similar.

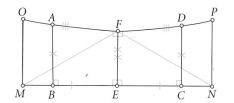 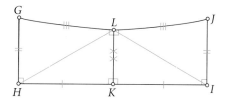

Extend *BC* and *AD* and cut $\overline{EM} = \overline{EN} = \overline{KH} = \overline{KI}$ and $\overline{FO} = \overline{FP} = \overline{LG} = \overline{LJ}$. Draw $\overline{OM}$, $\overline{PN}$, $\overline{FM}$, $\overline{FN}$, $\overline{LH}$, and $\overline{LI}$. Then $\triangle MFE \cong \triangle NFE \cong \triangle HLK \cong \triangle ILK$ by SAS so $\overline{MF} = \overline{NF} = \overline{HL} = \overline{IL}$, $\angle MFE = \angle NFE = \angle HLK = \angle ILK$, and $\angle FME = \angle FNE = \angle LHK = \angle LIK$. Thus, $\angle MFO = \angle NFP = \angle HLG = \angle ILJ$ so $\triangle MFO \cong \triangle NFP \cong \triangle HLG \cong \triangle ILJ$. Therefore, $\overline{OM} = \overline{PN} = \overline{GH} = \overline{JI}$ and $\angle FMO = \angle FNP = \angle LHG = \angle LIJ$. Thus, $\angle OME = \angle PNE = \angle GHK = \angle JIK = 90°$. We thus see that *OMNP* is a saccheri quadrilateral and is congruent to *GHIJ*. We are given that $\angle DAB < 90°$. Applying Lemma 5.15 to *ABCD*, it follows that $\overline{OM} > \overline{AB}$. We can then apply Lemma 5.14 to the saccheri quadrilateral *OMNP* to conclude that $\angle POM = \angle JGH < 90°$. Thus *GHIJ* also has acute summit angles.

Case 2: $\overline{KL} \neq \overline{EF}$: We illustrate the case in which $\overline{KL} < \overline{EC}$. The other case is similar.

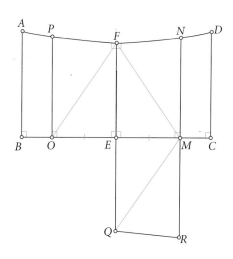 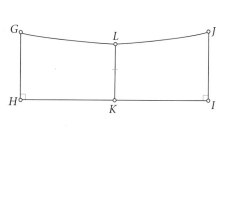

Cut $\overline{EM} = \overline{EO} = \overline{KL}$. Erect $MN \perp BC$ with *N* on $\overline{AD}$ and cut $\overline{FP} = \overline{FN}$. Draw $\overline{OP}$, $\overline{FO}$, and $\overline{FM}$. Then $\triangle OFE \cong \triangle MFE$ so $\overline{OF} = \overline{MF}$, $\angle FOE = \angle FME$, and $\angle OFE = \angle MFE$. Thus, $\angle OFP = \angle MFN$ so $\triangle OFP \cong \triangle MFN$. Then $\overline{OP} = \overline{MN}$ and $\angle FOP = \angle FMN$. Therefore, $\angle POE = \angle NME = 90°$ so *POMN* is a saccheri quadrilateral with midline $\overline{EF}$. By Case 1, *POMN* must have acute summit angles so $\angle FNM < 90°$.

Extend *FE* and *NM* and cut $\overline{EQ} = \overline{EF}$ and $\overline{MR} = \overline{MN}$. Draw $\overline{QR}$ and $\overline{MQ}$. Since $\triangle MEF \cong \triangle MEQ$, $\overline{MF} = \overline{MQ}$, $\angle EFM = \angle EQM$, and $\angle EMF = \angle EMQ$. Thus $\angle FMN = \angle QMR$ so $\triangle FMN \cong \triangle QMR$. Therefore, $\overline{FN} = \overline{QR}$ and $\angle MFN = \angle MQR$ and so $\angle EQR = \angle EFN = 90°$. Thus we see that *NFQR* is another saccheri quadrilateral with midline $\overline{EM}$ and acute summit angle $\angle FNM$. Since $\overline{EM} = \overline{KL}$, we can apply Case 1 to see that $GHIJ$ also has acute summit angles. □

▶ **Exercise 5.17.** Prove that if one saccheri quadrilateral has right summit angles, then every saccheri quadrilateral has right summit angles.

Corollary 5.10 stated that if one triangle exists with angle sum 180°, then a rectangle exists. We next state an extension of the converse of this:

Theorem 5.17. If one rectangle exists, then the sum of the angles in any triangle is 180°.

Corollary 5.18. If there exists a triangle whose angles sum to 180°, then all triangles have angle sum 180°.

Corollary 5.19. If there exists a triangle whose angles sum to less than 180°, then all triangles have angle sum less than 180°.

▶ **Exercise 5.18.** Prove Theorem 5.17.

▶ **Exercise 5.19.** Prove Corollary 5.18.

▶ **Exercise 5.20.** Prove Corollary 5.19.

As we noted previously, Saccheri ends Part 1 of his text by claiming that his Hypothesis of the Acute Angle "is absolutely false, because it is repugnant to the nature of a straight line." He seems to have recognized the unsatisfactory nature of his contradiction, since Part 2 of the book goes on to show other theorems that would be true if the Hypothesis of the Acute Angle were true, perhaps in search of a more substantive contradiction.

Johann Heinrich Lambert undertook similar studies to those of Saccheri without knowing of his predecessor's work. He based his study on what are called *Lambert quadrilaterals*: quadrilaterals $ABCD$ in which $\angle ABC = \angle BCD = \angle CDA = 90°$.

▶ **Exercise 5.21.** Prove that if $ABCD$ is a Lambert quadrilateral, then $\overline{AB} \geq \overline{DC}$ and $\overline{AD} \geq \overline{BC}$.

▶ **Exercise 5.22.** Prove that if one Lambert quadrilateral is a rectangle, then the angle sum of any triangle is 180°.

♦ 5.3 Alternate parallel postulates

In this section we consider some of the more notable statements that turn out to be logically equivalent to Euclid's Parallel Postulate 5. First, recall what Euclid claimed:

Euclid's Postulate 5. If a straight line falling on two straight lines make the interior angles on the same side less than two right angles, the two straight lines, if produced indefinitely, meet on that side on which are the angles less than two right angles.

and our restatement of this:

Postulate 5. [Parallel Postulate]. If two lines are cut by a transversal so that the sum of the interior angles on one side of the transversal is less than 180°, then the two lines meet on that side.

In his textbook of 1813, the Scottish mathematician John Playfair explains,

> The subject of parallel lines is one of the most difficult in the Elements of Geometry. . . . The methods by which Geometers have attempted to remove this blemish from the elements are of three kinds. 1. By a new definition of parallel lines. 2. By introducing a new Axiom concerning parallel lines, more obvious than Euclid's. 3. By reasoning merely from the definition of parallels, and the properties of lines already demonstrated, without the assumption of any new Axiom.[3]

He then gives several examples of each type, explaining where the arguments fail or where the proposed new definitions contain unstated assumptions that are in fact logically equivalent to the Parallel Postulate 5.

The most common form of the Parallel Postulate 5 in modern textbooks is the form called Playfair's Postulate, for the statement he proposed in place of Euclid's Parallel Postulate 5. Proclus, one of the early commentators on Euclid, mentioned this equivalent statement as early as 450 AD. In his 1813 geometry textbook, Playfair assumes as an axiom the following statement: "*Two straight lines which intersect one another, cannot be both parallel to the same straight line.*"[4] Note that Euclid's Proposition I.31 (which does not require the use of the Parallel Postulate 5) shows that there is at least one line that can be constructed through the given point and parallel to the given line. Thus, we know that we can construct one parallel line, and Playfair's Postulate claims that there is no other. We give the equivalent statement most commonly used in modern textbooks.

Alternate Postulate 5.1. [Playfair, 1813] Given a line and a point not on the line, exactly one line can be drawn through the given point and parallel to the given line.

Theorem 5.20. Playfair's Postulate is logically equivalent to the Parallel Postulate 5.

Proof: See Exercises 2.53 and 2.54. □

---

[3] Playfair, *Elements of Geometry*, p. 300.
[4] Playfair, *Elements of Geometry*, p. 21.

Any statement is logically equivalent to its contrapositive, so the next statement can also be considered as an alternate postulate:

Alternate Postulate 5.2. [Euclid's Proposition I.29] If two parallel lines are cut by a transversal, then the alternate interior angles are equal, each exterior angle is equal to the opposite interior angle, and sum of the interior angles on the same side of the transversal is 180°.

In order to show that two statements are logically equivalent, we must prove two things: that the first statement implies the second, and then that the second statement implies the first. When we assume that the Parallel Postulate 5 is true, then we can use any of Euclid's propositions, as well as any of the neutral theorems. However, when we are trying to prove the Parallel Postulate 5, we can only use Euclid's Propositions I.1–28 and 31 and the neutral Theorems 5.1–5.19. Often we show that a given statement is logically equivalent to Euclid's Proposition I.29 or Playfair's Postulate, rather than the Parallel Postulate 5 directly.

Proving that a statement is logically equivalent to another is more sophisticated than most of the theorems we have done thus far. Richard Trudeau in *The Non-Euclidean Revolution* calls these *metatheorems* and our development of this section owes much to that text. We are proving facts about the relationship between two statements, rather than proving facts about triangles or parallel lines.

One of the first attempts to prove the Parallel Postulate 5 was a proof by Poseidonios, ca. 100 B.C. The error in his proof is now called Poseidonios' Postulate:

Alternate Postulate 5.3. [Poseidonios] Parallel lines are equidistant.

Theorem 5.21. Poseidonios's Postulate is logically equivalent to the Parallel Postulate 5.

Proof: Part 1: Poseidonios's Postulate implies the Parallel Postulate 5.

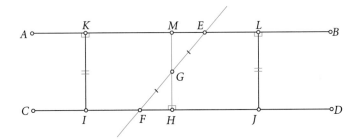

We assume that Poseidonios's Postulate is true and that we are given arbitrary lines $AB \parallel CD$ with transversal EF. Construct the midpoint G of $\overline{EF}$ and drop $GH \perp CD$. Choose an arbitrary point I on CD between C and H and cut $\overline{HJ} = \overline{HI}$. Then drop $IK \perp AB$ and $JL \perp AB$. Since we are assuming that parallel lines are equidistant, $\overline{IK} = \overline{JL}$. Thus, $IKLJ$ is a saccheri quadrilateral. Let M be the midpoint of $\overline{KL}$ and draw $\overline{HM}$. By Corollary 5.6, $\overline{HM}$ is the midline of $IKLJ$ and is perpendicular to both AB and CD. Thus, HM coincides with HG. By the Vertical Angle Theorem, $\angle EGM = \angle FGH$. Thus $\triangle EGM \cong \triangle FGH$ by AAS, and so $\angle GEM = \angle GFH$. Since the alternate interior angle property is true for this pair of arbitrary parallel lines, we now know that Proposition I.29 must be true. Therefore, its contrapositive, the Parallel Postulate 5, must also be true.

Part 2: Parallel Postulate 5 implies Poseidonios's Postulate.

We now assume that the Parallel Postulate 5 (and all theorems proven with it) is true. Consider an arbitrary pair of parallel lines $AB \parallel CD$ and choose any two points E and F on AB. Drop $EG \perp CD$ and $FH \perp CD$. By Proposition I.27, $EG \parallel FH$ and so $EGHF$ is a parallelogram. By Proposition I.34, $\overline{EG} = \overline{FH}$. This argument for any two points on AB shows that the lines must be equidistant, and Poseidonios' Postulate follows. □

Proclus pointed out Poseidonios' error in assuming that parallel lines are equidistant and went on to give a different 'proof' of the Parallel Postulate 5. However, he assumed the statement below in his proof.

Alternate Postulate 5.4. [Proclus] If a straight line intersects one of two parallel lines, it will also intersect the other.

Consider a situation in which Playfair's Postulate is false. We would then have the following situation: a point C not lying on line AB, and at least two lines DE and FG both passing through C and both parallel to AB.

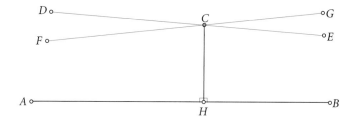

Theorem 5.22. Proclus' Postulate is logically equivalent to the Parallel Postulate 5.

▶ **Exercise 5.23.** Prove Theorem 5.22. First prove that if Playfair's Postulate is true, then Proclus' Postulate must also be true. Instead of proving Proclus's Postulate implies Playfair's Postulate, prove the contrapositive: show that if Playfair's Postulate is false, then Proclus' Postulate must be false.

If Playfair's Postulate is false it is easy to see that Euclid's Proposition I.30 is also false, since in this case $DE \parallel AB$ and $FG \parallel AB$ but DE intersects FG at C. Thus it is easy to prove that Euclid's Proposition I.30

is logically equivalent to Playfair's Postulate, which is logically equivalent to the Parallel Postulate 5 by Theorem 5.20.

Alternate Postulate 5.5. [Euclid's Theorem I.30] Lines that are parallel to the same line are parallel to each other.

In the picture illustrating the negation of Playfair's Postulate, note that this configuration of lines could be interpreted as an angle $\angle FCE$ and a line AB through a point H lying inside the angle.

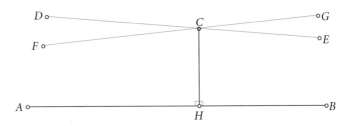

Alternate Postulate 5.6. [Lorenz, 1791] Every straight line through a point within an angle will meet at least one side of the angle.

▶ **Exercise 5.24.** Prove that Lorenz's Postulate is logically equivalent to Playfair's Postulate, and so to the Parallel Postulate 5.

In his book of 1733, Fr. Saccheri has a number of statements which he shows are equivalent to the Parallel Postulate 5. Among these, we have chosen four representatives.

Alternate Postulate 5.7. [Saccheri 1] In any saccheri quadrilateral the summit angles are right, and thus all saccheri quadrilaterals are rectangles.

Alternate Postulate 5.8. [Saccheri 2] At least one rectangle exists.

Alternate Postulate 5.9. [Saccheri 3 (Euclid's Proposition I.32)] The angle sum of any triangle is 180°.

Alternate Postulate 5.10. [Saccheri 4] At least one triangle with angle sum 180° exists.

Theorem 5.23. Saccheri's Postulates are logically equivalent to one another.

Proof:

S1 $\implies$ S2: It is clear that if all saccheri quadrilaterals have right summit angles, then they are rectangles, so Saccheri 1 implies Saccheri 2.

S2 $\implies$ S1: This is true by Exercise 5.17.

S2 $\Longrightarrow$ S3: This is true by Theorem 5.17.

S3 $\Longrightarrow$ S1 and S2: If Saccheri 3 is true, then any quadrilateral must have angle sum 360° and so any saccheri quadrilateral must be a rectangle.

S3 $\Longrightarrow$ S4: The proof is obvious.

S4 $\Longrightarrow$ S3: This is true by Corollary 5.18.

Thus, all four of the Saccheri Postulates are logically equivalent. □

Theorem 5.24. Saccheri's Postulates are logically equivalent to the Parallel Postulate 5.

Proof: Part 1: the Parallel Postulate 5 implies Saccheri's Postulates. Note that the Parallel Postulate 5 implies Euclid's Proposition I.32, which implies all of Saccheri's other postulates.

Part 2: Saccheri's Postulate 3 implies the Parallel Postulate 5.

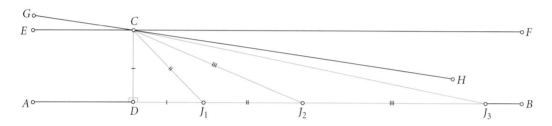

We assume that Saccheri's Postulate 3 is true and will try to deduce Playfair's Postulate. We are given a line AB and a point C not on AB. Drop $CD \perp AB$ and construct $CE \perp CD$. Extend $\overline{EC}$ to F and note that $EF \parallel AB$ by Proposition I.27. Let GH be another line through C, not the same as EF, and assume, without loss of generality, that H falls between $\overrightarrow{CF}$ and $\overrightarrow{CD}$. We wish to show that CH must intersect AB and thus cannot be parallel to AB. Cut $\overline{DJ_1} = \overline{DC}$ and draw $\overline{CJ_1}$. Then $\triangle CDJ_1$ is isosceles, so $\angle DCJ_1 = \angle DJ_1C$. By Saccheri's Postulate 3, $\angle DCJ_1 + \angle DJ_1C + 90° = 180°$, and so $\angle DJ_1C = 45°$. As supplementary angles, $\angle DJ_1C + \angle CJ_1B = 180°$, so $\angle CJ_1B = 135°$. Now cut $\overline{J_1J_2} = \overline{CJ_1}$ and draw $\overline{CJ_2}$. We again have an isosceles triangle so $\angle J_1CJ_2 = \angle J_1J_2C$. Applying Saccheri's Postulate 3 to $\triangle J_1CJ_2$, we have $\angle J_1CJ_2 + \angle J_1J_2C + \angle CJ_1J_2 = 180°$ and so $\angle J_1J_2C = 22.5°$ and thus $\angle CJ_2B = 157.5°$. Next, cut $\overline{J_2J_3} = \overline{CJ_2}$ and draw $\overline{CJ_3}$. As above, we compute $\angle J_2J_3C = 11.25°$ and $\angle CJ_3B = 168.75°$. Repeat this process, taking advantage of Archimedes' Axiom Postulate 10, to find J_n on AB so that $\angle DJ_nC < \angle FCH$.

Consider the case in which CH falls inside $\angle J_nCF$:

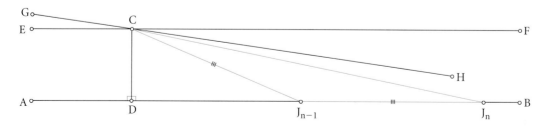

If so, we have $\angle J_nCD < \angle HCD$. Since $\angle DJ_nC < \angle FCH$, we know that $\angle J_nCD + \angle DJ_nC < \angle HCD + \angle HCF = 90°$. Therefore, $\angle J_nCD + \angle DJ_nC + \angle J_nDC < 180°$ and so the sum of the angles in $\triangle CDJ_n$ is less than 180°, which contradicts our assumption that Saccheri 3 is true. Thus CH cannot fall inside ∠J_nCF.

Thus, $\overrightarrow{CH}$ falls inside ∠J_nCD, and so CH must intersect AB by Pasch's Theorem 2.11.

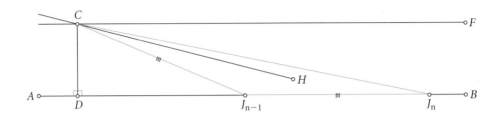

Therefore, any line through C other than EF must intersect AB and so there is only one line through C parallel to AB. Thus, Playfair's Postulate is true. By Theorem 5.20, this implies that the Parallel Postulate 5 must also be true. □

▶ **Exercise 5.25.** Consider the statement: *There is a saccheri quadrilateral whose diagonals bisect one another.* Prove that this statement is logically equivalent to the Parallel Postulate 5.

Alternate Postulate 5.11. [Pythagorean Theorem] In right-angled triangles the square on the side subtending the right angle is equal to the squares on the sides containing the right angle.

Theorem 5.25. The Pythagorean Theorem is logically equivalent to the Parallel Postulate 5.

▶ **Exercise 5.26.** Euclid showed that the Parallel Postulate 5 implies the Pythagorean Theorem. Finish the proof of Theorem 5.25 by showing that the Pythagorean Theorem implies Saccheri's Postulate 2. Begin with a right triangle △ABC with ∠BAC = 90°. Construct the associated saccheri quadrilateral BFGC as in Theorem 5.9. Show that the Pythagorean Theorem implies that $\overline{FG} = \overline{BC}$ and from this deduce that BFGC is a rectangle.

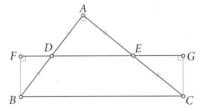

Yet another alternate postulate is due to a mistake by Giordano Vitale [1633–1711] in 1680:

Alternate Postulate 5.12. [Vitale] In any saccheri quadrilateral, there is a point on the summit so that the length of the perpendicular from this point to the base is equal to the sides of the saccheri quadrilateral.

▶ **Exercise 5.27.** Prove that Vitale's Postulate is logically equivalent to the Parallel Postulate 5.

Another attempt at proving the Parallel Postulate 5 was by English mathematician John Wallis. His proof is very much like an earlier proposition by Nasir-Eddin [1201–1274] and contains the same unjustified assumption about the existence of similar triangles.

Alternate Postulate 5.13. [Wallis, 1663] Given a line segment, one can construct on it a triangle similar to a given triangle.

Theorem 5.26. Wallis's Postulate is logically equivalent to the Parallel Postulate 5.

Proof: Part 1: Wallis's Postulate implies the Parallel Postulate 5.

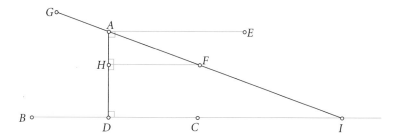

We assume that Wallis's Postulate is true, and will show that Playfair's Postulate must also be true. Thus, we assume the situation given by the hypotheses of Playfair's Postulate: we are given a line BC and a point A not on BC. Drop $AD \perp BC$ and construct $AE \perp AD$. By Proposition I.27, since the alternate interior angles are equal, $AE \parallel BC$. Now let GAF be another line through A, different from AE. Without loss of generality, we may assume that F falls between AE and BC. Drop $FH \perp AD$. By Wallis's Postulate, we may construct $\triangle ADI \sim \triangle AHF$ so that F and I lie on the same side of AD. We must show that point I lies on BC, as the illustration implies. By similarity, $\angle ADI = \angle AHF = 90°$. We already know that $\angle ADC = 90°$, so it follows that line DI and DC coincide. Therefore, $\angle DAF = \angle DAI$ and lines AF and AI also coincide. Thus, GF intersects BC at I. Therefore, GF, which was an arbitrary line through A, but different from AE, must intersect BC. Thus, AE is the only line through A parallel to BC. Playfair's Postulate is therefore true, and so the Parallel Postulate 5 is true.

Part 2: Parallel Postulate 5 implies Wallis's Postulate.

We assume that the Parallel Postulate 5 is true and must show that we can construct similar triangles of any size. We are given a triangle $\triangle ABC$ and line segment $\overline{DE}$, which will be equal to one side of the similar triangle we want to construct.

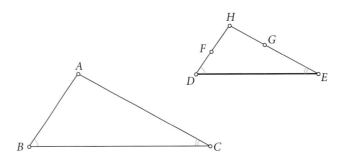

Construct $\angle FDE = \angle ABC$ and $\angle GED = \angle ACB$. By Proposition I.17, $\angle FDE + \angle GED = \angle ABC + \angle ACB < 180°$. Thus, by the Parallel Postulate 5, DF intersects EG at a point H, forming $\triangle HDE$. By Corollary 3.17 (AA similarity, which we can use since we are assuming the Parallel Postulate 5), $\triangle ABC \sim \triangle HDE$. □

Alternate Postulate 5.14. [Gauss 1, 1799] There exist similar triangles that are not congruent.

▶ **Exercise 5.28.** Prove that Alternate Postulate 5.14 is logically equivalent to the Parallel Postulate 5.

Some other notable statements that are also logically equivalent to the Parallel Postulate 5 follow, although we won't prove these equivalences.

Alternate Postulate 5.15. [Proclus 2] The distance between a pair of parallel lines is bounded.

Alternate Postulate 5.16. [Clavius, 1574] The set of all points on a given side of a line that are a given distance to the line forms a straight line.

Alternate Postulate 5.17. [Gauss 2, 1799] It is possible to construct a triangle with area greater than any given number.

Alternate Postulate 5.18. [Farkas Bolyai, 1800] Given three noncollinear points, there is a circle passing through all three.

The last mentioned alternate postulate is the same as our Corollary 3.27 (Euclid's Proposition IV.5). This is the mistake that Farkas (Wolfgang) Bolyai [1775–1856] made when he tried to prove the Parallel Postulate 5. His son, Janos Bolyai, was one of the founders of noneuclidean geometry, although his father tried very hard to discourage him from even attempting to work on the problem.

◆ 5.4 Projects

Project 5.1. Research some other alternate postulates and prove that they are logically equivalent to the Parallel Postulate 5.

Project 5.2. Read Fr. Giovanni Saccheri, *Euclides ab Omni Naevo Vindicatus*, and report precisely on the contradiction he claimed to have found.

Project 5.3. Research Omar Khayyam's mathematical work.

Project 5.4. Research Johann Lambert's attempts to prove the Parallel Postulate 5.

Project 5.5. Research Adrien Legendre's attempts to prove the Parallel Postulate 5.

Project 5.6. Research Alternate Postulate 5.15 by Proclus and present a proof that this is logically equivalent to the Parallel Postulate 5.

Project 5.7. Research Alternate Postulate 5.16 by Christopher Clavius and present a proof that this is logically equivalent to the Parallel Postulate 5.

Project 5.8. Research Alternate Postulate 5.17 by Carl F. Gauss and present a proof that this is logically equivalent to the Parallel Postulate 5.

Project 5.9. Research Alternate Postulate 5.18 by Farkas Bolyai and present a proof that this is logically equivalent to the Parallel Postulate 5.

Project 5.10. Discuss the influence of the scientific method as exemplified by Euclid and Newton on the culture of the 18th century. Examples of this influence can be seen in Samuel Johnson's *Dictionary* of 1755, which was the first to try to standardize definitions and spelling, and the American Declaration of Independence of 1776 (mainly written by Thomas Jefferson), which begins by setting out an axiom system or logical structure for sound government: "We hold these Truths to be self-evident: that all men are created equal, that they are endowed with certain unalienable Rights, that among these are Life, Liberty, and the Pursuit of Happiness."

6. Hyperbolic geometry

♦ 6.1 The history of hyperbolic geometry

The history of the discovery of noneuclidean geometry is full of near-misses. Many mathematicians worked on trying to prove the Parallel Postulate 5, but few of them ever considered the possibility that it was, in fact, unprovable. We will try to give a chronology of some of the more notable figures.

Carl Friedrich Gauss was one of the best mathematicians to have ever lived, and was probably the first to recognize that there were other valid geometries aside from that of Euclid. In 1813, he wrote that "In the theory of parallels we have advanced no further than Euclid."[1] Furthermore, he wrote in a letter from 1816, "It is easy to prove that, if Euclid's geometry is not the true one, then there are absolutely no similar figures." In 1817 he wrote, "I am becoming more and more convinced that the necessity of our [euclidean] geometry cannot be proved, at least not by the human intellect nor for human reason. Perhaps in another life we will be able to obtain insight into the nature of space, which is now unattainable. Until then we must place geometry not in the class with arithmetic, which is purely *a priori*, but with mechanics."[2] In the period after 1813, his letters show an increasing confidence and more advances in developing the theory of noneuclidean geometry, but he was careful not to publicize these. He even collected data from a general survey of Germany, working out the angles of a triangle formed by three distant mountains, perhaps hoping for an experimental proof of the type of geometry we live in. (The results were inconclusive, since the sum of the angles was close to 180°, within the margin of error.)

One of Gauss' students, Friedrich Ludwig Wachter [1792–1817], studied parallels and proved a deep result about noneuclidean geometry (that noneuclidean geometry can be considered as geometry on a sphere of infinite radius), but we are uncertain of his intent, whether to find a contradiction that would imply that the Parallel Postulate 5 must be true or if he accepted the potential truth of noneuclidean geometry. On April 3, 1817, Wachter vanished while on his customary evening walk and was never seen again. Gauss never published any of his work on noneuclidean geometry. His personal motto was *pauca sed matura* (few, but ripe). His writings were often long delayed while he polished them to perfection. He was also very disinclined to create controversy and thought that the announcement of noneuclidean geometry would do precisely that. Such a discovery would destroy the theory of Immanuel Kant, the preeminent European philosopher. Kant had written, "the concept of [euclidean] space is by no means of empirical origin, but is an inevitable necessity of thought." Noneuclidean geometry would show that this is not inevitable. Since euclidean geometry was his only example of a synthetic *a priori* truth, showing that it was only one of several possible geometries would wreak havoc on his theory of truth.

In 1818, there appears a brief one-page memorandum by a lawyer, Karl Schweikart [1780–1859], in which he states, "There exists a two-fold geometry—a geometry in the strict sense—the Euclidean geometry; and an astral geometry ... Triangles in the latter have the property that the sum of their three angles is not equal to two right angles."[3] But this short note contains no details or proofs. This is the first statement in print on the existence of noneuclidean geometries.

---

[1] Quoted in Dunnington, *Gauss—Titan of Science*, p. 179.
[2] Quoted in Dunnington, *Gauss—Titan of Science*, p. 180.
[3] Quoted in Bonola, *Non-Euclidean Geometry*, p. 76.

In 1820, Janos Bolyai [1802–1860], a Hungarian army officer, began to study parallels. His father, Farkas (or Wolfgang) Bolyai was a mathematician and had been a student and friend of Gauss at Göttingen. In the context of another mistaken proof of the Parallel Postulate 5, he was the originator of our Alternate Postulate 5.18. Farkas wrote to his son, trying to dissuade him, "Do not lose one hour on that. It brings no reward, and it will poison your whole life. Even through the pondering of a hundred great geometers lasting for centuries it has been utterly impossible to prove the [parallel] postulate without a new axiom. I believe that I have exhausted all imaginable ideas. . . . I can prove in writing that [Gauss] racked his brains on parallels. He averred both orally and in writing that he had meditated fruitlessly about it."[4] He also wrote,

> You must not attempt this approach to parallels. I know this way to its very end. I have traversed this bottomless night, which extinguished all light and joy from my life. I entreat you, leave the science of parallels alone. . . . I thought I would sacrifice myself for the sake of truth. I was ready to become a martyr who would remove the flaw from geometry and return it purified to mankind. I accomplished monstrous, enormous labors; my creations are far better than those of others and yet I have not achieved complete satisfaction. . . . I turned back when I saw that no man can reach the bottom of the night. I turned back unconsoled, pitying myself and all mankind. Learn from my example: I wanted to know about parallels, I remained ignorant, this has taken all the flowers of my life and all my time from me . . . I admit that I expect little from the deviation of your lines. It seems to me that I have been in these regions; that I have traveled past all reefs of this infernal Dead Sea and have always come back with broken mast and torn sail. The ruin of my disposition and my fall date back to this time.[5]

Janos persevered and in 1823 wrote to his father,

> I have now resolved to publish a work on the theory of parallels, as soon as I shall have put the material in order, and my circumstances allow it. I have not yet completed this work, but the road which I have followed has made it almost certain that the goal will be attained, if that is at all possible: the goal is not yet reached, but I have made such wonderful discoveries that I have been almost overwhelmed by them, and it would be cause of continual regret if they were lost. When you see them, you too will recognize it. In the meantime I can only say this: *I have created a strange new universe from nothing.* All that I have sent you till now is but a house of cards compared to the tower. I am fully persuaded that it will bring me honor, as if I had already completed the discovery.[6]

The tradition in mathematics is that discoveries are credited to the first to publish. Farkas prophetically replied to his son, "If you have really succeeded in the question, it is right that no time be lost in making it public, for two reasons: first, because ideas pass easily from one to another, who can anticipate its publication; and secondly, there is some truth in this, that many things have an epoch, in which they are found at the same time in several places, just as violets appear on every side in spring."[7]

In 1829 Janos finally finished writing up his results and sent them to his father. These were published in 1832 as an appendix to a textbook, the *Tentamen*, by his father. While Janos' treatise is usually referred to

---

[4]Quoted in Dunnington, *Gauss—Titan of Science*, p. 178.
[5]Quoted in Gray, *Janos Bolyai, Non-Euclidean Geometry, and the Nature of Space*.
[6]Quoted in Bonola, *Non-Euclidean Geometry*, p. 98.
[7]Quoted in Bonola, *Non-Euclidean Geometry*, p. 99.

as *The Appendix*, the formal title is *The Science of Absolute Space with a Demonstration of the Independence of the Truth or Falsity of Euclid's Parallel Postulate (Which Cannot Be Decided* a Priori*) and, in Addition, the Quadrature of the Circle in Case of its Falsity*, thus making it clear that Bolyai recognized the implication of his discovery to the currently held ideas in philosophy. A copy of this book was sent to Gauss, in the hopes that he would publicize these results and help establish Janos in the mathematical community. In a letter to Farkas, Gauss replied,

> If I commenced by saying that I am unable to praise this work, you would certainly be surprised for a moment. But I cannot say otherwise. To praise it, would be to praise myself. Indeed the whole contents of the work, the path taken by your son, the results to which he is led, coincide almost entirely with my meditations, which have occupied my mind partly for the last 30 or 35 years. . . . So far as my own work is concerned, of which up to now I have put little on paper, my intention was not to let it be published during my lifetime. . . . On the other hand it was my idea to write down all this later so that at least it should not perish with me. It is therefore a pleasant surprise for me that I am spared this trouble, and I am very glad that it is just the son of my old friend, who takes precedence of me in such a remarkable manner."[8]

Thus Gauss, although he had never published any of his work in noneuclidean geometry, clearly established his claim to have been the first discoverer. Nor would he recommend Janos for a university position, although a word from him would have established the young mathematician in a secure position. Janos was quite bitter about this and never again published his work. Much later, in 1851, Janos wrote,

> In my opinion, and as I am persuaded, in the opinion of anyone judging without prejudice, all the reasons brought up by Gauss to explain why he would not publish anything in his life on this subject are powerless and void; for science, as in common life, it is necessary to clarify things of public interest which are still vague, and to awaken, to strengthen and to promote the lacking or dormant sense for the true and right . . . . It is a fact that, among mathematicians, and even among celebrated ones, there are, unfortunately, many superficial people, but this should not give a sensible man a reason for writing only superficial and mediocre things and for leaving science lethargically in its inherited state. Such a supposition may be said to be unnatural and sheer folly; therefore I take it rightly amiss that Gauss, instead of acknowledging honestly, definitely and frankly the great worth of the Appendix and the Tentamen, and instead of expressing his great joy and interest and trying to prepare an appropriate reception for the good cause, avoiding all these, he rested content with pious wishes and complaints about the lack of adequate civilization. Verily, it is not this attitude we call life, work and merit.[9]

Nikolai Ivanovich Lobachevskii [1793–1856] was still trying to prove the Parallel Postulate when Janos Bolyai had already made significant progress in noneuclidean geometry. However, he was the first to publish an account of hyperbolic geometry. He wrote, "The fruitlessness of the attempts made, since Euclid's time, for the space of 2000 years, aroused in me the suspicion that the truth, which it was desired to prove, was not contained in the data themselves; that to establish it the aid of experiment would be needed, for example, of astronomical observations, as in the case of other laws of nature. When I had finally convinced myself of the justice of my conjecture and believed that I had completely solved this difficult

---

[8] Quoted in Gray, *Worlds out of Nothing*, p. 125.
[9] Quoted in Fejes Tóth, *Regular Figures*, p. 98–99.

question, I wrote, in 1826, a memoir on this subject."[10] While no trace of the 1826 paper remains, in 1829, Lobachevskii published an extended memoir on noneuclidean geometry in Russian in a provincial journal, the *Kazan Messenger*. This, not unexpectedly for such an obscure source, had absolutely no impact on the mathematical community. However, this deserves credit as the first detailed published account of hyperbolic geometry, which he calls "imaginary geometry." His attempts to find an audience failed for some time. One of his articles was rejected by the St. Petersburg Academy and in 1834 a reviewer wrote,

> Glory to Mr. Lobachevskii who took upon himself the labor of revealing, on the one hand, the insolence and shamelessness of false new inventions, and on the other the simpleminded ignorance of those who worship their new inventions. However, while I realize the full value of Mr. Lobachevskii's work, I cannot but hold it against him that, having failed to give his book an appropriate title, he forced us to think for a long time in vain. For instance, why not write, instead of *On the principles of geometry*, *A satire on geometry* or *A caricature of geometry* or a similar thing.[11]

Despite this humiliation, Lobachevskii continued to publish further findings through the University of Kazan. Finally, in 1837, he published an article in *Crelle's Journal*, the major European mathematics journal, and this is the article that first caught the attention of mathematicians. Of this article, in 1846 Gauss wrote, "Lobachevskii calls it imaginary geometry. You know that for 54 years (since 1792) I have shared the same views with some additional development of them that I do not wish to go into here; thus I have found nothing actually new for myself in Lobachevskii's work. But in developing the subject the author followed a road different from the one I took; Lobachevskii carried out the task in a masterly fashion and in a truly geometric spirit."[12] It is interesting how Gauss changes the time frame in which he claims to have first considered noneuclidean geometry.

It took another fifty years for noneuclidean geometry to be fully accepted. As Max Planck later wrote (about quantum mechanics), "A new scientific truth triumphs, not because it convinces its opponents and makes them see the light, but because the opponents eventually die and a new generation that is familiar with it grows up."[13]

♦ 6.2 Strange new universe

The "strange new universe" discovered by Janos Bolyai and Nikolai Lobachevskii is now called the *hyperbolic plane*. In this form of geometry, we assume Euclid's Postulates 1–4, Postulates 6–12, and add a statement equivalent to the negation of the Parallel Postulate 5.

Hyperbolic Postulate 5. Given a line *AB* and a point *C* not on the line, there are at least two lines through *C* that are parallel to *AB*.

Theorem 6.1. The negation of the Parallel Postulate 5 is logically equivalent to the Hyperbolic Postulate 5.

---

[10] Quoted in Bonola, *Non-Euclidean Geometry*, p. 92.
[11] Quoted in Rosenfeld, *A History of Non-Euclidean Geometry*, p. 209.
[12] Quoted in Gray, *Worlds out of Nothing*, p. 127.
[13] Quoted in T.S. Kuhn, *The Structure of Scientific Revolutions*, p. 150.

Proof: Assume that the Parallel Postulate 5 is false and consider an arbitrary line AB and a point C not on AB. □

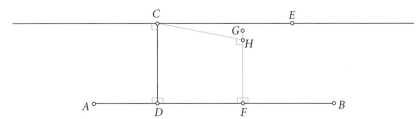

Drop $CD \perp AB$ and erect $CE \perp CD$. Then by Proposition I.28, $CE \| AB$. Let F be another point on AB and erect $FG \perp AB$. Drop $CH \perp FG$. Since $\angle DFH = 90° = \angle CHF$, CH is also parallel to AB. If CH coincides with CE, then $CDFH$ would be a rectangle. But from Theorem 5.24, we know that if the Parallel Postulate 5 is false, then rectangles do not exist. Thus, we have two distinct lines, CE and CH, both through C and both parallel to AB, so the Hyperbolic Postulate 5 is true.

The Hyperbolic Postulate 5 states that there is a line and a point such that there are at least two lines through the point parallel to the given line. This is the negation of Playfair's Postulate when combined with Proposition I.31, which guarantees the existence of at least one parallel line. Recall that Proposition I.31 was proven without the Parallel Postulate 5 and is thus a legitimate theorem in neutral geometry. By Theorem 5.20, we know that the negation of Playfair's Postulate is logically equivalent to the negation of the Parallel Postulate 5. Thus, the Hyperbolic Postulate 5 is logically equivalent to the negation of the Parallel Postulate 5. □

In this geometry, Euclid's Propositions I.1–28, Proposition I.31, Theorems 3.20, 3.21, and 3.24, Corollary 3.25, and all of the neutral Theorems 5.1–5.19 are true. Also, the negations of all of the alternate parallel postulates are true, since they are logically equivalent to the negation of the Parallel Postulate 5. Thus we have a list of ready-made theorems in hyperbolic geometry, some of which are listed below. Many of these could be proven directly (and usually quite simply) by using our new Hyperbolic Postulate 5, rather than negating the proofs given in Chapter 5.

Theorem 6.2. Negate Alternate Postulate 5.3: There are parallel lines that are not equidistant.

▶ **Exercise 6.1.** Using only the Hyperbolic Postulate 5 (rather than the results of Chapter 5), prove that there exist lines that are parallel but not equidistant.

Theorem 6.3. Negate Alternate Postulate 5.14: If two triangles are similar, then they must be congruent.

Theorem 6.4. Negate Alternate Postulate 5.10: The angles in any triangle sum to less than 180°.

Corollary 6.5. The angles in any convex quadrilateral sum to less than 360°.

Theorem 6.6. Negate Alternate Postulate 5.8: Rectangles do not exist.

Negating Alternate Postulate 5.7 gives us the statement that there is a saccheri quadrilateral with acute summit angles. Combining this with Theorem 5.16, we have the following theorem:

Theorem 6.7. The summit angles of any saccheri quadrilateral are acute.

Theorem 6.8. Negate Alternate Postulate 5.15: There is a pair of parallel lines so that the distance between them is unbounded.

Theorem 6.9. Negate Alternate Postulate 5.17: There is an upper bound for the possible area of a triangle.

▶ **Exercise 6.2.** Negate the statements of Alternate Postulates 5.1, 5.2, 5.4, 5.6, 5.9, 5.11, 5.12, 5.16, and 5.18.

The question naturally arises, can a space exist where all of these odd things are true, and, if so, what would it look like?

▷ **Activity 6.1.** Cut out a large number of paper equilateral triangles, about 2 inches on each side. Tape these together edge to edge so that seven meet at each vertex. Extend this to form a ripply surface at least 2 feet in diameter. (In practice, it is easier to take hexagons formed by six equilateral triangles of the suggested size, slit each of these open along one of the triangle edges, and then tape an extra triangle into the gap. Cut some extra triangles and tape these modified hexagons together so that there are seven triangles at each vertex. Better yet, use fabric instead of paper. A larger model can be made if students work in groups.)

a. Using a ruler, draw lines on your surface, smoothing the surface so that you have a straight flat bit to draw on. Draw a triangle with sides at least a foot long. Measure the angles and add together. What angle sum did you get?
b. Using a ruler and a protractor, construct a saccheri quadrilateral. Measure the summit angles. What do you find?
c. Construct two parallel lines. Consider the distance between them. What do you conclude?
d. Construct a quadrilateral with three right angles and measure the fourth angle. What do you find?

The preceding activity explains the use of the word hyperbolic for this type of geometry. "Hyperbolic" comes from the Greek for "excess" and as you have seen this geometry is constructed by joining more triangles together at each vertex than would fit in the euclidean plane.

By the Hyperbolic Postulate 5, there must be at least two lines through a point C that are parallel to a given line AB. On the other hand, we know how to construct one parallel line through C by Euclid's Proposition I.31. There are infinitely many lines through C that intersect AB (one for each point on AB). We need to study the relationships among all of these lines. Consider all of the lines through C. These can be divided into two groups: lines that intersect AB and lines that are parallel to AB. The Hyperbolic Postulate 5 guarantees at least two lines through C parallel to AB.

Lemma 6.10. Given a line AB and a point C not on AB, if there are two lines through C both parallel to AB, then there are infinitely many lines through C all parallel to AB.

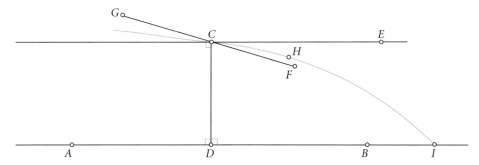

Proof: Begin with a straight line *AB* and a point *C* not on *AB*. Drop a perpendicular *CD* onto *AB*. Clearly *CD* intersects *AB* and so is not parallel. Erect *CE* ⊥ *CD*. By Euclid's Proposition I.28, *CE* must be parallel to *AB*. The Hyperbolic Postulate 5 guarantees us at least one other line *FG* through *C* parallel to *AB*. Without loss of generality we may assume that *F* lies inside ∠*ECD*. Consider a line *CH* falling inside ∠*ECF*. If *CH* is not parallel to *AB*, then *CH* will intersect *AB* at some point *I*. But then *CF* contains a point inside △*CDI* and so by Pasch's Theorem 2.11, *CF* must intersect △*CDI* one more time. Since *CF* cannot intersect $\overline{CD}$ or $\overline{CI}$ again, it must intersect $\overline{DI}$. But *CF* is known to be parallel to *AB*, so there is a contradiction. Thus, *CH* must be parallel to *AB*. Therefore, the Hyperbolic Postulate 5 implies that if there are two lines through *C* parallel to *AB*, then there must be infinitely many lines through *C* parallel to *AB*. □

Next, consider the illustration below. For any point *P* on *AB*, we can draw the line *CP* to get a line through *C* that intersects *AB*. Now, consider the angle ∠*DCP*. Any ray $\overrightarrow{CQ}$ on the *P*-side of *CD* so that ∠*DCQ* < ∠*DCP* must intersect *AB* by the Cross-bar Theorem 2.12. Thus, there are infinitely many lines through *C* that intersect *AB*, one for each point of *AB*. We know from Lemma 6.10 that *CF* and any ray *CR* so that ∠*DCF* < ∠*DCR* < 90° must be parallel to *AB*. The question is, what happens between the intersecting line *CP* and the parallel line *CF*?

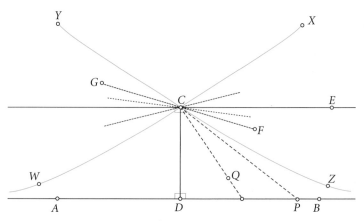

Once we have an angle at which a ray intersects line *AB*, the ray forming any smaller angle will also intersect. Similarly, once we have an angle at which a ray does not intersect *AB*, the ray at any larger angle will not intersect the line. We know that the set of measures of the angles at which a ray $\overrightarrow{CS}$ will intersect *AB* is bounded above because once we get to a non-intersecting angle, any ray at a larger angle will not intersect. In particular, a ray at the angle 90° will not intersect *AB*. At this point, we appeal to the Completeness Property for the real numbers, which says that every nonempty subset of the real numbers that is bounded above has a least upper bound: a smallest number that is greater than or equal to all of the elements of the set. This Completeness Property is another continuity axiom which we only use once, so we will not formally add it to our postulate list. The result is that we are guaranteed one specific angle measure so that any smaller angle will yield an intersecting ray while any larger angle will not. The existence of this angle matches our intuition and represents the angle where the intersecting rays turn into the nonintersecting rays. Note that what we have said so far is true in neutral geometry. In euclidean geometry, this least upper bound for the set of angles at which a ray intersects is 90°. In hyperbolic geometry, the existence of more than one line through *C* parallel to *AB* forces the least upper bound to be less than 90° on at least one side of *CD*. The question remains whether lines at these special angles will intersect *AB*.

Definition 6.11. Given a line *AB* and a point *C* not on *AB*. Drop a perpendicular *CD* ⊥ *AB*. Without loss of generality, assume *A* − *D* − *B*. Let *P* be an arbitrary point on the *B* side of *CD*. The *angle* *of parallelism* on the *B* side of *CD* is defined to be the least upper bound of the measures of the angles ∡*DCP* at which a ray $\overrightarrow{CP}$ will intersect *AB*.

We know that the angle of parallelism must be less than or equal to 90°. Thus, if the ray $\overrightarrow{CZ}$ at the angle of parallelism does not intersect *AB*, then neither will the line *CZ*.

Theorem 6.12. Given a line *AB* and a point *C* not on the line, there are two lines *XCW* and *YCZ* through *C* that are parallel to *AB* so that any line through *C* entering angle ∡*WCZ* will intersect *AB* and any line through *C* entering angle ∡*YCW* will be parallel to *AB*.

Proof: Let *CD* be the line through *C* perpendicular to *AB*. The lines *WCX* and *YCZ* are the extensions of the rays on each side of *CD* at the angles of parallelism. Their existence is explained above and depends on the Completeness Property, which we ask you to take on faith. We need to prove that these particular lines are indeed parallel to *AB*. Assume that *YCZ* intersects *AB* at some point *P*. The Betweenness Postulate 7(3) says that there is another point *Q* lying on the side of *P* opposite to *A*. But then *CQ* intersects *AB* and ∡*WCQ* = ∡*WCD* + ∡*DCQ* where ∡*DCQ* > ∡*DCZ*. This contradicts the assumption that ∡*DCZ* is the least upper bound of all of the angles at which a ray intersects *AB*. Thus, *YCZ*, and similarly *WCX*, must be parallel to *AB*.

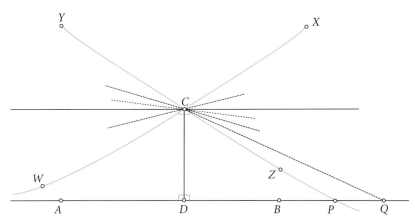

Furthermore, examining the proof of Lemma 6.10, we see that all of the lines through *C* that enter ∡*XCZ* (or ∡*YCW*) must be parallel to *AB*. Thus, *XCW* and *YCZ* separate the lines through *C* that intersect *AB* from the lines through *C* that are parallel to *AB*. □

We introduce some terminology for these special limiting parallel lines, since we will need to refer to them frequently. Intuitively, we think of these lines as the "first" lines through *C* that are parallel to *AB*.

Definition 6.13. Given the hypotheses of Theorem 6.12, the line *WX* is said to be *asymptotic parallel* to *AB* in the direction of *A* and *YZ* is asymptotic parallel to *AB* in the direction of *B*. Any other line through *C* that is parallel to *AB* is *divergent parallel*.

If *WX* is asymptotic parallel to *AB*, then we can refer to the *direction of parallelism* to indicate that they are asymptotic parallel in the direction of *A* and *W*. Similarly the direction of parallelism for the asymptotic parallels *YZ* and *AB* is the direction of *B* and *Z*. The angle of parallelism as defined in Definition 6.11 is denoted as $\Pi(\overline{CD})$ = ∡*WCD*, though it remains to be shown that ∡*WCD* = ∡*ZCD*.

Given line *AB* and point *C* not on *AB*, there are two asymptotic parallels, one in each direction, and infinitely many divergent parallels. Note that this terminology is not universal: In some other books asymptotic parallels are called limiting parallels, ultraparallels, critically parallel, or horoparallel. Similarly, divergent parallels are sometimes called parallel or hyperparallel.

♦ 6.3 Models of the hyperbolic plane

◇ 6.3.1 Physical models

Before proceeding to develop the theory of hyperbolic geometry, we want to give you some way to develop a feeling for this strange new universe. The floppy model of the hyperbolic plane made in Activity 6.1 is extremely useful in this. This paper model was popularized first by William Thurston and then by Jeffrey Weeks in his book *The Shape of Space*, who calls this model *Thurston paper*. Other models for hyperbolic space can be made similarly: use eight or more equilateral triangles instead of seven at each vertex. Another model, sometimes called the *hyperbolic soccer ball*, is formed by taping two hexagons and one heptagon together at each vertex. Depending on the size of the polygons and the sum of the angles at each vertex, one gets models of different hyperbolic planes, equally noneuclidean, but with different waviness. The quantity that determines this "waviness" is the angle sum at each vertex. If the angles sum to 360°, of course, the surface will lie flat. If the angles sum to less than 360°, the surface will close back on itself, forming a sphere-like polyhedron. And if the angle sum is greater than 360°, we will get one of the hyperbolic planes. We will return to this idea of classifying models by their angle sum in Chapter 7.

None of these polygonal representations are really quite a hyperbolic plane since each individual triangle is flat and euclidean. To represent hyperbolic phenomena, you must be careful that your lines and figures span several triangles. One can partially compensate for this deficiency by slightly curving the edges of the polygons. An even better and more durable model can be made of fabric polygons, sewn together instead of taped, since the fabric is more elastic than paper. It is even possible to crochet a model: For examples, see the article "Crocheting the Hyperbolic Plane" by David Henderson and Daina Taimina or Taimina's *Crocheting Adventures with Hyperbolic Planes*.

The first such model was made by Eugenio Beltrami [1835–1900] and was very important in convincing mathematicians that hyperbolic geometry was more than an artificial construct. He showed that hyperbolic space can be modeled on a surface called a *pseudosphere*.

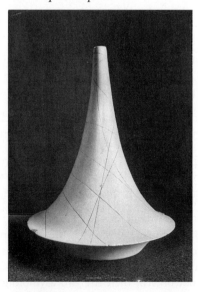

(*Source*: Time & Life Pictures/Getty Images, Inc.)

Beltrami also made several models much like the one we made in Activity 6.1, of thin cardboard using trapezoids with carefully curved edges. One that he gave to his friend Luigi Cremona in 1869 has been preserved at the Department of Mathematics of the University of Pavia. This model (and ours) can be folded and rolled up to form different sections of the pseudosphere.

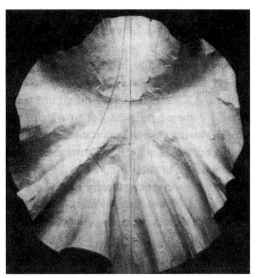

(*Source*: Beltrami model: Livia Giacardi, 'Scientific Research and Teaching Problems in Beltrami's Letters to Houel,' in Katz, Using History to Teach Mathematics/©The Mathematical Association of America)

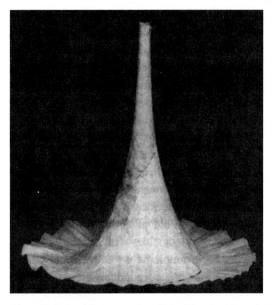

(*Source*: Beltrami model: Livia Giacardi, 'Scientific Research and Teaching Problems in Beltrami's Letters to Houel,' in Katz, Using History to Teach Mathematics/©The Mathematical Association of America)

Technically, the ideal model for hyperbolic space should have the same type of hyperbolic geometry at each point (*constant negative curvature*) and each line segment in the model should be extendable to a line (*geodesically complete*). In general, on surfaces (or manifolds, their higher dimensional analogs) a geodesic is a curve that minimizes the length between nearby points. Thus, a geodesic acts like a straight line on the surface, at least locally. A segment in a surface is then defined as a piece of a geodesic. With this terminology, a surface is geodesically complete if every segment can be extended to a complete geodesic. The model in Activity 6.1 formed by taping seven equilateral triangles together at each vertex has negative curvature because the sum of the vertex angles exceeds 360°. It does not have constant curvature because each triangle lies flat instead of curving, but is a reasonably good polygonal approximation of a hyperbolic plane.

In 1901, David Hilbert proved that there is no differentiable surface in $\mathbb{R}^3$ satisfying all three of these conditions. This means that we cannot construct a hyperbolic plane in the world we live in. However, we can construct large sections of hyperbolic space to experiment with. The interested reader is referred to Jeffrey Weeks' *The Shape of Space*.

These physical models of the hyperbolic plane are very useful for building intuition and experimenting with hyperbolic phenomena, but they aren't very convenient or portable. Furthermore, they are, naturally enough, finite, and so do not lend themselves to exploring the behavior of parallel lines at infinity. Thus we'd like to develop another way of representing hyperbolic geometry. An analogy could be drawn with the many different mapping projections used to represent our earth. None of these are ideal, since each distorts some property of the sphere, but they are very useful.

◇ 6.3.2 The Beltrami-Klein model

The first map for the hyperbolic plane that we give is the *Beltrami-Klein model*, for Eugenio Beltrami and Felix Klein [1849–1925]. To make such a map or model of a geometric space, we must tell you how to represent the plane and the lines and points on the plane. In this model the infinite hyperbolic plane is represented by the set of points inside a fixed circle. Points in the plane are then represented by points inside this disc and lines are represented by open chords of this circle:

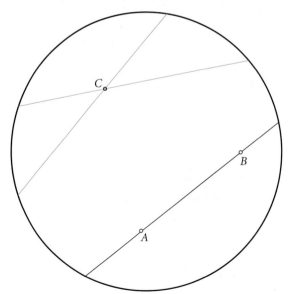

Points on the boundary of the circle are considered to be infinitely far way. Thus distance is measured in some new way. Imagine that as one approaches the boundary circle, one's legs get shorter and shorter, so that the same number of steps are needed to walk equal distances, although to an outside observer the distances look different.

In this model it is easy to see that through point C one can draw infinitely many lines parallel (not intersecting) line AB, as above. Below is pictured a line AB and a point C with the two asymptotic parallels through C to line AB, as well as several examples of divergent parallels, shown as dashed lines.

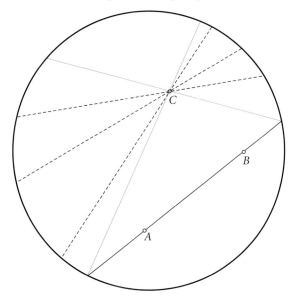

It thus appears that in this model, parallel lines need not be equidistant. To return to our analogy with maps of the earth, note that all mapping projections of the world involve distortions. One cannot represent a round earth on a flat piece of paper without giving up something. The Mercator projection, one of the oldest and most commonly used, badly distorts area and apparent distance. Notice how large Greenland and Antarctica appear:

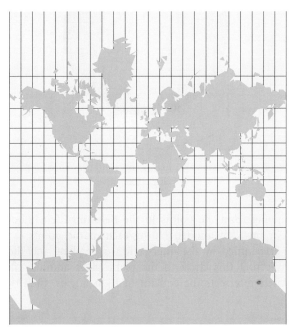

The importance of the Mercator projection is that it represents angle measure correctly. A sailor could draw a line from point A to point B on this map and from this know what compass heading to use to get the ship to point B. However, one wouldn't, without lengthy computations, know how long it would take to get there.

The advantage to the Beltrami-Klein model is that it is very simple and easy to draw, and it represents basic properties of parallel lines, both asymptotic and divergent, accurately. The disadvantage is that it distorts both distance and angle measure, so many figures are not at all what they appear. Our next model is necessarily more complicated, but eliminates one of these disadvantages.

◇ 6.3.3 The Poincaré disc model

The next map of the hyperbolic plane that we study is called the Poincaré disc model. This is an elaboration on the idea of Beltrami-Klein, due to Henri Poincaré [1854–1912], which correctly shows angle measure, though it still distorts distance.

The Poincaré disc model is related to the Beltrami-Klein model by *stereographic projection*: Let C be the circle that defines the Beltrami-Klein disc and consider a sphere with the same radius as C placed so the South Pole is at the center of the circle. Then a point P in the Beltrami-Klein model can be lifted straight up to give a point Q on the southern hemisphere of the sphere. Now consider the line from the North Pole N passing through Q and extending until it meets the plane that C lies on. This line will intersect the plane at a point P' which will lie within another circle C' with radius twice the radius of C.

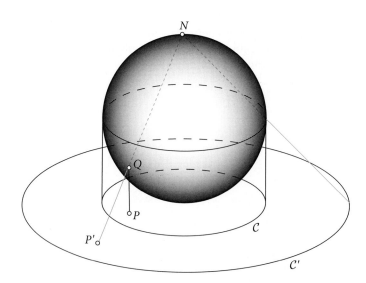

Thus, we have a mapping from the points in the Beltrami-Klein model to a new model. Next, consider a line in the Beltrami-Klein model. Lifting each of the points to the sphere in effect gives the circle on the sphere determined by the intersection of a plane and the sphere, restricted to the southern hemisphere. Projecting this circle to the plane gives another circle and it can be shown that since the spherical circle intersects the equator at right angles, then the projection will meet C' perpendicularly:

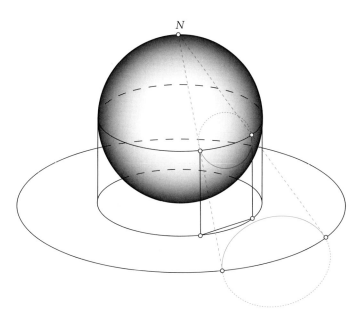

Thus, we define the Poincaré disc model for hyperbolic geometry to consist of the interior of a circle $\mathcal{C}'$, which may as well be chosen to have radius one. In order to illustrate hyperbolic geometry in terms of this model, we must provide interpretations of the basic primitive terms for the model. We define the following:

- point = a point inside the disc
- line = any diameter of the circle $\mathcal{C}'$, or the part lying inside $\mathcal{C}'$ of any circle $\mathcal{C}''$ which is perpendicular to the circle $\mathcal{C}'$.

Points on the boundary circle $\mathcal{C}'$ are called *ideal points* and are considered to be infinitely far away from each point inside the disc. Line segments are represented by segments of diameters or arcs of circles meeting the boundary circle at right angles.

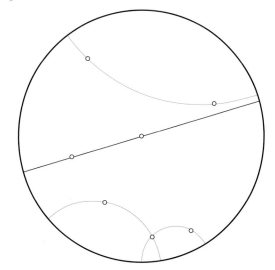

Note that circles are considered perpendicular to each other if their tangent lines are perpendicular. Exercise 3.10 shows that the tangent line at a point P to a circle is perpendicular to the radius from the center of the circle to P. Therefore, diameters meet the circle at right angles, which is why they are considered straight lines in the Poincaré model.

In order to draw pictures on the Poincaré disc model we must think how to construct such perpendicular circles. Providentially, we have already done this in Theorem 3.36 in the section on inversion in the circle. From this theorem we see that the following procedure is just what we want:

Constructing a hyperbolic line through interior points A and B:

- First invert point A through the circle $\mathcal{C}$ to get point A': Draw the ray $\overrightarrow{OA}$, erect $AD \perp OA$ where D lies on the circle, then erect $DE \perp OD$. DE intersects OA at A'.

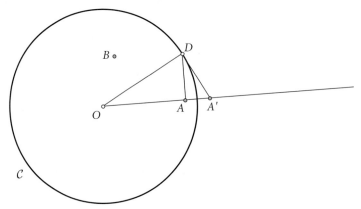

- Use Corollary 3.27 to construct the circle $\mathcal{C}'$ passing through points A, B, and A': Construct the perpendicular bisectors of any two sides of $\triangle ABA'$. These will meet at the circumcenter O' and the circle centered at O' through A will pass through all three points and will be perpendicular to circle $\mathcal{C}$.

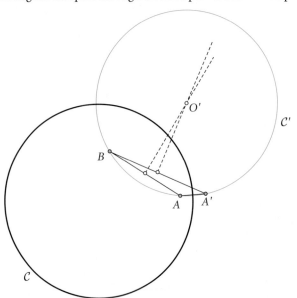

- The arc from the circle thus constructed and terminating at the intersection with $\mathcal{C}$ represents the line AB.

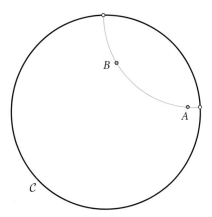

Note that the construction gives the same result if we chose to draw the circle determined by A, B, and B' (the inversion of B).

In Theorem 3.37 we proved that inversion in a circle preserves angle measure. This can be used to show that the Poincaré disc model represents angles correctly. This is the strong advantage to the Poincaré disc model. However, distance is badly distorted, as in the Beltrami-Klein model, although the formula for the distance is easier to formulate. We know another property that is invariant under inversion in the circle, the cross-ratio, by Theorem 3.39. We can measure distance in the Poincaré model using the formula

$$d(A,B) = \left| \ln \left(\frac{\overline{AY}}{\overline{AX}} \frac{\overline{BX}}{\overline{BY}} \right) \right|$$

where X and Y are the ideal points where the hyperbolic line determined by A and B meets the boundary circle.

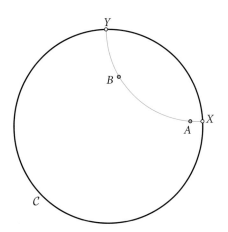

▶ **Exercise 6.3.** Show that the distance formula above has the following properties:
   a. $d(A,B) = d(B,A)$.
   b. $d(A,B) \geq 0$ for all points A, B.
   c. $d(A,B) = 0$ if and only if $A = B$.
   d. If A, B, and C are collinear points with B falling between A and C, then $d(A,C) = d(A,B) + d(B,C)$.

Below is a saccheri quadrilateral drawn in the Poincaré disc. Note that the angles at C and D are acute and equal. The distances AD and BC are equal by the metric defined above.

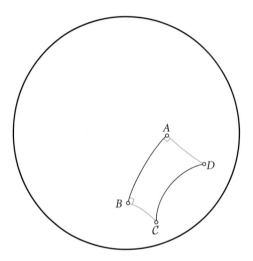

And here is an equilateral triangle:

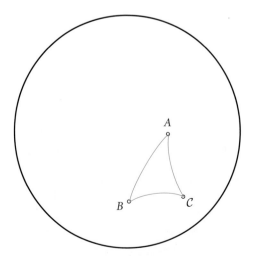

The Dutch artist M. C. Escher [1898–1972] was inspired by the Poincaré disc model to take some of his tilings and extend them to hyperbolic space. He made a series of prints, the *Circle Limit* pictures. Below is *Circle Limit III*. Note that all of the fish are congruent.

(*Source*: M.C. Escher's "Circle Limit III" © 2009 The M.C. Escher Company – Holland)

Experimenting with the Poincaré disc model will help build your intuition about hyperbolic geometry, but performing the ruler and compass constructions necessary to draw the pictures can be quite tedious. Therefore, the use of geometric software is very helpful. *Cinderella* has menus in which you can choose which geometry you wish to work in and which model to use. While *The Geometer's Sketchpad* does not have hyperbolic models built in to the basic installation, supplementary sketches and scripts are supplied which can be used for this. There is also an excellent interactive java program available online called *NonEuclid*, designed specifically to allow experimentation with the Poincaré disc. The use of software is almost essential in developing an intuition for the phenomena of hyperbolic geometry.

▷ **Activity 6.2.** Use geometric software to construct a hyperbolic triangle △*ABC*, measure each of the hyperbolic angles, and find their sum. Move the points *A*, *B*, and *C* around. What is the smallest angle sum you can get? What is the largest angle sum?

▷ **Activity 6.3.** Use geometric software to draw two parallel hyperbolic lines *AB* and *CD* as shown below. Draw the hyperbolic line segment $\overline{BC}$ that intersects the two parallels. Measure the hyperbolic angles ∠*ABC* and ∠*BCD*. What does this tell you about the alternate interior angle relationship between parallel lines?

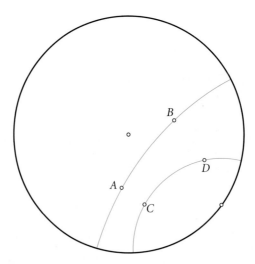

▷ **Activity 6.4.** Use geometric software to draw a hyperbolic line *AB* and then choose a point *C* within the disc but not on your line. Construct two hyperbolic lines through *C* that are both parallel to *AB*.

▷ **Activity 6.5.** Use geometric software to draw a hyperbolic line *AB* and a point *C* within the disc but not on *AB*. Drop a perpendicular $CD \perp AB$ and draw another line *CE* through *C* parallel to *AB*. Drag point *E* around to find how small you can make ∠*ECD* and still have *CE* parallel to *AB*. In this case *CE* is asymptotic parallel to *AB* through *C*, and so $\Pi(\overline{CD}) = \angle ECD$. Measure the angle ∠*ECD*. Move point *C* closer to *D*, make sure *CE* is still asymptotic parallel to *AB* and measure ∠*ECD* again. Repeat for various positions of *C* both farther from *D* and closer to *D*. Make a conjecture about the relationship between ∠*ECD* and the length of $\overline{CD}$.

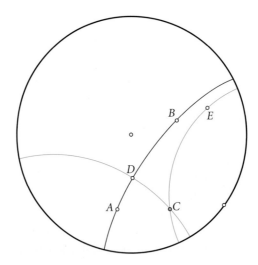

A circle is defined to be the set of all points a given distance from a center point (Definition 2.15). However, distances are not what they appear to be in the Poincaré disc model.

▷ **Activity 6.6.** Use geometric software to draw a hyperbolic circle and measure the distance from the center A to point B on the circle. Place another point C on the circle. Measure the distance from the center A to C and drag C around the circle. Convince yourself that this satisfies the definition of a circle.

▷ **Activity 6.7.** Use geometric software to draw a hyperbolic right triangle $\triangle ABC$. Measure the distances AB, BC, and AC. Compute $AB^2 + BC^2$ and AC^2. What does this tell you about the Pythagorean Theorem in hyperbolic geometry?

We have thus provided a way for you to draw reasonable pictures in hyperbolic space and a way to experiment with geometric properties. Be careful in using these. No piece of software is perfect, and no example constitutes a proof. Use the software to experiment and form conjectures, but you still have to prove your conjectures.

In general, our illustrations will be drawn freehand rather than in the Poincaré disc model. This will require that our straight lines sort of sag a bit to represent angles reasonably. Don't let that bother you. When we want more precise illustrations or want to experiment and form conjectures, we turn to the software.

♦ 6.4   Consistency of geometries

Noneuclidean geometry was not widely popularized until 1866 with a new generation of mathematicians and physicists such as Helmholtz, Riemann, Beltrami, and Klein. Most mathematicians accepted the new theory, especially after Gauss died and his correspondence was published showing that he had long been convinced of the impossibility of proving the Parallel Postulate 5. However, many other scientists and layman denied the truth of noneuclidean geometry. In part this was because of Kant's theory that euclidean geometry was a natural product of the way in which the human mind thought, and therefore was the only natural geometry: the one true geometry. Sermons were even preached against the new theory, such as the remarks in 1905 by Matthew Ryan, "The study of 'non-Euclidean' Geometry brings nothing to students but fatigue, vanity, arrogance, and imbecility ... 'Non-Euclidean' space is the false invention of demons, who gladly furnish the dark understanding of the 'Non-Euclideans' with false knowledge ... The 'Non-Euclideans,' like the ancient sophists, seem unaware that their understandings have become obscured by the promptings of the evil spirits."[14]

So we ask ourselves, which geometry is "true"? How can we decide if a mathematical theory is true? In mathematical terms, we want an axiomatic system to be *consistent*—does not contain any contradictions, and *complete*—all geometric statements can either be proven or disproven from the axioms. Euclid's original list of postulates was not complete, which is why we had to add new ones. As for consistency, that is precisely what a hundred generations of mathematicians have been trying to prove or disprove, without finding a contradiction. But how do we know if the next generation might succeed?

The modern approach to deciding whether an axiomatic system is consistent is to construct a *model* for the system: a physical interpretation of the primitive terms in which the axioms can be shown to be true. This is precisely what Beltrami did in providing physical models for hyperbolic space. For example, it

---

[14]Quoted in Martin, *The Foundations of Geometry and the Non-Euclidean Plane*, p. 215.

can be shown that in the Poincaré disc model Postulate 1 is true, since it is a theorem in euclidean geometry (which we used to draw the Poincaré disc) that two points and the inverse of one of them determine a unique circle. Thus there is a unique Poincaré line between any two distinct points inside the disc. Postulate 2 is true after we redefine how the distances will be measured. Similarly, Postulates 3, 4, and 6–12 are all true in the model. Furthermore, the Hyperbolic Postulate 5 is easily seen to be true. Thus, the model is a faithful representation of the axiom system for hyperbolic geometry. Therefore any hyperbolic theorems we might prove will also be true in the model. However, this model was constructed using euclidean ruler and compass constructions, so we are in the paradoxical position of having deduced that hyperbolic geometry is true if the euclidean geometry needed to construct the model is true. Therefore, if euclidean geometry is true, then hyperbolic geometry is equally true. We cannot prove Euclid's Parallel Postulate 5 using only neutral geometry, because the Parallel Postulate 5 would then also be true in hyperbolic geometry, where it is clearly in contradiction with the Hyperbolic Postulate 5.

Thus we have argued that euclidean and hyperbolic geometry are equally true. So the question remains whether euclidean geometry is "true." We have a familiar model of euclidean geometry: Descartes' cartesian plane, in which points are ordered pairs (x, y) and lines are solutions to equations of the form $\{(x, y) : ax + by = c\}$. We can thus reduce the question of the truth of euclidean geometry to the question of whether the algebra system of the real numbers is true. Whether this last question can ever be answered was addressed in the following famous theorem:

> **Gödel's Incompleteness Theorem:** *We cannot prove the logical consistency of any axiomatic system that is complicated enough to contain the real numbers without assuming a set of axioms of logic that is equally complicated.*

Thus, any attempt on the absolute truth of any but the simplest systems is doomed to failure. We must rest satisfied that our two geometries are equally true. Thus, Kant was wrong in viewing euclidean geometry as *a priori* knowledge. And amazingly, Euclid was right in claiming that it was necessary to assume the Parallel Postulate 5. Poincaré himself said,

> The geometric axioms are therefore neither synthetic *a priori* intuitions nor experimental facts. They are conventions. Our choice among all possible conventions is guided by experimental facts; but it remains free, and is only limited by the necessity of avoiding every contradiction ... In other words, axioms of geometry are only definitions in disguise. What, then, are we to think of the question: Is Euclidean geometry true? It has no meaning. We might as well ask if the metric system is true, and if the old weights and measures are false; if cartesian coordinates are true and polar coordinates false. One geometry cannot be more true that the other: it can only be more convenient.[15]

The mathematician De Sua further said, "Suppose we loosely define a religion as any discipline whose foundations rest on an element of faith, irrespective of any element of reason which may be present. Quantum mechanics for example would be a religion under this definition. But mathematics would hold the unique position of being the only branch of theology possessing a rigorous demonstration of the fact that it should be so classified."[16]

---

[15] Poincaré, *Science and Hypothesis*, p. 50.
[16] Frank De Sua, 'Consistency and Completeness—A Resume', *American Mathematical Monthly* 63(1956), p. 295–305.

As for what use hyperbolic geometry might be, when Albert Einstein was developing his theory of relativity, he found that some inequalities that Lobachevskii had developed for use in his imaginary geometry were precisely what he needed for his theory. This is often the way in mathematics: Theories developed for speculation alone often turn out to have unforeseen applications generations later.

♦ 6.5 Asymptotic parallels

In this section, we investigate the properties of asymptotic parallels. Every one of the proofs will use Theorem 6.12, so it is important that one thoroughly understand that result, the fundamental description of the asymptotic parallels.

Theorem 6.14. Given a line AB and a point C not on the line and asymptotic parallels WCX and YCZ, drop $CD \perp AB$. Then the angle of parallelism $\Pi(\overline{CD}) = \angle WCD = \angle ZCD < 90°$.

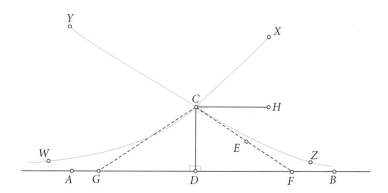

Proof: We are given two lines WCX and YCZ asymptotic parallel to AB. Drop $CD \perp AB$. We proceed to prove that $\angle WCD = \angle ZCD$ by contradiction. Let us assume $\angle WCD < \angle ZCD$. By Proposition I.23, we can construct $\angle ECD = \angle WCD$, so $\angle ECD < \angle ZCD$. Therefore, by Theorem 6.12, CE must intersect AB at some point F. Cut $\overline{DG} = \overline{DF}$ so that $G - D - F$ and draw $\overline{GC}$. Then $\triangle GCD \cong \triangle FCD$ by SAS. Therefore, $\angle GCD = \angle FCD = \angle WCD$. Thus GC must coincide with WC. This contradicts our assumption that G lies on AB while WC is known to be parallel to AB, so it must be true that $\angle WCD = \angle ZCD$.

Next we show that $\angle WCD = \angle ZCD < 90°$. Erect $CH \perp CD$. First note that if $\angle WCD = \angle ZCD > 90°$, then $\angle XCD = \angle YCD < 90°$, so CX lies below CZ and thus $\Pi(\overline{CD}) = \angle XCD$ rather than $\angle ZCD$. If $\angle WCD = \angle ZCD = 90°$, then XCW and YCZ coincide with CH, so there is only one line through C parallel to AB, contradicting the Hyperbolic Postulate 5. □

The next theorem verifies that the angle of parallelism depends on the distance from the point to the line, and not on the particular point chosen.

Theorem 6.15. $\Pi(\overline{CD})$ depends only on the length of $\overline{CD}$ and not on the point C or the line AB.

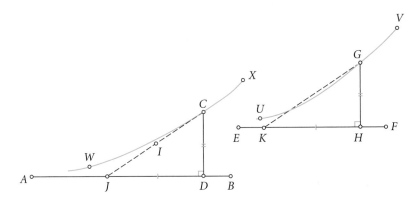

Proof: We are given $CD \perp AB$ and $GH \perp EF$ with $\overline{CD} = \overline{GH}$ and wish to show that $\Pi(\overline{CD}) = \Pi(\overline{GH})$. Consider WX asymptotic to AB at C in the A-direction and UV asymptotic to EF at G in the E-direction. We now need to show that $\angle WCD = \angle UGH$. Let us assume that $\angle WCD > \angle UGH$. By Proposition I.23, we can construct an angle on the A-side of CD so that $\angle ICD = \angle UGH$. Since $\angle WCD > \angle ICD$, by Theorem 6.12, CI intersects AB at J. Cut $\overline{HK} = \overline{JD}$ on the E-side of GH and draw $\overline{KG}$. Then $\triangle KGH \cong \triangle JCD$ by SAS, so $\angle KGH = \angle JCD = \angle UGH$. Therefore, KG must coincide with UG, so we have a contradiction. Therefore, $\Pi(\overline{CD}) = \angle WCD = \angle UGH = \Pi(\overline{GF})$. □

The next three theorems are technical in nature. The reader may choose to skim through these rather than to work through them carefully. The first theorem essentially guarantees that if a line is asymptotic parallel to a second line at one point, then it must be asymptotic parallel to that line at all points. Thus the property of being asymptotically parallel is well-defined. The corresponding theorem for divergent parallels will be proven in Theorem 6.28. Theorem 6.17 is the equivalent of Euclid's Proposition I.30 for asymptotic parallels and shows that the relation of asymptotic parallel is transitive, at least in the direction of parellelism. The corresponding statement for divergent parallels is not true. Finally, the last theorem justifies the term "asymptotic" for these parallels.

Theorem 6.16. If YZ is asymptotic parallel to AB at C in the direction of B and E is another point on YZ, then YZ is asymptotic parallel to AB at E in the direction of B.

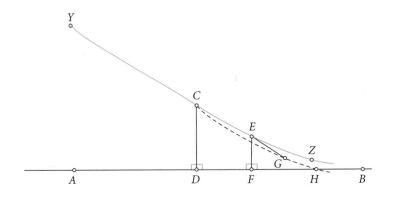

Proof: We are given *YZ* asymptotic parallel to *AB* at *C* and another point *E* on *YZ*, on the *Z* side of *C*. Drop $CD \perp AB$ and $EF \perp AB$. We must show that any line through *E* that enters angle ∠*FEZ* intersects *AB*, thus showing that *EZ* is asymptotic parallel to *AB*. Let *EG* be a line so that *G* lies inside ∠*FEZ*. Then *G* lies inside ∠*DCZ*. Draw *CG*. This line will lie inside ∠*DCZ* and so, by Theorem 6.12, must intersect *AB* at some point *H*.

Note that since *E* lies on *CZ* which is parallel to *AB*, *E* must lie outside ∠*HCD*, so *E* lies outside △*CDH*. Thus, by Pasch's Theorem 2.11, the line *EG* must then intersect △*CDH*. Since *G* lies inside ∠*DCZ*, *EG* must intersect *AB*. □

▶ **Exercise 6.4.** The proof of Theorem 6.16 above assumes that *E* is on the *Z*-side of *C*. Modify the proof for the case in which *E* lies on the *Y* side of *C*.

Theorem 6.17. *If AB is asymptotic parallel to CD in the direction of B and D and CD is asymptotic parallel to EF in the direction of D and F, then AB is asymptotic parallel to EF in the direction of B and F.*

Proof: We assume that *CD* lies between *AB* and *EF*. We must show that *AB* and *EF* are asymptotic parallel. We will do this by contradiction. First, assume that *AB* is not parallel to *EF* at all.

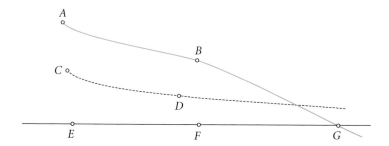

Then *AB* must intersect *EF* at some point *G*. But then *G* on *AB* lies on the *EF* side of line *CD*, while *A* lies on the opposite side of *CD*. Therefore, the Plane Separation Property Postulate 8 tells us that *AB* must intersect *CD*. This gives us a contradiction and so we know that *AB* must be parallel to *EF*.

It remains to show that *AB* is actually asymptotic parallel to *EF* and not divergent parallel. Drop $AH \perp EF$ and consider a line *AI* entering ∠*HAB*. We must show that *AI* is forced to intersect *EF* and the result follows.

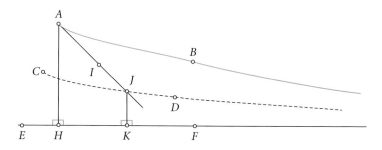

Since AB is asymptotic parallel to CD, we know by Theorem 6.12 that AI must intersect CD at some point J. Drop JK ⊥ EF. Note that AI (extended) enters ∠KJD. Since CD is asymptotic parallel to EF, Theorem 6.12 implies that AI must intersect EF. Thus AB must be asymptotic parallel to EF. □

▶ **Exercise 6.5.** Prove Theorem 6.17 for the case in which CD does not lie between AB and EF.

The next theorem shows that asymptotic parallels get arbitrarily close together in the direction of parallelism, and arbitrarily far apart in the other direction.

Theorem 6.18. Asymptotic parallels get arbitrarily close together in the direction of parallelism and arbitrarily far apart in the other direction.

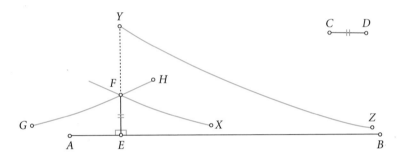

Proof: We are given YZ asymptotic parallel to AB in the B-direction, and an arbitrary length $\overline{CD}$. We need to find a point K on YZ so that the distance from K to AB is equal to the length $\overline{CD}$. Drop YE ⊥ AB. We assume that $\overline{CD} < \overline{YE}$. Cut point F on YE so that $\overline{EF} = \overline{CD}$. Draw FG asymptotic parallel to AB in the A-direction and extend this line in the other direction to H. We must first show that FH must intersect YZ. Draw FX asymptotic parallel to AB in the B direction. Since $\Pi(\overline{EF}) = \angle EFG = \angle EFX < 90°$ by Theorem 6.14 and ∠EFG = ∠YFH by the Vertical Angle Theorem, we have ∠YFH < 90° < ∠YFX. By Theorem 6.17, FX is asymptotic parallel to YZ and FH enters the angle of parallelism, so FH must intersect YZ at some point I.

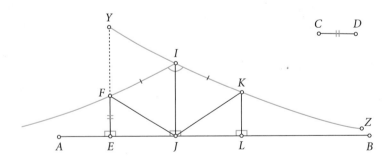

Drop IJ ⊥ AB, cut $\overline{IK} = \overline{IF}$, drop KL ⊥ AB, and draw $\overline{FJ}$ and $\overline{KJ}$. Note that $\Pi(\overline{IJ}) = \angle FIJ = \angle KIJ$. Thus △FIJ ≅ △KIJ by SAS. Therefore, $\overline{FJ} = \overline{KJ}$ and ∠IJF = ∠IJK. It follows that ∠FJE = ∠KJL, and so △FJE ≅ △KJL by AAS. Thus, $\overline{KL} = \overline{FE} = \overline{CD}$. □

▶ **Exercise 6.6.** Modify the proof of Theorem 6.18 above for the case in which $\overline{CD} > \overline{YE}$ as in the following illustration.

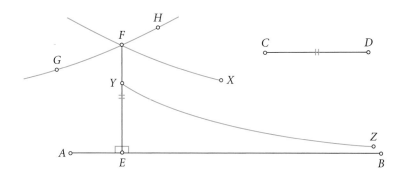

♦ 6.6 Biangles

There are figures in hyperbolic geometry that have no equivalent in euclidean geometry. One of these is a biangle, which can be thought of as a triangle with one vertex at infinity. Some texts call these figures singly asymptotic triangles or trilaterals.

Definition 6.19. If two lines AX and BY are asymptotic parallel in the direction of X and Y and they are cut by transversal AB, then $\square XABY$ is a *biangle*. The finite side $\overline{AB}$ is called the *base* of the biangle.

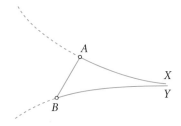

Definition 6.20. Two biangles $\square XABY$ and $\square WCDZ$ are *congruent* if their bases are equal, $\overline{AB} = \overline{CD}$, and their corresponding angles are equal, $\angle XAB = \angle WCD$ and $\angle YBA = \angle ZDC$. We then write $\square XABY \cong \square WCDZ$.

We prove two congruence theorems for biangles, and an analog to the Weak Exterior Angle Theorem (Proposition I.16) for triangles.

Theorem 6.21. [Angle-Base Congruence] Given biangles $\square XABY$ and $\square WCDZ$ with $\angle BAX = \angle DCW$ and $\overline{AB} = \overline{CD}$, then $\square XABY \cong \square WCDZ$.

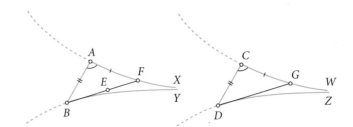

Proof: We are given biangles ☐*XABY* and ☐*WCDZ* with $\overline{AB} = \overline{CD}$ and ∠*BAX* = ∠*DCW*. We must show that the remaining angles are also equal. Assume that ∠*ABY* > ∠*CDZ*. Draw ∠*ABE* = ∠*CDZ*, so ∠*ABE* < ∠*ABY*. Since *BE* lies within the angle of parallelism, by Theorem 6.12, *BE* intersects *AX* at some point *F*. Next, cut $\overline{CG} = \overline{AF}$ and draw $\overline{DG}$. By SAS, △*FAB* ≅ △*GCD* and therefore, ∠*ABF* = ∠*CDG* = ∠*CDZ*. Thus the line *DG* coincides with *DZ*, and so *DZ* intersects *CW*. This provides us with a contradiction and so we deduce that ∠*ABY* = ∠*CDZ*. Therefore, ☐*XABY* ≅ ☐*WCDZ*. □

Theorem 6.22. [Exterior Angle Theorem for Biangles] Given biangle ☐*XABY*, extend *AB* to *C*. The exterior angle ∠*CBY* is greater than the opposite interior angle ∠*CAX*.

Proof: We are given a biangle ☐*XABY* with $\overline{AB}$ extended to *C*. We must show that ∠*CBY* > ∠*CAX*.

Case 1: Assume that ∠*CBY* < ∠*CAX*.

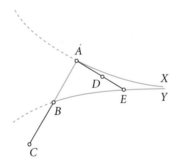

Draw ∠*DAB* = ∠*CBY* and note that ∠*DAB* < ∠*CAX*. By Theorem 6.12, *AD* intersects *BY* at *E* to form △*ABE*. By Proposition I.16, ∠*CBY* > ∠*EAB*. This is a contradiction, so we know that ∠*CBY* ≥ ∠*CAX*.

Case 2: Assume that ∠*CBY* = ∠*CAX*.

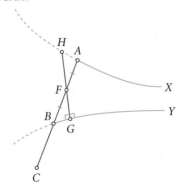

Construct F, the midpoint of $\overline{AB}$. Drop $FG \perp BY$, cut $\overline{AH} = \overline{BG}$, and draw $\overline{HF}$. We must first show that HFG forms a straight line. By Proposition I.13, $\angle XAB + \angle BAH = 180°$ and $\angle YBC + \angle YBA = 180°$. Thus, $\angle BAH = \angle YBA$, and so by SAS, $\triangle AHF \cong \triangle BGF$. Therefore, $\angle BFG = \angle AFH$. By Proposition I.13 again, $\angle BFG + \angle GFA = 180° = \angle AFH + \angle GFA$. Proposition I.14 then implies that GFH is a straight line. By the pair of congruent triangles, $\angle XHG = \angle HGB = 90°$. But by Theorem 6.14, we know that $\angle XHG < 90°$, thus giving us the desired contradiction.

Therefore, it must be true that $\angle CBY > \angle CAX$. □

Theorem 6.23. [Angle-Angle Congruence] Given biangles $\square XABY$ and $\square WCDZ$ with $\angle BAX = \angle DCW$ and $\angle ABY = \angle CDZ$, then $\square XABY \cong \square WCDZ$.

HW due 3/24

▶ **Exercise 6.7.** Prove Theorem 6.23. [Hint: Assume $\overline{AB} \neq \overline{CD}$, cut $\overline{AE} = \overline{CD}$, and draw an asymptotic parallel through E.]

The next theorem has no counterpart for triangles. It says that if a figure looks exactly like a known biangle, then it must be a biangle; i.e., the infinite sides must be asymptotic parallel.

Theorem 6.24. If $\square XABY$ is a biangle and $WCDZ$ is a figure formed so that $\overline{AB} = \overline{CD}$, $\angle BAX = \angle DCW$, and $\angle ABY = \angle CDZ$, then $\square WCDZ$ is a biangle.

▶ **Exercise 6.8.** Prove Theorem 6.24, showing that CW is asymptotic parallel to DZ in the direction of W and Z.

Previously, we defined the angle of parallelism $\Pi(\overline{CD}) = \angle ZCD$ where CZ is asymptotic parallel to AB and CD is perpendicular to AB. Thus $\square ZCDB$ is a biangle. We wish to investigate what happens to the angle of parallelism as the distance from C to the line AB increases. The following corollary to the Exterior Angle Theorem for Biangles 6.22 should confirm your observations from Activity 6.5.

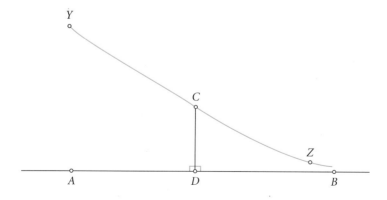

Corollary 6.25. The angle of parallelism decreases as the distance of C from the line AB increases.

▶ **Exercise 6.9.** Prove Corollary 6.25.

With the help of the Poincaré disc model, which we know represents angles correctly, we can compute the angle of parallelism. Since by Theorem 6.15, this angle depends only on the distance from the point to the line, and not on the location of the point or the line, we will choose the line to be a diameter OZ and a point A on the other diameter PQ perpendicular to the first. Let $d = \overline{AO}$, so we want to compute $\Pi(\overline{AO}) = \Pi(d)$. This will be equal to the angle formed by an asymptotic parallel through A to the diameter OZ. We measure this angle by measuring the angle between AO and the tangent line to the arc that represents the asymptotic parallel. Let B be the point where this tangent line intersects the diameter. Then $\Pi(d) = \angle BAO$.

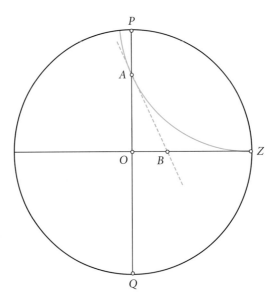

First, note that if O' is the center of the circle that forms the arc representing the hyperbolic line AZ as illustrated on the next page, then since $BA \perp AO'$ and $BZ \perp O'Z$, then $\triangle O'AB \cong \triangle O'ZB$. Thus, $\triangle BAZ$ is isosceles and $\angle BAZ = \angle BZA$.

Note that $\triangle AOB$ and $\triangle BAZ$ are euclidean, not hyperbolic: They are not drawn with hyperbolic lines and must follow the rules of euclidean geometry. We are using them to help us figure out the angles of the hyperbolic biangle formed by the angle of parallelism. Thus we can use the fact (from Proposition I.32) that $\angle ABO = \angle BAZ + \angle BZA = 2\angle BAZ$. Also, $\angle OAB + \angle ABO + 90° = 180°$, so $\angle BAZ = 45° - \frac{1}{2}\angle OAB = 45° - \frac{1}{2}\Pi(d)$. Now use the distance formula for the Poincaré disc model: $d = d(A, O) = \left| \ln\left(\frac{\overline{AQ}}{\overline{AP}} \frac{\overline{OP}}{\overline{OQ}}\right) \right|$. Note that if $\overline{AO} = d$ and the disc has radius r, then $\overline{AQ} = r + d$, $\overline{AP} = r - d$, and $\overline{OP} = \overline{OQ} = r$. Thus we have

$$d = \ln\left(\frac{r+d}{r-d}\right),$$

$$e^d = \frac{r+d}{r-d},$$

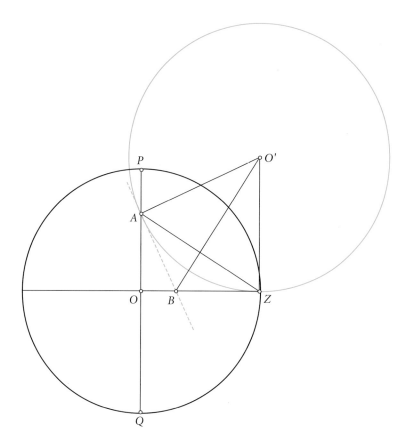

$$d = r\left(\frac{e^d - 1}{e^d + 1}\right).$$

On the other hand, considering $\triangle AOZ$, we see that $\tan(\angle AZO) = \tan(45° - \frac{1}{2}\Pi(d)) = \frac{\overline{AO}}{\overline{OZ}} = \frac{d}{r}$. Thus,

$$\tan\left(45° - \frac{1}{2}\Pi(d)\right) = \frac{e^d - 1}{e^d + 1},$$

and so we have:

Theorem 6.26. $\Pi(d) = 90° - 2\arctan\left(\frac{e^d-1}{e^d+1}\right)$.

If you remember your calculus, two simple computations give the following corollary:

Corollary 6.27. $\lim_{d\to\infty} \Pi(d) = 0$ and $\lim_{d\to 0^+} \Pi(d) = 90°$.

The result above can be used to construct hyperbolic triangles with any desired angle sum greater than 0 and less than 180°. To form a triangle with angle sum less than some chosen value of ϵ, draw AB

and *CD*, two perpendicular lines intersecting at *D*. Place point *C* so that $\Pi(\overline{CD}) = \frac{\epsilon}{4}$. Cut $\overline{DE} = \overline{DC}$ and draw $\overline{CE}$. Since *CE* intersects *AB*, $\angle DCE < \Pi(\overline{CD}) = \frac{\epsilon}{4}$. Since $\triangle CDE$ is isosceles, $\angle DEC = \angle DCE$. Now cut $\overline{DF} = \overline{DE}$ on the other side of *D* from *E* and draw $\overline{FC}$. Since $\triangle FCD \cong \triangle ECD$, we have $\triangle FCE$ with angle sum less than ϵ.

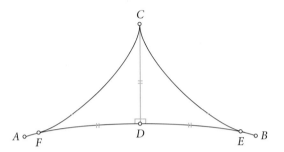

♦ 6.7 Divergent parallels

Recall that a divergent parallel is one of the infinitely many other parallel lines that is not one of the two asymptotic parallels. In this section, we discuss properties of these lines. The first theorem is the counterpart to Theorem 6.16, but for divergent parallels:

Theorem 6.28. If a line is divergent parallel to a given line at one point, then it is divergent parallel to that line everywhere.

▶ **Exercise 6.10.** Prove Theorem 6.28, using Theorem 6.16. Assume that we are given line *AB* and point *C* not on *AB* and let *CE* be divergent parallel to *AB*. Choose a point *F* at random on *CE*. Show that *FE* is divergent parallel to *AB*.

The following theorem relates the angle relationships (alternate interior angles are equal, the exterior angle is equal to the opposite interior angle, or the interior angles on one side of the transversal sum to 180°) we studied for euclidean parallel lines to the hyperbolic situation. Euclid's Propositions I.27 and I.28, which are also true in neutral and hyperbolic geometry, guarantee that if any of the eight angle relationships of Lemma 2.26 are true, then the lines must be parallel. The following theorem shows that they must be divergent parallel.

Theorem 6.29. If two lines are cut by a transversal so that the alternate interior angles are equal (or if the exterior angle is equal to the opposite interior angle or if the interior angles on one side of the transversal sum to 180°), then the lines are divergent parallel.

▶ **Exercise 6.11.** Prove Theorem 6.29.

The next two theorems are partial analogs of Euclid's Proposition I.29, which said that if two lines are parallel, then the eight angle relationships must be true. In particular, a line can be drawn perpendicular to both of the parallel lines. This is not true in general in hyperbolic geometry. However, for two divergent parallel lines, one can find one point where they do have a common perpendicular, and at this point, all eight angle relationships will hold true. However, when one moves away from this point, these properties will no longer hold. The proof of the first of these theorems is quite technical.

Theorem 6.30. If two lines are divergent parallel, then they have a common perpendicular.

Proof: AB and CD are divergent parallel. Choose E, F on AB and drop $EG \perp CD$ and $FH \perp CD$.

Case 1: $\overline{EG} = \overline{FH}$.

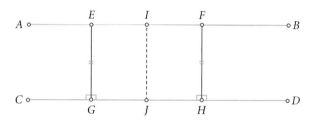

In this case, EGHF is a saccheri quadrilateral, and so the midline $IJ \perp AB$ and $IJ \perp CD$ by Corollary 5.6.

Case 2: Without loss of generality, we may assume that $\overline{EG} > \overline{FH}$.

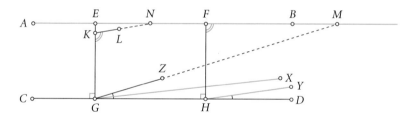

Cut $\overline{GK} = \overline{FH}$ and construct $\angle LKG = \angle BFH$. Draw GX asymptotic parallel to AB and HY asymptotic parallel to AB, both in the B direction, and construct $\angle ZGD = \angle YHD$. By Theorem 6.17, GX is asymptotic parallel to HY in the B direction. Therefore, $\sqsubset XGHY$ is a biangle and by Theorem 6.22, $\angle YHD > \angle XGD$. Thus $\angle ZGD > \angle XGD$, and so by Theorem 6.12, GZ must intersect AB at point M. Note that $\sqsubset MFHY$ is a biangle and $\angle KGD = \angle FHD = 90°$ and therefore $\angle KGM = \angle FHY$. Thus by Theorem 6.24, $\sqsubset LKGM$ is a biangle so KL is asymptotic parallel to GM. By Pasch's Theorem 2.11, KL must intersect $\triangle MEG$ along $\overline{EM}$ or along $\overline{GM}$. Since KL is parallel to GM, KL must intersect $\overline{EM}$ at point N.

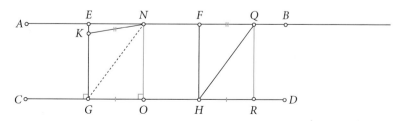

Drop $NO \perp CD$, cut $\overline{FQ} = \overline{KN}$ and $\overline{HR} = \overline{GO}$, and draw $\overline{GN}$, $\overline{QR}$, and $\overline{HQ}$. By SAS, $\triangle KGN \cong \triangle FHQ$ and so $\overline{NG} = \overline{QH}$ and $\angle NGK = \angle QHF$. Therefore, $\angle NGD = \angle QHD$. Thus $\triangle NGO \cong \triangle QHR$ by SAS.

Corresponding parts of congruent triangles gives $QR \perp CD$ and $\overline{QR} = \overline{NO}$. Thus, $NORQ$ is a saccheri quadrilateral, and the midline is perpendicular to both AB and CD. □

Corollary 6.31. If two lines are divergent parallel, then the parallel angle relationships hold at one point.

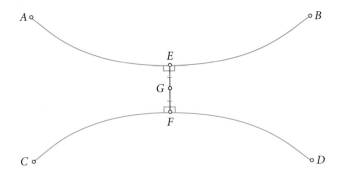

Proof: Let AB and CD be divergent parallel, with common perpendicular EF. Let G be the midpoint of $\overline{EF}$.

▶ **Exercise 6.12.** Finish the proof of Corollary 6.31 by showing that any transversal through G that intersects AB and CD must form equal alternate interior angles, and thus the other seven angle relationships of Lemma 2.26 will also be true. Note that you must first show that if such a line intersects one of AB and CD, then it must intersect the other.

The next theorem justifies the term "divergent" for this type of parallel:

Theorem 6.32. Divergent parallels get farther apart on either side of the common perpendicular.

▶ **Exercise 6.13.** Prove Theorem 6.32.

♦ 6.8 Triangles in hyperbolic space

Recall that in hyperbolic geometry, the negation of Wallis' Alternate Postulate 5.13 shows that noncongruent similar triangles do not exist. This gives an additional congruence theorem, which is only true in noneuclidean geometry, and we give a direct proof of this theorem.

Theorem 6.33. [AAA Congruence] If $\triangle ABC$ and $\triangle DEF$ have $\angle ABC = \angle DEF$, $\angle BAC = \angle EDF$, and $\angle ACB = \angle DFE$, then the triangles are congruent.

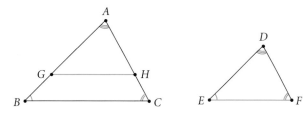

Proof: We are given △ABC and △DEF with ∠ABC = ∠DEF, ∠ACB = ∠DFE, and ∠BAC = ∠EDF. Assume that $\overline{AB} > \overline{DE}$. Cut $\overline{AG} = \overline{DE}$ and $\overline{AH} = \overline{DF}$, so △AGH ≅ △DEF by SAS. Then ∠AGH = ∠DEF = ∠ABC and ∠AHG = ∠DFE = ∠ACB. Since, by supplementary angles, ∠AGH + ∠BGH = 180° and ∠AHG + ∠CHG = 180°, it follows that ∠ABC + ∠BGH = 180° and ∠ACB + ∠CHG = 180°. Thus ∠ABC + ∠ACB + ∠BGH + ∠CHG = 360°. This contradicts Corollary 6.5. We deal with the case $\overline{AB} < \overline{DE}$ similarly. Therefore, $\overline{AB} = \overline{DE}$, and △ABC ≅ △DEF by SAS. □

▶ **Exercise 6.14.** Prove that the exterior angle of a triangle is greater than the sum of the opposite interior angles.

▶ **Exercise 6.15.** Prove that the summit of a saccheri quadrilateral is longer than the base, and the midline is shorter than the legs.

This last exercise gives us a way of a constructing a triangle that cannot be circumscribed (recall F. Bolyai's Alternate Postulate 5.18).

Theorem 6.34. There are triangles that cannot be circumscribed.

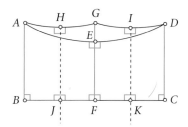

Proof: First construct a saccheri quadrilateral ABCD with midline $\overline{EF}$. By Exercise 6.15, $\overline{EF} < \overline{AB}$, so we can extend EF and cut it at G so that $\overline{GF} = \overline{AB}$. Draw $\overline{GA}$ and $\overline{GD}$. We will show that △GAD has the desired property. Note that ABFG and GFCD are saccheri quadrilaterals. Construct their midlines $\overline{HJ}$ and $\overline{IK}$. By Corollary 5.6, $HJ \perp AG$, $HJ \perp BC$, $IK \perp GD$, and $IK \perp BC$. By Proposition I.27, $HJ||IK||EF$, so these lines do not intersect. Considering the proof of Theorem 3.26, we see that we cannot draw a circle through the vertices of △GAD. □

In hyperbolic geometry, one has proper triangles, with which we are quite familiar, and the biangles discussed in Section 6.6. Biangles are sometimes called *singly asymptotic triangles* and can be viewed as triangles with one vertex at infinity.

▶ **Exercise 6.16.** An *isosceles biangle* is one in which the two base angles are equal. Prove that the perpendicular bisector of the base is asymptotic parallel to both of the other sides.

▶ **Exercise 6.17.** If two isosceles biangles have equal bases, prove that they are congruent.

▶ **Exercise 6.18.** Prove the following variation of Pasch's Theorem 2.11 for biangles: Let $\square XABY$ be a biangle and let ℓ be a line that intersects $\overrightarrow{AX}$ at some point D. Then ℓ must intersect either $\overline{AB}$ or $\overrightarrow{BY}$.

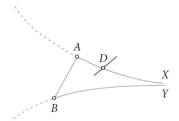

▶ **Exercise 6.19.** Prove that if $\square XABY$ and $\square WCDZ$ are biangles with $\angle XAB = \angle WCD$ and $\overline{AB} > \overline{CD}$, then $\angle ABY < \angle CDZ$.

▶ **Exercise 6.20.** Prove that if $\square XABY$ and $\square WCDZ$ are biangles with $\angle XAB = \angle WCD$ and $\angle ABY < \angle CDZ$, then $\overline{AB} > \overline{CD}$.

We can adapt the construction of the inscribed triangle (Corollary 3.25) to the case of a biangle as follows. Let $\square\ XABY$ be a biangle. Bisect $\angle XAB$ and $\angle YBA$. By Exercise 6.18 these angle bisectors will meet at a point C. Drop $CD \perp AX$ and $CE \perp BY$ and draw $\overline{DE}$. Drop $CF \perp DE$ and extend this line to Z. CZ is called the *midline* of the biangle, or more simply, the midline of the two asymptotic parallels AX and BY.

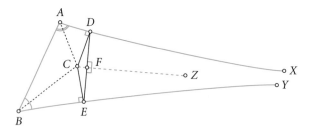

▶ **Exercise 6.21.** Prove that CZ is asymptotic parallel to both AX and BY. (Hint: Prove that $\square XDEY$ is an isosceles biangle.)

▶ **Exercise 6.22.** Prove that any point on the midline CZ is equidistant from AX and BY by choosing any point G on CZ and dropping $GH \perp AX$ and $GI \perp BY$ and then showing that $\overline{GH} = \overline{GI}$.

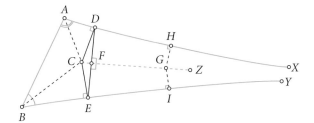

▷ **Activity 6.8.** Using geometric software and the Poincaré disc model, draw a biangle and construct the midline. Confirm the finding of Exercise 6.22.

▷ **Activity 6.9.** Using geometric software and the Poincaré disc model, draw a line *WZ* and a point *A* not on that line. Construct *AX* through *A* asymptotic parallel to *WZ* in the *W*-direction and another line *AY* asymptotic parallel to *WZ* in the *Z*-direction.

The figure of Activity 6.9 is called a *doubly asymptotic triangle*, a triangle with two vertices at infinity. By Exercise 6.23, we can denote this doubly asymptotic triangle as $\triangle AWZ$ without ambiguity.

▶ **Exercise 6.23.** Prove that point *A* and line *WZ* define a unique doubly asymptotic triangle.

In the construction of the doubly asymptotic triangle in Activity 6.9, we began with a line and a point not on that line. Alternatively, we could begin with the angle $\angle XAY$ and try to construct a line *WZ* that is asymptotic parallel to *AX* in the *X*-direction and simultaneously asymptotic parallel to *AY* in the *Y*-direction. This is called the *enveloping line* of the angle $\angle XAY$ and the next rather complicated theorem explains how to construct this line.

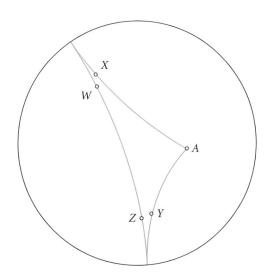

Theorem 6.35. Given any angle ∠*XOY*, there is a line that is asymptotic parallel to both *OX* and *OY*.

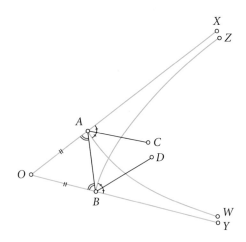

Proof: We are given ∠*XOY*. Choose point *A* on *OX* and cut $\overline{OB}$ on *OY* so that $\overline{OA} = \overline{OB}$. Draw $\overline{AB}$ and note that by Proposition I.5, ∠*XAB* = ∠*YBA*. Draw *AW* asymptotic parallel to *BY* and *BZ* asymptotic parallel to *AX*. Then the biangles ⊏ *XABZ* ≅ ⊏ *YBAW* by Theorem 6.21. Thus, ∠*ABZ* = ∠*BAW*, and it follows that ∠*YBZ* = ∠*XAW*. Let *AC* bisect ∠*XAW* and *BD* bisect ∠*YBZ*. Then ∠*DBY* = ∠*CAW*.

Case 1: AC intersects *BD* at *E*.

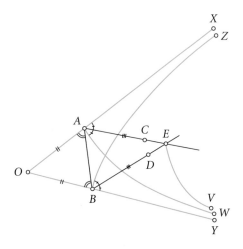

In this case, since we know that ∠*EAB* = ∠*EBA*, then $\overline{EA} = \overline{EB}$. Draw *EV* asymptotic parallel to *BY* and note that *EV* is asymptotic parallel to *AW* by Theorem 6.17. Thus, ⊏*VEAW* ≅ ⊏*VEBY* by Theorem 6.21 and so ∠*VEA* = ∠*VEB*. This gives us a contradiction, so we conclude that *AC*||*BD*. It remains to determine if *AC* is asymptotic parallel to *BD* or divergent parallel.

Case 2: AC is asymptotic parallel to BD.

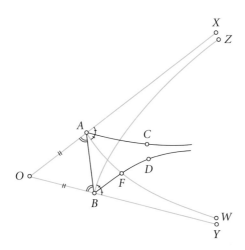

By Pasch's Theorem for biangles (Exercise 6.18), $\overrightarrow{AW}$ intersects $\overrightarrow{BD}$ at some point F. By vertical angles, $\angle AFD = \angle BFW$ and so $\square CAFD \cong \square YBFW$ by Theorem 6.23. Therefore, $\overline{BF} = \overline{AF}$, and so $\angle FAB = \angle FBA = \angle ZBA$. This again gives us a contradiction, and we conclude that AC is not asymptotic parallel to BD.

Case 3: Therefore, AC is divergent parallel to BD.

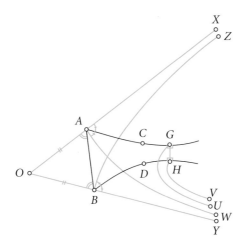

By Theorem 6.30, AC and BD have a common perpendicular GH. We want to show that GH is asymptotic parallel to AX and to BY. Assume that it is not asymptotic parallel to BY and construct GU and HV asymptotic parallel to BY. By Theorem 5.4, note that $\overline{AG} = \overline{BH}$. Therefore, $\square UGAW \cong \square VHBY$ by Theorem 6.21, and so $\angle AGU = \angle BHV$. Note that $\angle UGH = 90° - \angle AGU$, and so $\angle UGH + \angle VHG = 180°$. This contradicts Theorem 6.22, so we conclude that GH must be asymptotic parallel to BY. Similarly, HG is asymptotic parallel to AX. □

▷ **Activity 6.10.** Using geometric software and the Poincaré disc model, construct an angle and its enveloping line.

▶ **Exercise 6.24.** Given an acute $\angle ABC$ show that there is a point D on BC and a line $DE \perp BC$ so that DE is asymptotic parallel to BA in the A-direction.

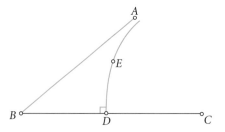

Let $\triangle AWZ$ be a doubly asymptotic triangle and choose a point B on WZ. Draw $\overline{AB}$, which makes two biangles ⌐ $XABW$ and ⌐ $YABZ$. Let CU be the midline for ⌐$XABW$ and DV be the midline for ⌐$YABZ$.

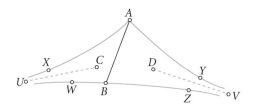

▶ **Exercise 6.25.** Show that CU intersects DV at some point E. Drop $EF \perp AX$, $EG \perp AY$, and $EH \perp WZ$. Show that $\overline{EF} = \overline{EG} = \overline{EH}$. Show that A lies on EH and that EH bisects $\angle XAY$. Finally, explain how to draw a circle inscribed in the doubly asymptotic triangle $\triangle AWZ$.

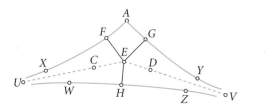

6.8. TRIANGLES IN HYPERBOLIC SPACE

An *ideal triangle* or *triply asymptotic triangle* has all three vertices at infinity. Note that it can be proven that all ideal triangles are congruent. Below is a picture of an ideal triangle drawn in the Poincaré disc model:

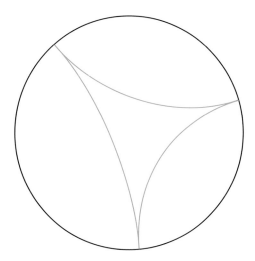

The construction of an ideal triangle involves finding the equivalent to the enveloping line, but for asymptotic parallels rather than for intersecting lines.

Theorem 6.36. Given two asymptotic parallel lines, there is a third line asymptotic parallel to both of the given lines at opposite ends.

Proof:

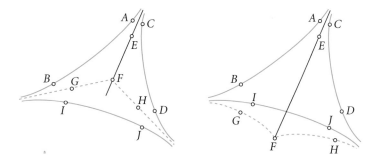

Let *AB* and *CD* be asymptotic parallel in the *A*-direction. Let *EF* be the midline of *AB* and *CD*. Draw *FG* asymptotic parallel to *AB* in the *B*-direction and *FH* asymptotic parallel to *CD* in the *D*-direction. Let *IJ* be the enveloping line for ∡*GFH*. Then by Theorem 6.17, *IJ* is asymptotic to both *AB* and *CD*. □

▷ **Activity 6.11.** Using geometric software and the Poincaré disc model, construct a triply asymptotic triangle.

▶ **Exercise 6.26.** Let *AB*, *CD*, and *EF* form an ideal triangle. Construct the midlines for each pair of sides of this triangle. Show that the three midlines meet at a single point equidistant from the three lines, and thus we can inscribe a circle inside the ideal triangle.

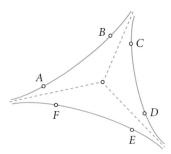

Finally we return to saccheri quadrilaterals in hyperbolic geometry. From Theorem 6.7 we know that the summit angles of any saccheri quadrilateral in hyperbolic geometry must be acute, and from Exercise 6.15 we know that the summit is longer than the base and the midline is shorter than the legs. The base and the summit are divergent parallel, as are the legs. We proved two congruence theorems for saccheri quadrilaterals, Theorem 5.12 (Base-Leg) and Theorem 5.13 (Base-Midline) which are true in neutral geometry and therefore true in both euclidean and hyperbolic geometry. Below are two other congruence theorems for saccheri quadrilaterals, Summit-Angle and Base-Angle, that are only true in hyperbolic geometry:

Theorem 6.37. [Summit-Summit Angle] If two saccheri quadrilaterals are such that the summit of one is equal to the summit of the other and the summit angles of the two saccheri quadrilaterals are equal, then the saccheri quadrilaterals are congruent.

Theorem 6.38. [Base-Summit Angle] If two saccheri quadrilaterals are such that the base of one is equal to the base of the other and the summit angles of the two saccheri quadrilaterals are equal, then the saccheri quadrilaterals are congruent.

▶ **Exercise 6.27.** Prove Theorem 6.37.

▶ **Exercise 6.28.** Prove Theorem 6.38.

♦ 6.9 Projects

Project 6.1. Research the life and work of Carl Friedrich Gauss.

Project 6.2. Research the life and work of Janos Bolyai.

Project 6.3. Research the life and work of Nikolai Lobachevskii.

Project 6.4. Research the life and work of Henri Poincaré.

Project 6.5. Build a model of the hyperbolic plane, either by crocheting or by quilting.

Project 6.6. Report on the upper half-plane model for hyperbolic geometry. Illustrate triangles, angle sums, asymptotic and divergent parallels, biangles, doubly asymptotic triangles, and ideal triangles in this model.

Project 6.7. Experiment with other models like the one you built in Activity 6.1, trying eight or nine equilateral triangles at each vertex, or try taping two hexagons and one heptagon together at each vertex. (This last suggestion is sometimes called the hyperbolic soccer ball. Why?)

Project 6.8. Prove that the distance measure given for the Poincaré disc model is indeed a metric by proving that the triangle inequality holds in general.

Project 6.9. Prove that the postulates of neutral geometry are true in the Poincaré disc model.

7. Other Geometries

♦ 7.1 Exploring the geometry of a sphere

In this section we explore the geometry of a sphere. Consider only the surface of a sphere: You are not allowed to dig tunnels or fly in space but must remain on the surface. In order to do geometry on a sphere, we must first consider what objects should take the place of straight lines, since clearly there is nothing straight about a sphere. In euclidean geometry, a line is sometimes called the path that forms the shortest distance between two points, though the closest Euclid comes to this idea is Proposition I.20, the triangle inequality. In general, the concept of a straight line on an arbitrary surface is somewhat more complicated and subtle, but the idea of finding the shortest path is a good a place to start.

▷ **Activity 7.1.** Find a ball and a piece of string. Any ball will do, but try to find one at least six inches in diameter. Choose two points on the ball. Use the string to help find the shortest path between the points. Repeat several times. Choose points that are close together. Choose points that are far apart. When you think you have the shortest distance marked with the string, rotate the ball so you are looking at it from a different angle. Be sure your path still looks like the shortest distance.

Note that a globe of the earth is ideal for this activity if you can find one. Instead of choosing random points, find your favorite major cities. What is the shortest route from Washington, D.C., U.S.A. to Beijing, China? From Sydney, Australia to Buenos Aires, Argentina? What path would an airplane take?

Now try extending your routes so they go the rest of the way around the ball. How would you describe these paths?

▷ **Activity 7.2.** For those interested in more hands-on explorations, think about all of the definitions and constructions from euclidean and hyperbolic geometry. Use your ball and string to decide what a circle should look like on the sphere and how it is different from a line. How would you define a line segment? A triangle? What would a saccheri quadrilateral look like? Can you construct midpoints? Perpendiculars? Parallels? Once you have some idea of how the geometry works, or if you don't mind some hints and answers, read on.

In the first activity, you should have found, by pulling a string taut from one point to another, that the shortest path between two points on the sphere seems to be a *great circle*: a circle like the equator or the lines of longitude that pass around the fattest part of the sphere. Another way of describing such a great circle is to say that it will be the intersection of the sphere with any plane that goes through the center of the sphere. Proving that this is indeed the shortest path on the sphere through two points requires techniques from differential calculus. *Differential geometry* is the field in mathematics in which calculus is joined with geometry and there are many excellent texts if you want to explore this area.

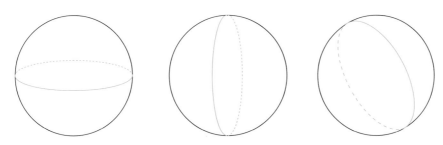

In *Experiencing Geometry on Plane and Sphere*, David Henderson and Daina Taimina suggest some other experiments to help convince you that such great circles act as the equivalent of a straight line on a sphere. Some of these are included in the following activity:

▷ **Activity 7.3.** Take a ball at least 3 inches in diameter.
   a. Stretch a rubber band around a ball. If the rubber band is placed around the sphere anywhere except along a great circle, it will slip off. Try to find an explanation for this phenomenon.
   b. Roll a ball along a chalked surface. Consider the imprint that the chalk makes on the ball. Try to find an explanation.
   c. Take a strip of paper about half an inch wide and wrap it as flat as you can around the sphere, first along a circle such as the Tropic of Cancer and then along the equator. Compare how the ribbon must be pleated or creased. Try to find an explanation.

A further thought experiment suggested by Henderson and Taimina is to imagine an insect walking around the sphere. If the insect walks along the Tropic of Cancer or the Arctic Circle, the legs on the equator side will have to take longer steps than the legs on the North Pole side. However, along the equator, and only along the equator, the legs on either side will take steps of equal length. All of these experiments should help convince you that great circles act as lines on the sphere.

Once you are accustomed to the idea of great circles as lines, rethink all of your geometric intuition. Explore this new geometry. Think about which postulates and propositions from euclidean or hyperbolic geometry should be true. While experimenting with a ball is a good beginning, you may also want to download the program *Spherical Easel* developed by David Austin and William Dickinson. This program allows us to draw on a virtual sphere and to manipulate the results.

▷ **Activity 7.4.** Considering great circles as lines on the sphere, consider Postulate 1: Any two points determine a unique line. Is this true for your lines on the ball? Are there any exceptional points?

▷ **Activity 7.5.** Is Postulate 2 true on the sphere? That is, can any line segment be extended indefinitely? Explain.

▷ **Activity 7.6.** Is the Plane Separation Property Postulate 8 true on the sphere? Is the Line Separation Property true on the sphere? Explain.

We have decided to consider a great circle as the natural counterpart of a straight line. Two points on the sphere are called *antipodal* or *opposite* if they form the intersection of the sphere with any diameter, like the North and South Poles. Two distinct points will determine a unique line (great circle) as long as they are not antipodal. We can justify this using euclidean geometry since the two points and the center of the sphere are three noncollinear points that determine a unique euclidean plane, and great circles are the intersection of the sphere with a euclidean plane through the center of the sphere. If the two points are antipodal, the center of the sphere is collinear with them and a unique plane is not established. Thus, two antipodal points determine infinitely many lines, like the lines of longitude from the North Pole to the South Pole.

Another peculiarity of spherical geometry concerns length. Euclidean and hyperbolic lines can be extended indefinitely. One can follow a great circle indefinitely but not without repeatedly covering the same ground. Lines have finite length in spherical geometry and there is, in a sense, a natural way to measure that length. Thus on the sphere there is a canonical standard for length: the circumference of your ball, which depends on the radius. It is striking that all of the lines defined by the great circles have the same length, $2\pi r$ for a ball with radius r.

In contrast, consider that in euclidean and hyperbolic geometry, as well as on the sphere, there is a natural idea of angle measure, based on the idea of a straight line. If we assign the straight line an angle measure (180° is as good a number as any), then perpendicular lines form an angle measuring half of that of the straight line, and so on. These angle measures carry over to spherical geometry as well. Thus, we have a natural unit of measure for angles in each of the geometries we have studied, but only on the sphere have we found a canonical measure for length.

If we consider a line or great circle as the intersection of the sphere with a plane that passes through the center of the sphere, then there is a unique diameter that is perpendicular to the plane. The intersection of this diameter with the sphere is a pair of antipodal points called the *poles* of the line. For example, the North and South Poles are the poles of the line formed by the equator.

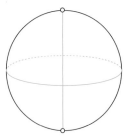 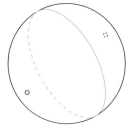

There is an additional difficulty to be navigated when attempting to define line segments: the Line Separation Property does not hold. A point on a great circle does not separate the line into two pieces.

Given two points A and B on a great circle, these points break this circle up into two pieces we call *arcs*. In our treatment of euclidean geometry, we defined a line segment $\overline{AB}$ as the set of points lying between A and B. In spherical geometry, it takes two points to break the great circle into two arcs. We can no longer simply describe a line segment by $\overline{AB}$ but must indicate which of the two arcs we intend. In the illustration below, A and B divide the circle into two arcs, one containing C and the other containing D. We will denote the two arcs as $\widehat{ACB}$ and $\widehat{ADB}$. In this situation, we say that A and B *separate* C and D. In the next section, we will give formal axioms for the primitive term "separate", so for now we merely ask you to familiarize yourself with the concept.

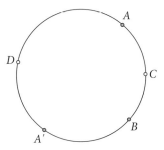

If A and B are not antipodal, we can define the line segment $\widehat{AB}$ to be the shorter arc of the great circle through A and B. An alternate definition is to define $\widehat{AB}$ to be the arc that does not contain the point A' antipodal to A. Note that there is a maximal length: $\widehat{AB}$ must have length less than πr. If A and A' are antipodal points that lie on a given great circle, they divide the great circle into two arcs that are equal in length. We can denote these two arcs as $\widehat{ABA'}$ and $\widehat{ADA'}$. Note that in this case, $\widehat{ABA'} = \widehat{ADA'} = \pi r$.

▷ **Activity 7.7.** Let A, B, and C be three points on a line on the sphere so that none of these are antipodal to any of the others. Is it always true that $\widehat{AB} + \widehat{BC} = \widehat{AC}$? Explain.

▷ **Activity 7.8.** Let A, B, and C be three points on a line on the sphere so that none of these are antipodal to any of the others. If $\widehat{AB} = \widehat{BC}$, is B necessarily the midpoint of $\widehat{AC}$? Explain.

▷ **Activity 7.9.** Consider two lines (great circles) on your sphere. What can you say about their intersection? Is the Parallel Postulate 5 true on the sphere? Is Playfair's Postulate true on the sphere? Explain.

We are thus faced with a geometry in which several of the neutral postulates fail dramatically. This is another type of noneuclidean geometry, in a sense, even more noneuclidean than hyperbolic geometry. The Greeks were quite familiar with the geometry on a sphere. Euclid wrote a text on astronomy, his *Phaenomena*, which includes the basics of spherical geometry. He defines a great circle as the intersection of the sphere with a plane through the center of the sphere.

Another Greek mathematician who studied spherical geometry was Menelaus of Alexandria [ca. 70–140], who wrote a book called *Sphaerica* devoted to the theory of spherical triangles, in which he develops the theory of spherical geometry in terms analogous to Euclid's treatment of the plane in *The Elements*. The Indian astrologers also knew a great deal about spherical geometry and trigonometry. Thus, in many cultures, spherical geometry was never seen as intractable. Oddly enough, during the great debates about the validity of noneuclidean geometry, no one seems to have mentioned the fact that we are all walking about on a noneuclidean space because of the failure of several of the euclidean postulates.

A further question is what a circle would look like on the sphere. This is more easily answered. Choose a point on the surface of the sphere and a radius, represented as a length of string. A circle is defined to be the set of points a given distance away from the chosen point (Definition 2.15). Using the string to measure this distance gives us a figure that is recognizable as a circle:

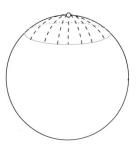

Such a circle can also be described as the intersection of the sphere with a plane that does not necessarily pass through the center of the sphere.

▷ **Activity 7.10.** In neutral geometry, a circle is uniquely described by its center and radius. Show that on the sphere the same circle can be given by two different centers and two different radii. Describe the relationship between the centers and the radii.

▷ **Activity 7.11.** Considering circles as described above, is Postulate 3 true on the sphere? Explain. What bounds should be set on the possible radii a circle on the sphere can have?

Consider a segment $\widehat{AB}$ and construct a circle with center A that passes through point B and a second circle with center B that passes through A. Recall that this construction was used in Euclid's first proposition to construct an equilateral triangle. At the time, we noted the use of the Circular Continuity Principle to assert that the two circles intersect, after showing that the second circle connected a point outside the first circle and a point inside the circle. In spherical geometry, the Circular Continuity Principle remains true, but some unexpected developments occur. Consider the illustrations below of the first circle, of the second circle, and finally of their relationship:

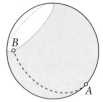 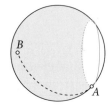 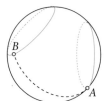

If the length of $\widehat{AB}$ is greater than one-third of the circumference of the sphere, the two circles will not intersect. If $\widehat{AB} = \frac{2\pi r}{3}$, the circles are tangent. We get two points of intersection only if one circle contains both a point inside and a point outside the other circle, which implies that $\widehat{AB} < \frac{2\pi r}{3}$. If the radii of the circles are too big, this won't happen.

▷ **Activity 7.12.** Give an example to show that, while one always can drop a perpendicular line from a given point to a given line, this perpendicular need not be unique.

▷ **Activity 7.13.** Is the Exterior Angle Theorem (Proposition I.32) true in spherical geometry? Try to find a counterexample on your ball. What about the Weak Exterior Angle Theorem (Proposition I.16)?

▷ **Activity 7.14.** Find two triangles on the sphere which are not congruent but satisfy the AAS criterion.

▷ **Activity 7.15.** Construct a saccheri quadrilateral on your ball. What can you say about the summit angles?

▷ **Activity 7.16.** Experiment with triangles on your ball. What angle sums can you form? Find maximum and minimum values for the angle sum of a triangle.

There is a figure in spherical geometry with no euclidean equivalent: a *lune* (which some books call a biangle or a bilateral) is the figure one gets from intersecting two great circles at antipodal points. Its double or shadow lies on the opposite side of the sphere:

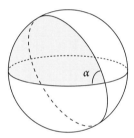

The area of the lune is determined by the angle α, which identifies how much of the sphere the lune covers. The standard formula for the surface area of a sphere is $A = 4\pi r^2$, where r is the radius of the sphere. We will use these concepts to find the area of some spherical triangles.

▷ **Activity 7.17.** Consider the triangle formed on the surface of a sphere by the equator and two longitudinal circles meeting at the north pole at an angle of 90°. This spherical triangle thus has three angles, each measuring 90°. Find the sum of the angles and the area of this spherical triangle (in terms of the radius r).

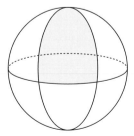

▷ **Activity 7.18.** Consider the spherical triangle formed on the surface of a sphere by the equator and two longitudinal circles meeting at the north pole at an angle of 30°. Find the sum of the angles and the area of this spherical triangle.

▷ **Activity 7.19.** Consider the spherical triangle formed on the surface of a sphere by the equator and two longitudinal circles meeting at the north pole at an angle of 120°. Find the sum of the angles and the area of this spherical triangle.

▶ **Exercise 7.1.** Using the examples of Activities 7.17–7.19, conjecture a formula for the area of a spherical triangle in terms of the angle sum and r, the radius of the sphere.

▶ **Exercise 7.2.** Use your conjecture of Exercise 7.1 to find the area of a spherical triangle with angles 45°, 60°, 90°.

▶ **Exercise 7.3.** Use your conjecture of Exercise 7.1 to find the area of a spherical triangle with angles 61°, 62°, 63°.

◆ 7.2 Elliptic geometry

In this section, we study the geometry of the sphere more formally. We begin with constructing a model for this geometry that eliminates some of the difficulties we have encountered in Section 7.1. After developing this model, we continue with a formal treatment of the axioms for elliptic geometry and some of the more significant elementary theorems.

◇ 7.2.1 The projective plane

One of several difficulties with spherical geometry is the failure of Postulate 1: Through any two points there exists a unique straight line. This postulate is not true if the points happen to be antipodal. Furthermore, a circle on the sphere can be described as having either of two centers, which are antipodal points. In many of the activities of Section 7.1, you should have found that antipodal points tend to give exceptional behavior.

Felix Klein suggested a way to avoid this difficulty by using the *single elliptic model*, obtained by identifying opposite points on the sphere. Think about taking a sphere and gluing each point to its antipode. This procedure can be carried out in stages: first, mark at least three points on the sphere and their antipodes. Cut the sphere in half along a great circle, avoiding the marked points. We now have two hemispheres. Turn one inside out and rotate it so the points are aligned and glue the lower hemisphere to the upper hemisphere. The points on the dividing circle remain to be glued together. This cannot be done in the three-dimensional space in which we live, so we have to remember to think of the pairs of opposite points as being identified. The object we have (almost) constructed is called a *projective plane*.

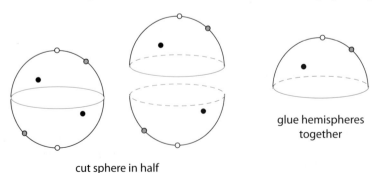

cut sphere in half

glue hemispheres together

It is an interesting exercise to make a disc out of clay and try to identify opposite points along the equator circle that forms the boundary. Alternatively, cut a circle of cloth and try to sew the equator points together so each point on the equator is sewn to its antipode, for example, A to A' and B to B' below, etc.

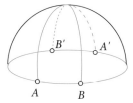

In this model, lines are represented by arcs of great circles drawn on the hemisphere that end at a pair of antipodal points. There are, of course, many lines through any point, including a pair of antipodal points that are considered to be a single point:

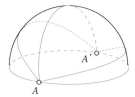

If we identify pairs of antipodal points on any great circle on the sphere, we get a circle with half of the original circumference:

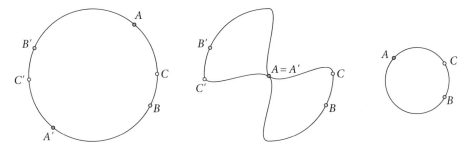

Since all great circles on the sphere have length $2\pi r$, all lines on the projective plane have length πr.

◇ 7.2.2 Formal elliptic geometry

The formal geometry of this modified sphere or projective plane is called *single elliptic geometry* and is identical to the spherical geometry we have discussed except the difficulties posed by pairs of antipodal points have been removed. We wish to devise postulates that describe the behaviors we observed on the sphere, but suitably modified for the projective plane model. Remember that what seems obvious on the model must be postulated or proven in the formal system. We begin our treatment of elliptic geometry by assuming that our primitive terms of points, lines, and planes exist. We picture the plane as the hemispherical projective plane described above, the points as points on this projective plane, and the lines as arcs through a pair of antipodal points on the boundary of the hemisphere. Most of the postulates carry over from neutral geometry with suitable modifications.

Postulate 1. Given two distinct points, there exists exactly one line through them.

Note that the advantage to working with single elliptic geometry lies in Postulate 1: on the sphere, any two points determine at least one line, rather than a unique line. In particular, on the sphere there are infinitely many lines through a pair of antipodal points, but if the points are not antipodal, then there is exactly one line through them. In single elliptic geometry, any two points determine a unique line. We picture the lines of Postulate 1 as circles on the sphere with all pairs of antipodal points identified. As we remarked above, identifying antipodal points on a great circle gives us a another circle with half of the circumference of the original. When we talk of lines in elliptic geometry, this is a convenient visual image to bear in mind.

Postulate 2 refers to line segments which we defined in euclidean geometry in terms of the primitive notion of betweenness with the properties specified in Betweenness Postulate 7. Betweenness fails dramatically in elliptic geometry. We replace the entire concept with one based on a different primitive term.

In elliptic geometry, no single point separates a line (think of a great circle with antipodal points identified). It takes two points to break a line into two *arcs*. In the illustration below, A and B divide the line into two arcs, one containing C and the other containing D. We denote the two arcs as $\overgroup{ACB}$ and $\overgroup{ADB}$ and say that A and B *separate* C and D. In place of the concept of betweenness used in neutral geometry, we accept the idea of *separation* as a primitive term and postulate the following properties of this separation relation:[1]

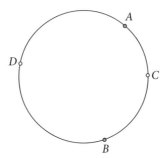

Elliptic Postulate 7. [Separation]

1. If A and B separate C and D, then A, B, C, and D are distinct points and lie on a line.
2. If A and B separate C and D, then C and D separate A and B, and B and A separate C and D.
3. If A and B separate C and D, then A and C do not separate B and D.
4. Given any four distinct points A, B, C and D on a line, then A and B separate C and D, or A and C separate B and D, or A and D separate B and C.
5. Given three distinct points A, B, and C on a line, then there is a point D so that A and B separate C and D.
6. Given any five distinct points A, B, C, D, and E on a line, if A and B separate D and E, then either A and B separate C and D or A and B separate C and E.

---

[1] From Greenberg's *Euclidean and Non-Euclidean Geometries*.

The concept of separation can then be used to define arcs and segments:

Definition 7.1. Given three collinear points A, B, and C, the *arc* $\widehat{ACB}$ is the set consisting of the points A, B, C, and all points E on the line so that A and B do not separate C and E.

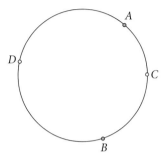

By Postulate 1, given distinct points A and B, there is a unique line through these points. Thus, there are exactly two arcs connecting them. If these two arcs are unequal in length, then we define the *segment* $\widehat{AB}$ to be the shorter of the two. In the illustration above, $\widehat{AB} = \widehat{ACB}$. If the two arcs are equal in length, then we need an additional point in order to specify which of the two we want. In this case, as shown below, when the two arcs given by the points A and B are equal, we denote the two segments as $\widehat{ACB}$ and $\widehat{ADB}$.

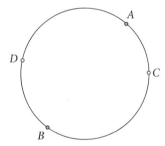

Now that we have a definition for line segment, we have:

Elliptic Postulate 2. One can extend a given segment to form a line.

In single elliptic geometry, note that we can extend any arc to a line, but this line will be finite in extent. Thus, Euclid's original Postulate 2 is literally true, though the interpretation is quite different from that to which we are accustomed. Recall the distance between antipodal points on the sphere is πr. Since those points are identified in single elliptic geometry, all lines are exactly πr long. (Note: We will prove that all lines have equal length later in Corollary 7.5. We mention it here in our development of the axioms for elliptic geometry since we want our axioms to reflect behavior we have observed to be true on the projective plane. All of the restrictions on length we adopt in the following postulates mimic the known properties of the projective plane and will be proved to be true in our theoretical single elliptic geometry in Theorem 7.3 and Corollaries 7.4 and 7.5.)

In Postulate 3, note that Euclid's original statement is true but from the observations just made, the radius of any circle must be less than $\frac{\pi r}{2}$.

Elliptic Postulate 3. Given a point and a length less than $\frac{\pi r}{2}$, there exists exactly one circle with the given point as center and radius equal to the given length.

Postulate 4. All right angles are congruent.

Postulate 6. [Incidence]

1. There exists at least one (elliptic) plane.
2. Every plane contains at least three noncollinear points.
3. Every line contains at least two points.

In Incidence Postulate 6, the plane referred to is an elliptic or projective plane: a sphere with antipodal points identified.

Elliptic Postulate 8. [Pasch's Postulate] If A, B, and C are noncollinear points and a line ℓ intersects $\widehat{AB}$ at a single point D between A and B, then ℓ must intersect exactly one of the following: $\widehat{BC}$ between B and C, $\widehat{AC}$ between A and C, or the vertex C.

We have substituted Pasch's Postulate for the Plane Separation Postulate 8 of neutral geometry. In our study of euclidean geometry in Chapter 2, we mentioned that these statements are logically equivalent (see Project 2.5) but that proof relies on the context of neutral geometry in which the Betweenness Postulate 7 is true. On the sphere, any great circle divides the sphere into two disjoint parts, but when we identify opposite points to form the projective plane, these two regions become one. Thus, in single elliptic geometry, the Plane Separation Property fails and we replace it with Pasch's Postulate.

Postulate 9. [Congruence]

1. Congruence or equality in length is an additive equivalence relation on arcs.
2. Congruence or equality in measure is an additive equivalence relation on angles.
3. Equality in area is an additive equivalence relation on polygons.
4. Congruent triangles have equal areas.

As in our discussion of line segments and arcs on the sphere, we must be cautious in dealing with arcs and segments, since the behavior we noticed in Activities 7.7 and 7.8 on the sphere carries over to the projective plane model. Note that the notation $\widehat{ADB}$ may denote an arc or a segment, depending on the circumstances. One must also be cautious in applying the additivity of lengths of arcs, as given by Congruence Postulate 9(1). For example, consider the line determined by A and B shown below. The arcs $\widehat{AD}$, $\widehat{DB}$, and $\widehat{AB}$ are segments, but

$$\widehat{AD} + \widehat{DB} \neq \widehat{AB}.$$

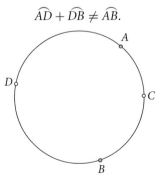

Before we can state the next axiom, we need to reformulate the terms ray and angle. Since lines have finite length and are not separated by a point, the ray $\overrightarrow{AB}$ simply gives a direction on a line. Choose the segment $\widehat{AB}$ and continue from A through B.

▷ **Activity 7.20.** Define angle and the terms interior and exterior for angles, triangles, and circles in single elliptic geometry, if you can. We know that lines do not separate the elliptic plane. Do circles separate it? Justify your answer.

Elliptic Postulate 10. [Archimedes' Axiom] If a and b are positive real numbers with $a < b$, then there exists $n \in \mathbb{N}$ so that $na > b$. In particular,

1. Given arcs $\widehat{AB}$ and $\widehat{CD}$ with $\widehat{CD}$ longer than $\widehat{AB}$, there exists $n \in \mathbb{N}$ and a point X on $\overrightarrow{CD}$ so that $\widehat{CDX} = n \cdot \widehat{AB}$ or $\widehat{CDX}$ exceeds πr.
2. Given angles $\angle ABC$ and $\angle DEF$, there exists $n \in \mathbb{N}$ and a ray $\overrightarrow{EX}$ so that $\angle XEF = n \cdot \angle ABC$ and $\angle XEF > \angle DEF$.

Elliptic Postulate 11. [Circular Continuity Principle]

1. An arc with one endpoint outside a given circle and the other endpoint inside the circle will intersect the circle exactly once.
2. A circle passing through a point inside a given circle and a point outside that circle will intersect the given circle twice.

Postulate 12. [SAS] If two triangles have two sides equal to two sides respectively and have the angles contained by the equal sides equal, then the triangles will be congruent.

Comparing our list of postulates with those in euclidean and hyperbolic geometries, we have adjusted some of the postulates and replaced others. The only postulate left to address is the Parallel Postulate 5. Euclid's Postulate 5, that if the sum of the angles formed by two lines and a transversal is less than 180°, then the lines must intersect, is technically true in elliptic geometry, since any two great circles must intersect regardless of the sum of the angles. However, Playfair's Postulate, that given a line and a point not on that line, there is exactly one line through the point parallel to the given line, is clearly false for our model since any two great circles intersect twice on the sphere and so any two lines will intersect once on the projective plane. This may seem contradictory since in Theorem 5.20 we claimed that Postulate 5 and Playfair's Postulate are logically equivalent. However, that proof was done in the context of neutral geometry, i.e., assuming that Postulates 1–4 and 6–12 are true. But we have shown above that several of these axioms fail, and others are only true when greatly restricted. Therefore, we must remember that the theorems of neutral geometry are not necessarily true in elliptic geometry. We replace the Parallel Postulate 5 by a statement that implies that no parallel lines exist in the elliptic plane.

Elliptic Postulate 5. Any two distinct lines in a plane meet in exactly one point.

Some of Euclid's early propositions in *The Elements* and the neutral theorems we proved in Chapter 5 are true in elliptic geometry, while others are false. Essentially, one has to go through each proof line by line and verify that the constructions either work in elliptic geometry or can be modified to remove any difficulties. To avoid repetition, we give some theorems that we are willing to accept without proof. This list is not meant to be inclusive, but rather is chosen from those we will need to prove the succeeding results and other theorems of interest. All of these can be proved by suitable modifications to the proofs given for neutral geometry. The reader is welcome to reconstruct these proofs.

Theorem 7.2. The following statements are true in elliptic geometry:

1. Given a point on a line and a length less than or equal to $\frac{\pi r}{2}$, one can cut off a segment of the given length.
2. The base angles of an isosceles triangle are equal.
3. If the base angles of a triangle are equal, then the triangle is isosceles.
4. Vertical angles are equal.
5. Two adjacent angles form a line if and only if their sum is 180°.
6. Any arc can be bisected.
7. Any angle can be bisected.
8. Given a line and a point on the line, a unique perpendicular can be constructed to the given line through the given point.
9. Given a line and a point not on that line, at least one perpendicular can be constructed to the given line through the given point.
10. Given an angle and a segment, one can construct on that segment at an endpoint an angle equal to the given angle.
11. If two triangles satisfy the SSS criterion, then they are congruent.
12. If two triangles satisfy the ASA criterion, then they are congruent.
13. The summit angles of any saccheri quadrilateral are equal and the midline is perpendicular to both the base and the summit.

We are more interested in the ways in which elliptic geometry differs from euclidean. Consider the equator and the lines of longitude. All of the lines of longitude are perpendicular to the equator. This behavior carries over to single elliptic geometry.

Theorem 7.3. Let $\mathcal{C}$ be a line. There is a point P, called the *pole* for $\mathcal{C}$, such that every straight line connecting P to a point on $\mathcal{C}$ is perpendicular to $\mathcal{C}$. The distance from P to any point X on $\mathcal{C}$ is called the *polar distance* and is the same for any X on $\mathcal{C}$.

Proof: We are given straight line $\mathcal{C}$, which we can, without loss of generality, picture as the equator. Let X be an arbitrary point on $\mathcal{C}$. By Theorem 7.2(1), we can construct points A and B on $\mathcal{C}$ so that $\widehat{XA} = \widehat{XB}$. By Theorem 7.2(8), we can erect perpendiculars $\mathcal{C}'$ and $\mathcal{C}''$ at A and B. By Elliptic Postulate 5, $\mathcal{C}'$ and $\mathcal{C}''$ must intersect at some point P. Choose points C_A and D_A on $\mathcal{C}'$ which separate the points A and P. Similarly, choose points C_B and D_B on $\mathcal{C}''$ separating B and P. Below we have on the left an illustration of the situation drawn on the upper hemisphere model of the projective plane as well as a bird's eye view of the same picture.

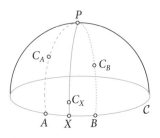
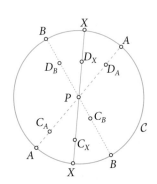

Consider $\triangle PAB$ and note that $\angle PAB = \angle PBA = 90°$, so by Theorem 7.2(3), $\widehat{PC_A A} = \widehat{PC_B B}$. Now, draw PX, choosing points C_X and D_X separating X and P. Note that $\triangle PXA \cong \triangle PXB$ by SSS. Thus, $\angle PXA = \angle PXB$ so $\widehat{PC_X X} \perp \mathcal{C}$. Considering $\triangle PXA$, by Theorem 7.2(3), $\widehat{PC_X X} = \widehat{PC_A A} = \widehat{PC_B B}$.

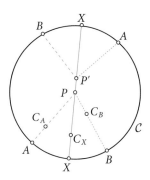

We next show that $\widehat{PC_X X} = \widehat{PD_X X}$. If this is not true, then we can cut a point P' on PX so that $\widehat{PC_X X} = \widehat{P'X}$. Draw $P'A$ and $P'B$. By SAS, $\triangle P'AX \cong \triangle P'BX \cong \triangle PBX \cong \triangle PAX$, so $\angle P'AX = \angle P'BX = \angle PBX = \angle PAX = 90°$. Thus $P'A$ and $P'B$ are perpendicular to $\mathcal{C}$. Thus by Theorem 7.2(8), the lines $P'A$ and PA must coincide, as do $P'B$ and PB. Since two lines intersect in exactly one point, $P = P'$ and thus the points P and X divide the line PX into two equal segments.

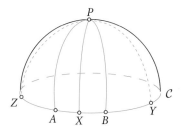 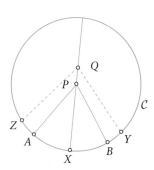

Now let Y be another point on $\mathcal{C}$. Cut point Z so that $\widehat{XZ} = \widehat{XY}$. Erect lines at Z and Y perpendicular to $\mathcal{C}$ and let Q be the intersection of these lines. By Theorem 7.2(3), $\triangle QYZ$ is isosceles, so $\widehat{QY} = \widehat{QZ}$. Draw $\widehat{QX}$ and note that by SAS, $\triangle QZX \cong \triangle QYX$. Thus, $\angle QXZ = \angle QXY = 90°$. Therefore, Q must lie on the previously defined line PX. As above, Q and X must divide PX into two equal segments, so Q must coincide with the point P. Thus, $\widehat{PY} = \widehat{PZ} = \widehat{PX} = \widehat{PA} = \widehat{PB}$. Therefore, the length $\widehat{PY}$ is the same for all Y on $\mathcal{C}$. Since PY is unique by Postulate 1, it follows that PY is perpendicular to $\mathcal{C}$ for any point Y. □

Conversely, if P is a point, there is a line $\mathcal{C}_P$, called the *polar* of P, so that every line from P to a point of $\mathcal{C}_P$ is perpendicular to $\mathcal{C}_P$.

▶ **Exercise 7.4.** Given a point P, construct the polar $\mathcal{C}_P$ of P and prove that every line from P to a point of $\mathcal{C}_P$ is perpendicular to $\mathcal{C}_P$.

Let X be a point on line $\mathcal{C}$ with pole P and polar distance k, so $\widehat{PX} = k$. Note that by Theorem 7.3, the points P and X divide the line PX into two equal arcs, which will both have length k. Thus, the length of the line PX is $2k$, and similarly, the length of any line through P is also $2k$.

Corollary 7.4. All lines have the same polar distance, defined to be $k = \frac{\pi r}{2}$.

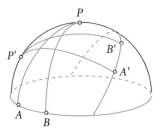

Proof: Let $\mathcal{C}$ be a line with pole P and polar distance k. Choose points A and B on $\mathcal{C}$. By Theorem 7.3, $\widehat{PA} = \widehat{PB} = k$. Now consider any other line $\mathcal{C}'$ with pole P' and polar distance k'. Choose a point A' on $\mathcal{C}'$ and construct B' on $\mathcal{C}'$ with $\widehat{AB} = \widehat{A'B'}$ by Theorem 7.2(1). Then $\triangle PAB \cong \triangle P'A'B'$ by ASA (Theorem 7.2(12)). Thus, $\widehat{P'A'} = k' = \widehat{PA} = k$, which we define to be $k = \frac{\pi r}{2}$, in analogy with the distance from the North Pole to the equator on the sphere. □

Corollary 7.5. All lines have the same length, πr.

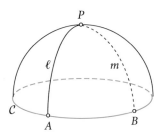

Proof: Let $\mathcal{C}$ be a line with pole P and polar distance $\frac{\pi r}{2}$. Let ℓ and m be two lines through P which are perpendicular to each other. Since ℓ must intersect $\mathcal{C}$ at some point A by Elliptic Postulate 5, ℓ is also perpendicular to $\mathcal{C}$ by Theorem 7.3. Furthermore, $\widehat{PA} = \frac{\pi r}{2}$, so the length of ℓ is πr. Similarly, line m intersects $\mathcal{C}$ at B and is perpendicular to $\mathcal{C}$, and $\widehat{PB} = \frac{\pi r}{2}$ and the length of m is πr. Note that A is thus the pole for line m, so m has polar distance $\widehat{AB} = \widehat{AP} = \frac{\pi r}{2}$. Therefore, the length of $\mathcal{C}$ is also πr. By Corollary 7.4, any other line $\mathcal{C}'$ will also have length πr. □

Theorem 7.6. Consider a right triangle $\triangle ABC$ with $\angle ABC = 90°$. Then $\angle ACB < 90°$ if and only if $\widehat{AB} < \frac{\pi r}{2}$, $\angle ACB = 90°$ if and only if $\widehat{AB} = \frac{\pi r}{2}$, and $\angle ACB > 90°$ if and only if $\widehat{AB} > \frac{\pi r}{2}$.

Proof: Case 1: $\widehat{AB} = \frac{\pi r}{2}$.

In this case, by Corollary 7.4, A must be a pole for the line BC, and so $\angle ACB = 90°$. Conversely, assume that $\angle ACB = \angle ABC = 90°$. By Theorem 7.3, A must be a pole, and so $\widehat{AB} = \frac{\pi r}{2}$.

Case 2: $\widehat{AB} < \frac{\pi r}{2}$.

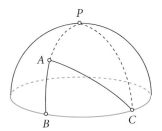

Let P be the pole of line BC. Then $\triangle PBC$ has two right angles, $\angle PBC = \angle ABC$ and $\angle PCB$. Since $\widehat{AB} < \widehat{PB} = \frac{\pi r}{2}$, P does not lie on $\widehat{AB}$ and so $\angle ACB < \angle PCB = 90°$. Conversely, if $\angle ACB < 90°$, then A is not the pole of BC. Thus, $\widehat{AB} \neq \frac{\pi r}{2}$. If the arc $\widehat{AB} > \frac{\pi r}{2}$, then the pole P lies on $\widehat{AB}$ and thus $90° = \angle PCB < \angle ACB$, contradicting our assumption.

Case 3: $\widehat{AB} > \frac{\pi r}{2}$ follows in a similar manner. □

Theorem 7.7. In elliptic geometry, the summit angles of a saccheri quadrilateral are obtuse.

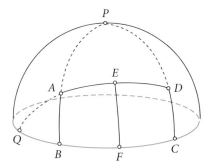

Proof: Let $ABCD$ be a saccheri quadrilateral with $\angle ABC = \angle DCB = 90°$. By Theorem 7.2(13), the midline $\widehat{EF}$ is perpendicular to both AD and BC. Let P be the pole of line BC and Q the pole of line EF. Then by Theorem 7.3, $\widehat{QAE} = \widehat{QBF}$, so $\widehat{QB} < \frac{\pi r}{2}$. Considering the right triangle $\triangle QBA$, by Theorem 7.6, $\angle QAB < 90°$. Applying Theorem 7.2(5) and (13), $\angle DAB = \angle ADC > 90°$. □

Lemma 7.8. In elliptic geometry, the sum of the angles in a right triangle is greater than $180°$.

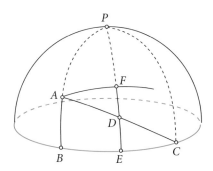

Proof: Let $\triangle ABC$ have $\angle ABC = 90°$. If either $\angle ACB$ or $\angle BAC$ were greater than or equal to $90°$, then the result follows immediately, so we assume that both angles are acute. Thus, $\widehat{AB} < \frac{\pi r}{2}$ and $\widehat{BC} < \frac{\pi r}{2}$. Let D be the midpoint of $\widehat{AC}$ and drop $DE \perp BC$. Before proceeding, we must show that E falls on $\widehat{BC}$. Let P be the pole of line BC. Then PE is perpendicular to BC by Theorem 7.3, so DE coincides with PE. Draw PB. Since P is the pole of BC, $PB \perp BC$ so AB coincides with PB. By Pasch's Postulate, since PD intersects $\triangle ABC$ at D, it must intersect $\widehat{AB}$ or $\widehat{BC}$. Since PD and PB already intersect at P, they cannot intersect at a second point. Therefore, PD must intersect $\widehat{BC}$ and so the point E is on $\widehat{BC}$.

Note that $\angle DBE = \angle DBC < \angle ABC = 90°$, so $\widehat{DE} < \frac{\pi r}{2}$. Extend ED, cut $\widehat{DF} = \widehat{DE}$, and draw $\widehat{AF}$. By Theorem 7.2(4), $\angle FDA = \angle EDC$. Thus, $\triangle DEC \cong \triangle DFA$ by SAS, so $\angle DFA = 90°$ and $\angle DCB = \angle DAF$. Construct the saccheri quadrilateral with base $\widehat{BC}$, leg $\widehat{AB}$, midline $\widehat{EF}$, and summit angle $\angle FAB$. Note that by Theorem 7.7, $90° < \angle FAB = \angle FAC + \angle CAB = \angle BCA + \angle CAB$. Thus, the sum of the angles in $\triangle ABC$ is greater than $180°$. □

Theorem 7.9. *In elliptic geometry, the sum of the angles in a triangle is greater than $180°$.*

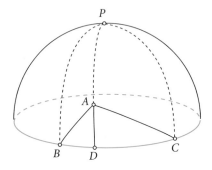

Proof: Let $\triangle ABC$ be given. If at least two of the angles are right or obtuse, then the result follows immediately. Thus, we need only consider the case when both $\angle ABC$ and $\angle ACB$ are acute. Let P be the pole of BC. Draw $\widehat{PB}, \widehat{PA}$, and $\widehat{PC}$, all of which will be perpendicular to BC. Let D be the intersection of PA and BC. Since $\angle ABC$ and $\angle ACB$ are acute, A lies inside $\triangle PBC$ and so D lies between B and C, forming two right triangles $\triangle ADB$ and $\triangle ADC$. By Lemma 7.8, $\angle ABC + \angle BAD + 90° > 180°$ and $\angle ACB + \angle CAD + 90° > 180°$. Thus $\angle ABC + \angle BAC + \angle ACB = \angle ABC + (\angle BAD + \angle CAD) + \angle ACB > 180°$. □

Theorem 7.10. **[AAA Congruence]** *If $\triangle ABC$ and $\triangle DEF$ have $\angle ABC = \angle DEF$, $\angle BAC = \angle EDF$, and $\angle ACB = \angle DFE$, then the triangles are congruent.*

▶ **Exercise 7.5.** Prove Theorem 7.10.

◊ 7.2.3 The Riemann Disc Model

The next model for single elliptic geometry is a variation on the hemisphere. We consider the hemisphere model with antipodal points on the equator identified to each other. Assuming this hemisphere was made of some infinitely flexible material, it can be flattened out to form a disc. This is the *Riemann disc model* for single elliptic geometry in which pictures are drawn on the disc, while remembering that opposite points on the boundary circle are to be considered as the same point in elliptic space.

This model for single elliptic geometry has the advantage that it can be modeled using geometric software, so one no longer needs to carry a sphere or hemisphere around for a model. Lines on this model are the images of the great circles on the sphere. Thus, we have interpretations for the primitive terms:

- point = a point inside the disc or a pair of opposite points on the boundary circle.
- line = any diameter of the circle, or any arc from a circle which intersects the circle at two antipodal points.

Some lines in this model are illustrated below:

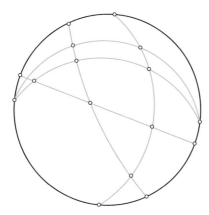

Another way to derive the Riemann model is by stereographic projection. Consider the sphere centered at the origin, with the *xy*-plane cutting through it along the equator. Any point P on the sphere is projected onto the plane by drawing the line from the North Pole, N, through P and following this line to the point Q where it intersects the *xy*-plane. Points in the upper hemisphere will fall outside the equator circle, and points in the lower hemisphere will fall inside the circle. Since P and P' are to be considered as the same point, we must also consider Q and Q' the same. Thus every point can be considered as represented by a point on or inside the circle. Alternately, we could consider the projection of the lower hemisphere alone, to get a flat model for the hemisphere model.

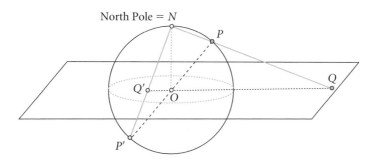

Notice that by Thales' Theorem 3.1, since the line from P to P' passes through the origin O, the angle formed by $\angle PNP'$ is $90°$. If the sphere has radius r, use the Pythagorean Theorem:

$$\overline{OQ'}^2 + \overline{NO}^2 = \overline{NQ'}^2,$$

$$\overline{NO}^2 + \overline{OQ}^2 = \overline{NQ}^2,$$
$$\overline{NQ'}^2 + \overline{NQ}^2 = \overline{Q'Q}^2.$$

Note that $\overline{Q'Q} = \overline{OQ'} + \overline{OQ}$. Thus,

$$(\overline{OQ'}^2 + \overline{NO}^2) + (\overline{NO}^2 + \overline{OQ}^2) = (\overline{OQ'} + \overline{OQ})^2,$$
$$\overline{OQ'}^2 + r^2 + r^2 + \overline{OQ}^2 = \overline{OQ'}^2 + 2\overline{OQ'} \cdot \overline{OQ} + \overline{OQ}^2,$$
$$\overline{OQ'} \cdot \overline{OQ} = r^2.$$

Therefore, Q and Q' are inverses in the circle as in Section 3.4, but Q and Q' are on opposite sides of the center O. This insight allows us to give a procedure for constructing a line through any two points in the Riemann disc model.

Constructing an elliptic line through points A and B: First invert the point A through the circle as in Section 3.4 to get point A', then construct the point A'' so that A'' is on the line OA, on the opposite side of A from A', and so that $\overline{OA'} = \overline{OA''}$.

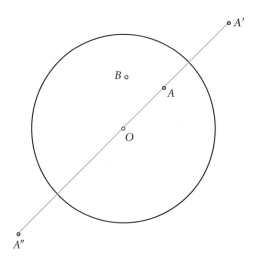

Construct the circle C' passing through points A, B, and A'', the inverse of A through the circle C, but on the opposite side of the center O. We do this by finding the perpendicular bisectors of any two sides of $\triangle AA''B$. These intersect at point O', which forms the center of the circle containing A, B, and A''. (Note that one could use $\triangle ABB''$ instead, where B'' is the point opposite the inverse B' of B.) The portion of the circle so constructed that lies inside the given circle represents the line through A and B in the model.

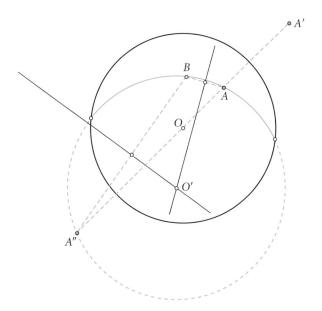

▷ **Activity 7.21.** Use geometric software to draw a lune in the Riemann disc model for single elliptic geometry.

The angles between two elliptic lines in the Riemann disc model are measured by finding the angle between their tangent lines.

▷ **Activity 7.22.** Use geometric software to draw a right-angled lune in the Riemann disc model.

▷ **Activity 7.23.** Use geometric software to draw an equilateral triangle in the Riemann disc model, and give the sum of its angles.

▷ **Activity 7.24.** Use geometric software to draw a saccheri quadrilateral in the Riemann disc model and measure the summit angles, and the lengths of its base and summit.

▷ **Activity 7.25.** Use geometric software to draw an elliptic circle in the Riemann disc model. Measure the distance from the center to at least three points on the circle to verify that this is indeed a circle.

▷ **Activity 7.26.** Use geometric software to draw a quadrilateral with four equal sides and four equal angles in the Riemann disc model.

♦ 7.3 Comparative geometry

Our purpose in this section is to clarify and review the properties of the different geometries we have studied. In this text, we have discussed three main types of geometry: euclidean and hyperbolic geometry in some depth and elliptic geometry more lightly. We ask you to summarize the characteristics of each of these:

▶ **Exercise 7.6.** Fill out the table below as instructed.

   a. In the first row, give how many points of intersection two distinct lines can have.

b. In the second row, give the number of lines parallel to a given line and passing through a given point not on that line.
c. In the third row, describe the summit angles of a saccheri quadrilateral in each geometry as acute, right, or obtuse.
d. In the fourth row, describe the angle sum for an arbitrary triangle as an equation or inequality.
e. In the fifth row, answer whether the triangles are similar or congruent.

	Euclidean	Hyperbolic	Elliptic
Two lines intersect at			
Given a line and a point, there are? parallel lines.			
Summit angles of a saccheri quadrilateral are			
The angle sum of a triangle is			
If two triangles satisfy AAA, they are			

The next exercise is another review question. Think carefully about your answers in each geometry.

▶ **Exercise 7.7.** Answer True (always true) or False (not always true) for each statement for each of the three geometries, euclidean, hyperbolic, and (single) elliptic. Justify your answers.

a. If $AB \perp CD$ and $CD \perp EF$, then $AB \| EF$.
b. If a statement is true in neutral geometry, then it is true.
c. Given point P not on line AB, there is a unique line through P perpendicular to AB.
d. Parallel lines are equidistant.
e. Any three noncollinear points lie on some circle.
f. The opposite sides of a parallelogram are equal.
g. A saccheri quadrilateral is a parallelogram.
h. If lines ℓ and m are parallel, then there are two points on ℓ that are equidistant from m.
i. If lines ℓ and m intersect, then there are two points on ℓ that are equidistant from m.
j. If lines $\ell \| m$ and $m \| n$, then $\ell \| n$.
k. Similar triangles are congruent.
l. A line divides the plane into two regions.
m. A point divides a line into two parts.
n. A circle divides the plane into two regions.

As you recall, the angle sum of a triangle in euclidean geometry is always 180°, but in hyperbolic and elliptic geometry the angle sum varies. In fact, for small triangles in either of these geometries, the angle sum will be close to 180°. This might explain why no one noticed that the world is not flat for a long time. It also explains the still current assumption that the universe is euclidean, since anything close enough to measure is small in terms of the size of the universe.

One of Gauss' later students, Georg Friedrich Bernhard Riemann [1826–1866] changed how mathematicians thought about geometry. His theory is that we can only know the local geometry of a space. He defined the concept of a *metric* or distance measure, which in turn gives rise to the idea of a *geodesic*, a curve that acts like a locally straight line or the shortest path between two nearby points (if such a path exists, though it should be noted that exceptions occur where the geodesic is not the shortest path). Riemann also resurrected interest in elliptic geometry. He pointed out the distinction between boundlessness (such as the great circles that act as lines on the sphere, which have no beginning and no end) and the infiniteness of the euclidean and hyperbolic planes.

Riemann generalized Gauss' concept of the *curvature* of a space, considering this as a function depending on the position of a point. This allows a space to have variable curvature and thus variable geometry. In order to study how the geometry varies from point to point, one must apply techniques from both geometry and calculus. This field is called *differential geometry*.

Reality and real objects rarely fall into neat categories: euclidean, hyperbolic, or elliptic. Instead, many things have mixed geometries. The angle sum of a triangle is a convenient characterization to use. The following activity was suggested by James Casey in *Exploring Curvature*:

▷ **Activity 7.27.** Cut the two enlarged copies of the following 45° angle out of cardboard, and staple or tape them together at the indicated points as shown. This forms a triangle with two fixed 45° angles at the base and a 90° angle formed by the two overlapping flaps at the apex.

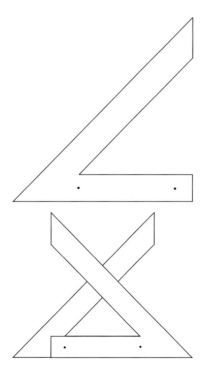

▷ **Activity 7.28.** Choose a partner and fit your triangle to various body parts. The two flaps should flex to form various angles depending on the local geometry. Measure the angle formed by the flaps and use this to characterize the local geometry. (Some suggestions: the top of the head, under the chin, shoulder, the back, but experiment to find several portions that have each of the three geometries.)

▷ **Activity 7.29.** Describe (as euclidean, hyperbolic, elliptic, or mixed) the geometry of a cylinder.

▷ **Activity 7.30.** Find three objects that typify each of the three geometries. Hint: Vegetables work well.

Felix Klein devised the single elliptic model for elliptic geometry which solves some of the difficulties we faced in spherical geometry. He also showed that all the geometries we have studied can be viewed as special cases of a more general theory, called *projective geometry*. Klein made many major contributions to geometry and mathematics, but is best known for what is now called the *Erlanger Program*. This was a new approach to geometry, in which spaces are characterized by the types of transformations, also called symmetries or rigid motions, they allow and the invariants of these transformations. We will begin the study of these transformations in Chapter 8.

♦ 7.4 Area and defect

We studied Euclid's treatment of area in Propositions I.34–46, in which he discusses the relative areas of parallelograms and triangles in the same parallel. At that point, we discussed the more modern formulae for the area of some common polygons. However, there is a fundamental difficulty in extending these ideas to noneuclidean geometry, since all of the area formulae are in terms of square units, i.e., in comparison to a standard square 1 unit on each side. In hyperbolic and elliptic geometry, there are no such standard squares to use as a basis for measurement of area. Thus, an entirely new approach to area needed to be devised, one that, in some ways, is closer to Euclid's approach than the algebraic method with which we are more familiar.

In this section, we reconsider the question of area in all of the geometries we have studied. We begin by recalling and extending our previous findings in euclidean geometry, then introduce the neutral geometry idea of relating area and decomposition of polygons. We then discuss the application of this concept to hyperbolic geometry. The fundamental result of this section is the Gauss-Bonnet Theorem for hyperbolic and elliptic geometry.

Throughout this section, we have mixed our geometries. Be very careful to observe in which geometry each theorem is true and in, doing the exercises, pay attention to the context. We indicate which geometries each theorem applies to.

◊ 7.4.1 Area in euclidean geometry

In second or third grade, you were introduced to the concept of area, with illustrations such as

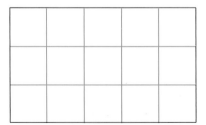

to justify the formula $A = bh$ for the area of a rectangle in square units, where b denotes the width and h the height of the rectangle. The question is: what can one do with area in hyperbolic or elliptic geometry where squares do not exist? Implicit in Euclid's *The Elements* are the following assumptions, which we adopted as axioms. Congruence Postulate 9 is a part of neutral geometry and is also true in elliptic geometry:

Postulate 9. [Congruence]

3. Equality in area is an additive equivalence relation on polygons.
4. Congruent triangles have equal areas.

Postulate 13. [Area] The area of a rectangle is the product of the lengths of its base and height.

Area Postulate 13, with its reference to rectangles, is only true in euclidean geometry. In modern notation, Euclid's Propositions I.35 and I.37 (proved using the Parallel Postulate 5) and Area Postulate 13 can be combined to show that the area of a parallelogram of base b and height h is bh and that the area of a triangle is $\frac{1}{2}bh$.

▶ **Exercise 7.8.** Prove that in euclidean geometry, the area of a rhombus is one-half the product of the two diagonals.

We prove another area formula, called Heron's formula. Heron of Alexandria [ca. 10–75], of course, proved this theorem geometrically, but modern algebra makes it much more accessible. For a triangle $\triangle ABC$ with sides a, b, and c, the perimeter is $a + b + c$. The quantity $s = \frac{1}{2}(a + b + c)$ is called the *semiperimeter*.

Theorem 7.11. [Heron's formula] (euclidean geometry) $A(\triangle ABC) = \sqrt{s(s-a)(s-b)(s-c)}$, where $s = \frac{1}{2}(a + b + c)$.

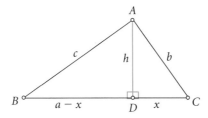

Proof: Drop $AD \perp BC$, dividing side $\overline{BC}$ into lengths of x and $a - x$. The Pythagorean Theorem for $\triangle ABD$ and $\triangle ADC$ gives us:

$$c^2 = h^2 + (a - x)^2,$$
$$h^2 = b^2 - x^2.$$

Combining these gives $c^2 = b^2 + a^2 - 2ax$ and so $x = \frac{a^2 + b^2 - c^2}{2a}$. Substituting this into the second equation above gives

$$h^2 = b^2 - \left(\frac{a^2 + b^2 - c^2}{2a}\right)^2 = \left(b - \frac{a^2 + b^2 - c^2}{2a}\right)\left(b + \frac{a^2 + b^2 - c^2}{2a}\right)$$
$$= \frac{2ab - a^2 - b^2 + c^2}{2a} \cdot \frac{2ab + a^2 + b^2 - c^2}{2a}.$$

Thus,
$$\frac{1}{4}a^2h^2 = \frac{1}{16}\left(c^2 - (a-b)^2\right)\left((a+b)^2 - c^2\right) = \frac{1}{16}(c - (a-b))(c + (a-b))(a+b-c)(a+b+c),$$
$$\frac{1}{4}a^2h^2 = \frac{b+c-a}{2}\frac{a+c-b}{2}\frac{a+b-c}{2}\frac{a+b+c}{2}.$$

Since $A(\triangle ABC) = \frac{1}{2}ah$, we have
$$A(\triangle ABC) = \sqrt{s(s-a)(s-b)(s-c)}.$$

□

In Proposition XII.2, Euclid uses a method called *exhaustion*, due to Eudoxus, to show that the area of a circle depends only on the square of the radius. Note that he never states the formula $A = \pi r^2$. In fact, he never refers to the number π, though he was probably aware of some of its properties. Archimedes extended this result to derive the approximation $\frac{223}{71} < \pi < \frac{22}{7}$, or $3.14084507 < \pi < 3.142857143$. Subsequently, mathematicians have continued to work out approximations for π, even in the current day.

In proving Proposition XII.2, Euclid first considers a circle $\mathcal{C}$ with an inscribed and circumscribed regular polygon, so that the vertices of the inscribed polygon are the points of tangency for the circumscribed polygon. For example, below, we have drawn such a circle with an inscribed octagon $P_8 = A_1A_2A_3\ldots A_8$ and a circumscribed octagon $P_8' = B_1B_2B_3\ldots B_8$:

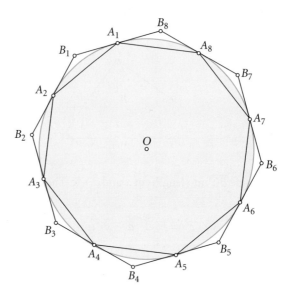

▷ **Activity 7.31.** Using geometric software, draw a circle and an inscribed hexagon. Next, draw a circumscribed hexagon so that the vertices of the inscribed hexagon are the points of tangency for the circumscribed hexagon.

By bisecting the central angles such as $\angle A_1OA_2$, we construct points such as C_1 shown below. Note that if $P_n = A_1A_2\ldots A_n$ is a regular n-sided polygon inscribed in the circle, then $P_{2n} = A_1C_1A_2C_2\ldots A_nC_n$

is a regular $2n$-sided polygon which is also inscribed in the circle $\mathcal{C}$. Constructing the tangent to the circle at C_1 and intersecting it with the sides of the regular n-sided circumscribed polygon $P'_n = B_1 B_2 \ldots B_n$ gives the vertices D_1, etc., so that $P'_{2n} = D_1 D_2 \ldots D_{2n}$ is a regular $2n$-sided circumscribed polygon.

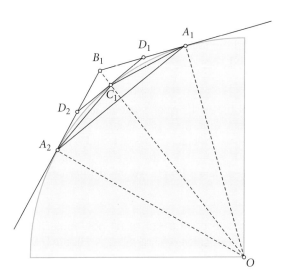

▷ **Activity 7.32.** Follow the procedure above to modify your drawing of Activity 7.31 to construct an inscribed and a circumscribed dodecagon.

Note that

$$A(P_n) < A(P_{2n}) < A(\mathcal{C}) < A(P'_{2n}) < A(P'_n)$$

Euclid shows that $A(\triangle A_1 D_1 C_1) + A(\triangle C_1 D_2 A_2) < \frac{1}{2} A(\triangle A_1 B_1 A_2)$. Thus, each time we double the number of sides of polygons, the difference between the area of the circumscribed polygon and the inscribed polygon is cut in half. By Archimedes' Axiom Postulate 10, this difference can be made as small as we like. Thus, the area of $\mathcal{C}$ can be approximated by the area of either P_n or P'_n for large n.

▶ **Exercise 7.9.** Prove the assertion above, that $A(\triangle A_1 D_1 C_1) + A(\triangle C_1 D_2 A_2) < \frac{1}{2} A(\triangle A_1 B_1 A_2)$.

This proof is interesting because it involves a limiting procedure 2000 years before the invention of calculus. The modern approach to areas inside curves is to use calculus and the Riemann integral. With a little trigonometry and one fact from calculus, we can derive the formula for the area of a circle in the following exercise:

▶ **Exercise 7.10.** Let C be a circle of radius r and P_n a regular n-sided polygon as above.

   a. Find the angle $\angle A_1 O A_2$ in radians.

b. Using trigonometry, find the height of $\triangle A_1OA_2$, i.e., drop a perpendicular from O to A_1A_2 and find the length of this.
c. Using trigonometry again, find $\overline{A_1A_2}$.
d. Find the perimeter of P_n.
e. Find the area of $\triangle A_1OA_2$.
f. Find the area of P_n.
g. By Euclid's argument summarized above, $\lim_{n \to \infty} A(P_n) = A(C)$. Use this to find a formula for $A(C)$. [You will need the following fact, commonly proved in a first semester calculus class: $\lim_{n \to \infty} n \sin\frac{1}{n} = 1$.]

Further Saccheri proved that the existence of a square or rectangle is logically equivalent to Euclid's parallel postulate. Without a square unit, there is no natural way to measure area and Area Postulate 13 fails. However, Congruence Postulate 9 remains reasonable and is the only valid knowledge we have about area that is true in all geometries. We will develop an alternate theory of area for use in noneuclidean geometry, following in the steps of Janos Bolyai. We need to clarify what properties an area measure must have.

Definition 7.12. An *area function* on the set of all polygons in a space is a function which associates to every polygon P a positive real number $A(P)$, such that:

1. Congruent triangles have equal areas: If $\triangle_1 \cong \triangle_2$, then $A(\triangle_1) = A(\triangle_2)$.
2. If the interiors of two polygons, P and Q, do not overlap, then the area of their union is equal to the sum of their areas: $A(P \cup Q) = A(P) + A(Q)$.

The second condition guarantees that any area function is additive. It also suggests that cutting a polygon up into bits might prove useful in other contexts. This concept is generalized in the following definition:

Definition 7.13. A polygon P is *decomposed* into subpolygons $P_1, P_2, \ldots, P_n$ if $P = \bigcup_{i=1}^{n} P_i$ with two subpolygons P_i and P_j intersecting at either a vertex or an edge or not at all. We then write $P = P_1 + P_2 + \ldots + P_n$.

It is clear from the definition of an area function (if we ever find one) that if P is decomposed as $P = P_1 + P_2 + \ldots + P_n$ as in this definition, then

$$A(P) = A(P_1) + A(P_2) + \ldots + A(P_n).$$

When we talk about area in neutral and elliptic geometry, we are referring to any area function as defined above.

Definition 7.14. Polygons P and Q are *equivalent by decomposition* if $P = P_1 + P_2 + \ldots + P_n$ and $Q = Q_1 + Q_2 + \ldots + Q_n$, with $P_i \cong Q_i$. We then write $P \equiv Q$.

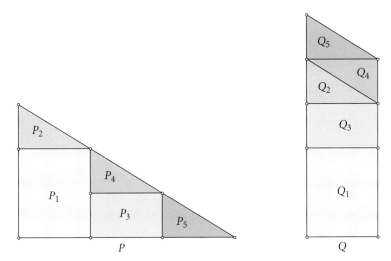

Theorem 7.15. (all geometries) Equivalence by decomposition is an equivalence relation.

Proof: It is quite easy to show that this equivalence is reflexive and symmetric. Proving transitivity is far more difficult and we omit the proof. The interested reader is referred to Hilbert's *Foundations of Geometry*, Theorem 43. □

Theorem 7.16. (all geometries) If triangles △ABC and △DEF are equivalent by decomposition, then $A(\triangle ABC) = A(\triangle DEF)$.

▶ **Exercise 7.11.** Prove Theorem 7.16.

The procedure of Theorem 5.9 can be used to show that any triangle is equivalent by decomposition to the associated saccheri quadrilateral in neutral geometry, or a rectangle in euclidean geometry. In that theorem, we showed that given △ABC, we can bisect $\overline{AB}$ and $\overline{AC}$ at D and E and draw DE. Drop perpendiculars $\overline{AH}, \overline{BF},$ and $\overline{CG}$ to DE. Then BFGC is a saccheri quadrilateral and furthermore, ∠ABC + ∠ACB + ∠BAC = ∠FBC + ∠GCB. In proving Theorem 5.9, we showed that △AHD ≅ △BFD and △AHE ≅ △CGE, so it is clear that △ABC ≡ BFGC. (See also the cases proved in Exercise 5.9.)

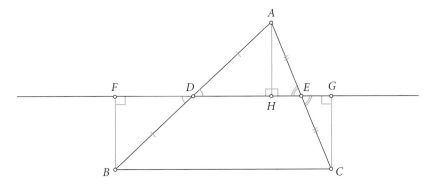

Theorem 7.17. (euclidean geometry) Any parallelogram is equivalent by decomposition to a rectangle.

▶ **Exercise 7.12.** Prove Theorem 7.17, considering the two cases for the parallelogram *ABCD* pictured below:

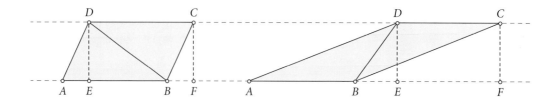

Lemma 7.18. (euclidean geometry) If *ABCD* is a rectangle with another length $\overline{AE}$ so that $\overline{AB} < \overline{AE} < 2\overline{AB}$, then *ABCD* is equivalent by decomposition to a rectangle with $\overline{AE}$ as one side.

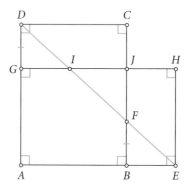

Proof: We are given the rectangle *ABCD* and line segment $\overline{AE}$. Draw *DE*, which intersects *BC* at *F*. Cut $\overline{DG} = \overline{BF}$. Construct rectangle *AEHG* and points *I* and *J* where *GH* intersects *DE* and *BC*. By Proposition I.29, $\angle ADI = \angle BFE$. Also note that $\angle DGI = \angle FBE = 90°$, so $\triangle DGI \cong \triangle FBE$ by ASA, so $\overline{GI} = \overline{BE}$. Thus

$$\overline{DC} = \overline{GJ} = \overline{GI} + \overline{IJ} = \overline{IJ} + \overline{BE} = \overline{IJ} + \overline{JH} = \overline{IH}.$$

Since $\angle CDI = \angle HIF$ and $\angle IHE = \angle DCF = 90°$, it follows that $\triangle DCF \cong \triangle IHE$. Therefore,

$$ABCD \equiv \triangle DGI + \triangle DCF + AGIFB,$$

$$AEHG \equiv \triangle FBE + \triangle IHE + AGIFB,$$

and so $ABCD \equiv AEHG$. □

▶ **Exercise 7.13.** Draw the illustration for Lemma 7.18 when $\overline{AE} \geq 2\overline{AB}$ and show how the proof fails.

Theorem 7.19. (euclidean geometry) If *ABCD* is a rectangle and a line segment $\overline{EF}$ is given, then *ABCD* is equivalent by decomposition to a rectangle with $\overline{EF}$ as one side.

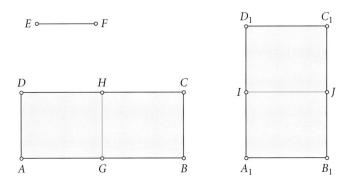

Proof: We consider the case in which $\overline{AB} > \overline{EF}$. Let G and H be the midpoints of $\overline{AB}$ and $\overline{DC}$ and draw GH. Construct rectangle $A_1B_1C_1D_1$ as shown so that $A_1B_1JI \cong AGHD \cong GBCH \cong IJC_1D_1$ and note that $\overline{A_1B_1} = \frac{1}{2}\overline{AB}$. Clearly $A_1B_1C_1D_1 \equiv ABCD$. Continue thus, using Archimedes's Axiom 10, until you obtain a rectangle $A_nB_nC_nD_n \equiv ABCD$ with $\overline{A_nB_n} \le \overline{EF} < 2\overline{A_nB_n}$. Then we apply Lemma 7.18 to construct a rectangle with side $\overline{EF}$ equivalent to $ABCD$. □

▶ **Exercise 7.14.** Prove Theorem 7.19 for the case in which $\overline{EF} > 2\overline{AB}$.

Corollary 7.20. (euclidean geometry) Given a line segment $\overline{EF}$, any polygon is equivalent by decomposition to some rectangle with side $\overline{EF}$.

Proof: First subdivide the polygon into a finite number of triangles. Apply the procedure of Theorem 5.9 to decompose each triangle into an equivalent rectangle, and then use Theorem 7.19 to decompose each rectangle into an equivalent one with side $\overline{EF}$. These rectangles can then be stacked to form a larger rectangle with side $\overline{EF}$ that is clearly equivalent to the original polygon. □

Theorem 7.21. [Bolyai-Gerwien] (euclidean geometry) Two polygons P and Q are equivalent by decomposition if and only if they have the same area.

Proof: We have proved half of this theorem in Theorem 7.16. If P and Q are polygons with equal areas, then by Corollary 7.20, P is equivalent by decomposition to a rectangle with one side of length 1 and height a, and similarly, polygon Q is equivalent to a 1 by b rectangle. By Area Postulate 13, $A(P) = 1 \cdot a = A(Q) = 1 \cdot b$, so $a = b$ and the two rectangles are congruent. It follows that P is equivalent to Q. □

◊ **7.4.2 Defect**

The series of theorems above were proven in the context of euclidean geometry, and are dependent on the existence of rectangles. To develop a similar theory for noneuclidean geometries, we must look for a connection with other concepts. Recall that the Parallel Postulate 5 is true if and only if the sum of the angles in a triangle is 180°.

Definition 7.22. If ABC is a triangle, then the *defect* of $\triangle ABC$ is

$$\text{def}(\triangle ABC) = 180° - \angle A - \angle B - \angle C$$

The defect of a quadrilateral $ABCD$ is $360° - \angle A - \angle B - \angle C - \angle D$.

▶ **Exercise 7.15.** Define the defect of an *n*-sided polygon.

The defect of a polygon measures how badly it deviates from the euclidean measure, and so measures how noneuclidean it is. The next theorem is true in euclidean, hyperbolic, and elliptic geometries, and is suggested by the idea of dividing any polygon into smaller pieces.

Theorem 7.23. (all geometries) The defect is additive. I.e., if $\triangle ABC$ is a triangle, and ABC is divided into polygons P_i, for $i = 1, 2, \ldots, n$, then

$$\text{def}(\triangle ABC) = \sum_{i=1}^{n} \text{def}(P_i).$$

Proof: Case 1: $\triangle ABC$ is divided into two parts by a line from one vertex, say vertex A, to a point D on the opposite side $\overline{BC}$, as pictured below:

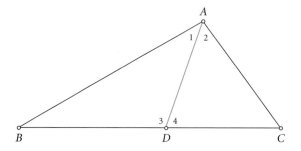

Then

$$\begin{aligned}
\text{def}(\triangle ABC) &= 180° - \angle A - \angle B - \angle C \\
&= 180° - (\angle 1 + \angle 2) - \angle B - \angle C - (\angle 3 + \angle 4 - 180°) \\
&= (180° - \angle 1 - \angle B - \angle 3) + (180° - \angle 2 - \angle C - \angle 4) \\
&= \text{def}(\triangle ABD) + \text{def}(\triangle ADC).
\end{aligned}$$

Case 2: $\triangle ABC$ is divided by a line DE that cuts through two of the sides forming the triangle $\triangle ADE$ and the quadrilateral $DECB$, as shown below:

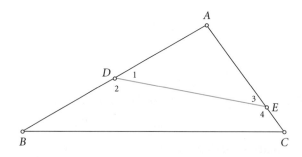

$$\begin{aligned}
def(\triangle ABC) &= 180° - \angle A - \angle B - \angle C \\
&= 180° - \angle A - \angle B - \angle C + (180° - (\angle 1 + \angle 2)) + (180° - (\angle 3 + \angle 4)) \\
&= (180° - \angle A - \angle 1 - \angle 3) + (360° - \angle 2 - \angle 4 - \angle B - \angle C) \\
&= def(\triangle ADE) + def(DECB).
\end{aligned}$$

Case 3: If $\triangle ABC$ is obtained by subdividing into several parts then note that the subdivision can be done one step at a time by applying case (1) and case (2) repeatedly. We omit the details, but the interested reader is referred to a similar proof (for areas) in Hilbert's *Foundations of Geometry*. □

Theorem 7.24. (all geometries) If triangles $\triangle ABC$ and $\triangle DEF$ are equivalent by decomposition, then $def(\triangle ABC) = def(\triangle DEF)$.

▶ **Exercise 7.16.** Prove Theorem 7.24.

▶ **Exercise 7.17.** Prove, in the context of hyperbolic geometry, that if $\triangle ABC$ is an equilateral triangle and D, E, and F are the midpoints, respectively, of $\overline{AB}$, $\overline{BC}$, and $\overline{CA}$, then $\triangle DEF$ is also equilateral and $def(\triangle DEF) < def(\triangle ABC)$.

◇ **7.4.3 Area in hyperbolic geometry**

Theorems 7.16 and 7.24 suggest a similarity in the behavior of the defect and area functions. The precise relationship will be given in Theorem 7.27 below. The following theorems are proven in the context of hyperbolic geometry.

Lemma 7.25. (hyperbolic geometry) If $\triangle ABC$ and $\triangle DEF$ are two triangles such that $\overline{AB} = \overline{DE}$ and $def(\triangle ABC) = def(\triangle DEF)$, then $\triangle ABC$ has the same area as $\triangle DEF$.

▶ **Exercise 7.18.** Prove Lemma 7.25 by constructing the associated saccheri quadrilaterals and using Theorem 6.37.

Theorem 7.26. (hyperbolic geometry) If $\triangle ABC$ and $\triangle DEF$ are two triangles such that $def(\triangle ABC) = def(\triangle DEF)$, then $\triangle ABC$ has the same area as $\triangle DEF$.

Proof: If any side of $\triangle ABC$ is equal to a side of $\triangle DEF$, then apply Lemma 7.25 above. Otherwise, we may assume that $\overline{FE} > \overline{BC}$. Let G and H be the midpoints of $\overline{AC}$ and $\overline{BC}$, and draw $\overline{GH}$. Let I be the midpoint of $\overline{FE}$. Consider a circle centered at B with radius equal to $\overline{EI}$. Since $\overline{BH} < \overline{EI}$, point H lies within this circle. By the Circular Continuity Principle 11, line GH must intersect the circle at a point J, and so $\overline{BJ} = \overline{EI}$. Extend BJ to K with $\overline{BK} = \overline{EF}$ and draw $\overline{AK}$, intersecting GH at L.

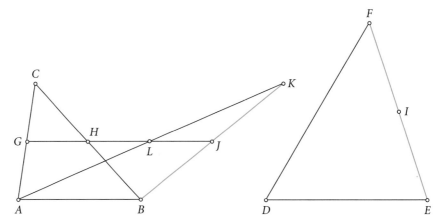

▶ **Exercise 7.19.** Finish the proof of Theorem 7.26 above by showing first that *L* is the midpoint of $\overline{AK}$. Then show that $def(\triangle ABC) = def(\triangle ABK)$ and $\triangle ABC \equiv \triangle ABK$. Use Lemma 7.25 to prove that $\triangle ABK$, and thus $\triangle ABC$, has the same area as $\triangle DEF$.

Theorems 7.16 and 7.24 suggest that the converse of Theorem 7.26 is true, though this is not rigorous since we lack a precise definition of area in hyperbolic geometry. We will not prove the following very difficult theorem. We have shown that the defect acts like an area function as defined in Definition 7.12. The proof depends on showing that area functions are unique up to a constant, and so the defect is some constant multiple of the area function. Note that the following theorem is easily seen to be true in euclidean geometry by letting $\kappa = 0$. If κ is positive, the underlying space must be hyperbolic, while if κ is negative, the space must be elliptic.

Theorem 7.27. [Gauss-Bonnet Theorem] (all geometries) If $\triangle ABC$ is a triangle, then

$$def(\triangle ABC) = \kappa\, A(\triangle ABC)$$

for some constant κ. Note that the constant κ depends only on the space and not on a particular triangle.

◊ **7.4.4 Area in elliptic geometry**

We can prove the counterpart of Theorem 7.27 for spherical geometry, where we do know something about area, confirming your conjecture and computations of Activities 7.17–7.19 and Exercise 7.1. First, recall the figure known as a lune. This is one component of the figure formed by two great circles. The double or shadow of the lune is a duplicate.

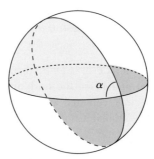

A lune is determined by two great circles meeting at antipodal points with angle α. Recall, from your calculus class, the formula for the area of a sphere (we will give Archimedes' proof of this formula in Chapter 10):

$$A(sphere) = 4\pi r^2.$$

▶ **Exercise 7.20.** From the formula for the area of a sphere, deduce the area of an α-lune.

Now consider a spherical triangle. This is formed by the intersection of three lunes of angles α, β, and γ.

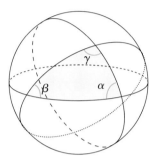

The lunes and their doubles cover the entire sphere, except that the triangle (and its double on the back side of the sphere) are in each of the lunes and are therefore covered several times over. The result of the next exercise is called Girard's Theorem, for Albert Girard [1595–1632]. This is the Gauss-Bonnet Theorem for the case of a spherical triangle.

▶ **Exercise 7.21.** Give a formula for the area of the spherical triangle in terms of the radius and the three angles. Prove that your formula is valid.

▶ **Exercise 7.22.** Using the fact that any polygon with n sides can be cut up into $(n-2)$ triangles, find a general formula for the area of a spherical polygon with n sides in terms of the angle sum.

▶ **Exercise 7.23.** If a triangle on a sphere of unknown radius is surveyed and it is found that its area is 120 units and that its angles measure 90°, 60°, and 60°, find the radius of the sphere.

Theorem 7.26 and Exercise 7.21 give quite similar results for hyperbolic and elliptic geometry:

$$def(\triangle ABC) = \kappa \, A(\triangle ABC),$$

where κ is a constant, called the curvature of the space, that depends on the space itself. This constant depends on the radius in the case of elliptic geometry and a quantity that behaves similarly for hyperbolic space. If you recall the discussion of building models of hyperbolic space in Section 6.3.1, the constant k is related to how the paper models are built. The more the circumference of a disc centered at a vertex exceeds $2\pi r$, the greater the constant. From this equation, we deduce that very small triangles will have angle sums very close to 180°. Thus, both hyperbolic and elliptic geometry are approximately euclidean in the small.

♦ 7.5 Taxicab geometry

Analytic geometry, invented by René Descartes, gave mathematicians a way of relating purely geometric objects to algebraic equations. A point P on the plane is represented by an ordered pair of real numbers (x_1, y_1). A line is then represented by the set of all points satisfying an equation of the form $ax + by = c$. A circle with center (a,b) and radius r is the set of all points satisfying the equation $(x-a)^2 + (y-b)^2 = r^2$. The distance between two distinct points $P(x_1, y_1)$ and $Q(x_2, y_2)$ is defined by the formula:

$$d_E(P,Q) = \sqrt{(x_2 - x_1)^2 + (y_2 - y_1)^2}$$

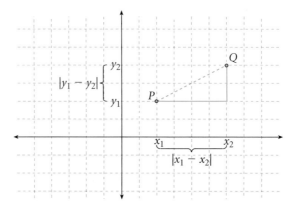

The distance formula in this case is the length of the line segment from P to Q. Note that the formula above is based on the Pythagorean Theorem for the right triangle with vertices at P and Q. Thus, this geometry is euclidean, since by Theorem 5.25 the Pythagorean Theorem is logically equivalent to Postulate 5. The distance formula or metric in effect determines the geometry.

In general, a metric is defined as in Definition 2.27 in the section on distance and cartesian coordinates, which we recall:

Definition 2.27. Let X be a nonempty set of points. A *metric* or distance function on X is a function $d: X \times X \to \mathbb{R}$ such that for all $P, Q, R \in X$,

1. $d(P,Q) \geq 0$.
2. $d(P,Q) = 0$ if and only if $P = Q$.
3. $d(P,Q) = d(Q,P)$.
4. $d(P,Q) \leq d(P,R) + d(R,Q)$.

Property (4) is called the triangle inequality, since it says in effect that the length of the line segment $\overline{PR}$ plus the length of the line segment $\overline{RQ}$ is longer than or equal to the line segment $\overline{PQ}$. In other words, it is shorter to go straight from P to Q than to detour by way of R.

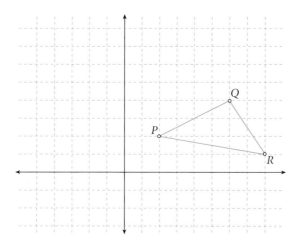

Let us investigate another metric and the geometry it generates. We still consider points as ordered pairs on the cartesian plane, but we will change the metric. The euclidean metric d_E above assumes that the shortest distance between two points can cut diagonally from one point to another. The new metric is called the *taxicab* or *Manhattan metric*, and is modeled after an urban geography. In a city whose streets are laid out in a rectangular grid, one cannot travel as a bird would, but must travel from point P to point Q by going so many blocks east or west and then so many blocks north or south. In these terms, the travelling distance from P to Q is defined by the function

$$d_T(P,Q) = |x_2 - x_1| + |y_2 - y_1|$$

▶ **Exercise 7.24.** Show that d_T defines a metric.

▷ **Activity 7.33.**

a. Using graph paper, plot the points $A(0,0)$, $B(5,0)$, $C(2,3)$, and $D(3,4)$.
b. Find $d_E(A,B)$, $d_E(A,C)$, and $d_E(A,D)$.
c. Find $d_T(A,B)$, $d_T(A,C)$, and $d_T(A,D)$.
d. Does $d_E(P,Q) = d_E(R,S)$ imply that $d_T(P,Q) = d_T(R,S)$?
e. Does $d_T(P,Q) = d_T(R,S)$ imply that $d_E(P,Q) = d_E(R,S)$?

▶ **Exercise 7.25.** When is $d_E(P,Q) = d_T(P,Q)$?

a. Find an algebraic criterion by setting the equations for the distances with the two metrics equal to one another and solving to find an equation that states when the two distance measures are equal.
b. In general, what is the algebraic relation (i.e., $<$, $=$, or $>$) between $d_E(P,Q)$ and $d_T(P,Q)$?
c. Give a geometric criterion that describes the position of two points when the two distance measures are equal.

While the points are the same as in euclidean geometry, changing the metric can change other geometric figures. While we have drawn the line from $P(x_1, y_1)$ to $Q(x_2, y_2)$ as if it were the euclidean line, in fact the metric implies that $\overline{PQ}$ is a path that travels east from $P(x_1, y_1)$ to (x_2, y_1) and then north from (x_2, y_1)

to $Q(x_2,y_2)$, or alternatively, north from $P(x_1,y_1)$ to (x_1,y_2) and then east from (x_1,y_2) to $Q(x_2,y_2)$, or any other path from P to Q that involves travel only in the east-west and north-south directions. In taxicab geometry, we picture a line segment as we do in euclidean geometry on the coordinate plane, but more formally, define the line segment $\overline{PQ}$ as the equivalence class of paths from P to Q that travel only east-west and north-south. Alternatively, one can dictate that to travel from P to Q, one must first travel east-west from $P(x_1,y_1)$ to (x_2,y_1) and then travel north-south from (x_2,y_1) to $Q(x_2,y_2)$.

Recall that a euclidean circle with center P and radius r can be described as the set $\{Q : d_E(P,Q) = r\}$.

▷ **Activity 7.34.** Consider the points $A(0,0)$, $B(3,0)$, $C(2,1)$, $D(1,2)$, and $E(1.5, 1.5)$. Find $d_T(A,B)$, $d_T(A,C)$, $d_T(A,D)$, and $d_T(A,E)$.

▷ **Activity 7.35.** Draw a taxicab circle centered at $A(0,0)$ with radius 3.

▷ **Activity 7.36.** Draw a taxicab circle centered at $P(3,4)$ with radius 2.

▶ **Exercise 7.26.**

  a. Find a formula for the circumference (in the taxicab metric) of a taxicab circle.
  b. Find a formula for the area of a taxicab circle.

Exercise 7.26 leads us to the recognition that there is no single number in taxicab geometry that plays the role that π does in euclidean geometry, in which the circumference of a circle is $C = 2\pi r$ and the area is $A = \pi r^2$.

In euclidean geometry, the set of points equidistant from two given points is given by the perpendicular bisector of the line segment connecting the two points: If M is the midpoint of $\overline{PQ}$ and $RM \perp PQ$, it is easy to see that $\triangle PRM \cong \triangle QRM$ so $\overline{PR} = \overline{QR}$.

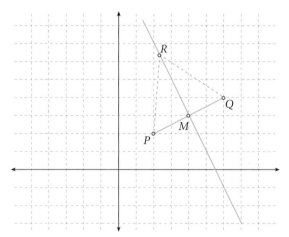

The situation in taxicab geometry is considerably different and changes depending on the relative positions of the points. In the following activities, be careful to test a number of points to be sure that you find all equidistant points.

▷ **Activity 7.37.** Consider $P(2,2)$ and $Q(6,4)$. Find the set of all points equidistant from P and Q using the taxicab metric.

▷ **Activity 7.38.**

a. Find the set of points equidistant from $A(0,0)$ and $B(6,0)$.
b. Find the set of points equidistant from $A(0,0)$ and $C(4,2)$.
c. Find the set of points equidistant from $A(0,0)$ and $D(2,4)$.
d. Find the set of points equidistant from $A(0,0)$ and $E(3,3)$.

Now let us address the question of what kind of geometry we have. Since Postulates 1 and 2 refer only to lines and line segments and we have not redefined the lines themselves, these postulates must be true in taxicab geometry. Postulate 3 is also true, although we have a new type of circle. We haven't changed the concept of angle measure, so Postulate 4 is still true. Similarly, Parallel Postulate 5 will remain true. All of our added Postulates 6–13 also remain true, with one notable exception.

Theorem 7.28. The SAS Postulate 12 is not true in the taxicab geometry.

Proof: We prove this theorem by exhibiting a counterexample. Consider the following right triangles $\triangle ABC$ and $\triangle DEF$:

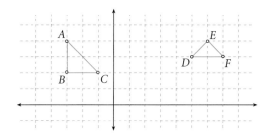

Measuring the taxicab lengths of the sides, we find that $d_T(A,B) = 2 = d_T(D,E)$, $d_T(B,C) = 2 = d_T(E,F)$, and, since we have not redefined angle measure, $\angle ABC = 90° = \angle DEF$, so SAS would imply that $\triangle ABC$ should be congruent to $\triangle DEF$. However, $d_T(A,C) = 4 \neq d_T(D,F) = 2$ so these triangles are not congruent. □

▷ **Activity 7.39.** Find an example of two noncongruent triangles that satisfy the ASA criterion.

▷ **Activity 7.40.** Find an example of two noncongruent triangles that satisfy the AAS criterion.

▷ **Activity 7.41.** Find an example of two noncongruent triangles that satisfy the SSS criterion.

We have a geometry which satisfies all of the euclidean postulates except SAS. This proves that SAS is logically independent of the other axioms. It further implies that many of Euclid's theorems which rely on congruent triangles will fail.

▷ **Activity 7.42.** Find an isosceles triangle in which the base angles are not equal.

The last counterexample we ask you to find is more difficult. One way to approach this is to remember that since we have not redefined angle measure, the sides of the triangle should be equal in euclidean distance but not in taxicab geometry.

▷ **Activity 7.43.** Find a triangle with equal base angles that is not isosceles.

♦ 7.6 Finite geometries

In this section we will investigate *finite geometries*. This area of geometry is a relatively recent mathematical innovation. In the second half of the 19th century, the belief in a single undisputable geometric system had already been brought into question (and discarded by most) by the noneuclidean geometries we have studied in Chapters 6 and 7. Even as Hilbert was working to construct his complete and independent set of axioms for euclidean geometry, other mathematicians were exploring new geometries based on changing some of the assumptions in Euclid's original set of postulates. As the name finite geometry implies, there are only finitely many objects in such a geometry. The first finite geometry was introduced by Gino Fano [1871–1952] in 1892.

Finite geometries are rather artificial mathematical systems consisting of a finite number of points. These have the advantage of being easy to study and the further advantage of being useful to show that some of our axioms are independent. We begin our exploration of finite geometry as we began our study of euclidean geometry: with the introduction of primitive terms. Our undefined terms are "point," "line," "plane," and "on." Points, lines, and planes are objects while on is a relation between points, lines, and planes. Thus we say that a point is on a line or on a plane or a line is on a plane and we may also say that a line is on a point or a plane is on a line or point. The standard model of points and lines that worked well in Euclidean geometry may have drawbacks in visualizing finite geometries, leading us to think that things are true that are not guaranteed by the axioms. In using primitive terms, one must be careful to assume only those properties explicitly given in the axioms, rather than those that intuition implies.

We begin our study of finite geometry with one of the simplest cases:

Three Point Geometry #1.

> **Axiom 1:** *There exists exactly one plane on exactly three distinct points.*
> **Axiom 2:** *For any two distinct points, there is exactly one line on the plane so that this line is on both of the points.*
> **Axiom 3:** *Not all points are on the same line.*
> **Axiom 4:** *For any two distinct lines in the plane, there is at least one point on both of them.*

▷ **Activity 7.44.** Construct a drawing to represent the three point geometry. Your figure should have three lines, and each line should be on exactly two points.

▶ **Exercise 7.27.** Using only the four axioms, prove that there are exactly three lines and that each line is on exactly two points.

We can use these four axioms to prove a stronger version of Axiom 4 as a theorem:

▶ **Exercise 7.28.** Prove that in Three Point Geometry #1, any two distinct lines are both on exactly one point.

Note that in this geometry, as in the other finite geometries we will study, lines do not have infinitely many points, but rather a specific and finite number of points. We can still say that two lines *intersect* if there is a point that lies on each of the lines. Two lines are *parallel* if they lie on the same plane and there is no point that lies on both of the lines. *Skew* lines are lines which do not lie in the same plane and do not intersect.

7.6. FINITE GEOMETRIES

Four Point Geometry #1.

Axiom 1: *There exists exactly one plane on exactly four distinct points.*
Axiom 2: *For any two distinct points, there is exactly one line on the plane so that this line is on both of the points.*
Axiom 3: *Every line lies on exactly two points.*

Three possible graphical representations for this geometry are given below. These models are simply different rearrangements of the same four points and the same four lines. Note that in the last picture, we assume the thc two diagonal lines pass each other without intersecting, since we have not drawn a point of intersection.

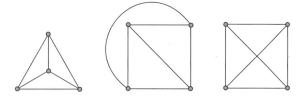

▶ **Exercise 7.29.** Prove that Four Point Geometry #1 has exactly six lines.

▶ **Exercise 7.30.** Prove that each point in Four Point Geometry #1 has exactly three lines on it.

▶ **Exercise 7.31.** Prove that given a line and a point not on the line in Four Point Geometry #1, there is exactly one line on the point that is parallel to the given line.

It is often convenient to represent a finite geometry with a drawing as we have done above. However, such drawings must be used with caution, since they often imply the existence of more points than really exist. Another way to represent a finite geometry is as a collection of sets. We can list the set of all points. Furthermore, we can list the planes and lines as sets by telling which points lie on which planes and lines. For example, for Four Point Geometry #1, we have four points we designate w, x, y, and z. We have one plane $P = \{w, x, y, z\}$, and six lines $\ell_1 = \{w, x\}$, $\ell_2 = \{w, y\}$, $\ell_3 = \{w, z\}$, $\ell_4 = \{x, y\}$, $\ell_5 = \{x, z\}$, and $\ell_6 = \{y, z\}$. Two lines intersect if they have at least one point in common, so ℓ_1 intersects ℓ_5 since $\ell_1 \cap \ell_5 = \{x\}$. Similarly, ℓ_1 is parallel to ℓ_6 since they lie on the same plane and $\ell_1 \cap \ell_6 = \emptyset$.

Finite geometries are sometimes called *incidence geometries* and can be used to study the logical independence of Postulate 1 and the separate clauses of our Incidence Postulate 6. In general, one way to show that two particular axioms, Axiom A and Axiom B, are logically independent is to exhibit two mathematical systems so that in the first system, Axiom A and B are both true, while in the second Axiom A is true but Axiom B is false. This proves that Axiom B is not a consequence of Axiom A. Note that Betweenness Postulate 7 and Plane Separation Postulate 8 make no sense in finite geometry since we are explicitly assuming that lines are discrete and so are not continuous entities. We recall the postulates we wish to study. We give the extension of Incidence Postulate 6 to planes that we will use when we study three-dimensional geometry in Chapter 10.

Postulate 1. If one is given two distinct points, then one can draw exactly one straight line on them.

Postulate 6. Incidence

1. There exists at least one plane.
2. Every line is on at least two points.
3. Every plane is on at least three noncollinear points.
4. For any three noncollinear points, there is exactly one plane on these points.
5. If a plane is on two points of a line, then every point on the line lies on the plane.
6. If two planes P_1 and P_2 have one point in common, then they have at least one other point in common.

We also wish to study the existence of parallel lines in our finite geometries. Recall that in Theorem 5.20 we proved that Euclid's Postulate 5 is logically equivalent to Playfair's Postulate. Euclid's original parallel postulate does not make sense in finite geometry, where we have no sense of angle measure. Thus, we will make use of Playfair's Postulate, which asserts, given a line on a plane and a point not on that line, there exists exactly one line on the given point and parallel to the given line. If this is false, then there are two possibilities: that there are no lines on the given point that are parallel to the given line, or that there are at least two lines on the given point that are parallel to the given line. These three possibilities give us three possible axioms to chose from:

Playfair's Postulate: Given a line on the plane and a point on the plane but not on the given line, exactly one straight line can be drawn on the plane that is on the given point and parallel to the given line.

Elliptic Parallel Postulate: Any two lines on a plane intersect.

Hyperbolic Parallel Postulate: Given a line on the plane and a point on the plane but not on the given line, more than one straight line can be drawn on the plane that are on the given point and parallel to the given line.

Definition 7.29. Let ℓ be a line and P a point not on ℓ. The pair (P, ℓ) is said to have the *euclidean property* if there is exactly one line through P that is parallel to ℓ. The pair (P, ℓ) has the *elliptic property* if no lines through P are parallel to ℓ. The pair (P, ℓ) has the *hyperbolic property* if more than one line through P are parallel to ℓ.

In Three Point Geometry #1 that we introduced above, Axiom 2 implies Postulate 1. Parts (1), (2), (3), (4), and (5) of the Incidence Postulate are also easy to verify, while Incidence Postulate (6) is vacuously true. In this geometry, Axiom 4 implies that the Elliptic Parallel Postulate is true.

A different three-point geometry can be defined by considering the plane $\mathcal{P} = \{x, y, z\}$ and defining the lines to be $\ell_1 = \{x, y\}$, $\ell_2 = \{x, z\}$, $\ell_3 = \{y, z\}$, and $\ell_4 = \{x\}$. Again Postulate 1 is true as are Incidence Postulate parts (1), (3), (4), (5), and (6). Incidence Postulate (2) obviously fails. Comparing this geometry with Three Point Geometry #1 proves that Incidence Postulate (2) is independent of the other clauses of the Incidence Postulate. Notice that the pair (x, ℓ_3) has the euclidean parallel property, but the pair (z, ℓ_1) satisfies the elliptic parallel property.

▷ **Activity 7.45.** Invent a (very small) geometry in which Postulate 1 is true and Incidence Postulate parts (1), (2), (4), (5), and (6) are true but Incidence Postulate (3) fails, proving that part (3) is independent of the other incidence postulates.

Next, we return to possible four-point geometries with points w, x, y and z.

▶ **Exercise 7.32.** Consider Four Point Geometry #1 that we investigated earlier.
   a. Is Postulate 1 true?
   b. Check each part of the Incidence Postulate.
   c. Is Playfair's Postulate true?

▶ **Exercise 7.33.** Consider a different four-point geometry in which there are four planes, $P_1 = \{w,x,y\}$, $P_2 = \{w,x,z\}$, $P_3 = \{w,y,z\}$, and $P_4 = \{x,y,z\}$. The lines are $\ell_1 = \{w,x\}$, $\ell_2 = \{w,y\}$, $\ell_3 = \{w,z\}$, $\ell_4 = \{x,y\}$, $\ell_5 = \{x,z\}$, and $\ell_6 = \{y,z\}$.
   a. Is Postulate 1 true?
   b. Check each part of the Incidence Postulate.
   c. Check the parallel properties of all pairs (P, ℓ) with P not on ℓ.
   d. Draw a model for this geometry.

▶ **Exercise 7.34.** Another four-point geometry is defined with planes $P_1 = \{w,x,y\}$, $P_2 = \{w,y,z\}$, and $P_3 = \{x,y,z\}$. The lines are $\ell_1 = \{w,x,z\}$, $\ell_2 = \{w,y\}$, $\ell_3 = \{x,y\}$, and $\ell_4 = \{y,z\}$.
   a. Is Postulate 1 true?
   b. Check each part of the Incidence Postulate.
   c. Check the parallel properties of all pairs (P, ℓ) with P not on ℓ.
   d. Draw a model for this geometry.

▶ **Exercise 7.35.** Prove that in Four Point Geometry #1, if two lines intersect, then they have exactly one point in common. Is this true in the geometries of Exercises 7.33 and 7.34?

▶ **Exercise 7.36.** Consider a universe consisting of exactly five points, $v, w, x, y,$ and z with one plane $P = \{v,w,x,y,z\}$. The lines of this geometry are $\ell_1 = \{v,w\}$, $\ell_2 = \{v,x\}$, $\ell_3 = \{v,y\}$, $\ell_4 = \{v,z\}$, $\ell_5 = \{w,x\}$, $\ell_6 = \{w,y\}$, $\ell_7 = \{w,z\}$, $\ell_8 = \{x,y\}$, $\ell_9 = \{x,z\}$, and $\ell_{10} = \{y,z\}$.
   a. Is Postulate 1 true?
   b. Check each part of the Incidence Postulate.
   c. Is Playfair's Postulate true? What about the Hyperbolic Parallel Postulate?
   d. Draw a model for this geometry.

▶ **Exercise 7.37.** Consider two universes consisting of exactly five points, $v, w, x, y,$ and z. The first geometry has planes $P_1 = \{v,w,x,y\}$, $P_2 = \{v,w,x,z\}$, $P_3 = \{v,y,z\}$, $P_4 = \{w,y,z\}$, and $P_5 = \{x,y,z\}$ and lines $\ell_1 = \{v,w,x\}$, $\ell_2 = \{v,y\}$, $\ell_3 = \{v,z\}$, $\ell_4 = \{w,y\}$, $\ell_5 = \{w,z\}$, $\ell_6 = \{x,y\}$, $\ell_7 = \{x,z\}$, and $\ell_8 = \{y,z\}$. The second geometry has planes $Q_1 = \{v,w,x,y\}$, $Q_2 = \{v,w,x,z\}$, $Q_3 = \{w,x,y,z\}$, and $Q_4 = \{v,y,z\}$ and the same lines as the first. Use these two models to prove the independence of one of the parts of the Incidence Postulate.

The next example is a modification of the one that Fano introduced in 1892. This version is based on a single plane and is scaled down from his more complex original model.

Fano's Geometry

Axiom 1: *There exists exactly one plane.*
Axiom 2: *There exists at least one line on the plane.*

Axiom 3: *Every line on the plane has exactly three points on it.*
Axiom 4: *Not all points on the plane are on the same line.*
Axiom 5: *For any two distinct points on the plane, there is exactly one line on both of them.*
Axiom 6: *Any two lines on a plane intersect.*

Note that Axioms 4–6 of Fano's geometry appear in Three Point Geometry #1 above. Axiom 5 also occurs in Four Point Geometry #1 along with a two-point version of Axiom 3. Axiom 6 is the Elliptic Parallel Postulate. Before exploring the properties of Fano's geometry, we introduce Young's geometry for comparison. John Wesley Young changes only the sixth axiom, replacing the elliptic axiom with Playfair's Postulate.

Young's Geometry

Axiom 1: *There exists exactly one plane.*
Axiom 2: *There exists at least one line on the plane.*
Axiom 3: *Every line on the plane has exactly three points on it.*
Axiom 4: *Not all points on the plane are on the same line.*
Axiom 5: *For any two distinct points on the plane, there is exactly one line on both of them.*
Axiom 6: *Given a line on the plane and a point on the plane but not on the given line, exactly one straight line can be drawn on the plane that is on the given point and parallel to the given line.*

▷ **Activity 7.46.** Build as many models as you can using only Axioms 1–5 of Fano's and Young's geometries. How many points must you have as a minimum? What relationships can you establish, if any, between the points and lines? (Hint: Begin with Axioms 1 and 2 so there exists at least one line. Axiom 3 gives us at least three points on that line. Axiom 4 gives us a fourth point not on the original line, and so on.) Can you find a model that satisfies Axioms 1–5 and, in addition, the Hyperbolic Parallel Postulate?

▶ **Exercise 7.38.** In Fano's geometry, prove there exist exactly seven points and seven lines.

▶ **Exercise 7.39.** In Young's geometry, prove there exist exactly nine points and twelve lines.

▶ **Exercise 7.40.** In Young's geometry, prove that two distinct lines each parallel to a third line are parallel to each other.

▷ **Activity 7.47.** In both Fano's and Young's geometries, suppose we replace Axiom 3: "Every line has exactly three points on it" with "Every line has exactly four points on it" or even "Every line has exactly n points on it." How many points and lines will the new geometries have?

We conclude this section with the finite analogs of some well-known classical geometries.

Finite Affine Geometry

Axiom 1: *There exists exactly one plane.*
Axiom 2: *There exists at least one line on the plane.*
Axiom 3: *Every two distinct points lie on a unique line.*
Axiom 4: *There is a set of four points so that no three of these points lie on the same line.*
Axiom 5: *Given a line on the plane and a point on the plane but not on the given line, exactly one straight line can be drawn on the plane that is on the given point and parallel to the given line.*

▶ **Exercise 7.41.** In the Finite Affine Geometry, prove that if two distinct lines are each parallel to a third line, then they are parallel to each other.

▶ **Exercise 7.42.** In the Finite Affine Geometry, prove that all the lines are on the same number of points and all points are on the same number of lines.

We adopt the convenient convention that a line is considered to be parallel to itself. Accepting this, we get the following computations:

▶ **Exercise 7.43.** In the Finite Affine Geometry, prove that there is an integer $n > 0$ such that:

  a. The total number of points is n^2.
  b. Each line is parallel to n lines.
  c. The total number of lines is $n^2 + n$.
  d. Each line intersects n^2 lines.

Note that Axiom 5 of the Finite Affine Geometry is Playfair's Postulate. Replacing this with a strong form of the Elliptic Parallel Postulate gives us our final example. This is again a finite version of an important class of geometry:

Finite Projective Geometry

  Axiom 1: *There exists exactly one plane.*
  Axiom 2: *There exists at least one line on the plane.*
  Axiom 3: *Every two distinct points lie on a unique line.*
  Axiom 4: *There is a set of four points so that no three of these points lie on the same line.*
  Axiom 5: *Every two lines intersect at a unique point.*

▶ **Exercise 7.44.** In the Finite Projective Geometry, prove that given a point, all of the other points lie on lines that lie on the given point.

▶ **Exercise 7.45.** In the Finite Projective Geometry, prove that there is an integer $n > 0$ such that:

  a. Every point belongs to $n + 1$ lines.
  b. Every line contains $n + 1$ points.
  c. The total number of points is $n^2 + n + 1$.
  d. The total number of lines is $n^2 + n + 1$.

There is a way to construct a projective geometry from any affine geometry. We start with an affine geometry. With the convention that a line is considered to be parallel to itself, we define a relation on the set of lines by:

$$\ell_1 \sim \ell_2 \text{ if and only if } \ell_1 \text{ is parallel to } \ell_2.$$

▶ **Exercise 7.46.** Show that $\sim$ is an equivalence relation on the set of lines.

Now we can use this equivalence relation to define a projective geometry:

- The plane of the projective geometry is the plane of the affine geometry.
- The points of the projective geometry are the points of the affine geometry plus one point P_ℓ for each equivalence class of lines $[\ell]$. These additional points are called the *points at infinity*.
- The lines of the projective geometry are the lines of the affine geometry plus one line that lies on all the points at infinity.
- The points lie on the lines as in the affine geometry, but additionally P_ℓ lies on ℓ where P_ℓ is the point at infinity determined by the equivalence class of line ℓ.

▶ **Exercise 7.47.** Show that if we begin with a finite affine geometry, the model constructed above is a finite projective geometry.

♦ 7.7 Projects

Project 7.1. Describe a full proof of the Bolyai-Gerwein theorem.

Project 7.2. Research Archimedes' method for finding the area under a parabola.

Project 7.3. Study Hilbert's proof of the transitivity of the equivalence relation defined by decomposition. See Theorem 7.15 above.

Project 7.4. Derive the explicit formula for stereographic projection.

Project 7.5. Investigate the details of performing the analogs of the ruler and compass constructions on the spherical model for elliptic geometry.

Project 7.6. We call two models of a finite geometry *isomorphic* if there are bijections between the corresponding sets of points and lines that preserve incidence.

1. Find all the models, up to isomorphism, that contain three points and satisfy Postulate 1 and Incidence Postulate 6.
2. Find all the models, up to isomorphism, that contain four points and satisfy Postulate 1 and Incidence Postulate 6.
3. Find all the models, up to isomorphism, that contain five points and satisfy Postulate 1 and Incidence Postulate 6.

III. Symmetry

8. Isometries

♦ 8.1 Transformation geometry

Throughout this section, we assume euclidean geometry, so Postulates 1–13, Propositions I.1–I.48, and all of the results from Chapters 3 and 5 are true. Let us reconsider Euclid's original proof of Proposition I.4 from *The Elements*, the SAS criterion for congruence of triangles. This is commonly assumed as a postulate by modern geometers, but there is still an important idea that can be gleaned from studying his proof.

Proposition I.4. *If two triangles have two sides equal to two sides respectively, and have the angles contained by the equal straight lines equal, they will also have the base equal to the base, the triangle equal to the triangle, and the remaining angles will be equal to the remaining angles respectively, namely those which the equal sides subtend.*

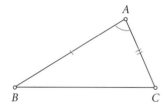 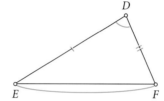

Proof: Let ABC, DEF be two triangles having two sides AB, AC equal to the two sides DE, DF respectively, namely AB to DE and AC to DF, and the angle BAC equal to the angle EDF.
 I say that the base BC is also equal to the base EF, the triangle ABC will be equal to the triangle DEF, and the remaining angles will be equal to the remaining angles respectively, namely those which the equal sides subtend, that is, the angle ABC to the angle DEF, and the angle ACB to the angle DFE.
 For, if the triangle ABC be applied to the triangle DEF and if the point A be placed on the point D and the straight line AB on DE, then the point B will also coincide with E, because AB is equal to DE.
 Again, AB coinciding with DE, the straight line AC will also coincide with DF, because the angle BAC is equal to the angle EDF; hence the point C will also coincide with the point F, because AC is again equal to DF.
 But B also coincided with E; hence the base BC will coincide with the base EF. [For if, when B coincides with E and C with F, the base BC does not coincide with the base EF, two straight lines will enclose a space: which is impossible. Therefore, the base BC will coincide with EF] and will be equal to it.
 Thus the whole triangle ABC will coincide with the whole triangle DEF, and will be equal to it. And the remaining angles will also coincide with the remaining angles and will be equal to them, the angle ABC to the angle DEF, and the angle ACB to the angle DFE.

 This proof depends on the idea of superposition: picking up triangle △ABC and placing it on top of △DEF, so that point A lies on top of D and the line AB on top of DE. This is an unprecedented action and cannot be justified by Euclid's postulates or common notions, none of which address the idea of moving objects. Euclid uses this concept of superposition once more, in the proof of Proposition I.8: SSS

Congruence. He does not use it in Proposition I.5 to show that the base angles of an isosceles triangle are equal, although there is a tradition that Thales' original proof of this theorem was done using superposition.

The idea of moving triangles around requires rather more justification than Euclid gave, and will be the object of this section. The approach we have used thus far is to decide whether two figures are congruent by measuring the lengths of sides and the size of the angles. Thus, the traditional approach is to say that figures are congruent because they have the same measurements. Felix Klein introduced a new approach to geometry: the *Erlanger Program*, named for his university. He presented a unifying idea: to study geometry in general by studying certain families of functions, referred to as groups of transformations or isometries. In Klein's view, measurements are the same because figures are congruent, and figures are defined to be congruent if they can be superimposed. In other words, two figures are considered to be congruent if there is a function taking the first to the second, but not just any function can be used, only functions belonging to a certain family. Different fields of mathematics use different types of functions: In a calculus class one studies differentiable functions, in linear algebra one studies linear functions represented by matrices, and so on. Different families or groups of functions give rise to different geometries, and we then study the properties that the functions preserve. We will not follow Klein's full program of reinventing geometry, but will explore the idea of geometric functions.

Let us first consider the particular type of function that will give the same concept of congruence we are used to in euclidean geometry. The ideal geometric function should preserve all geometric properties, such as length, angle measure, congruence, parallelism of lines, and area.

▷ **Activity 8.1.**

a. Find the ways to move an object in a plane so that length, angle measure, congruence and area do not change. You may have seen some of these motions in other classes. Describe as many different types of motion as you can.

b. Cut a scalene triangle from an index card. Place your triangle on a sheet of paper and trace around it. Now, hold the triangle at least eight inches above the paper and let it go. Trace the new position. Can you describe the relationship between your two drawings in terms of the motions you described above? If not, add to your list and then drop the triangle again. You should have described four different types of motion to complete this activity.

The standard terms for the four functions that preserve geometric properties on the euclidean plane are rotation, reflection, translation, and glide reflection. We will study these in depth, but here is a brief description of each.

- *Rotation about a point:* Rotate the object about a point by a chosen angle:

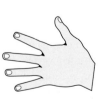

- *Reflection across a line:* Reflect the object across a chosen line:

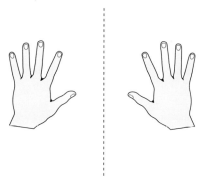

- *Translation in a given direction by a given distance:* Slide the object in a chosen direction by a chosen distance:

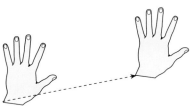

- *Glide reflection across a line by a given distance:* Reflect the object across a chosen line and then slide it along that line by a chosen distance:

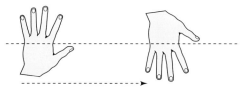

The previous and the following activities are designed to allow you to play with these motions before analyzing them more formally.

▷ **Activity 8.2.** In these activities, consider the white triangle as the original and the shaded triangle the result of one of the motions discussed above.

  a. Construct the point about which the object is rotated.

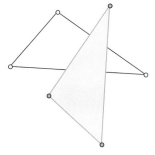

b. Construct the point about which the object is rotated.

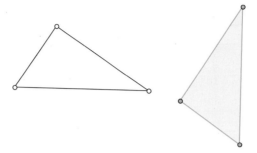

c. Construct the line about which the object is reflected.

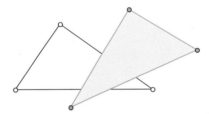

d. Construct the line about which the object is reflected.

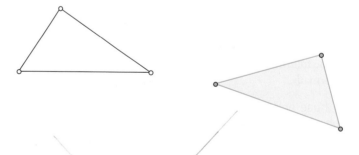

e. Construct the direction and distance of translation.

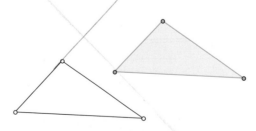

f. Construct the line of reflection and the distance of translation for this glide reflection.

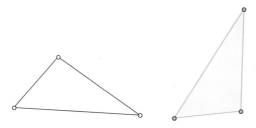

As mentioned above, one example of a function that preserves geometric properties is rotation: let us rotate the plane, including the triangle $\triangle ABC$ shown, about the origin by $45°$ to get a new triangle $\triangle A'B'C'$:

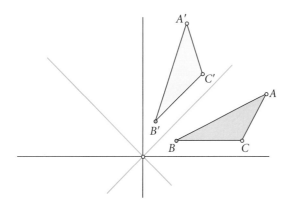

Functions such as the rotation above are called *isometries* (from the Greek words for "same measure"), or *rigid motions*.

Definition 8.1. An *isometry* is a function T defined on the plane so that for any pair of points P and Q, the distance from P to Q is the same as the distance from $T(P)$ to $T(Q)$. I.e.,

$$\overline{PQ} = \overline{T(P)T(Q)}.$$

Theorem 8.2. If T is an isometry, then T is one-to-one and T takes lines to lines.

Proof: To show that a function is one-to-one, we must show that if $T(A) = T(B)$ for points A and B, then $A = B$. If $T(A) = T(B)$, then $\overline{T(A)T(B)} = 0$. Since T is an isometry, $\overline{T(A)T(B)} = 0 = \overline{AB}$, so $A = B$. We next show that T takes lines to lines. Let A, B, and C be three points on a line and assume that B lies between A and C, so $\overline{AC} = \overline{AB} + \overline{BC}$. Since T is an isometry, $\overline{T(A)T(C)} = \overline{T(A)T(B)} + \overline{T(B)T(C)}$. If $T(A)$, $T(B)$ and $T(C)$ were not collinear, then the triangle inequality (Euclid's Proposition I.20) would imply that $\overline{T(A)T(C)} < \overline{T(A)T(B)} + \overline{T(B)T(C)}$. Therefore, $T(B)$ must lie on the line determined by $T(A)$ and $T(C)$ and furthermore, must lie between those two points. Thus, T takes all of the points on the line AB to points on the line $T(A)T(B)$. □

Exercise 8.1. Let $\triangle ABC$ be a triangle and T be an isometry of the plane, with $A' = T(A)$, $B' = T(B)$, and $C' = T(C)$. Prove that the triangles $\triangle ABC$ and $\triangle A'B'C'$ are congruent.

Corollary 8.3. If T is an isometry, then T preserves angle measure.

▶ **Exercise 8.2.** Prove Corollary 8.3.

▶ **Exercise 8.3.** Let $\mathcal{C}$ be a circle and T an isometry of the plane. Prove that $T(\mathcal{C})$ is a circle with the same radius as $\mathcal{C}$.

We next prove that isometries preserve the property of being parallel, at least in the context of euclidean geometry.

Corollary 8.4. Let ℓ_1 and ℓ_2 be parallel lines on the plane and T an isometry of the plane. Then $T(\ell_1)$ is parallel to $T(\ell_2)$.

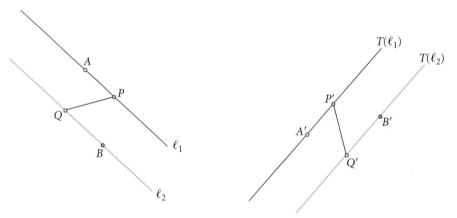

Proof: We are given $\ell_1 \| \ell_2$. Choose points A and P on ℓ_1 and Q and B on ℓ_2 and draw the transversal PQ. Since the lines are parallel, by Euclid's Proposition I.29, the alternate interior angles are equal, so $\angle APQ = \angle PQB$. Apply the isometry T and let $A' = T(A)$, $P' = T(P)$, $Q' = T(Q)$, and $B' = T(B)$. Since $A, P \in \ell_1$, we know that $A', P' \in T(\ell_1)$ and similarly $Q', B' \in T(\ell_2)$. By Corollary 8.3 above, $\angle A'P'Q' = \angle APQ$ and $\angle P'Q'B' = \angle PQB$. Therefore, $\angle A'P'Q' = \angle P'Q'B'$, and so by Euclid's Proposition I.27, $T(\ell_1)$ is parallel to $T(\ell_2)$. □

Next, let us investigate the types of isometries more formally.

◊ 8.1.1 Translation

Translation involves moving a figure a certain distance in a certain direction. We indicate this distance and direction by giving a beginning point P and an ending point Q. In other words, we must pay attention not only to the line segment $\overline{PQ}$, but also to its direction. (Thus, a transformation is defined by a vector.)

Definition 8.5. Given two distinct points P and Q, for any point A on the euclidean plane, define the *translation* $T_{P,Q}(A)$ of A to be the unique point so that $\overline{AT_{P,Q}(A)} = \overline{PQ}$ and the ray $\overrightarrow{AT_{P,Q}(A)}$ is parallel to $\overrightarrow{PQ}$ and points in the same direction.

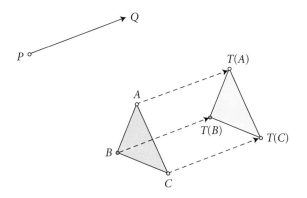

Note that if P' and Q' were any other two points satisfying $PQ \| P'Q'$, $\overline{PQ} = \overline{P'Q'}$ and the rays $\overrightarrow{PQ}$ and $\overrightarrow{P'Q'}$ point in the same direction, then P' and Q' would define the same translation. Thus, under these conditions $T_{P,Q} = T_{P',Q'}$.

Theorem 8.6. Translation is an isometry.

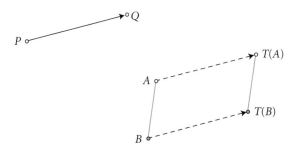

Proof: Let P and Q be the given points that define the translation $T_{P,Q}$, and let A and B be any two distinct points. Then we know that $\overline{AT_{P,Q}(A)} = \overline{PQ} = \overline{BT_{P,Q}(B)}$, that $AT_{P,Q}(A) \| PQ \| BT_{P,Q}(B)$, and that $\overrightarrow{PQ}, \overrightarrow{AT(A)}$, and $\overrightarrow{BT(B)}$ all point in the same direction. Thus $AT_{P,Q}(A)T_{P,Q}(B)B$ is a parallelogram by Proposition I.33, and so by Proposition I.34, $\overline{AB} = \overline{T_{P,Q}(A)T_{P,Q}(B)}$. Therefore, $T_{P,Q}$ is an isometry. □

▷ **Activity 8.3.** Sketch the direction and distance of the translation below.

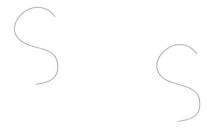

▶ **Exercise 8.4.** If you know an isometry *T* is a translation and you know the location of a single point *A* and its image *T(A)*, can you completely identify the translation (i.e., find suitable points *P* and *Q*)? If so, describe how. If not, explain why not.

The inverse of a function is the operation that puts everything back where you found it. Not all functions have inverses, but the isometries do. If T is an isometry, then we denote its inverse as T^{-1} and then $T^{-1}(T(A)) = A = T(T^{-1}(A))$ for every point A.

▶ **Exercise 8.5.** Let $T_{P,Q}$ be a translation. Describe T^{-1}.

◇ **8.1.2 Rotation**

Rotation is another example of an isometry. A rotation of the plane is defined by giving the center of rotation O and the angle θ (measured counterclockwise).

Definition 8.7. Given a center O and an angle θ, for any point A in the plane, define the *rotation* $R_{O,\theta}(A)$ of A to be the unique point such that $\overline{OA} = \overline{OR_{O,\theta}(A)}$ and $\angle AOR_{O,\theta}(A) = \theta$, where θ is measured counterclockwise from $\overrightarrow{OA}$ to $\overrightarrow{OR_{O,\theta}(A)}$.

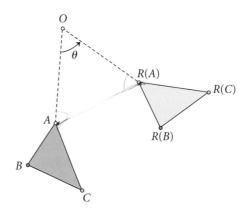

Theorem 8.8. Rotation is an isometry.

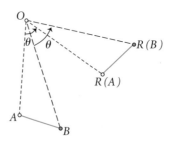

Proof: Let O be the center of rotation and θ the angle for the rotation $R_{O,\theta}$. Let A and B be two points on the plane and assume that neither A nor B coincide with O. Then $\overline{OA} = \overline{OR_{O,\theta}(A)}$ and $\overline{OB} = \overline{OR_{O,\theta}(B)}$. We also know that $\angle AOR_{O,\theta}(A) = \theta = \angle BOR_{O,\theta}(B)$. Note that

$$\angle AOB = \angle AOR_{O,\theta}(A) - \angle BOR_{O,\theta}(A) = \angle BOR_{O,\theta}(B) - \angle BOR_{O,\theta}(A) = \angle R_{O,\theta}(A)OR_{O,\theta}(B).$$

Therefore, $\triangle AOB$ is congruent to $\triangle R_{O,\theta}(A)OR_{O,\theta}(B)$ by SAS, so $\overline{AB} = \overline{R_{O,\theta}(A)R_{O,\theta}(B)}$, and $R_{O,\theta}$ is an isometry. □

▶ **Exercise 8.6.** Prove Theorem 8.8 in the case when the rotation center O coincides with A and in the case in which $\angle AOB > \angle AOR_{O,\theta}(A)$.

▷ **Activity 8.4.** Find the center of rotation and the angle for the rotation pictured below. Explain how you found these.

▶ **Exercise 8.7.** If you know an isometry R is a rotation and you know the location of a single point A and its image $R(A)$, can you completely identify the rotation (i.e., find the center and the angle of rotation)? If so, describe how. If not, explain why not.

▶ **Exercise 8.8.** Let $R_{O,\theta}$ be the rotation about O by angle θ. Describe R^{-1}.

◊ 8.1.3 Reflection in a line

Reflection is our third example of an isometry. A reflection of the plane is defined by the line ℓ (called the *line of reflection* or the *mirror line*) we wish to reflect across.

Definition 8.9. Given a line ℓ in the plane, for any point A which does not lie on ℓ, define the *reflection* $F_\ell(A)$ of A to be the unique point found by dropping a perpendicular from A to ℓ and extending the perpendicular to the point $F_\ell(A)$ an equal distance on the other side of ℓ. Thus, ℓ is the perpendicular bisector of $\overline{AF_\ell(A)}$. If A lies on the line of reflection ℓ, then define $F_\ell(A) = A$.

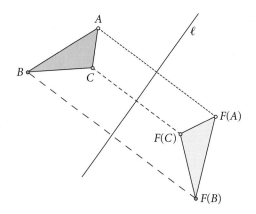

Theorem 8.10. Reflection across a line is an isometry.

▶ **Exercise 8.9.** Prove Theorem 8.10.

▷ **Activity 8.5.** Find the mirror line for the reflection pictured below:

▶ **Exercise 8.10.** If you know an isometry F is a reflection and you know the location of a single point A and its image $F(A)$, can you completely identify the reflection (i.e., find the line of reflection)? If so, describe how. If not, explain why not.

▶ **Exercise 8.11.** Let F_ℓ be the reflection across line ℓ. Describe F^{-1}.

◇ 8.1.4 Glide reflection

The fourth and last of the euclidean isometries is the only one to require two steps to produce. A glide reflection involves a reflection across a line, followed by a translation in the direction of the line for a given distance (or vice versa, the order does not matter in this case). Although this sounds like a composition, we delete the result of the intermediate step:

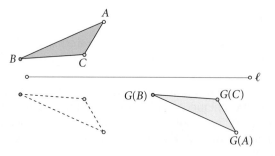

Definition 8.11. Given a line ℓ in the euclidean plane and two points P and Q so that PQ is parallel to ℓ, for any point A, define the *glide reflection* $G_{\ell,P,Q}(A)$ of A to be the unique point found by reflecting A across ℓ and then translating by $\overline{PQ}$.

Theorem 8.12. Glide reflection is an isometry.

▶ **Exercise 8.12.** Prove Theorem 8.12.

▶ **Exercise 8.13.** If you know an isometry G is a glide reflection and you know the location of a single point A and its image $G(A)$, can you completely identify the glide reflection (i.e., find the line of reflection and the direction and distance of the translation)? If so, describe how. If not, explain why not.

▶ **Exercise 8.14.** Let G be the glide reflection across line ℓ with a translation given by P and Q. Describe G^{-1}.

▷ **Activity 8.6.** Identify which of the four isometries was used to generate each of the pictures below. If the isometry is a rotation, identify the center of rotation. For a reflection, identify the line of reflection. For a translation, find the direction and distance of translation. For a glide reflection, find the line of reflection and the direction and distance of translation.

a.

b.

c.

d.

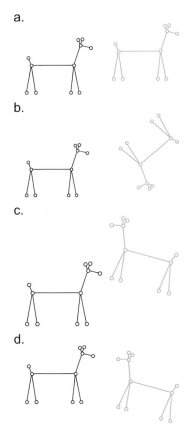

Some geometric software packages, such as *The Geometer's Sketchpad*, *Cabri*, and *GeoGebra*, have built-in isometry actions, although *Cinderella* and *Geometry Playground* do not. Of those that do, some do not have a command for the glide reflection. One can perform any of the isometries by appropriate ruler and compass constructions, but this can be tedious. If you have access to any of the software packages that include isometries, you can use these to explore their effects.

▷ **Activity 8.7.** Using geometric software, create a non-symmetric object. For each of the four isometries, we want to learn how to move your object appropriately.

  a. Place a point to use as the rotation center and figure out how to rotate your object by a chosen angle. Experiment by moving the rotation center and varying the rotation angle.
  b. Draw a line to use as a mirror line and figure out how to reflect your object across the chosen line. Experiment with moving the line.
  c. Draw two points and figure out how to translate your object by the direction and distance given by the chosen points. Experiment by changing the position of the given points.
  d. Figure out how to get the result of a glide reflection of your object. Note that you may have to hide the results of intermediate steps.

◇ **8.1.5 Orientation**

A property preserved by the action of an isometry is called an *invariant*. By the definition of an isometry, distance is invariant. In Exercise 8.1 and Corollary 8.3, you showed that congruence and angle measure are also invariants. Corollary 8.4 proved that parallelism is an invariant property.

One property that isometries may not preserve is called *orientation*. A triangle has clockwise orientation if, when one traverses the circumference in the clockwise direction, the vertices are labeled in alphabetic order. Below on the left is a triangle with clockwise orientation, while the triangle on the right has counterclockwise orientation. Of course, in a sense the labeling of a triangle, and thus its orientation, are arbitrary, so what we are interested in is whether this orientation changes under the action of an isometry.

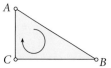 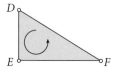

We must determine whether the triangle retains its orientation after being acted on by one of the isometries. This concept of the change in orientation does not depend on the particular orientation chosen for the original triangle. An isometry T is called *orientation-preserving* if $\triangle ABC$ and $\triangle T(A)T(B)T(C)$ have the same orientation, either both clockwise or both counterclockwise. If $\triangle ABC$ and $\triangle T(A)T(B)T(C)$ have opposite orientations, one clockwise and the other counterclockwise, then T is said to be *orientation-reversing*.

Another way of thinking about orientation is to consider the action on a hand. An orientation-preserving isometry will take a left hand to a left hand, while a orientation-reversing isometry will change a left hand into a right hand.

▶ **Exercise 8.15.** Which of the four isometries (translation, rotation, reflection, and glide reflection) preserve orientation, and which reverse orientation?

An isometry that preserves orientation is called a *direct isometry*, while one that reverses orientation is called *indirect* or *opposite*. Some properties, such as congruence and parallelism, are preserved by any isometry. Orientation is one example of an invariant that is preserved only by certain types of isometries and may therefore be useful in classifying the type of an isometry.

◇ 8.1.6 Composing isometries

Like other functions, isometries can be combined by *composition* or product. We define $S \circ T(A) = S(T(A))$. In other words, $S \circ T$ gives us instructions to first perform isometry T, then apply isometry S to the result.

▶ **Exercise 8.16.** Prove that the composition of two isometries is an isometry.

▶ **Exercise 8.17.** Prove that the composition of two translations is a translation. Do not assume that your translations are parallel. Does the order in which you perform the two translations matter? Justify your answer.

▶ **Exercise 8.18.** Prove that the composition of two rotations about the same center is a rotation. What is the angle of this rotation? Does the order in which you perform the two rotations matter? Justify your answer.

▷ **Activity 8.8.** Explore the composition of two rotations about different centers. Which isometry is the result? Does the order in which you perform the two rotations matter? Justify your answer.

▷ **Activity 8.9.** In the exercises above, we composed two translations to get a different translation. We composed two rotations about the same center to get a different rotation about that center. To finish exploring composition, there are several more ways to pair isometries. There are four different types of functions so there are 16 ordered pairs. Exercise 8.17 and Activity 8.8 discussed two of those pairs. For each of the remaining pairs, use geometric software to determine what type of isometry is the result of the composition. Then consider whether the order of composition matters. For example, if R is a rotation and F is a reflection, is $R \circ F = F \circ R$ in general? Write your results in the table below. If order does change the composition, give an example to demonstrate that composition is not commutative.

Composition	Result	Order matters?
TT	T	no
TR, RT		
TF, FT		
TG, GT		
RR	R	no
RF, FR		
RG, GR		
FF		
FG, GF		
GG		

By Exercise 8.17, we know that the composition of two translations is a translation. Suppose we know only the final translation. Can we recover the original isometries that were composed to get the end result? In general, the answer to this question is no—there are infinitely many different compositions that could be combined to give the same result. Compositions give a unique result, but decompositions do not. However, there is an interesting idea to be pursued by means of decompositions.

Lemma 8.13. Any translation can be written as a composition of two reflections.

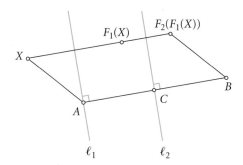

Proof: Let $T_{A,B}$ be a translation in the direction from A to B and let ℓ_1 be the line through A perpendicular to AB. Let C be the midpoint of $\overline{AB}$ and let ℓ_2 be the line through C perpendicular to AB. Let F_1 denote reflection across ℓ_1 and F_2 reflection across ℓ_2. Let X be any point on the plane and construct $F_1(X)$ and $F_2(F_1(X))$. Draw $\overline{XA}$, $\overline{F_2(F_1(X))B}$, and $\overline{XF_2(F_1(X))}$. Since $\ell_1 \perp AB$ and $XF_1(X) \perp \ell_1$, and $\ell_2 \perp AB$ and $F_1(X)F_2(F_1(X)) \perp \ell_2$, $XF_1(X)F_2(F_1(X))$ is a straight line parallel to AB. Since $B = F_2(F_1(A))$, $XABF_2(F_1(X))$ is a parallelogram and so $\overline{XF_2(F_1(X))} = \overline{AB}$.

Thus $T_{A,B}(X) = F_2 \circ F_1(X)$ for an arbitrary point X. Therefore, the translation $T_{A,B} = F_2 \circ F_1$. Note that this representation is not unique. □

▷ **Activity 8.10.** What can you say about the relationship between the direction of the translation $T_{A,B}$ and the two lines of reflection ℓ_1 and ℓ_2 for A, B, ℓ_1, and ℓ_2 as in Lemma 8.13? What can you say about the relationship between the distance $\overline{AB}$ of the translation and the distance d between the two lines of reflection?

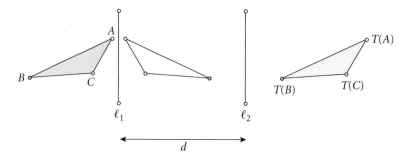

▶ **Exercise 8.19.** Find two different lines of reflection which compose to give the same translation as the example above. Note that the intermediate white figure will not be in the same position, but the original figure $\triangle ABC$ and final shaded figure $\triangle T(A)T(B)T(C)$ should end up in the same positions as above.

▶ **Exercise 8.20.** Prove that any glide reflection can be written as the composition of reflections in three lines.

▷ **Activity 8.11.** What isometry do you get when you first reflect across the line ℓ_1 and then across the perpendicular line ℓ_2 in the diagram below?

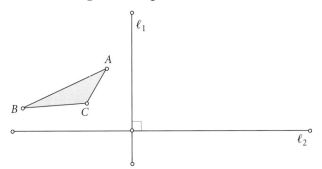

▷ **Activity 8.12.** What isometry do you get when you first reflect across the line ℓ_1 and then across ℓ_2 in the diagram below?

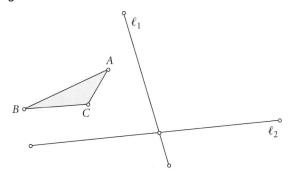

▷ **Activity 8.13.** Activities 8.11 and 8.12 should have shown you that rotations can also be expressed as the composition of two reflections. What can you say about the relationship between the center of rotation and the two lines of reflection? What can you say about the relationship between the angle of rotation and the two lines of reflection?

Lemma 8.14. If $R_{O,\theta}$ is a rotation with center O and angle $\theta \neq 0$, then $R_{O,\theta} = F_2 \circ F_1$, where F_1 and F_2 are reflections in lines ℓ_1 and ℓ_2 respectively for ℓ_1 and ℓ_2 that intersect at point O and so that the angle formed between ℓ_1 and ℓ_2 is $\frac{\theta}{2}$.

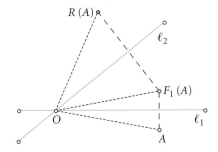

▶ **Exercise 8.21.** Prove Lemma 8.14.

The idea of writing an isometry as a composition of reflections is central to the proof that we have found all possible isometries of the euclidean plane.

◇ 8.1.7 Classification of isometries

We next prove two *rigidity theorems*. A point A is called a *fixed point* for a transformation T if $T(A) = A$. For example, if $R_{O,\theta}$ is a rotation with rotation center O and $0 < \theta < 360°$, then O is the only fixed point and every other point on the plane moves under the action of the rotation.

Theorem 8.15. If an isometry T fixes two distinct points A and B, then T fixes all points on the line AB.

Proof: If T fixes A and B, then $T(A) = A$ and $T(B) = B$. Let C be another point on line AB. We wish to show that $T(C) = C$. There are three cases to consider:

Case 1: C lies between A and B: Note that $\overline{AC} + \overline{CB} = \overline{AB}$ since C lies on line AB between A and B. Since T is an isometry, $\overline{AC} = \overline{T(A)T(C)} = \overline{AT(C)}$ and $\overline{CB} = \overline{T(C)T(B)} = \overline{T(C)B}$. Thus $\overline{AB} = \overline{AC} + \overline{CB} = \overline{AT(C)} + \overline{T(C)B}$. Therefore, $T(C)$ must lie between A and B or we would have a contradiction to the Triangle Inequality (Proposition I.20). Furthermore, since there is only one point at distance $\overline{AC}$ from A in the direction of B, $T(C)$ must coincide with C, so $T(C) = C$.

Case 2: B lies between A and C: Note that $\overline{AB} + \overline{BC} = \overline{AC}$ since B lies on line AC between A and C. Since T is an isometry, $\overline{BC} = \overline{T(B)T(C)} = \overline{BT(C)}$. Thus $\overline{AC} = \overline{AB} + \overline{BC} = \overline{AB} + \overline{BT(C)}$. Therefore, $T(C)$ must lie on the far side of B from A. Furthermore, since there is only one point at distance $\overline{AC}$ from A in the direction of B, $T(C)$ must coincide with C, so $T(C) = C$.

Case 3: A lies between B and C. The proof is similar to Case 2. □

▶ **Exercise 8.22.** Describe the fixed points for an arbitrary rotation, reflection, translation, and glide reflection of the plane.

▶ **Exercise 8.23.** A line is fixed by an isometry S if $S(\ell) = \ell$. Thus points on this line are mapped to other points on the same line. Describe the fixed lines for an arbitrary rotation, reflection, translation, and glide reflection of the plane.

The *identity transformation* is the function T so that $T(X) = X$ for every point X. The identify transformation can be considered as a translation of distance 0, or a rotation of angle 0.

Theorem 8.16. If an isometry T of the plane has three noncollinear fixed points, then T is the identity transformation.

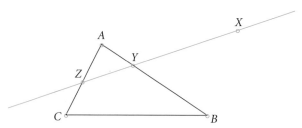

Proof: Let A, B, and C be the noncollinear fixed points of T. By Theorem 8.15, T fixes all points on the lines AB, AC, and BC, and so fixes $\triangle ABC$. Let X be any point on the plane. If X happens to lie on AB, AC, or BC, then $T(X) = X$ by Theorem 8.15. Assume that X does not lie on AB, AC, or BC and choose another point Y on edge $\overline{AB}$ of the triangle between A and B. By Pasch's Theorem 2.11, the line XY must intersect $\triangle ABC$ at one other point, Z. Since T fixes the triangle, both Y and Z are fixed points and T must fix all points on the line YZ, including point X. Thus, $T(X) = X$ for any point X. □

Corollary 8.17. If T and S are isometries of the plane, and there are three noncollinear points A, B, and C so that $T(A) = S(A)$, $T(B) = S(B)$, and $T(C) = S(C)$, then $T = S$.

▶ **Exercise 8.24.** Prove Corollary 8.17.

Theorem 8.18. Any isometry of the plane can be expressed as the composition of at most three reflections.

Proof: We are given an isometry T. Choose three noncollinear points A, B and C and find $A' = T(A)$, $B' = T(B)$, and $C' = T(C)$. Consider the triangles $\triangle ABC$ and $\triangle A'B'C'$. We proceed by cases determined by the number of fixed points for T:

Case 1: T fixes all three points: If $A' = A$, $B' = B$, and $C' = C$, then by Theorem 8.16, T is the identity transformation.

Case 2: T fixes exactly two of the points, $A' = A$ and $B' = B$: By Theorem 8.15, T fixes the line $\ell = AB = A'B'$. Also, note that $F_\ell(A) = A$ and $F_\ell(B) = B$. Drop $CX \perp AB$. Since X lies on ℓ, $X = T(X)$ and thus $\overline{CX} = \overline{T(C)T(X)} = \overline{C'X}$. Draw $\overline{C'X}$. Note that X might coincide with A or B, so choose a point D on ℓ distinct from X and note that $D = T(D)$. By Corollary 8.3, $\angle CXD = 90° = \angle T(C)T(X)T(D) = \angle C'XD$. Hence, $C'XC$ is a straight line by Proposition I.14 and ℓ is the perpendicular bisector of $\overline{CC'}$. Therefore, $C' = T(C) = F_\ell(C)$ and by Corollary 8.17, $T = F_\ell$.

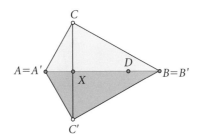

Case 3: T fixes exactly one of the points, $A' = A$: Let ℓ_1 be the perpendicular bisector of $\overline{BB'}$ and let F_1 denote reflection across ℓ_1, so $F_1(B) = B'$ and $F_1(B') = B$. Since $\overline{T(A)T(B)} = \overline{A'B'} = \overline{AB'} = \overline{AB} = \overline{AF_1(B)}$, $A = A'$ must lie on ℓ_1 and thus $F_1(A) = A = A'$. Thus $\triangle F_1(A)F_1(B)F_1(C) = \triangle AB'F_1(C)$. Consider $F_1 \circ T$ and note that $F_1(T(A)) = F_1(A) = A$ and $F_1(T(B)) = F_1(B') = B$. Hence $F_1 \circ T$ fixes at least two points, A and B. If $F_1 \circ T$ also fixes C, then by Case 1, $F_1 \circ T$ is the identity, and then $T = F_1^{-1} = F_1$. If $F_1 \circ T$ does not fix C, then by Case 2, $F_1 \circ T = F_2$ for some reflection F_2, and then $T = F_1^{-1} \circ F_2 = F_1 \circ F_2$. Thus if T fixes one point, then T is the composition of at most two reflections.

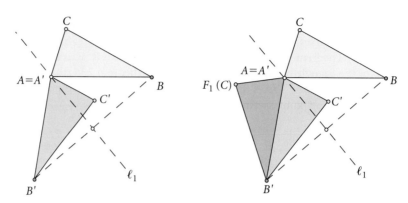

Case 4: T doesn't fix any of the points: Construct the perpendicular bisector ℓ_1 of $\overline{AA'}$. Denote reflection across this line as F_1 and note that $F_1(A) = A'$. Thus $\triangle F_1(A)F_1(B)F_1(C) = \triangle A'F_1(B)F_1(C)$ has at least one vertex in common with $\triangle A'B'C'$. Consider $F_1 \circ T$ and note that $F_1 \circ T(A) = F_1(A') = A$. Thus $F_1 \circ T$ fixes at least one point. If $F_1 \circ T$ fixes exactly one of the points, then by Case 3, $F_1 \circ T = F_2 \circ F_3$ for some reflections F_2 and F_3, so $T = F_1 \circ F_2 \circ F_3$. If $F_1 \circ T$ fixes two points, then by Case 2, $T = F_1 \circ F_2$ for some reflection F_2. Finally, if $F_1 \circ T$ fixes all three points, then $T = F_1$.

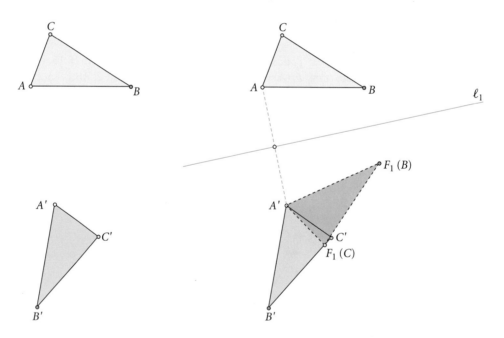

Thus, in all cases T is the composition of at most three reflections. □

We must next show that we have identified all possible isometries of the euclidean plane. We do this by showing that if an isometry is the composition of at most three reflections, as we have shown all isometries must be, then it is either the identity (if zero reflections), a reflection, a rotation or a translation (if two reflections), or a glide reflection (if three reflections).

Theorem 8.19. A euclidean isometry S is either a rotation or a translation if and only if S can be written as the composition of two reflections in distinct lines ℓ_1 and ℓ_2.

Proof: In Lemma 8.13, we showed that if S is a translation, then S can be written as the composition of two reflections in two parallel lines. Lemma 8.14 proved that any rotation can be written as the composition of reflections in two intersecting lines.

Conversely, assume that S is the composition of two reflections, so $S = F_1 \circ F_2$ for two distinct lines ℓ_1 and ℓ_2. If ℓ_1 intersects ℓ_2 at some point O at angle α, then $S(O) = F_1 \circ F_2(O) = F_1(O) = O$. For any point X, $\overline{OX} = \overline{OF_1(X)} = \overline{OF_2(F_1(X))}$ and $\angle F_2(F_1(X))OX = 2\angle \alpha$ as in Lemma 8.14. Thus $S = R_{O,2\alpha}$. If $\ell_1 \| \ell_2$, choose any point A on ℓ_1 and drop $AC \perp \ell_2$. Extend AC and cut $\overline{CB} = \overline{AC}$ so C lies between A and B. As above, for any point X, $XABF_2(F_1(X))$ is a parallelogram and so $S = F_2 \circ F_1 = T_{A,B}$. □

Theorem 8.19 tells us that if two lines intersect, then the composition of reflections in both lines gives a rotation, while if the two lines are parallel, the composition is a translation. Next, we prove the analogous result about glide reflections and reflections in three lines. The following proof is adapted from Venema's *Foundations of Geometry*.

Theorem 8.20. A euclidean isometry S is a glide reflection if and only if S can be written as the composition of three reflections in distinct lines ℓ_1, ℓ_2, and ℓ_3.

Proof: By Exercise 8.20, if S is a glide reflection, then S can be written as the composition of three reflections. Conversely, assume that $S = F_3 \circ F_2 \circ F_1$ where F_i is a reflection across line ℓ_i. We break the proof down into cases, determined by the whether the lines intersect and, if so, how they intersect.

Case 1: ℓ_1, ℓ_2 and ℓ_3 all pass through point O: In this case, we will prove that $F_3 \circ F_2 \circ F_1$ can be rewritten as a reflection across a single line n—i.e., a glide reflection with translation distance 0.

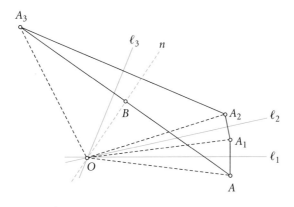

Let A be a point on the plane and construct $A_1 = F_1(A)$, $A_2 = F_2 \circ F_1(A)$ and $A_3 = F_3 \circ F_2 \circ F_1(A)$. Since O lies on all three lines, $F_1(O) = F_2(O) = F_3(O) = O$. Since F_1, F_2, and F_3 are isometries, by Definition 8.1, $\overline{OA} = \overline{OA_1} = \overline{OA_2} = \overline{OA_3}$. Draw $\overline{AA_3}$ and drop $OB \perp AA_3$. Then $\triangle OA_3B \cong \triangle OAB$ by the Hypotenuse-Side Theorem 3.20 and so $\overline{BA} = \overline{BA_3}$. Thus, the line $n = OB$ is the perpendicular bisector of

$\overline{AA_3}$ and therefore $A_3 = F_n(A)$, the reflection of A across the line $n = OB$. Clearly, $F_3 \circ F_2 \circ F_1(O) = F_n(O)$ since all of these reflections fix the point O. We claim that $F_3 \circ F_2 \circ F_1(B) = B$.

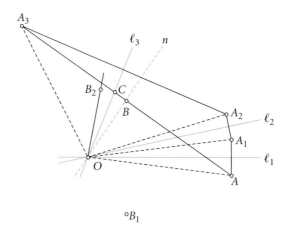

Let $B_1 = F_1(B)$ and $B_2 = F_2 \circ F_1(B)$. We must show that $F_3(B_2) = B$. Since F_1 and F_2 are isometries, $\overline{OB} = \overline{OB_1} = \overline{OB_2}$. By Lemma 8.14, $F_2 \circ F_1$ is the rotation about O by 2θ where θ is the angle formed by ℓ_1 and ℓ_2 at O. Thus, $\angle AOA_2 = \angle BOB_2 = 2\theta$. Let C denote the intersection of ℓ_3 with A_3A. Considering $\triangle AOA_3$, we see that

$$\angle A_3OC + \angle COB = \angle BOA_2 + \angle A_2OA = \angle BOA_2 + 2\theta.$$

Note that since $A_3 = F_3(A_2)$, ℓ_3 is the perpendicular bisector of $\overline{A_2A_3}$. From $\triangle A_2OA_3$, we have

$$\angle A_3OC = \angle COB + \angle BOA_2.$$

Combining these equations, we find that $\angle COB = \theta$. Since $B_2 = F_2 \circ F_1(B) = R_{O,2\theta}(B)$, it follows that $\angle COB = \theta = \angle B_2OC$, so $\triangle BOC \cong \triangle B_2OC$ by SAS. Therefore, ℓ_3 is the perpendicular bisector of $\overline{BB_2}$. Thus, $F_3(B_2) = B$ and $F_3 \circ F_2 \circ F_1(B) = B = F_n(B)$ for reflection across line n. Since $F_3 \circ F_2 \circ F_1 = F_n$ at the three points A, O, and B, by Corollary 8.17, $F_3 \circ F_2 \circ F_1 = F_n$ everywhere. $F_3 \circ F_2 \circ F_1$ can thus be considered as a glide reflection with a translation of length 0.

Case 2: Two of the lines ℓ_1, ℓ_2 and ℓ_3 intersect at a point O: There are three subcases. When two lines of reflection intersect, we can use Lemma 8.14 to rewrite their composition as a rotation, which we can change into the composition of two different reflections with the same enclosed angle. The key in each case is to replace the reflections in lines ℓ_1, ℓ_2 and ℓ_3 with reflections in lines that are either perpendicular or parallel to one another. In Exercise 8.20, you should have found that since a glide reflection can be considered as a reflection followed by a translation (or vice versa), and a translation can be written as two reflections in parallel lines, then a glide reflection can be written as a reflection followed by two reflections in lines perpendicular to the first line of reflection, or as two parallel reflections followed by a reflection perpendicular to those lines.

Case 2(a): ℓ_2 *and* ℓ_3 *intersect at O:*

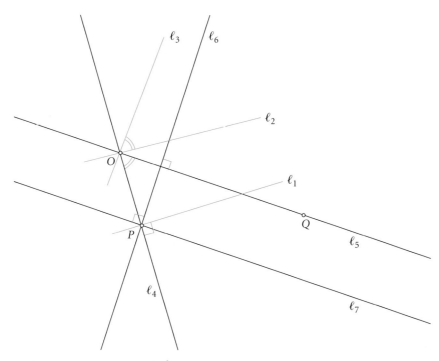

Let ℓ_2 and ℓ_3 intersect at O at angle $\frac{\theta}{2}$. By Lemma 8.14, $F_3 \circ F_2 = R_{O,\theta}$, the rotation centered at O with angle θ. From O, drop $\ell_4 \perp \ell_1$ and let P be the point where the lines ℓ_1 and ℓ_4 intersect. Construct $\angle POQ = \frac{\theta}{2}$ and label OQ as ℓ_5. Note that ℓ_4 and ℓ_5 intersect at O and form angle $\frac{\theta}{2}$ and therefore $F_5 \circ F_4 = R_{O,\theta} = F_3 \circ F_2$. Drop ℓ_6 through P perpendicular to ℓ_5 and then construct ℓ_7 through P perpendicular to ℓ_6. Note that ℓ_7 is thus parallel to ℓ_5. Consider $F_4 \circ F_1$. Since ℓ_4 intersects ℓ_1 at P at right angles, $F_4 \circ F_1 = R_{P,180°} = F_7 \circ F_6$. Therefore, $F_3 \circ F_2 \circ F_1 = F_5 \circ F_4 \circ F_1 = F_5 \circ F_7 \circ F_6$. Since $\ell_5 \parallel \ell_7$ and both are perpendicular to ℓ_6, $F_5 \circ F_7$ is a translation in the ℓ_6 direction, and F_6 is a reflection perpendicular to both of these. Therefore, $F_3 \circ F_2 \circ F_1 = F_5 \circ F_7 \circ F_6$ is a glide reflection along ℓ_6.

Case 2(b): ℓ_1 *and* ℓ_2 *intersect:*

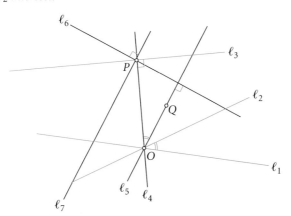

Let O be the point where ℓ_1 and ℓ_2 intersect at angle $\frac{\theta}{2}$. By Lemma 8.14, $F_2 \circ F_1 = R_{O,\theta}$, the rotation centered at O. Drop $\ell_4 \perp \ell_3$ and let P be the point where these lines intersect. Construct $\angle POQ = \frac{\theta}{2}$ and label OQ as ℓ_5. Note that ℓ_4 and ℓ_5 intersect at O and form angle $\frac{\theta}{2}$ and therefore $F_4 \circ F_5 = R_{O,\theta} = F_2 \circ F_1$. Construct ℓ_6 through P perpendicular to ℓ_5 and then construct ℓ_7 through P perpendicular to ℓ_6, so ℓ_7 is parallel to ℓ_5. Consider $F_3 \circ F_4$. Since ℓ_4 intersects ℓ_3 at P at right angles, $F_3 \circ F_4 = R_{P,180°} = F_6 \circ F_7$. Therefore, $F_3 \circ F_2 \circ F_1 = F_3 \circ F_4 \circ F_5 = F_6 \circ F_7 \circ F_5$. Since $\ell_5 \| \ell_7$ and both are perpendicular to ℓ_6, $F_7 \circ F_5$ is a translation in the ℓ_6 direction. Therefore, $F_3 \circ F_2 \circ F_1 = F_6 \circ F_7 \circ F_5$ is a glide reflection along ℓ_6.

Case 2(c): ℓ_1 and ℓ_3 intersect: In this case, either $\ell_2 \| \ell_1$, in which case ℓ_3 must also intersect ℓ_2 and the proof follows from Case 2(a), or ℓ_2 intersects ℓ_1 and then the proof follows from Case 2(b).

Case 3: ℓ_1, ℓ_2, and ℓ_3 are parallel:

▶ **Exercise 8.25.** If ℓ_1, ℓ_2, and ℓ_3 are three distinct parallel lines, prove that $F_3 \circ F_2 \circ F_1$ is a reflection (a glide reflection with translation of length 0).

Thus in all cases $F_3 \circ F_2 \circ F_1$ is a glide reflection. □

Theorem 8.21. [Classification of Euclidean Isometries] Every isometry of the euclidean plane is either the identity, a reflection, a translation, a rotation, or a glide reflection.

Proof: Let T be an isometry of the plane. By Theorem 8.18, T is a composition of at most three reflections. For zero or one reflections, we get T is the identity or a reflection. If T is the composition of two reflection, then by Theorem 8.19, T is either a translation or a rotation. If T is the composition of three reflections, then by Theorem 8.20, T is a glide reflection. □

Definition 8.22. An isometry is *even* if it can be expressed as the composition of an even number of reflections, and an isometry is *odd* if it can be written as the composition of an odd number of reflections.

▶ **Exercise 8.26.** Prove that any even isometry preserves orientation, while any odd isometry reverses orientation.

One reason why we are interested in isometries in general and reflections in particular is the following postulate and theorem:

Isometry Postulate: For every line ℓ, there is a function F_ℓ defined on the plane, called the reflection across ℓ. This function takes lines to lines, preserves distance and angle measure, has the property that for any point A on ℓ, $F_\ell(A) = A$, and also has the property that for points B not on ℓ, B and $F_\ell(B)$ lie on opposite sides of ℓ.

Theorem 8.23. In the context of neutral geometry, the Isometry Postulate is logically equivalent to the SAS Postulate 12.

Proof: Part 1: The Isometry Postulate implies SAS: Let $\triangle ABC$ and $\triangle DEF$ have $\angle BAC = \angle EDF$, $\overline{AB} = \overline{DE}$, and $\overline{AC} = \overline{DF}$. Rereading the proof of Theorem 8.18 shows that this not only proved the desired result, but is a constructive proof, giving an explicit procedure to construct T, a composition of reflections with only those properties given by the Isometry Postulate, so that $T(A) = D$ and $T(B) = E$. Since $\angle BAC = \angle EDF$, the line $T(AC)$ coincides with DF. Since $\overline{AC} = \overline{T(A)T(C)} = \overline{DF}$, $T(C) = F$. Thus by Postulate 1, the line $T(BC)$ coincides with EF. Therefore, T takes $\triangle ABC$ to $\triangle DEF$ and since by the properties of isometries, $\overline{AB} = \overline{T(A)T(B)} = \overline{DE}$, $\overline{AC} = \overline{T(A)T(C)} = \overline{DF}$, $\overline{BC} = \overline{T(B)T(C)} = \overline{EF}$, $\angle ABC = \angle T(A)T(B)T(C) = \angle DEF$, $\angle ACB = \angle T(A)T(C)T(B) = \angle DFE$, and $\angle BAC = \angle T(B)T(A)T(C) = \angle EDF$, $\triangle ABC \cong \triangle DEF$.

Part 2: SAS implies the Isometry Postulate: Definition 8.9 of a reflection is valid in neutral geometry and so defines a reflection with the desired properties for every line ℓ in the plane. □

◆ 8.2 Rosette groups

Definition 8.24. An isometry which takes a figure to itself, though some points may be rearranged, is called a *symmetry* of the figure.

For example, consider the equilateral triangle $\triangle ABC$. Let R be the rotation counterclockwise around the center of the triangle by $120°$. Then $R(\triangle ABC) = \triangle ABC$, although $A' = R(A) = B$, $B' = R(B) = C$, and $C' = R(C) = A$. Thus every point of the triangle moves, but the triangle itself is invariant under this isometry. We say that the equilateral triangle has *rotational symmetry*, with a rotation of $120°$, or a *3-fold rotation*, since three turns will return every point to its original position.

 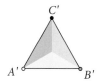

We will use some convenient shorthand to designate the images of the points of the triangle after applying the rotation, suppressing the prime notation. If we label the original configuration of the triangle as 1 (for the identity transformation) and the configuration after a $120°$ counterclockwise rotation as R, then, adopting multiplication notation, R^2 is the composition of two successive $120°$ counterclockwise rotations, or a $240°$ counterclockwise rotation. Then it is clear that $R^3 = 1$, since the composition of three counterclockwise $120°$ rotations is the same as a $360°$ rotation, returning the triangle to the original configuration.

 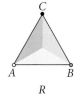
      1   R R^2 $R^3 = 1$

▶ **Exercise 8.27.** One could also rotate the triangle $120°$ clockwise and we denote this isometry as R^{-1}, since this would undo the action of R. The image of this rotation is the same as which of the configurations already pictured?

The equilateral triangle also has three lines of reflection, shown below:

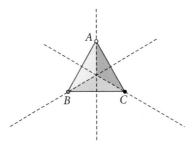

Let us denote the reflection across the vertical line by F. Two reflections across the vertical line returns every point to where it started, so $F^2 = 1$.

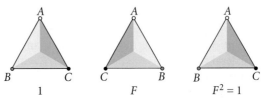

These isometries, rotation and reflection, are symmetries of the figure. One can also compose these rotations and reflections. For example, FR means first rotate the triangle by $120°$ counterclockwise, then reflect it across the vertical line, using the standard order for composition of functions.

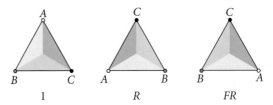

Note that this is a different configuration than those already pictured above. Similarly, we can find the image of the triangle under FR^2:

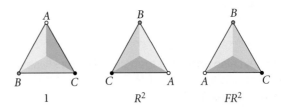

Since there are six ways to rearrange three symbols and we have found six different ways of labeling the vertices of the triangle, we must have found all of the possible configurations and thus all of the symmetries of the equilateral triangle. In summary, here are our results:

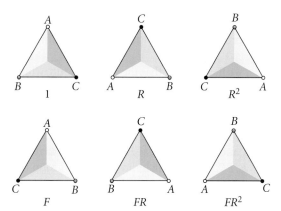

If you have studied abstract algebra, what we have done is show geometrically that the symmetries of the equilateral triangle form a *group*. A group is a set of elements that has an associative operation so that the set is closed under this operation and contains an identity element and inverses for every element. If you are not familiar with this concept, don't worry—the exposition in this text is meant to be self-contained.

▷ **Activity 8.14.** Reflecting the original triangle across the diagonal line that goes through vertex *B* should turn out to be the same as one of the configurations we have listed. Which one?

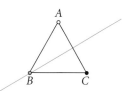

▷ **Activity 8.15.** Reflecting the triangle across the diagonal line that goes through vertex *C* also turns out to be the same as one of the above configurations. Which one?

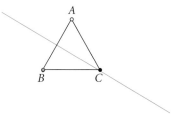

▶ **Exercise 8.28.** Recall that *FR* means to rotate the original triangle by 120° counterclockwise and then reflect across the vertical line. Thus, *RF* would mean to reflect the original triangle across the vertical line first and then rotate 120° counterclockwise. Which configuration of those pictured above is the same as *RF*? R^2F? *RFR*?

We have chosen F to denote flipping the triangle across the vertical line. This decision was arbitrary, since there are other lines of reflection, but one must be consistent. In the six different configurations of this triangle, $1, R, R^2, F, FR$, and FR^2, the choice always to rotate first then flip was arbitrary, but again one must be consistent.

Definition 8.25. The *rosette group* of a finite figure is the set of all symmetries of the figure.

For the equilateral triangle, we have shown that the rosette group, which is denoted by D_3 and called the *dihedral group with a 3-fold rotation*, is $D_3 = \{1, R, R^2, F, FR, FR^2\}$. The term 'dihedral' and the letter D are used to indicate that the figure has a line of reflection and the subscript 3 in D_3 indicates that there is a 3-fold rotation. However, any rosette group contains more information than just the list of elements, since there is also information on how the symmetries can be combined. We can build a multiplication table for the rosette group D_3. Be careful, and remember that $FR \neq RF$. Note that each entry in the table is one of our six configurations: $1, R, R^2, F, FR$, or FR^2, which shows that the group is closed under the operation of composition.

D_3	1	R	R^2	F	FR	FR^2
1	1	R	R^2	F	FR	FR^2
R	R	R^2	1	FR^2	F	FR
R^2	R^2	1	R	FR	FR^2	F
F	F	FR	FR^2	1	R	R^2
FR	FR	FR^2	F	R^2	1	R
FR^2	FR^2	F	FR	R	R^2	1

To use the table, for example to figure out which configuration is given by FRF, we write $FRF = FR \cdot F$ so that it is a product of two of our chosen entries. Then find the row that begins with FR and the column headed by F. From the table, we find the intersection of the FR row and the F column to get $FRF = R^2$. Notice that each of our six configurations occurs exactly once in each row and column. After you are about halfway through filling in the table, you can use what you already know to figure out the rest.

▶ **Exercise 8.29.** For the symmetries of the equilateral triangle, use the table above to simplify each of the following to one of the elements of $D_3 = \{1, R, R^2, F, FR, FR^2\}$.

a. FR^2FR
b. $RFRFRFRF$
c. FR^5FR^{10}
d. $R^{-2}FR^2$
e. $R^2FR^{-3}FR^2$

The table for D_3 and the rules we have found are true only for the equilateral triangle and any figures that have precisely the same symmetries as an equilateral triangle: a rotation by 120° counterclockwise that takes the figure back onto itself and a reflection across a mirror line. Actually, if there is a 3-fold rotation and one mirror line, there must be three reflection lines, since the rotations must take a mirror line to another mirror line.

The rosette group of a figure is the set of symmetries of the figure, the set of isometries that preserve the figure as a whole, though individual points of the figure may move. Each rosette group has a multiplication table which lays out all possible compositions of these symmetries. Note that all such figures have rotations, at least by 0°, and may or may not have reflectional symmetry. Therefore, the rosette groups will consist of rotations alone or rotations in combination with reflections. Thus, all rosette groups are of two types: *cyclic*, denoted $C_n = \{1, R, R^2, \ldots, R^{n-1}\}$ where $R^n = 1$ for a figure with an n-fold rotation and no reflectional symmetries, and *dihedral*, denoted $D_n = \{1, R, R^2, \ldots, R^{n-1}, F, FR, FR^2, \ldots, FR^{n-1}\}$ where $R^n = 1$ and $F^2 = 1$ for a figure with an n-fold rotation and n lines of reflection. This fact is named in honor of Leonardo da Vinci, who first showed that these are the only possibilities.

Theorem 8.26. [Leonardo's Theorem] The rosette group of a finite figure is either a cyclic group or a dihedral group.

Proof: First note that the symmetries of a finite figure must leave the center O of the figure fixed. Therefore, the only isometries that can be symmetries of the figure are those that fix at least one point. By Exercise 8.22, the symmetries must be rotations or reflections. If there is a rotation, let ϕ be the smallest angle of rotation so that $R_{O,\phi}$ is a symmetry of the figure. Then $R_{O,\phi} \circ R_{O,\phi} = R_{O,2\phi}$, so the rotation by 2ϕ is also a symmetry. Similarly, $R_{O,3\phi}, \ldots, R_{O,n\phi}$ are symmetries of the figure for all $n \in \mathbb{Z}$. We must show that $n\phi = 360°$ for some n. If not, there is an integer k so that $k\phi < 360° < (k+1)\phi$. From this we deduce that $360° < (k+1)\phi < 360° + \phi$, and thus $0° < (k+1)\phi - 360° < \phi$. Since rotation by $(k+1)\phi$ and $360°$ are both obviously symmetries of the figure, $(k+1)\phi - 360°$ must be a symmtery, but this contradicts the assumption that ϕ was the smallest angle of rotation. Therefore there is an integer n so that $n\phi = 360°$ and so $\phi = \frac{360°}{n}$. Thus, if the rosette group contains only rotations, it must be C_n for some n.

If the rosette group contains a reflection F, then $FR, FR^2, \ldots, FR^{n-1}$ give $(n-1)$ other lines of reflection, so the figure has n lines of reflection. If there were other lines of reflection, by Theorem 8.19 we would have other rotations. Thus there can only be n lines of reflection, so the rosette group must be D_n. □

For the activities below that ask for constructions, you may do these either by hand or use geometric software.

▷ **Activity 8.16.** Construct another figure that has the same symmetries, and so the same rosette group D_3, as the equilateral triangle.

Consider the figure below, which has a rotation but no reflectional symmetry. The rosette group is denoted by C_6, indicating that it is cyclic with a six-fold rotation and no reflections.

▷ **Activity 8.17.** List the elements of the rosette group C_6 of the figure above and work out the multiplication table for this group.

▷ **Activity 8.18.** Construct another figure that has the same symmetries, and so the same rosette group C_6, as the one above.

▷ **Activity 8.19.** Construct a figure that has 8-fold rotational symmetry but no lines of reflection, and thus has rosette group C_8.

▷ **Activity 8.20.** Construct a figure that has reflectional symmetry and an 8-fold rotation. This will have rosette group D_8 since there is a reflection. How many lines of reflection does your figure have? Can you find another figure with the same rotational symmetry but with a different number of lines of reflection?

▶ **Exercise 8.30.** Construct a figure that has a line of reflection but no rotational symmetry. List the elements of the rosette group D_1 of the figure and work out the multiplication table for this group.

▷ **Activity 8.21.** Construct figures that have the following rosette groups: C_1, C_2, and D_2.

▶ **Exercise 8.31.** Describe (in words) the symmetries of a square. List the elements of the rosette group D_4 of the square and work out the multiplication table for this group.

▷ **Activity 8.22.** Construct another figure that has the same rosette group D_4 as the square.

▶ **Exercise 8.32.** Describe the symmetries and identify the rosette groups of each of the following patterns.

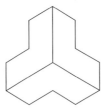 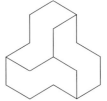

▶ **Exercise 8.33.** Describe the symmetries and identify the rosette groups of each of the following patterns.

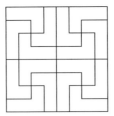 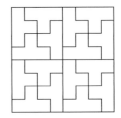

◆ 8.3 Frieze patterns

In 1891, the Russian mineralogist Yevgraf Stepanovich Fedorov [1853–1919] published an exhaustive analysis of the 230 ways of filling space with basic building blocks, and thus introduced the field of mathematical crystallography. In the same paper, he showed that there are 17 ways to fill the plane with regular repetitions of a basic two-dimensional figure. These are called the 17 wallpaper groups and will be discussed in the next section. While Fedorov's classification of three-dimensional symmetries attracted notice, his work on wallpaper groups was forgotten. These were rediscovered independently by George Polya [1887–1985] and Paul Niggli [1888–1953], both in 1924. In 1926, Paul Niggli applied the same methods to frieze patterns, proving that there are only seven such patterns. Frieze and wallpaper patterns are not only of interest to crystallographers and mathematicians, but have also been used by archaeologists and anthropologists to identify ornamental patterns on objects such as baskets, pottery, and weavings and so to trace contacts between various cultures. In particular, you are urged to look at Washburn and Crowe's *Symmetries of Culture*, a collaboration between an anthropologist and a mathematician.

Frieze, or border or strip, patterns are formed by an infinite number of repetitions of a motif along a line. Think of a pattern made with a rubber stamp. Each repetition of the basic motif must be exactly the same in size and shape as the original figure. Imagine an infinite strip of paper with a dotted line running lengthwise down the middle. Use a motif without any symmetry of its own, so that any symmetry comes from applying one of the four isometries studied in Section 8.1: translation, rotation, reflection, and glide reflection.

Since we want to create a border or strip pattern from these four moves, we can use them only in ways compatible with the linear character of the strip. The images of the motif under each isometry must be confined to the strip. Thus, we must use only the following versions of the isometries:

Translation is only allowed in the direction of the strip. This isometry is denoted by T. All friese patterns will have translations.

Reflection across a line can be used only for lines running along the center of the strip or lines perpendicular to the strip. We will denote the reflection across the center line by $F_{||}$ and a reflection across any line perpendicular to the center line by $F_{\perp}$.

Glide reflection must combine a reflection across the centerline of the strip and a translation in the direction of the strip. This is denoted by G.

Rotation about a point on the center line by $180°$ will rotate the figure in the direction of the strip. We will denote this isometry by R.

Any other isometries, other angles of rotation or other lines of reflection, would make the copies of the original motif migrate off the strip. Let us first consider the action of each of these isometries alone.

If a frieze pattern has only translational symmetry, then we get the pattern shown below, which we can name familiarly as *hop*, or more formally, *p111* (we will explain this notation shortly):

Reflection along the midline is the isometry $F_{||}$. Reflecting our motif across the center line gives us two feet, a left and a right, and then we can translate this doubled motif along the center line to get another frieze pattern *p1m1*, called *jump*. (Note that we moved the pattern off the center line only to make the symmetry more obvious and to fit the common name. The symmetry would be the same if the original pattern crossed the center line but the pattern would overlap itself making the symmetry harder to identify.)

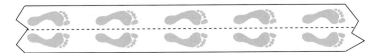

By considering the shaded feet below, note that the *jump* pattern also has glide reflection symmetry.

To apply the isometry $F_\perp$, we reflect the motif in a vertical line perpendicular to the dotted line, and then translate the doubled motif along the strip, to get *pm11: sidle* or sidestep. We have changed the orientation of the feet to match the name.

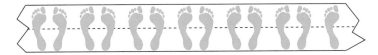

A glide reflection G along the center line gives the frieze pattern *p1g1*, called *step*:

To respect the linear symmetry of the strip itself, we only allow rotations R by 180°, or 2-fold rotations around points on the dotted center line. A fifth pattern is given by rotating the motif by 180° and then translating to get *p112* or *spinning hop*. Note that there are only left feet.

We have now generated five patterns using each of the isometries in isolation, though in the case of *p1m1* we accidentally involved a second isometry. We claim that there are exactly seven frieze patterns that can be formed using these five isometries, R, $F_\parallel$, $F_\perp$, T, and G, and combinations of them. All frieze patterns have translational symmetry, but there are many ways to compose the remaining four to try to create new patterns. We study these, adapting the approach used by sarah-marie belcastro and Thomas Hull in their article "Classifying Frieze Patterns Without Using Groups." Let us first consider ways to combine any two of the symmetries. There are six ways of combining any two of these symmetries, $F_\parallel F_\perp$, $GF_\parallel$, $F_\parallel R$, $F_\perp G$, $F_\perp R$, and GR, but of course, we must also consider the order in which they are composed. However, we already know that $GF_\parallel$ gives us the pattern we have already found and named *jump* or *p1m1*.

If we apply first $F_\parallel$ and then apply $F_\perp$ to the result, we get the following pattern, which we will call *pmm2* or *spinning jump*. Note that the final pattern contains a glide reflection (since we have a horizontal reflection and a translation) and also a 180° rotation, around the centers marked with little dots. Also note that we would get the same pattern if we reflected over the vertical line first and then reflected across the horizontal line.

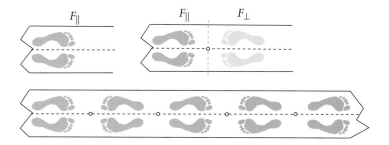

▶ **Exercise 8.34.** Investigate the compositions $RF_\|$ and $F_\| R$ to see if you get a new frieze pattern.

Next, consider the composition $GF_\perp$. Again, a rotation appears, and we call this pattern *spinning sidle* or *pmg2*.

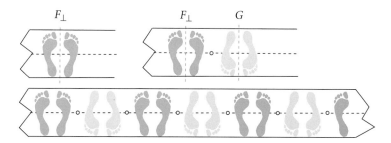

▶ **Exercise 8.35.** Investigate the compositions $RF_\perp$ and $F_\perp R$ to see if you get a new frieze pattern.

▶ **Exercise 8.36.** Investigate the compositions GR and RG to see if you get a new frieze pattern.

▶ **Exercise 8.37.** Explain why the previous examples and exercises show that no new patterns can emerge from taking the isometries three or four at a time.

By checking all possible combinations of the allowed isometries, T, $F_\|$, $F_\perp$, R and G, we have proved the following theorem:

Theorem 8.27. [Classification of Frieze Patterns] There are exactly seven different symmetry groups for frieze patterns.

The hop-step-jump terminology above is due to John H. Conway. Frieze patterns have been studied extensively by crystallographers, chemists who analyze how crystals are shaped. The notation they have evolved to denote each type of pattern is called the *international crystallographic symbol* and is of the form *p* ___ ___ ___. The *p* stands for primitive cell, which we will discuss in more detail in the next section on wallpaper patterns. The first blank is filled in by *m* if there is a vertical mirror, and 1 otherwise. The second blank is filled in by *m* if there is a horizontal mirror, *g* if there is a glide reflection, and 1 otherwise. The last blank is filled by 2 if there is a 180° (or 2-fold) rotation, and 1 otherwise. Since *jump* has no vertical mirror, a horizontal mirror, and no rotation, its symbol is *p1m1*. *Spinning sidle* has a vertical mirror, no horizontal mirror but a horizontal glide reflection, and a 2-fold rotation, so its symbol is *pmg2*. In classifying a frieze pattern with this notation, use the following outline:

- The first character is always *p*.
- The second character is:
    - *m* if there is a reflection perpendicular to the center line, so $F_\perp$ is a symmetry.
    - *1* if there is no reflection perpendicular to the center line: i.e., if $F_\perp$ is not a symmetry.
- The third character is
    - *m* if there is a reflection along the center line, so $F_\parallel$ is a symmetry.
    - *g* if there is not a reflection but there is a glide reflection along the center line, so G is a symmetry.
    - *1* if there is no reflection or glide reflection along the center line.
- The fourth character is
    - *2* if there is a 180° rotation, so R is a symmetry.
    - *1* if there is no 180° rotation, so R is not a symmetry.

In all cases, "1" designates "no" to the question asked for each position. One easy way to determine whether a motif is reflected or rotated is to trace the pattern on a bit of tracing paper, then reflect (turn the paper over, being careful to distinguish between a horizontal or vertical reflection) or rotate and see whether the traced pattern lines up with the original. Alternatively, you can use a mirror to check for reflection lines. All of the frieze patterns use a translation to fill up the (infinite) strip of paper, so the notation makes no mention of this isometry.

▶ **Exercise 8.38.** Draw a flow chart or outline based on the crystallographic notation to help classify the seven frieze patterns.

▶ **Exercise 8.39.** Classify the following patterns:

a. DDDDDDDDDDDDD
b. XXXXXXXXXXXX
c. ZZZZZZZZZZZZ
d. YYYYYYYYYYYY
e. ∪∩∪∩∪∩∪∩∪∩
f. GGGGGGGGGGGGG
g. bpbpbpbpbpbp
h. bdbdbdbdbdbd
i. bqbqbqbqbqbq
j. pqpqpqpqpqpq
k. pdpdpdpdpdpd
l. dqdqdqdqdqdq

▶ **Exercise 8.40.** Classify the following patterns. Note that we have only drawn a finite segment of what should be an infinite pattern. You must imagine the patterns continued to both the left and the right to accurately identify the symmetry.

▶ **Exercise 8.41.** Classify the following patterns:

300 • 8. ISOMETRIES

b.
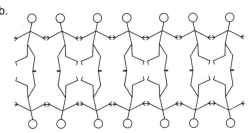

c.

d.

e.

f.

g.

▷ **Activity 8.23.** Using geometric software, draw the motif shown below. Then draw the seven frieze patterns formed from this motif.

▷ **Activity 8.24.** Draw your own examples of each of the seven frieze patterns.

▷ **Activity 8.25.** How would you classify the pattern formed by the footprints of someone skipping?

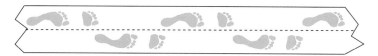

♦ 8.4 Wallpaper patterns

Wallpaper patterns are formed by repetitions of a motif in such a way as to cover the plane. There are exactly 17 different wallpaper patterns. This doesn't mean that if you have 18 rooms in your new mansion that you must have the same wallpaper in two rooms. We often hear that there must be more than 17 different patterns, and in the artistic sense, there are. What this means is that the paper in two of your 18 rooms will have the same pattern of repetitions. The paper may have flowers or birds or abstract swirls. It is the way that the motif repeats itself that identifies what mathematicians call the wallpaper pattern.

All wallpaper patterns are generated by a basic motif which is acted on by the isometries we have studied: translations, rotations, reflections, and glide reflections. These patterns are then studied and classified by their symmetries. Since the wallpaper is considered to cover the entire plane, there must be translations in two different directions.

We begin by seeing which wallpaper patterns we can generate by stacking copies of the frieze patterns. The simplest frieze pattern was *hop* or *p111*:

If we take copies of this row pattern and translate them in a new direction, we get the wallpaper pattern shown below, denoted **p111**. The frieze pattern and wallpaper notations are very similar. To distinguish between them, we will use the italic form *p111* for a frieze pattern and the bold form **p111** for a wallpaper pattern. We will explain the further differences between the frieze and wallpaper pattern notations shortly.

Another of our frieze patterns was *spinning hop* or *p112*. This frieze pattern was generated by translation and 180° rotation. We can mark the rotation centers with a ○:

If we again translate this pattern in a second direction, we get the wallpaper pattern **p211**:

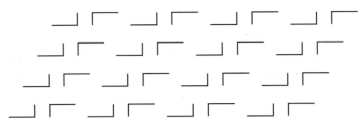

▷ **Activity 8.26.** For the pattern **p211** above, mark all of the 2-fold (180°) rotation centers with a ○. Be careful, in addition to the ones we noticed in the frieze pattern, there are others. What do you notice about the alignment of the rotation centers?

▷ **Activity 8.27.** Use geometric software to generate your own frieze patterns or use patterns you may have from the last section. Stack copies of the different patterns as we did for *hop* and *spinning hop*. See how many different types of symmetry you can create this way before reading further.

Neither of the patterns we have discussed has any reflectional symmetry. We also chose the two translation directions so that they were not perpendicular, to emphasize that they need not be.

The pattern of reflection lines, glide reflection lines, and rotation centers of various orders completely determine the wallpaper pattern (with only two exceptions). We will use the following symbols to indicate elements of symmetry:

- direction and length of translation = ⟶
- line of reflection = - - - -
- line of glide reflection = ----
- center of a 2-fold rotation (180°) = ○
- center of a 3-fold rotation (120°) = △
- center of a 4-fold rotation (90°) = ×
- center of a 6-fold rotation (60°) = ∗

We will soon explain why these are the only possible rotations. If we generate a wallpaper pattern with a reflection, then the directions of translation must conform to that line of reflection, since the translation of a line of reflection gives another line of reflection. For example, if we take copies of the frieze pattern *jump* or *p1m1* and translate in direction $\overline{PQ}$ to form a planar pattern, we get a pattern like this:

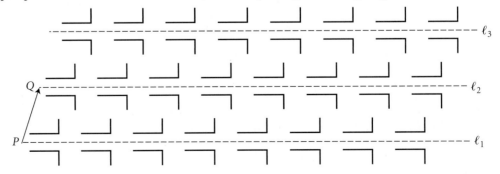

In this illustration, ℓ_1 is the line of reflection for the original frieze pattern. Translating ℓ_1 in the direction of $\overline{PQ}$ gives us another line of reflection ℓ_2, and another translation by $\overline{PQ}$ gives ℓ_3. However, ℓ_2 is not a line of reflection for the whole pattern since the motifs along ℓ_1 are not reflected to the corresponding motifs along ℓ_3. Thus, the planar pattern does not have reflectional symmetry in this case. In order for the planar pattern to have reflectional symmetry, PQ must be perpendicular to ℓ_1. This gives us the wallpaper pattern **p1m1**:

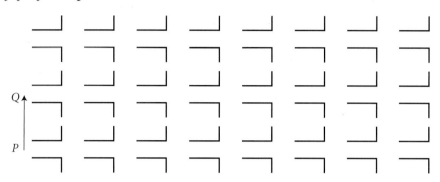

Note that the vertical translation introduces additional lines of reflection, midway between the ones for the original frieze pattern and its translates:

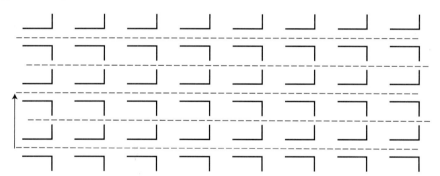

▶ **Exercise 8.42.** What wallpaper pattern do you get by lining up rows of the frieze pattern *sidestep* or *pm11*?

Similarly, copies of the frieze *spinning jump* or *pmm2*, give us another rectangular lattice wallpaper pattern, **p2mm**:

▶ **Exercise 8.43.** For the wallpaper pattern **p2mm** draw the lines of reflection and mark any rotation centers with the appropriate symbol.

Similarly, copies of the frieze pattern *step* or *p1g1* can be replicated to form a wallpaper pattern, **p1g1**:

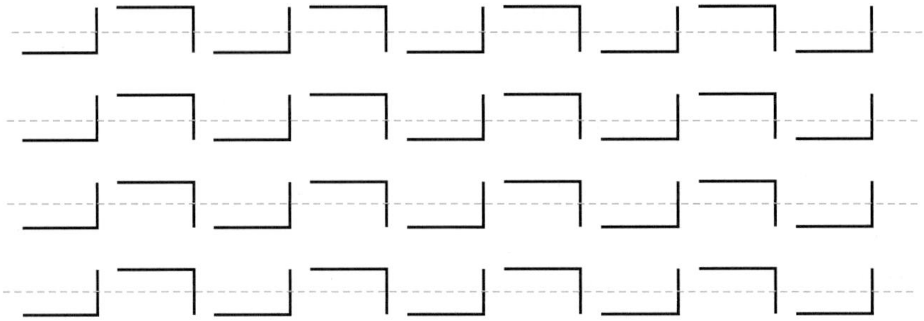

▶ **Exercise 8.44.** For the wallpaper pattern **p1g1** mark any rotation centers with the appropriate symbol.

On the other hand, if we take the frieze pattern *step* and translate it in the perpendicular direction using a glide reflection, we get a different pattern, called **p2gg**:

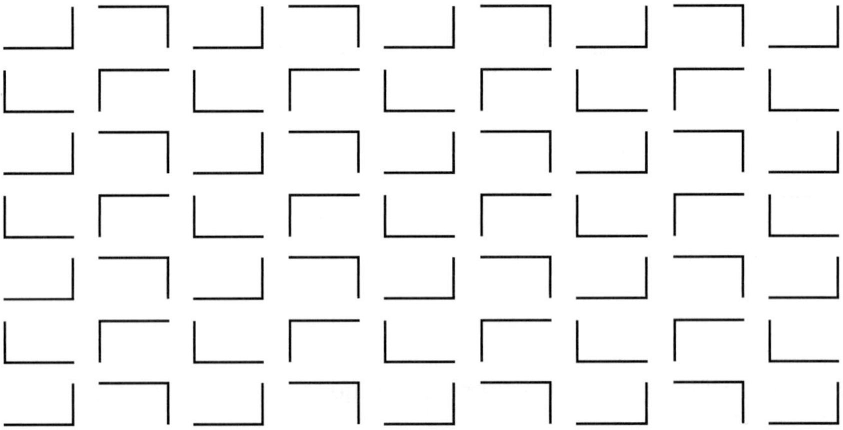

▶ **Exercise 8.45.** For the wallpaper pattern **p2gg** draw the lines of glide reflection and mark any rotation centers with the appropriate symbol.

Stacking copies of the frieze pattern *spinning sidestep* or *pmg2*, gives us the wallpaper pattern **p2mg**:

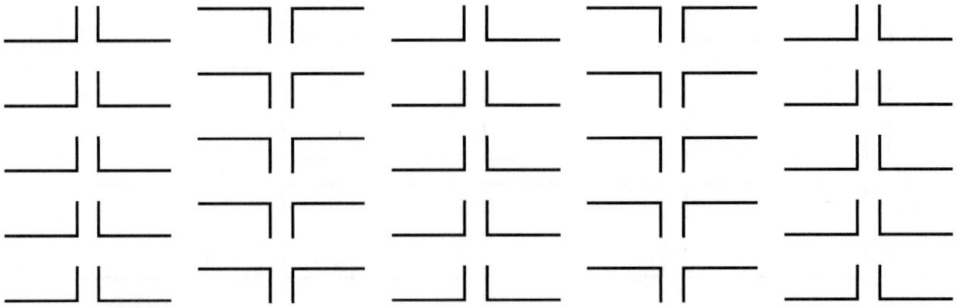

▶ **Exercise 8.46.** For the wallpaper pattern **p2mg**, draw the lines of reflection and glide reflection and mark any rotation centers with the appropriate symbol.

As you can see from Exercises 8.43–8.46, each of the wallpaper patterns we have considered has a unique pattern of lines of reflection, lines of glide reflection, and rotation centers. These are the clues by which you can differentiate among the seventeen wallpaper patterns.

In the patterns we have analyzed so far, note that the translation translates, not only the motif, but also any lines of reflection or glide reflection. You should also have noticed the recurrence of additional lines of reflection and glide reflection midway between the lines given by the frieze patterns forming the rows of the patterns. The following theorem formalizes this observation.

Theorem 8.28. Let ℓ be a line of reflection for a wallpaper pattern, and let $T_{P,Q}$ be a translation for the pattern which is not parallel to ℓ, so $T_{P,Q}(\ell)$ is another line of reflection. Let m be the line parallel to ℓ and $T_{P,Q}(\ell)$ and lying midway between the two. Then

1. If PQ is perpendicular to ℓ, then m is a line of reflection for the pattern.
2. If PQ is not perpendicular to ℓ, then m is a line of glide reflection.

Proof:

Case 1: PQ is perpendicular to ℓ.

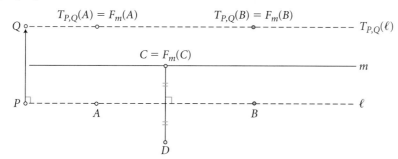

Denote reflection across line ℓ by F_ℓ and reflection across m by F_m. Choose points A and B on ℓ and construct $T_{P,Q}(A)$ and $T_{P,Q}(B)$ on $T_{P,Q}(\ell)$, so $\overline{PQ} = \overline{AT_{P,Q}(A)} = \overline{BT_{P,Q}(B)}$ and $PQ \| AT_{P,Q}(A) \| BT_{P,Q}(B)$. Note that $F_m(A) = T_{P,Q}(A) = T_{P,Q} \circ F_\ell(A)$ and $F_m(B) = T_{P,Q}(B) = T_{P,Q} \circ F_\ell(B)$, since $F_\ell(A) = A$, $F_\ell(B) = B$ and m lies half way between ℓ and $T_{P,Q}(\ell)$. Thus, F_m and $T_{P,Q} \circ F_\ell$ agree at two points. Now consider an arbitrary point C on m. Drop a perpendicular from C to ℓ and extend this segment an equal amount on the other side of ℓ to point D, so ℓ is the perpendicular bisector of $\overline{CD}$. Then $F_\ell(C) = D$ and $T_{P,Q}(D) = C$. Thus, $T_{P,Q} \circ F_\ell(C) = C = F_m(C)$. Therefore the isometries F_m and $T_{P,Q} \circ F_\ell$ agree at the three points A, B, and C, and so by Corollary 8.17, they must agree everywhere. Since F_ℓ and $T_{P,Q}$ are isometries of the given wallpaper pattern, so is F_m. Thus m is a line of reflection for the wallpaper pattern.

Case 2: PQ is not perpendicular to ℓ.

Choose a point A on ℓ and construct $A' = T_{P,Q}(A)$ on $T_{P,Q}(\ell)$. Erect a perpendicular to ℓ from point A, which will intersect m at B and $T_{P,Q}(\ell)$ at C. Drop a perpendicular from $A' = T_{P,Q}(A)$ to ℓ, intersecting m at D. Note that since $PQ \| AA'$ and $\overline{PQ} = \overline{AA'}$, $T_{P,Q} = T_{A,A'}$.

Consider the lines AC and $CT_{P,Q}(A) = CA'$. The translation $T_{P,Q}$ can be regarded as the composition of two translations: $T_{P,Q} = T_{A,A'} = T_{C,A'} \circ T_{A,C}$. Thus, $T_{P,Q} \circ F_\ell = T_{C,A'} \circ T_{A,C} \circ F_\ell$. Since $AC \perp \ell$, we can apply Case 1 to conclude that $T_{A,C} \circ F_\ell = F_m$ and so

$$T_{P,Q} \circ F_\ell = T_{C,A'} \circ T_{A,C} \circ F_\ell = T_{C,A'} \circ F_m.$$

Since $\overline{CA'} = \overline{BD}$ and $CA' \| BD$, we see that $T_{P,Q} \circ F_\ell = T_{C,A'} \circ F_m = T_{B,D} \circ F_m$ is a reflection in m followed by a translation along m. Thus m is a line of glide reflection for the wallpaper pattern. □

This theorem gives us some insight into the next two wallpaper patterns. Recall the pattern **p1m1**, generated by stacking copies of the frieze pattern *jump*. In Exercise 8.42 we noted that **p1m1** can also be generated by stacking copies of *sidestep* or *pm11*. These could be stacked in a different staggered pattern, while still respecting the lines of reflection, as shown below:

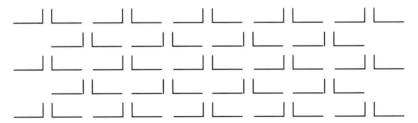

This gives a pattern known as **c1m1**, where the **c** stands for *centered*. The crystallographers who invented the notation we are using could find no adequate way of describing this pattern with a combination of "1"s and "m"s and this is their solution. There are only two centered-type wallpaper patterns, both closely related to patterns we have already discussed and both constructed similarly, by offsetting rows of one of the frieze patterns in such a way as to preserve the lines of symmetry.

▶ **Exercise 8.47.** For the wallpaper pattern **c1m1**, draw the lines of reflection and glide reflection.

Similarly, the wallpaper pattern **c2mm** looks much like **p2mm**, but offset:

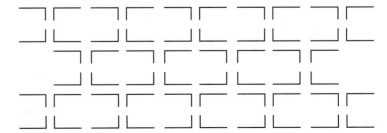

▶ **Exercise 8.48.** For the wallpaper pattern **c2mm** draw the lines of reflection and glide reflection and mark any rotation centers with the appropriate symbol.

The wallpaper patterns we have studied thus far have had at most a 2-fold rotation. We next study what other rotations are possible. It turns out that only a few types of rotations can occur in wallpaper patterns. We first prove a lemma:

Lemma 8.29. Let $T_{A,B}$ denote the translation defined by two distinct points A and B on the plane. If S is any other isometry of the plane, then $S \circ T_{A,B} \circ S^{-1} = T_{S(A),S(B)}$.

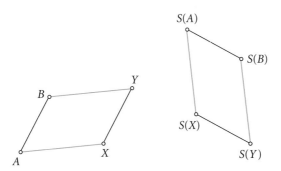

Proof: Let X be another point on the plane that is not collinear with A and B and let $Y = T_{A,B}(X)$. Then $AXYB$ is a parallelogram. Since an isometry must preserve both angle measure and length, $S(A)S(X)S(Y)S(B)$ will also form a parallelogram congruent to $AXYB$. Note that it follows that $S(Y) = T_{S(A),S(B)}(S(X))$. Therefore,

$$S \circ T_{A,B} \circ S^{-1}(S(X)) = S \circ T_{A,B}(X) = S(Y) = T_{S(A),S(B)}(S(X)).$$

Since X was an arbitrary point, we have shown that $S \circ T_{A,B} \circ S^{-1} = T_{S(A),S(B)}$. □

Theorem 8.30. [The Crystallographic Restriction] The only possible angles of rotation for a wallpaper pattern on the plane are 180°, 120°, 90°, and 60°.

Proof: Part 1: We assume that a given wallpaper pattern has a rotation of the form $R_{O,\theta}$ with rotation center O and angle θ where $0 < \theta < 90°$. Note that $R_{O,-\theta}$ will also be an isometry for this wallpaper pattern. Let A and B be two rotation centers for the pattern chosen so that A and B are the closest pair of rotation centers. Note that since A and B are rotation centers, $R_{A,\theta}$ and $R_{B,\theta}$ are both isometries for the pattern. Furthermore, the translation $T_{A,B}$ is also an isometry for this pattern. Thus, the composition $S_1 = R_{A,\theta} \circ T_{A,B} \circ R_{A,-\theta}$ must be another isometry for this wallpaper. Let $B' = R_{A,\theta}(B)$ and note $A = R_{A,\theta}(A)$. By the Lemma 8.29, $S_1 = T_{A,B'}$, so $B' = S_1(A) = T_{A,B'}(A)$.

Similarly, note that $T_{B,A}$ is also an isometry for this wallpaper pattern and $T_{B,A}(B) = A$. Note that $B = R_{B,-\theta}(B)$ and denote $A' = R_{B,-\theta}(A)$. Then $S_2 = R_{B,-\theta} \circ T_{B,A} \circ R_{B,\theta} = T_{B,A'}$. Note that $A' = S_2(B) = T_{B,A'}(B)$. Then A, B, B', and A' lie as shown below for different values of θ:

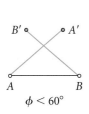 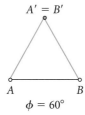 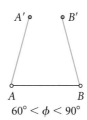

Since A' and B' are translations of rotation centers, they are also rotation centers. But if $0 < \theta < 90°$ and $\theta \neq 60°$, then A' and B' are closer that A and B were, contradicting our choice of A and B. Thus the only valid rotation angle less than 90° for wallpaper is 60°.

Part 2: Note that any angle of rotation for wallpaper must divide the region surrounding a rotation center into equal sectors, each having measure of the angle of rotation. Thus any valid angle of rotation for wallpaper must be of the form $\theta = \frac{360°}{n}$. By Part 1, we know that $\theta = \frac{360°}{n} \geq 60°$, and so $n \leq 6$. For $n = 6$, we have a rotation of 60°. For $n = 5$, we get $\theta = \frac{360°}{5} = 72°$, which fails by Part 1. For $n = 4$, $\theta = 90°$, and $n = 3$ gives $\theta = 120°$, and finally $n = 2$ gives $\theta = 180°$. □

The 17 wallpaper patterns are classified, as were the frieze patterns of the previous section, using standard international crystallographic notation. In analyzing a pattern, one looks for lines of reflection (mark these in black with dashed lines), lines of glide reflection (blue dashed lines), and rotation centers, marked appropriately with ∘, △, ×, or ∗.

The international crystallographic notation is ___ ___ ___ ___.

- The first blank is filled by **p** or **c** (for primitive or centered). There are only two wallpaper patterns of type **c**.
- The second blank is filled in with **n** for the highest order of rotation: this will be 1 (no rotation), 2 when there is a 2-fold rotation, 3 for a 3-fold rotation, 4 for a 4-fold rotation, or 6 for a 6-fold rotation.
- The third and fourth blanks are filled by **m** if there are lines of reflection, **g** if there is no line of reflection but there is a glide reflection, and **1** otherwise. The angle at which these lines of symmetry meet also determines the pattern.

Below is a list of all 17 forms. Some books use the abbreviated forms so we have also given those.

long form	short form
p111	p1
p211	p2
p1m1	pm
p1g1	pg
c1m1	cm
p2mm	pmm
p2mg	pmg
p2gg	pgg
c2mm	cmm
p411	p4
p4mm	p4m
p4gm	p4g
p311	p3
p3m1	p3m1
p31m	p31m
p611	p6
p6mm	p6m

8.4. WALLPAPER PATTERNS • 309

Here is a flowchart to help identify the wallpaper patterns, modeled on the one from Washburn and Crowe's *Symmetries of Culture*:

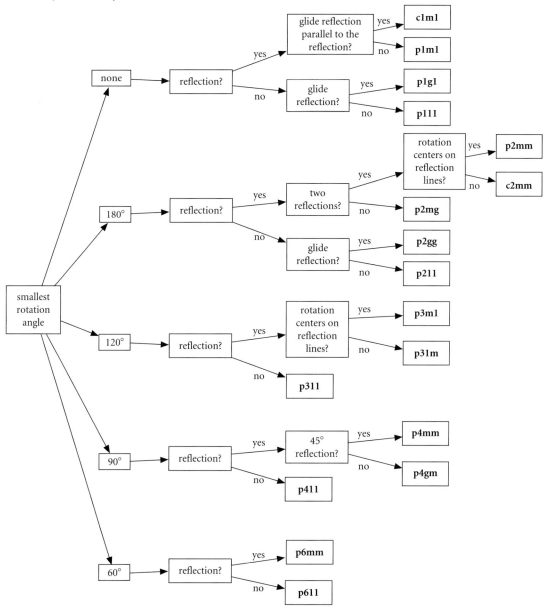

We have already found all of the wallpaper patterns with a 180° rotation. There are three possible wallpaper patterns with a 90° rotation. The first has a 4-fold rotation but no reflections or glide reflections, and you will draw this in the next activity:

▷ **Activity 8.28.** Here is the original motif and a central 4-fold center of rotation marked with a × for the pattern **p411**. Fill in the pattern first surrounding the ×, then translate this section of wallpaper to fill a space four blocks wide and three blocks tall. There are other rotation centers on the border of the cell. Mark them with the appropriate symbols. You may want to use software for greater accuracy with your pictures.

▶ **Exercise 8.49.** Here is a **p4gm** pattern. Draw the reflection lines as black dashed lines and the glide reflection lines in blue. You should find families of reflection lines in two directions and glide reflection lines in four directions. Indicate the centers of rotation appropriately.

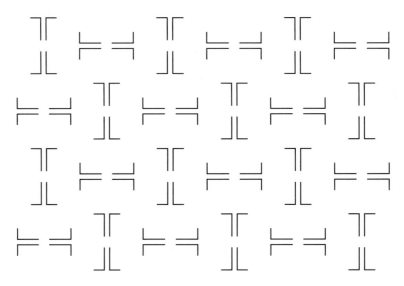

▷ **Activity 8.29.** Show what happens when you reflect the motif across the dashed lines, which meet at a 45° angle. Be sure to get all the repetitions of the figure.

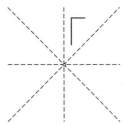

▶ **Exercise 8.50.** A **p4mm** will have 4-fold centers of rotation and reflection lines forming squares and the diagonals of the squares. It can also be described by reflecting a motif through each of the sides of a right isosceles triangle. Draw such a pattern. You may want to use software for greater accuracy with your pictures.

The remaining wallpaper patterns, **p311**, **p3m1**, **p31m**, **p611** and **p6mm**, will have translation directions making a 60° angle.

▷ **Activity 8.30.** A **p311** pattern will have 3-fold rotation centers, translations at an angle of 60°, and no lines of reflection or glide reflection. Here is a motif with a 3-fold rotational symmetry. Use this to generate a **p311** wallpaper pattern. You may want to use software for greater accuracy with your pictures.

▷ **Activity 8.31.** The patterns **p3m1** and **p31m** are often confused, even in some textbooks. Below are examples of each. Draw the reflection lines in red and the glide reflection lines in blue. Indicate the centers of rotation appropriately. One key difference between the two is that **p3m1** has all rotation centers on the reflection lines, while **p31m** has some on the reflection lines and some off the lines. Figure out which is which.

▷ **Activity 8.32.** Draw a **p611** pattern, which will have a 6-fold rotation and no other symmetries. Use software for greater accuracy with your pictures.

▷ **Activity 8.33.** Show what happens when you reflect the motif across the dashed lines, which meet at a 30° angle. Be sure to get all the repetitions of the figure.

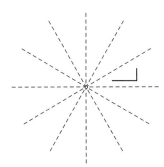

▷ **Activity 8.34.** A **p6mm** will have a hexagonal lattice, 6-fold centers of rotation, and mirror lines perpendicular to the *x*-axis and at an angle of 60° to the *x*-axis. It can also be described by reflecting a motif through each of the sides of a 30°-60°-90° right triangle. Draw such a pattern.

We state, but do not prove, the following theorem, which states that our enumeration of wallpaper patterns is complete. The proof could be done in two ways. The first of these would be to conduct an exhaustive analysis of all possible combinations of the isometries which preserve such patterns, the five allowable rotations (by 0°, 180°, 120°, 90°, and 60° by Theorem 8.30), translations, and reflections and glide reflections in various direction. This is the approach we used in analyzing frieze patterns for Theorem 8.27. Since there are almost 300 different combinations of the isometries in the case of wallpaper patterns, we choose not to attempt this. The second approach utilizes group theory, to which we do not assume that the reader has had much exposure.

Theorem 8.31. [Classification of Wallpaper Patterns] There are exactly seventeen different symmetry groups for wallpaper patterns.

▶ **Exercise 8.51.** Classify the patterns below:

8.4. WALLPAPER PATTERNS • 313

▷ **Activity 8.35.** Draw your own examples of each of the seventeen wallpaper patterns. It may be helpful to use geometric software to get accurate pictures. We encourage you to be artistic in your efforts.

♦ 8.5 Isometries in hyperbolic geometry

In Theorem 8.21 of Section 8.1, we proved that every isometry of the euclidean plane is either the identity, a reflection, a rotation, a translation, or a glide reflection. In hyperbolic geometry, things are a little more complicated since there are more possibilities for the relationships between lines. In addition, we must rewrite the definition of translation, since this assumes euclidean properties of parallelism, and try to modify the proofs of Theorems 8.6, 8.12, and 8.16 and Corollary 8.17 to apply to hyperbolic geometry. The other definitions and theorems of Section 8.1 assume only neutral geometry and so are true in hyperbolic geometry. In particular, the definitions of rotation and reflection are fine just as they are.

▷ **Activity 8.36.** Using geometric software and the Poincaré disc model, draw a hyperbolic line and a triangle. Construct the hyperbolic reflection of the triangle across the line.

▷ **Activity 8.37.** Using geometric software and the Poincaré disc model, draw a point and construct a hyperbolic angle. Draw a hyperbolic triangle and construct the hyperbolic rotation of the triangle about the point for the chosen angle.

The software package *NonEuclid* is particularly convenient since it has a menu command for a hyperbolic reflection. We shall see that any hyperbolic isometry can be written as a composition of reflections, providing a hyperbolic counterpart to Theorem 8.18.

Let A and B be two points on the hyperbolic plane. For any point X we want to figure out how to translate X in direction AB. Let us first discuss why the most obvious candidate for the definition does not work. Let us drop $XC \perp AB$. Draw a line through X perpendicular to XC and construct a point Y so that $\overline{XY} = \overline{AB}$. We would then have the following picture in hyperbolic geometry:

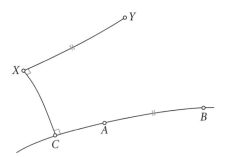

Note that although $\overline{XY} = \overline{AB}$ as seems right, $\overline{AX} \neq \overline{BY}$, so if $T(A) = B$ and $T(X) = Y$, then T is not an isometry since $\overline{AX} \neq \overline{T(A)T(X)}$. Thus this idea fails. Instead, we define translation in such a way that Theorem 8.19 will be true, so that a translation is the composition of two reflections.

Definition 8.32. Given two distinct points A and B, construct the midpoint C of $\overline{AB}$ and erect perpendiculars $\ell_1 \perp AB$ at A and $\ell_2 \perp AB$ at C. For any point X, define the *hyperbolic translation* $T_{A,B}(X) = F_{\ell_2} \circ F_{\ell_1}$.

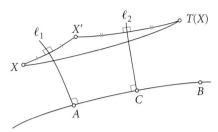

The lines ℓ_1 and ℓ_2 have common perpendicular AB, so ℓ_1 and ℓ_2 are divergent parallel. If $X' = F_{\ell_1}(X)$, note that there is no reason to think that X, X', and $T_{A,B}(X)$ lie on a straight line. Furthermore, $\overline{XT_{A,B}(X)} > \overline{AB}$, which certainly seems weird. However, as defined, $T_{A,B}$ is an isometry.

▶ **Exercise 8.52.** Prove that $T_{A,B}$ is an isometry. Let X and Y be two points in the hyperbolic plane and let $X' = F_{\ell_1}(X)$ and $Y' = F_{\ell_1}(Y)$. Show that $\overline{XY} = \overline{X'Y'} = \overline{T_{A,B}(X)T_{A,B}(Y)}$.

▷ **Activity 8.38.** Using geometric software and the Poincaré disc model, draw a hyperbolic line with two points A and B and a triangle. Construct the hyperbolic translation $T_{A,B}$ of the triangle.

With this definition of translation, Theorem 8.6 is true in hyperbolic geometry and the definition of a glide reflection as the composition of a hyperbolic translation and a reflection also makes sense.

▶ **Exercise 8.53.** Modify Definition 8.11 of a glide reflection for hyperbolic geometry.

Thus we have hyperbolic versions of translation (reflection in two divergent parallel lines), rotation (reflection in two intersecting lines), reflection, and glide reflection. There is one additional isometry in hyperbolic geometry. The name of this isometry comes from a term used by some texts: what we have called asymptotic parallel lines are sometimes called horoparallel and then divergent parallels are called hyperparallel.

Definition 8.33. If ℓ and m are asymptotic parallel, then $H_{\ell,m} = F_\ell \circ F_m$ is a *horolation*.

Some texts use the term *parallel displacement* instead of horolation. With this definition, it follows immediately that $H_{\ell,m}$ is an isometry. The horolation is illustrated below in the Poincaré disc model:

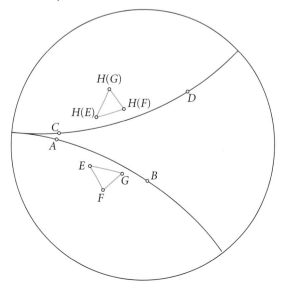

▷ **Activity 8.39.** Using geometric software and the Poincaré disc model, draw two asymptotic parallel lines AB and CD and a triangle. Construct the horolation of the triangle.

▶ **Exercise 8.54.** Does horolation preserve or reverse orientation?

▶ **Exercise 8.55.** Describe the fixed points of horolation.

The following theorem follows directly from the definitions of hyperbolic translation and horolation and by modifying the proof of Theorem 8.19:

Theorem 8.34. A hyperbolic isometry S is either a rotation, a translation, or a horolation if and only if S can be written as the composition of two reflections in distinct lines ℓ_1 and ℓ_2.

Proof: If S is a rotation, then by Lemma 8.14, there are distinct intersecting lines ℓ_1 and ℓ_2 so that $S = F_2 \circ F_1$. If S is a translation from A to B, by Definition 8.32, $S = F_2 \circ F_1$ for reflections across divergent parallel lines ℓ_1 and ℓ_2. If S is a horolation, then by Definition 8.33, $S = F_2 \circ F_1$ for reflections across asymptotic parallel lines ℓ_1 and ℓ_2.

Conversely, assume that $S = F_1 \circ F_2$ for two distinct lines ℓ_1 and ℓ_2. If ℓ_1 intersects ℓ_2 at O at angle α, then we can show that S is a rotation as in Theorem 8.19. If ℓ_1 and ℓ_2 are divergent parallel, then S is a translation by Definition 8.32. If ℓ_1 and ℓ_2 are asymptotic parallel, then S is a horolation by Definition 8.33.
□

Theorem 8.35. A hyperbolic isometry S is a glide reflection if and only if S can be written as the composition of three reflections in distinct lines ℓ_1, ℓ_2, and ℓ_3.

Proof: If S is a glide reflection, we can write it as a reflection followed by a hyperbolic translation (or vice versa), so it can be written as the composition of reflections in three distinct lines. Conversely, consider $S = F_3 \circ F_2 \circ F_1$. As in Theorem 8.20, we break the proof into cases determined by how the three lines intersect or are parallel.

Case 1: ℓ_1, ℓ_2 and ℓ_3 all pass through point O: The proof of this case is precisely the same as Case 1 in Theorem 8.20 for euclidean geometry.

Case 2a: ℓ_2 and ℓ_3 intersect and *Case 2b: ℓ_1 and ℓ_2 intersect:* The proof of Theorem 8.20 carries over without modification. *Case 2c* requires extensive rewriting and will be replaced by Case 5 below.

Case 3: ℓ_1 and ℓ_2 are divergent parallel:

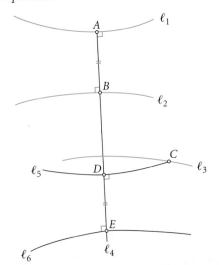

Lines ℓ_1 and ℓ_2 have a common perpendicular ℓ_4 intersecting ℓ_1 at A and ℓ_2 at B. Then $F_2 \circ F_1$ is a translation in the direction from A to B. Choose a point C on ℓ_3 and drop ℓ_5 from C perpendicular to ℓ_4 at D. Since ℓ_1, ℓ_2, and ℓ_5 are divergent parallel and share common perpendicular ℓ_4, Exercise 8.25 can be extended to show that there exists a line ℓ_6 so that $F_5 \circ F_2 \circ F_1 = F_6$ and furthermore, $\ell_6 \perp \ell_4$. Therefore, $F_3 \circ F_2 \circ F_1 = F_3 \circ F_5 \circ F_5 \circ F_2 \circ F_1 = F_3 \circ F_5 \circ F_6$. Since ℓ_3 intersects ℓ_5 at C, $F_3 \circ F_2 \circ F_1$ is a glide reflection by Case 2b.

Case 4: ℓ_1, ℓ_2, and ℓ_3 are asymptotic parallel:

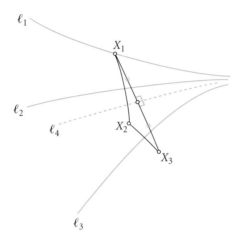

Let X_1 be any point on ℓ_1 and construct $X_2 = F_2(X_1)$ and $X_3 = F_3 \circ F_2(X_1)$. Consider $\triangle X_1 X_2 X_3$. Since ℓ_2 is the perpendicular bisector of $\overline{X_1 X_2}$ and ℓ_3 is the perpendicular bisector of $\overline{X_2 X_3}$, this is a triangle that cannot be circumscribed (recall Theorem 6.34). The perpendicular bisector ℓ_4 of $\overline{X_1 X_3}$ is thus also asymptotic parallel to both ℓ_2 and ℓ_3. Let F_4 denote reflection across ℓ_4 and note that $F_3 \circ F_2 \circ F_1(X_1) = X_3 = F_4(X_1)$. Let Z be the midpoint of $\overline{X_1 X_3}$ so Z also lies on ℓ_4.

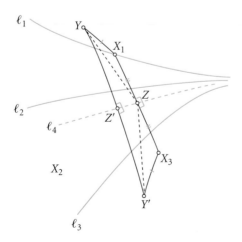

Let Y be an arbitrary point on the plane and let $Y' = F_3 \circ F_2 \circ F_1(Y)$. We want to show that $Y' = F_4(Y)$. Note that $\overline{X_1 Y} = \overline{F_3 \circ F_2 \circ F_1(X_1) F_3 \circ F_2 \circ F_1(Y)} = \overline{X_3 Y'}$. We also know that

$\angle X_3X_1Y = \angle F_3 \circ F_2 \circ F_1(X_3)F_3 \circ F_2 \circ F_1(X_1)F_3 \circ F_2 \circ F_1(Y) = \angle X_1X_3Y'$. Thus $\triangle ZX_1Y \cong \triangle ZX_3Y'$ by SAS, so $\overline{ZY} = \overline{ZY'}$ and $\angle X_1ZY = \angle X_3ZY'$. Let Z' be the intersection of $\overline{YY'}$ and ℓ_4. Therefore, $\angle YZZ' = \angle Y'ZZ'$ so $\triangle YZZ' \cong \triangle Y'ZZ'$. Thus $\overline{YZ'} = \overline{Y'Z'}$ and $\angle YZ'Z = \angle Y'Z'Z$. Therefore, ℓ_4 is the perpendicular bisector of $\overline{YY'}$. It follows that $F_3 \circ F_2 \circ F_1(Y) = Y' = F_4(Y)$. Since Y was an arbitrary point, $F_3 \circ F_2 \circ F_1 = F_4$ so $F_3 \circ F_2 \circ F_1$ is a reflection or a glide reflection of distance 0.

Case 5: ℓ_1 and ℓ_3 intersect: If ℓ_1 and ℓ_2 also intersect, the proof follows from Case 2b. If ℓ_1 and ℓ_2 are divergent parallel, then $F_3 \circ F_2 \circ F_1$ is a glide reflection by Case 3. Thus we assume the ℓ_1 and ℓ_2 are asymptotic parallel.

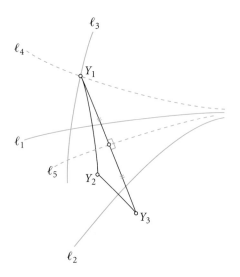

Choose a point Y_1 on ℓ_3 and construct ℓ_4 through Y_1 asymptotic parallel to ℓ_1 and ℓ_2. Let $Y_2 = F_1(Y_1)$ and $Y_3 = F_2 \circ F_1(Y_1)$. As in Case 4, we can construct a line ℓ_5 that is the perpendicular bisector of $\overline{Y_1Y_3}$ and as above, $F_4 \circ F_5 = F_2 \circ F_1$. Then $F_4 \circ F_2 \circ F_1 = F_4 \circ F_4 \circ F_5 = F_5$. Thus $F_3 \circ F_2 \circ F_1 = F_3 \circ F_4 \circ F_4 \circ F_2 \circ F_1 = F_3 \circ F_4 \circ F_5$. Since ℓ_3 and ℓ_4 intersect at Y_1, apply Case 2b and the result follows.

Thus in all cases, $F_3 \circ F_2 \circ F_1$ is a glide reflection. □

Once we have Theorems 8.34 and 8.35, the proof of Theorem 8.21 carries over to hyperbolic geometry without change, and so we get the classification theorem in hyperbolic geometry:

Theorem 8.36. [Classification of Hyperbolic Isometries] Every isometry of the hyperbolic plane is either the identity, a reflection, a translation, a rotation, a horolation, or a glide reflection.

Note that any hyperbolic isometry on the Poincaré Disc model extends to the boundary circle. There is a characterization of the isometries with respect to the fixed points of their extensions on the closed disc:

- An isometry of the Poincaré disc is called *elliptic* if it has fixed points inside the open disc. There are two types of elliptic isometries: the elliptic isometries that preserve orientation are the rotations (with fixed point the center of the rotation) and the ones that reverse orientation are the reflections (fixing the axis of reflection).

- An isometry of the Poincaré disc is called *hyperbolic* if it has no fixed points inside the open disc but its extension has two fixed points on the boundary circle. As in (1), there are two types of hyperbolic isometries. These are the translations which preserve orientation and the glide reflections that reverse orientation. In both cases, the fixed points are the two points of the hyperbolic line on the boundary circle.
- Finally, an isometry of the Poincaré disc is called *parabolic* if it has no fixed points in the open disc and it fixes one point on the boundary circle. These isometries are precisely the horolations. The fixed point is the point on the boundary circle where the two reflection lines intersect.

♦ 8.6 Projects

Project 8.1. Build a kaleidoscope and explain how it works.[1]

Project 8.2. Read "The Geometry of Folding Paper Dolls," by Brigitte Servatius. Replicate each of the frieze patterns by rolling and folding a strip of paper and cutting out an asymmetric pattern.

Project 8.3. For students who have had a course in group theory: Investigate the role of group theory in the classification of frieze and/or wallpaper patterns.[2]

Project 8.4. Draw several Islamic-style lattice patterns and classify as wallpaper patterns.[3]

Project 8.5. Report on how frieze and wallpaper patterns are used by anthropologists.[4]

Project 8.6. Read John H. Conway, *The Symmetries of Things*, and explain his notation for the classification of frieze and/or wallpaper patterns.

---

[1] A possible reference is Kinsey and Moore, *Symmetry, Shape, and Space*.
[2] A possible reference is David W. Farmer, *Groups and Symmetry*.
[3] Some references are Kinsey and Moore, *Symmetry, Shape, and Space*, Branko Grünbaum and G.C. Shephard, "Interlace Patterns in Islamic and Moorish Art," A.K. Dewdney, "Computer Recreations: Imagination meets geometry in the crystalline realm of latticeworks," or David Hutton, *Islamic Designs*.
[4] One place to start is Donald W. Crowe and Dorothy K. Washburn, *Symmetries of Culture*.

9. Tilings

♦ 9.1 Tilings on the plane

The art of tiling has been of interest to both laymen and mathematicians since antiquity. Outstanding examples can be found in the Moorish Alhambra in Granada, Spain and in the work of the Dutch artist M. C. Escher.

Definition 9.1. A *tesselation* or *tiling* of the plane is a pattern of repeated copies of figures covering the plane so that the copies do not overlap and leave no gaps uncovered. The figures are called the *tiles*.

The most common example of a tiling is the covering of the plane by squares.

Definition 9.2. A tiling is *regular* if it consists of repeated copies of a single regular polygon, meeting edge to edge so that at every vertex the same number of polygons meet.

Thus for the tiling by squares, four squares must meet at each vertex, and since each square contributes a 90° angle, they join to form a 360° angle, thus lying flat on the plane. The figure below on the right is not a regular tiling, since the junction of two of the squares forces a vertex at the midpoint of the edge of the adjoining square:

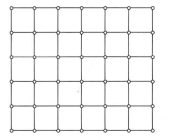 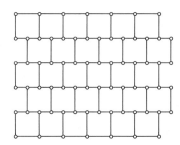

▷ **Activity 9.1.** Which other regular polygons can tile the plane?

▶ **Exercise 9.1.** Recall that the vertex angle of an n-sided regular polygon measures $\frac{(n-2)180°}{n}$. Use this fact to prove that there are only three regular tilings of the plane. If k of these n-sided polygons meet at a vertex, then $k\frac{(n-2)180°}{n} = 360°$. Find the integer solutions of this equation.

▶ **Exercise 9.2.** Classify the three regular tilings as wallpaper patterns.

◊ 9.1.1 Semiregular tilings

Next we analyze tilings with more than one type of polygon.

Definition 9.3. A *semiregular* or *Archimedean* tiling is a tiling in which each tile is a regular polygon and each vertex is identical.

There are exactly three regular tilings and eight semiregular tilings. An example of a semiregular tiling is one in which two regular octagons and a square, all with the same edge length, meet at each vertex. Each octagon has a $135°$ vertex angle and the square a $90°$ angle. Since $2 \times 135° + 90° = 360°$, the pattern lies flat and furthermore can be extended to form a tiling of the plane.

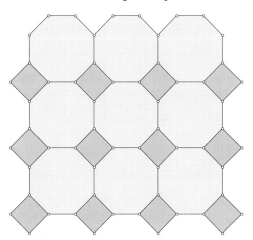

We want to find all of the semiregular tilings. This will be done by first deriving a list of rules to which such a tiling must conform, then making a list of candidates for tiling, and finally checking which candidates yield a tiling of the plane.

▶ **Exercise 9.3.** Fill in the following table.

Regular polygon	# sides	vertex angle
Triangle	3	
Square	4	
Pentagon	5	
Hexagon	6	
Heptagon	7	
Octagon	8	
Nonagon	9	
Decagon	10	
Dodecagon	12	
Pentakaidecagon	15	
Octakaidecagon	18	
Icosagon	20	
Tetrakaicosagon	24	

It is convenient to find notation by which we can specify a particular tiling. Since each vertex is identical, we denote a tiling by describing the number of sides of the polygons meeting at a vertex. At this point, we have mentioned four tilings: the three regular tilings found in Activity 9.1 and the octagon-square shown above. The regular tiling of squares will be denoted as 4.4.4.4, since each vertex is surrounded by four 4-sided figures. The semiregular tiling by squares and octagons will be denoted as 4.8.8, since each vertex is surrounded by one square and two octagons. It could also be denoted by 8.4.8 or 8.8.4, but it is traditional to begin with the number of sides of the smallest polygon. Also note that it does not matter whether one reads off the polygons in a clockwise or counterclockwise manner.

Since we are attempting to tile the plane, the arrangement of polygons around each vertex must lie flat. From this observation we get the following rule.

Rule 1: *In a tiling of the euclidean plane, the sum of the vertex angles of the polygons meeting at each vertex must be exactly 360°.*

▷ **Activity 9.2.** What is the largest number of regular polygons that can fit around a vertex? What is the smallest number of regular polygons that can fit around a vertex?

▷ **Activity 9.3.** Can there be four different regular polygons at a vertex? Explain your reasoning.

Activities 9.2 and 9.3 give us the following rules for semiregular tilings:

Rule 2: *A semiregular tiling must have at least three and no more than five polygons meeting at each vertex.*

Rule 3: *No semiregular tiling can have four different polygons meeting at a vertex. Thus, if a semiregular tiling has four polygons at a vertex, there must be some duplicates.*

We have set some parameters for our search for semiregular tilings. We will continue in two stages. First we will identify candidates for tilings so that a single vertex satisfies Rules 1, 2, and 3. You should find 14 such candidates. Then we will check each one to see if it can be extended to a tiling of the plane.

We have two examples given previously in the text: 4.4.4.4 and 4.8.8. Another candidate is 3.7.42, clearly satisfying Rules 2 and 3. A simple computation shows that it satisfies Rule 1: the vertex angle for a regular heptagon is $(128\frac{4}{7}°)$, and for a 42-gon the angle is $(171\frac{3}{7}°)$. Since $60 + 128\frac{4}{7} + 171\frac{3}{7} = 360°$, Rule 1 is satisfied. In the next exercise, you will find the remaining candidates.

▶ **Exercise 9.4.** Make a list of semiregular tiling candidates that satisfy Rules 1, 2, and 3. To make the task easier, we have indicated the number of polygons meeting at a vertex.

Symbol	# polygons
	5
	5
	4
	4
	4
4.8.8	3
3.7.42	3
	3
	3
	3
	3
	3
	3
	3

In the exercise above we found 14 ways to fit regular polygons around a single vertex using at least two different types of polygons. Unfortunately, not all of these extend to tilings of the plane. A little additional thought gives us further conditions that must be met in order to form a tiling of the plane. Note that if we try to arrange one equilateral triangle and copies of two other different polygons at each vertex to make pattern 3.n.m, they would be arranged as below:

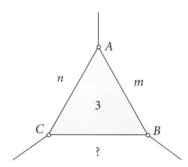

At vertex A we have the vertex configuration 3.n.m. A triangle and an m-sided polygon meet at B, so the polygon marked '?' ought to have n sides. But at vertex C there are already a triangle and an n-sided polygon, so the polygon marked '?' should have m sides. Thus we cannot have a semiregular tiling with a vertex configuration of the form 3.n.m if $n \neq m$.

▶ **Exercise 9.5.** Generalize the result of the paragraph above to address the case of a vertex configuration of the form k.n.m where k is odd.

This observation gives us Rule 4 below, and a similar argument gives us Rule 5.

Rule 4: *No semiregular tiling can have vertex configuration k.n.m where k is odd and $n \neq m$.*

Rule 5: *No semiregular tiling can have vertex configuration 3.k.n.m unless k = m.*

▶ **Exercise 9.6.** Prove Rule 5 is valid.

▶ **Exercise 9.7.** Note that we could generalize Rule 5 as we did Rule 4 to j.k.n.m for j odd and $k \neq m$. Why is this generalization not useful for semiregular tilings?

We can use Rule 4 to eliminate 3.7.42, since it is of the form 3.n.m with $n \neq m$. To see this geometrically, try to draw this tiling. A triangle, a heptagon, and a 42-gon meet at each vertex. If we arrange this correctly at vertex A and vertex B, then we are forced to have the triangle and *two* heptagons at vertex C. Thus the vertex pattern fails at C and therefore, this pattern cannot be extended to a semiregular tiling of the plane.

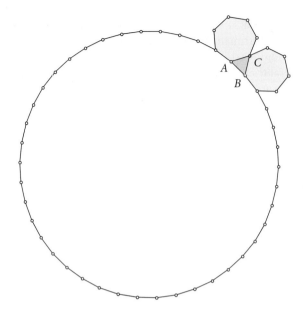

We have twelve other vertex configurations to check. Rules 4 and 5 will eliminate many of the candidates. For the vertex configurations that pass all five rules, you will draw sections of the tiling. These drawings must be extensive enough to be convincing that the tiling could be extended indefinitely. Thus, a single vertex or a row of polygons won't do. Your pictures should extend in all directions. To create these, one can fit together cardboard polygons, use tracing paper to trace illustrations of the polygons, or use geometric software.

▶ **Exercise 9.8.** There are nine candidates from Exercise 9.4 with three polygons at a vertex. We have shown that 4.8.8 tiles the plane and 3.7.42 does not. Use Rule 4 to eliminate all but two of the remaining candidates. Draw sections of the two tilings of this type.

When we turn to the candidates with four and five polygons at a vertex, there is an additional difficulty. One of the candidates from Exercise 9.4 is 3.4.4.6. There are two different ways that a triangle, two squares, and a hexagon can be arranged around a single vertex. We designate these by 3.4.4.6 and 3.4.6.4, shown below:

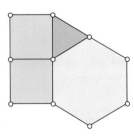 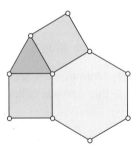

Note that the pattern 3.4.4.6 violates Rule 5, but 3.4.6.4 does not.

▶ **Exercise 9.9.** Draw a section of the tiling 3.4.6.4. Draw enough of 3.4.4.6 to show that it does not extend to a semiregular tiling, giving geometric confirmation of Rule 5.

▶ **Exercise 9.10.** Check the remaining tiling candidates from Exercise 9.4 with four polygons at a vertex, considering all possible rearrangements. Explain which ones fail and draw a section of the one tiling that results.

▶ **Exercise 9.11.** Both of the patterns found in Exercise 9.4 with five polygons around a vertex can be extended to tilings. One of these gives two different tilings, depending on the order of the polygons around a vertex. Draw sections of these three tilings.

You should now have found and drawn the all of the regular and semiregular tilings of the plane.

Theorem 9.4. [Classification of Euclidean Tilings] There are three regular tilings and eight semiregular tilings of the euclidean plane.

▶ **Exercise 9.12.** Classify the eight semiregular tilings as wallpaper patterns.

◇ **9.1.2 Dual tilings**

To each polygonal tiling we can associate another *dual tiling*, which might not be regular or semiregular. For example, below is the tiling 4.8.8 and its dual tiling, which is a tiling by 45°–45°–90° triangles:

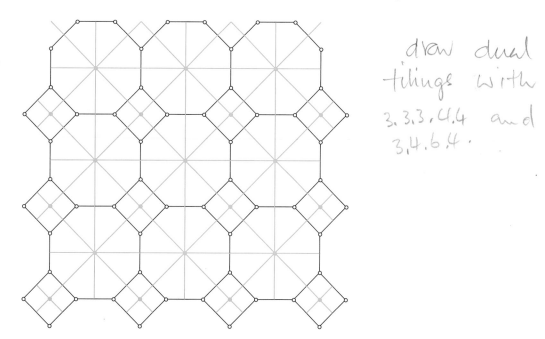

draw dual tilings with 3.3.3.4.4 and 3.4.6.4.

The process of forming the dual is as follows: place a vertex at the center of each polygon in the original tiling. Whenever two polygons share an edge in the original tiling, draw a dual edge connecting the new vertices at the centers of those polygons. In order to carry out this procedure, you will need to locate the centers for various regular polygons.

▶ **Exercise 9.13.** Figure out a general procedure for finding the center for any regular polygon. (Note that your procedure may have to be different depending on the number of sides.)

▶ **Exercise 9.14.** Suppose somewhere in a tiling there are two regular polygons that share an edge, one with *n* sides and the other with *k* sides. Where will the line connecting the centers of the two polygons intersect the shared edge? What angle will this line make with the shared edge? Prove your answers.

▷ **Activity 9.4.** Describe the dual tilings for each of the three regular tilings.

Here is the tiling 3.4.6.4 and its dual. Note that the dual contains only one type of tile, an irregular quadrilateral.

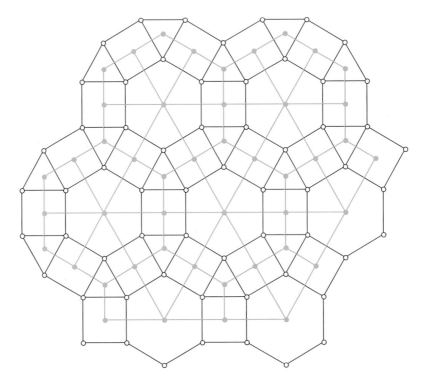

To find the angles of this irregular quadrilateral, note that it has three types of vertices: Two of the vertices are at the centers of squares of the original tiling, one of the vertices is at the center of the hexagon, while the last is at the center of the equilateral triangle. Since the lines of the dual tiling divide the square into four equal sectors, the two angles in the squares must be $\frac{360°}{4} = 90°$. Since the dual tiling divides the triangle into three equal sectors, those angles must be $\frac{360°}{3} = 120°$. The angles formed at the center of the hexagon must be $\frac{360°}{6} = 60°$. Thus, the irregular quadrilateral has angles 120°, 90°, 90°, 60°, which sum to 360° as they should for a quadrilateral.

▷ **Activity 9.5.** Draw the dual tiling for each of the tilings found in Exercises 9.8, 9.10 and 9.11. Describe the types of polygons formed and find their vertex angles.

▶ **Exercise 9.15.** Explain why the dual tiling of a regular or semiregular tiling will always have only one type of polygon, though it may be irregular.

▶ **Exercise 9.16.** Of the examples and exercises above, which tilings have duals formed by triangles? quadrilaterals? pentagons? Find a rule for the type of polygon in the dual. Justify your answer.

▶ **Exercise 9.17.** Explain the relationship between the wallpaper type of the original tiling and the wallpaper type of the dual tiling.

♦ 9.2 Tilings by irregular tiles

In tiling, we need not restrict ourselves to regular polygons. It is easy to tile the plane with copies of any rectangle or parallelogram:

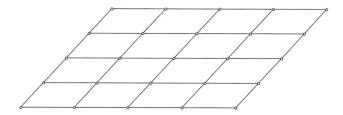

▶ **Exercise 9.18.** Prove that any triangle can tile the plane.

▶ **Exercise 9.19.** Prove that any trapezoid can tile the plane.

In fact, any quadrilateral will tile the plane if the copies are put together right. Label the angles α, β, γ, and δ. Since the sum of the angles in any quadrilateral must be $360°$, $\alpha + \beta + \gamma + \delta = 360°$. Arrange four copies of the quadrilateral so that the sides of the same length match up and so that α, β, γ, and δ surround the center vertex. Then repeat this basic building block to create a tesselation:

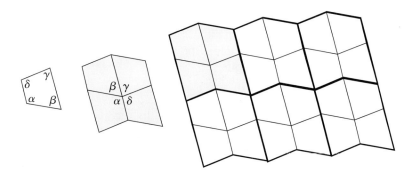

Note that the quadrilateral above is convex.

▷ **Activity 9.6.** Demonstrate, either on paper or using geometric software, that the above technique works for nonconvex quadrilaterals. Your drawing should contain at least 12 repeats of the quadrilateral.

◇ 9.2.1 Reptiles

A *reptile* (short for repeating tile) is a tile that can be arranged to form a larger copy of itself. This larger copy must be an exact scaled replica of the original. For example, a square is easily seen to be a reptile, since four squares (or, for that matter, 9 squares, or 16, etc.) can be arranged to form a larger square:

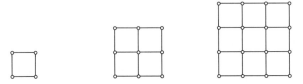

▶ **Exercise 9.20.** Below is a trapezoidal reptile. Find the angles of the trapezoid.

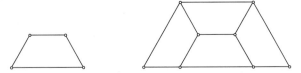

▶ **Exercise 9.21.** Below is a shaded rectangular reptile such that two repeats of the original rectangle give a larger similar rectangle. Find the dimensions of the shaded rectangle if the longer side has length 1.

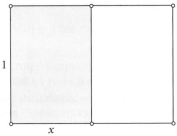

▷ **Activity 9.7.** Below is a 30°–60°–90° triangle. Show how to put four copies of this triangle together to make a larger similar triangle. Show how to put three copies of this triangle together to make a larger similar triangle.

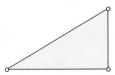

▶ **Exercise 9.22.** There is only one triangular reptile such that two repeats of the original triangle form a larger similar one. Which triangle is this?

▷ **Activity 9.8.** Show that each of the following is a reptile:

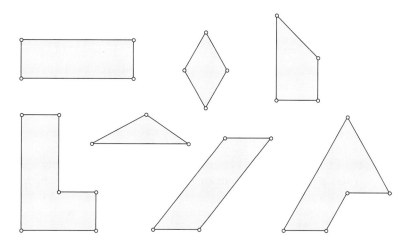

◇ 9.2.2 Escher-style tilings

Maurits Cornelis Escher was not a mathematician, but a mathematical artist. He was a Dutch graphic artist and created many tiling pictures. His early work was representational, etchings and engravings of landscapes and people, though with a strong geometric flavor. After visiting the Alhambra in Granada, Spain in 1936 and being inspired by the floor and wall tiles there, he began a systematic study of tilings and the art of generating and embellishing them. Many of his pictures involved repetitions of pictures of stylized animals, plants, and people rather than simply polygons. After working in this vein for a couple of years, he showed some of his work to his elder half-brother, Beer Escher, a geologist, who recommended some articles in German journals on crystallography, saying "It is all very theoretical, but the illustrations may be of some use to you." The journalist's-Gravesande, after interviewing Escher in 1940, reported,

> Most of the articles were far too difficult for a layman, too dry and too theoretical, but among them there was one in particular by a certain Polish professor in Zürich [George Polya, actually Hungarian] entitled "Über die Analogie der Kristallsymmetrie in der Ebene" ["On the analogy of crystal symmetry in the plane"] which Escher thoroughly read and studied. It contained many nice illustrations which he copied along with the entire article.[1]

---

[1] In *De vrije bladen*, 1940, quoted in Schattschneider, *Visions of Symmetry*, p. 22.

The article by Polya is the same one we mentioned earlier in discussing the discovery of the 17 wallpaper groups and the seven frieze groups. Escher studied this article and the accompanying illustrations and developed his own system of cataloging patterns, not only in terms of 17 wallpaper groups, but also in terms of relationship between motif and underlying grid (for example, he distinguished between a **p211** pattern drawn with perpendicular translations and a **p211** pattern on a grid of parallelograms), and in terms of coloring, developing a rule that adjacent figures must have different colors. Escher called a tiling a *regular division of the plane*. He remarks,

> In mathematical quarters, the regular division of the plane has been considered theoretically, since it forms part of crystallography. Does this mean that it is an exclusively mathematical question? In my opinion, it does not. Crystallographers have put forward a definition of the idea, they have ascertained which and how many systems or ways there are of dividing a plane in a regular manner. In doing so, they have opened the gate leading to an extensive domain, but they have not entered this domain themselves. By their very nature they are more interested in the way in which the gate is opened than in the garden lying behind it. To develop this metaphor: a long time ago, I chanced upon this domain [of regular division of the plane] in one of my wanderings; I saw a high wall and as I had a premonition of an enigma, something that might be hidden behind the wall, I climbed over with some difficulty. However, on the other side I landed in a wilderness and had to cut my way through with a great effort until—by a circuitous route—I came to the open gate of mathematics. From there, well-trodden paths lead in every direction, and since then I have often spent time there. Sometimes I think I have covered the whole area, I think I have trodden all the paths and admired all the views, and then I suddenly discover a new path and experience fresh delights.[2]

We are fortunate that his notebooks have survived, explaining his classification scheme and showing preliminary sketches for some of his etchings and other drawings that were never published. The interested reader is referred to Schattschneider's *Visions of Symmetry*. In these notebooks one sees the construction of a grid underlying many of these pictures. In this section, we will outline some procedures for generating such tilings by irregular figures.

We begin by laying out a tiling or grid, which can be a regular or semiregular tiling or a tiling by irregular polygons. Then the basic tile or tiles are modified by an isometry, which must be chosen to be compatible with the tiling, and the modified tile is then used to tile the plane. We will restrict our investigation to tilings that use a single type of tile, though the methods of this section may be used for other tilings.

Modifying a tile using translation
To use this technique, the original tile must have at least one pair of parallel sides of equal length. Thus, consider tilings such as squares, rectangles, parallelograms, or hexagons. We will demonstrate the technique with a tiling by parallelograms. One tile has been shaded.

---

[2]Escher, *Regelmatige vlakverdeling (Regular division of the plane)*, translated and reprinted in Bool, Kist, Locher, and Wierda, *M.C. Escher, His Life and Complete Graphic Work.*

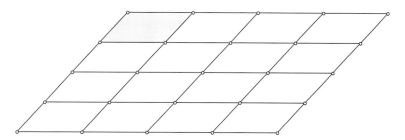

We can cut the shaded tile out of cardboard. Take this tile and cut a section out of one side, leaving the vertices intact. Tape this piece onto the parallel side, using a translation. (Note that it is not really necessary to leave the vertices intact, but this does make the result easier to analyze.)

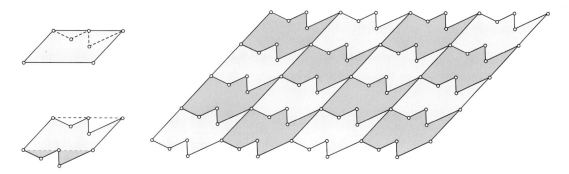

Modifying one tile changes all of the others, generating a new tiling of irregular tiles. The process can be repeated for the other pair of parallel sides. It becomes increasingly difficult to see the original underlying grid of parallelograms that formed the original tiling, but the translational symmetry of that tiling is preserved in the modified tiling.

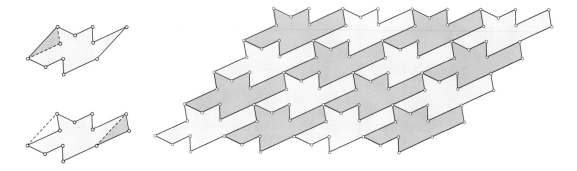

Escher used the technique of translation as well as other similar methods in generating many of his pictures. Below is one of his tilings of winged horses. In order to analyze this drawing as a tiling, note that the smallest repeating unit (ignoring the coloring) is a single horse. The horses are colored alternately light and dark to help the viewer figure out where one horse stops and another starts, but tracing will show you that the light and dark horses are indeed identical.

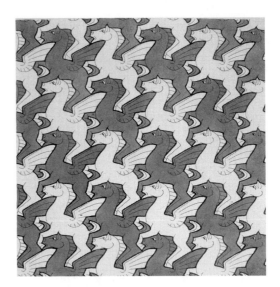

(*Source*: M.C. Escher's "Symmetry Drawing E105" © 2009 The M.C. Escher Company – Holland)

We wish to find a single polygonal tile which generates the tiling of horses and to describe how to modify that tile to form a single copy of a horse. Thus each polygonal tile must contain enough parts to be assembled into one horse. To find the grid or underlying tiling, connect some easily identifiable feature: Below on the left we have chosen the horse's eyes. Connecting the eyes as shown gives a grid of squares. In the picture below on the left there is a square tile containing the torso of a dark horse. The front legs are cut off, but the square contains the front legs of a white horse to compensate. The square is missing most of the back legs of the dark horse, but the square contains precisely those parts of a white horse. The upper left corner of the square cuts the horse's head off, and in fact the grid slices the head up into four pieces. Duplicates of the missing three pieces can be found in the other three corners of the square. Thus the square contains exactly enough parts to make up a single horse. The horse is derived from the square by parallel translation on the

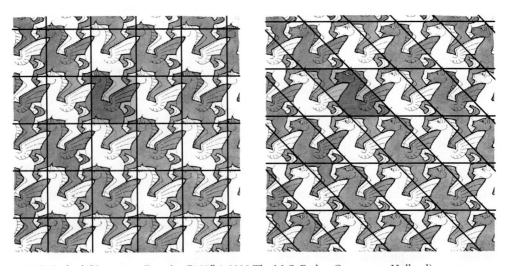

(*Source*: M.C. Escher's "Symmetry Drawing E105" © 2009 The M.C. Escher Company – Holland)

top and bottom sides, and on the left and right sides. Note that a different choice for the grid lines, as shown above on the right, leads to a different tile. There are commonly several ways to generate tilings such as this.

▷ **Activity 9.9.** Using either paper and pencil or geometric software, create a tiling from a rectangular grid, using translation of both pairs of parallel sides.

▷ **Activity 9.10.** Using either paper and pencil or geometric software, create a tiling from a hexagonal grid, using translation of all three pairs of parallel sides.

Modifying a tile using glide reflection

Another way to modify a tiling is to use glide reflection. Again, this requires a tiling in which the tiles have at least one pair of parallel sides of equal length. Take the original tile and cut something off of one of the parallel sides. Flip the cut piece over and tape it onto the parallel side. Then tile the plane with the new tile. You will have to turn over every other tile, so we have drawn the tile with two types of shading, the darker denoting copies of the basic tile in its original position, and white ones showing the tiles flipped over:

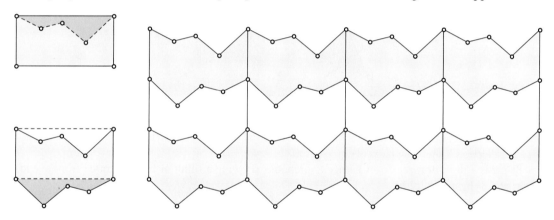

Here is an Escher-style tiling of fish that utilizes glide reflection. Drawing lines connecting the eyes, we get a grid of parallelograms. Each parallelogram contains enough parts to make a single fish. Using a glide reflection on the top and bottom edge and a translation on the left and right edges changes the parallelogram into a fish:

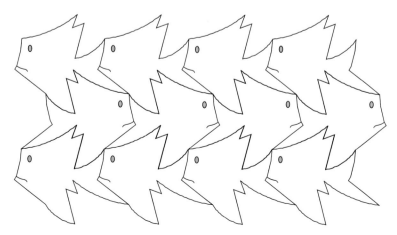

▷ **Activity 9.11.** Using either paper and pencil or geometric software, create an irregular tiling from a rectangular grid, using glide reflection on both pairs of parallel sides.

Modifying a tile using midpoint rotation

Another of Escher's techniques for modifying tilings is *midpoint rotation*. To use this method, one can use a basic grid formed by any tiling. Take one side of a tile and find its midpoint. Cut a pattern out of the tile that begins at one endpoint and ends at the midpoint of the same side. Then take the piece that was cut out, rotate it by 180° about the midpoint, and tape it back along the edge on the other side of the midpoint. Then tile the plane with the new tile.

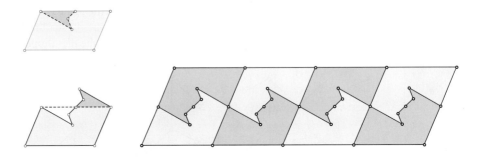

This maneuver can be repeated on the other sides of the basic unit or can be combined with translation or glide reflection if there are pairs of equal parallel sides.

Below is a draft drawing from Escher's notebooks showing a tiling of fish generated from a grid of quadrilaterals by midpoint rotation.

(*Source*: M.C. Escher's "Symmetry Drawing E89" © 2009 The M.C. Escher Company – Holland)

▷ **Activity 9.12.** Create an irregular tiling from an equilateral triangle grid, using midpoint rotation on each side.

▷ **Activity 9.13.** Create an irregular tiling from a parallelogram grid, using a translation on one pair of parallel sides and a midpoint rotation on each of the other sides.

Modifying a tile using side rotation

The fourth of Escher's techniques we investigate, *side rotation*, requires that the original tile has a pair of adjacent sides of equal length, such as a grid of squares, equilateral triangles, rhombi, or regular hexagons. Cut a pattern out that begins at one vertex of one of the equal edges of the tile and ends at the other vertex. Then take the piece you cut out, rotate it about the vertex connecting the two equal sides, and tape it to the adjacent equal edge. Then tile the plane with the new modified tile.

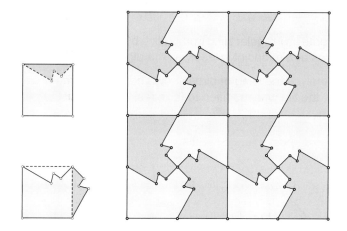

This maneuver can be repeated on the other sides of the original tile or can be combined with the other modifications.

Here is another Escher-style tiling of turtles, generated by side rotation and drawn using the program *TesselMania*. There are some fairly obvious rotation points: where the heads of four turtles meet, and again where the back feet of four turtles meet. Draw a set of horizontal and vertical lines through the points where the four heads meet and another through the back feet points. The head lines meet the feet lines at another, less obvious, rotation center where the tails of two turtles and the front feet of two other turtles meet. All of these lines give a square grid. Each square contains the body of one turtle. For example, the square outlined below contains most of a dark turtle. The back foot and tail of this turtle lie outside the square, but a back foot and tail belonging to a light turtle compensate for that loss. The head and part of the front foot is cut off, but a light turtle's foot and head make up for that. A side rotation about the head point rotates the light turtle head and front foot around to where we want the dark turtle head and foot to be. A side rotation around the back foot point rotates the light turtle's foot and tail into the correct position for the dark turtle's rear.

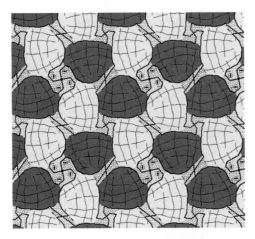
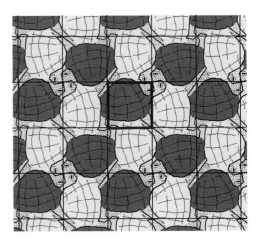

▷ **Activity 9.14.** Create an irregular tiling from a tiling by equilateral triangles, using side rotation on one pair of sides and midpoint rotation on the other side.

▶ **Exercise 9.23.** Analyze the tiling of birds shown below. Find the underlying polygonal tiling and describe how the tile was modified to form a single copy of the bird.

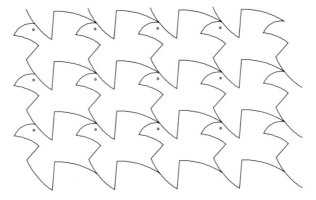

▶ **Exercise 9.24.** Analyze the tiling of birds shown below. Find the underlying polygonal tiling and describe how the tile was modified to form a single copy of the bird.

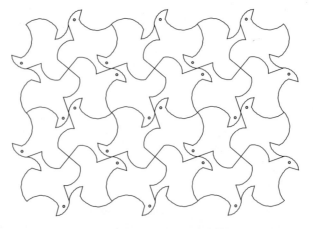

▶ **Exercise 9.25.** Analyze the tiling of people shown below. Find the underlying polygonal tiling and describe how the tile was modified to form a single person.

▶ **Exercise 9.26.** Analyze the tiling shown in M. C. Escher's *Reptiles*. Find the underlying polygonal tiling and describe how the tile was modified to form a single reptile.

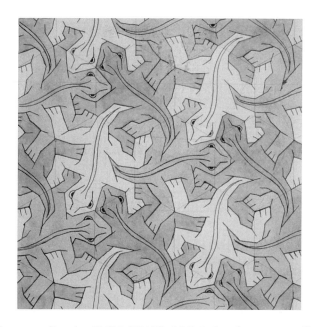

(*Source*: M.C. Escher's "Symmetry Drawing E25" © 2009 The M.C. Escher Company – Holland)

(*Source*: M.C. Escher's "Reptiles" © 2009 The M.C. Escher Company – Holland)

Many of Escher's tiling pictures make use of more than one tile. For example, in the drawing below there are both angels and devils. Thus, the basic repeating unit must contain one angel and one devil.

(*Source*: M.C. Escher's "Symmetry Drawing E45" © 2009 The M.C. Escher Company – Holland)

◊ 9.2.3 The Conway criterion

There are many shapes that cannot tile the plane, such as regular pentagons. Here is one rule that, when satisfied, implies that a figure will tile the plane. However, there are many figures that do not fulfill these criteria but succeed in tiling. This rule was proved by John H. Conway, the mathematician who developed the terminology we used for frieze patterns. A *simple region* is a subset of the plane that contains no punctures or holes and that has a boundary that ends where it begins (such as a circle or polygon) and does not intersect itself.

Theorem 9.5. [Conway Criterion] A simple region will tile the plane if the boundary can be divided into six sections by six points labeled *A*, *B*, *C*, *D*, *E*, and *F* in order as one travels around the boundary, and

1. The curve *AB* from *A* to *B* is the translation of the curve *ED*.
2. The curves *BC*, *CD*, *EF*, and *FA* have rotational symmetry about their midpoints.

Here is an example of a tile fitting the Conway criterion. The points marked with a ○ are the midpoints and centers of rotation.

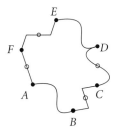

▷ **Activity 9.15.** Creat a tile that satisfies the Conway criterion and fit copies together to cover a roughly square region to show a piece of the tiling.

Proof of Theorem 9.5: Consider a tile satisfying the Conway Criterion. Label the vertices of this tile as $A_1B_1C_1D_1E_1F_1$. Let $A_2B_2C_2D_2E_2F_2$ be another copy of this tile and join these two tiles so $E_2 = A_1$ and $D_2 = B_1$, noting that since curve A_1B_1 is congruent to E_2D_2, these curves join without gaps. Take another

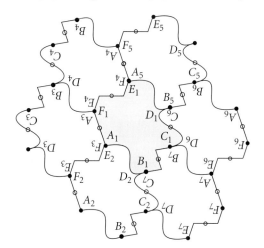

tile $A_3B_3C_3D_3E_3F_3$, rotate by 180°, and join this to the other two so that $A_3 = F_1$, $F_3 = A_1 = E_2$, and $E_3 = F_2$. Since AF and EF have two-fold symmetry, these curves fit together. Take a fourth tile $A_4B_4C_4D_4E_4F_4$, rotate by 180°, and join this to the other three so that $F_4 = E_1$, $E_4 = F_1 = A_3$, and $D_4 = B_3$. Since EF has two-fold symmetry and ED is congruent to AB, these curves fit together. A fifth tile tile $A_5B_5C_5D_5E_5F_5$ is joined so that $F_5 = A_4$, $A_5 = E_1 = F_4$, and $B_5 = D_1$. A sixth tile $A_6B_6C_6D_6E_6F_6$ is rotated by 180° and joined so that $B_6 = C_5$, $C_6 = D_1 = B_5$, and $D_6 = C_1$. A seventh tile $A_7B_7C_7D_7E_7F_7$ is rotated by 180° and joined so that $A_7 = E_6$, $B_7 = C_1 = D_6$, $C_7 = B_1 = D_2$, and $D_7 = C_2$. Thus we have surrounded the first tile by copies of itself. Repeat for the second tile, and so on.

▶ **Exercise 9.27.** Show that the tile below fits the Conway criterion by marking the points A, B, C, D, E, and F, and the centers of rotation for the curves BC, CD, EF, and FA. Trace copies of the tile and fit them together to surround the original tile.

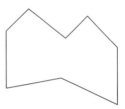

▶ **Exercise 9.28.** Show that the tile below fits the Conway criterion by marking the points A, B, C, D, E, and F, and the centers of rotation for the curves BC, CD, EF, and FA. Trace copies of the tile and fit them together to surround the original tile.

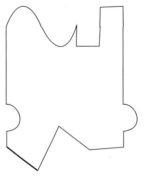

▷ **Activity 9.16.** Find or create a figure that tiles the plane but does not satisfy the Conway criterion.

An extension of the Conway criterion allows some of the points A, B, C, D, E, and F to be the same, as long as there are at least three distinct points. Of course, if $A = B$, then the arc AB is a single point, and then we must have $D = E$. In this case there are no translated sides, and we have a quadrilateral $ACDF$ with a midpoint rotation on each side, such as the one pictured below:

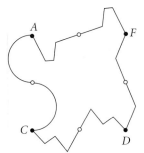

▷ **Activity 9.17.** Trace copies of the tile above and fit them together to surround the original tile.

▶ **Exercise 9.29.** Describe the situation in which a tile satisfies the Conway criterion with $A = B = C$ and $D = E$.

▶ **Exercise 9.30.** Describe the situation in which a tile satisfies the Conway criterion with $B = C$ and $E = F$.

♦ 9.3 Tilings of noneuclidean spaces

The same concepts and techniques we used to study regular and semiregular tilings of the euclidean plane can be adapted to study tilings in other geometries. We do not intend an in-depth analysis of these, but rather a sampling of results to demonstrate the more notable similarities to and differences from the euclidean case.

◇ 9.3.1 Spherical tilings

Here is an example of the spherical tiling 3.5.3.5.

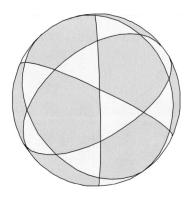

As in our study of euclidean tilings, there are only finitely many regular and semiregular tilings of the sphere. Tilings of the sphere are more commonly thought of as polyhedra, in which case one considers the tiles to be flat polygons. Here is a picture of the tiling 3.5.3.5 considered as a polyhedron. In this form, it is called the icosidodecahedron:

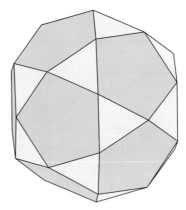

Any such polyhedron can be inflated to form a spherical tiling, but doing so will change the vertex angles. In the illustration of the spherical tiling, the angles at each vertex will sum to 360°. In the polyhedron, the angles are those formed by two equilateral triangles and two regular pentagons, for an angle sum of $2 \cdot 60° + 2 \cdot 108° = 336°$.

▶ **Exercise 9.31.** Modify the five rules of Section 9.1 for tiling to the case of semiregular spherical polyhedra with flat faces.

The next chapter is devoted to the study of three-dimensional geometry and so we will reserve further discussion of polyhedra until then.

◊ ## 9.3.2 Hyperbolic tilings

Here is an example of the hyperbolic tiling 3.7.3.7, drawn using the program *KaleidoTile* written by Jeffrey Weeks. This shows a view of the hyperbolic plane as a curved space, much like the sphere.

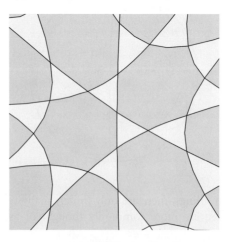

A section of the same tiling is shown below in the Poincaré disc model.

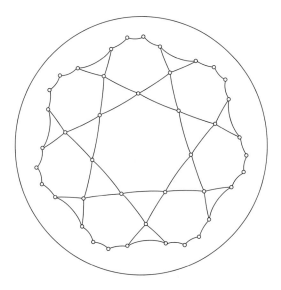

Here is a section of the 3.4.3.4.3.4 tiling in the Poincaré disc model:

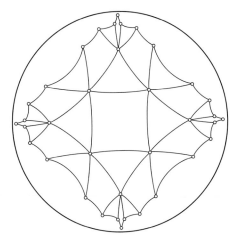

A drawing of a hyperbolic tiling similar to this formed the inspiration for a series of Escher's prints, such as *Circle Limit IV* below. In a 1958 letter to his son, he remarks,

> I'm engrossed again in the study of an illustration which I came across in a publication of the Canadian professor H.S.M. Coxeter, of Ottawa (whom I met in Amsterdam some time ago), *A Symposium on Symmetry*. I am trying to glean from it a method for reducing a plane-filling motif which goes from the centre of a circle out to the edge, where the motifs will be infinitely close together. His hocus-pocus text is no use to me at all, but the picture can probably help me to produce a division of the plane that promises to become an entirely new variation of my series of divisions of the plane. A circular regular division of the plane, logically bordered on all sides by the infinitesimal, is something truly beautiful, almost as beautiful as the regular division of the surface of a sphere.[3]

---

[3] Reprinted in Bool, Kist, Locher, and Wierda, *M.C. Escher, His Life and Complete Graphic Work*, p. 91.

Note that all of the hundreds of angels pictured are congruent and thus have equal area in hyperbolic geometry, as do the devils.

(*Source*: M.C. Escher's "Circle Limit IV" © 2009 The M.C. Escher Company – Holland)

Escher sent a print of a similar picture to the mathematician H.S.M. Coxeter [1907–2003] whose illustration had provided the initial impetus for the development of this series of engravings. He wrote his son again, "I had an enthusiastic letter from Coxeter about my colored fish, which I sent him. Three pages of explanation of what I actually did . . . It's a pity that I understand nothing, absolutely nothing of it."[4]

The software package *NonEuclid* is particularly well suited to drawing tilings in the Poincaré disc model, since it has a menu command for hyperbolic reflection.

▷ **Activity 9.18.** Using geometric software, draw the 3.3.3.3.3.3.3 tiling in the Poincaré disc model.

▷ **Activity 9.19.** Using geometric software, draw the 4.4.4.4.4.4 tiling in the Poincaré disc model.

▶ **Exercise 9.32.** Explain why there are infinitely many regular tilings of hyperbolic space. (Consider tilings by equilateral triangles.)

---

[4]Reprinted in Bool, Kist, Locher, and Wierda, *M.C. Escher, His Life and Complete Graphic Work*, pp. 100–101.

▶ **Exercise 9.33.** Modify the five rules of Section 9.1 for tiling to the case of semiregular hyperbolic polyhedra.

▶ **Exercise 9.34.** In Section 9.1, we showed that on the euclidean plane there are only eight possible semiregular tilings. What happens in hyperbolic space? Explain your answer.

Using basic facts about angle sums in euclidean, spherical, and hyperbolic geometry, we have considered the number and types of possible tilings in each space. In euclidean geometry, where the angle sum for a triangle is exactly 180°, there are the fewest possibilities. In spherical geometry, where the angle sum for a triangle lies between 180° and 540°, there are more, but still finitely many possibilities. And in hyperbolic geometry, with angle sums lying between 0° and 180°, the possibilities are endless.

One can also tile the hyperbolic plane using biangles or ideal triangles. Below is part of a tiling of ideal triangles. Note that there should be infinitely many triangles meeting at each ideal vertex and all of these triangles are congruent and equal in area.

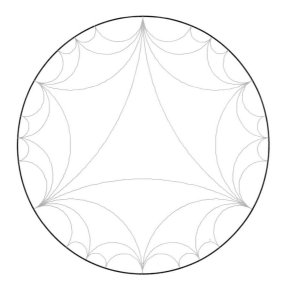

◆ 9.4 Penrose tilings

◇ 9.4.1 The Extension Theorem

We began this chapter with the most restrictive tilings: the regular tilings, in which all tiles must be regular polygons and must be identical. We then discussed the semiregular tilings which allow more than one type of tile (though always a finite number of distinct tiles) and each tile must still be a regular polygon. In order to fit together correctly, the sides of one tile must have the same length as the sides of any other tile. When we introduced irregular tilings, we allowed irregular polygons where side length could differ and might even have curved edges. To simplify things, we reverted to using only one type of tile (up to rotation and reflection), but this certainly isn't necessary. Escher frequently used more than one animal tile in his drawings. Even then, a single basic tile consisting of one copy of each could have been used.

In fact, all of the tilings we have discussed have a fixed pattern that is repeated over and over again in a predictable way to cover the plane. It is intuitively obvious that we can tile the entire plane with such tiles. For more complicated tilings, we will need the following theorem. For the proof, see *Tilings and Patterns* by Grünbaum and Shephard. A tile is a *closed topological disc* if it is connected (i.e., has only one piece), has no holes or punctures in it, and if removing an edge does not disconnect it into separate pieces. The regular polygons and modified tiles we have experimented with are all closed topological discs.

Theorem 9.6. [The Extension Theorem] Let S be any finite set of tiles, each of which is a closed topological disc. If S tiles arbitrarily large circular discs, then S admits a tiling of the plane.

In other words, if you have a finite number of distinct tiles (so that each tile is connected, has no holes, and has no sections where taking off an edge would disconnect the tile) and you can cover an arbitrarily large circular area with copies of these tiles, then you can use your tiles to tile the whole plane.

For example, below are illustrated patches of a tiling, based on one of the reptiles you saw in Section 9.2, that cover larger and larger discs:

▶ **Exercise 9.35.** Prove that any reptile will tile the plane.

◊ **9.4.2   Aperiodic tiles**

In this section, we explore some more complicated tilings which do not cover the plane by using translations alone.

Definition 9.7. A tiling is *periodic* if there are translations in two nonparallel directions which leave the tiling unchanged.

Such a translation will shift the tiling but still have it match up perfectly. Another way to visualize periodicity is that if you can make a rubber stamp out of some tiles and use it to cover the rest of the plane without rotating or reflecting the stamp and with no gaps or overlap, then you have a periodic tiling. The regular tilings of the plane by squares and by hexagons are periodic, and the simplest rubber stamp would be a single tile. The regular tiling by triangles is also periodic, but the rubber stamp would be made of two triangular tiles (a parallelogram) since you are not allowed to rotate the stamp. In general, if a tiling is periodic, your rubber stamp will need at least one of each type of tile in your tiling and more than one if there are rotations or glide reflections in your pattern. The tiling of the plane by an arbitrary quadrilateral in Section 9.2 would have four tiles in the stamp. (You could also use 8 or 12 or 16, etc., but we prefer the smallest repeating region.)

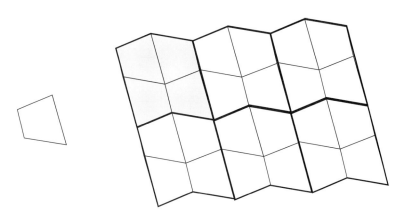

▶ **Exercise 9.36.** Find a smallest repeating region for the tilings you made for Exercise 9.8 and Activities 9.11 and 9.12.

Note that these regions are not unique. For the octagon-square tiling, we could choose several different regions for the rubber stamp, as the figures below on the right indicate.

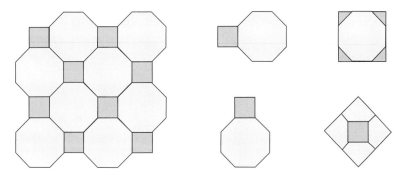

The question arises whether all tilings are periodic. Certainly all of the ones we have studied so far have been. However, the answer is no: Consider the figure below on the left. While the pattern is very symmetric, there is no one section that can be copied and translated to cover the whole plane because there is only one point that has rotational symmetry. However, it is possible to tile the plane in a periodic way using the same tiles as the tiling shown below on the right demonstrates. Thus this tile allows both a periodic and a nonperiodic tiling.

▷ **Activity 9.20.** Find a nonperiodic tiling that does not have rotational symmetry.

The next logical question is whether there is a set of tiles that can tile the plane only in a nonperiodic way. We adopt the language of Grünbaum and Shephard in *Tilings and Patterns* and call such a set of tiles aperiodic. Note that many of the results in this section were first reported in that text or in Martin Gardner's *Penrose Tiles to Trapdoor Ciphers*.

Definition 9.8. A tiling is *nonperiodic* if it is not periodic. A set of tiles is called *aperiodic* if any tiling of the plane by copies of the tiles is nonperiodic.

For a long time, people thought that such aperiodic tiles did not exist, though no proof was known. Thus, most mathematicians were surprised when Robert Berger produced a set of aperiodic tiles in 1964. On the other hand, his original set contained over 20,000 different tiles. He continued to work on the problem and later produced a set of 104 tiles that could cover the plane only in a nonperiodic way.

Mathematicians continued to refine Berger's results. The most well-known set of aperiodic tiles are the *Penrose tiles*, introduced by Roger Penrose in 1974. This set of aperiodic tiles contains only two tiles along with a set of rules for how the tiles must be put together. The tiles are known as the *kite* and the *dart*, names suggested by John H. Conway, who has contributed a great deal to what is known about Penrose tilings. These tiles are descended, after several generations, from Penrose's first attempts using pentagons. From Chapter 4.1, we might thus expect to find the golden ratio somewhere, and indeed it makes several appearances here.

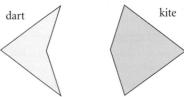

The kite and dart are cut from a rhombus with sides of length the golden ratio ϕ and with main diagonal length equal to $\phi + 1$. To form the tiles, connect the vertices at the obtuse angles to the main diagonal at a distance of the golden ratio from one of the acute angles as shown below. The large piece is the kite; the small piece is the dart. Notice that the kite is made of two isosceles triangles with the golden ratio for the length of the equal sides. The length of the third side of these triangles is equal to one. Consequently, the kite is made of two isosceles golden triangles.

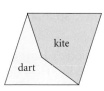 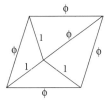

▶ **Exercise 9.37.** Find the proportion of the area of a kite to the area of a dart.

The rules for constructing a nonperiodic tiling with the Penrose tiles are simple. First, as with most tilings, only sides of the same length can be put together. This rule ensures that no vertex for one tile can occur in the middle of the side of another tile. Additionally, a second rule is stipulated, requiring a certain direction along the sides of the tiles. To enforce this direction rule, one can put notches and bumps on the tiles, or use dots or holes, or reshape the tiles to fit only in the correct way. Inspired by some of Escher's prints and using the methods we explored in Section 9.2, Penrose reshaped the tiles as chickens, as below.

(*Source*: Courtesy Roger Penrose)

The shaping of the fat and skinny chickens forces the tiles to fit together correctly. However, instead of asking you to cut out innumerable chickens of various body types, we will simply put dots on our tiles and ask you to follow certain rules about matching these dots. To work with actual tiles, we suggest you cut lots of tiles following the pattern below:

 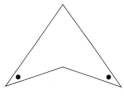

The following activities explore how the kite and darts can be fit together, first obeying only the rule that the sides of equal lengths fit together, and then experimenting with the positions of the dots. After this exercise, once we have more familiarity with the tiles and how they can interact, we will begin enforcing Penrose's rules.[5]

▷ **Activity 9.21.** Think about how the tiles fit together, so that only sides of the same length can be put together.

  a. Tile the plane using only darts. Is your tiling periodic? How do the dots match?
  b. Tile the plane using only kites. Is your tiling periodic? How do the dots match?
  c. Can you tile the plane using only darts if you must put dots next to dots? Justify your answer by showing how or proving it cannot be done.
  d. Can you tile the plane using only kites if you must put dots next to dots? Justify your answer.
  e. If you do not allow dots to be put next to dots, can you tile the plane periodically using some of each kind of tile in a way other than a rhombus tiling (where a rhombus is made from one kite and one dart)?

To begin tiling the plane with these tiles, the first question is how can the tiles be put together at a single vertex. In order to address this question for the kite and dart, we will label the vertices of the tiles by kite head (KH), kite tail (KT), kite wing (KW), dart head (DH), dart tail (DT) and dart wing (DW) as shown below.

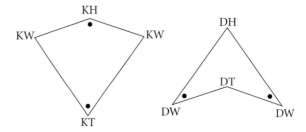

▶ **Exercise 9.38.** Determine the angles at the vertices of the tiles. Use that information to conjecture and prove the answers to the following questions.

  a. What is the maximum number of kite heads that can be put at any vertex in the plane? of kite tails?
  b. What is the maximum number of dart heads that can be put at any vertex in the plane? of dart tails?
  c. What is the maximum number of kite wings that can be put at any vertex in the plane? of dart wings? Be careful here. You need to think about more than just angle measures.

---

[5]This exercise was suggested by Robert Fathauer in a lesson plan accompanying the puzzle *Kites and Darts* by Tesselations.

To create a proper Penrose nonperiodic tiling with the kite and dart, the rules below must be followed. We obey these rules for the rest of this section.

- Only sides of the same length can be put together.
- Tiles must be joined so that dots are next to dots and blanks next to blanks.

With these rules, there are only seven ways tiles can be put around a vertex. One way is to put all dart heads together. This figure is called *the Star*. Another, called *the King*, puts two kite wings and three dart heads together. Notice that the two kites must be adjacent in order for the other pieces to fit.

The Star

The King

▶ **Exercise 9.39.** Find the other five ways that tiles can be put around a vertex following the Penrose rules. Prove that there are no others.

In building a tiling, as you work out from a vertex, sometimes there is only one way to place a tile at a certain point, and sometimes there are several choices. Sometimes it seems as if you have a choice, but farther out in your tiling you will find a place where no tile will fit and you will have to go back and change your original choice. If there is only one choice for fitting a tile around a vertex grouping, we say that tile is in the *empire* of the vertex.

If you begin a tiling with the grouping from your solution to Exercise 9.39 which has five kite tails meeting at the vertex (this pattern is called *the Sun*), you can choose your next tiles in several ways. Therefore, the empire of the Sun is empty.

The Sun

However, if you start with the Star grouping, there is only one way to add the next tile. (Try it!) There is no room for another dart, and a kite can be placed in only one way to satisfy the Penrose rules. In fact, you are forced to fit ten kites around the star as the figure below shows before you can choose any other tile. After these ten kites, there are several ways to fit tiles. Hence, the star has ten tiles in its empire.

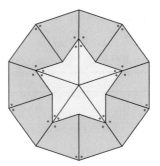

Empire of the Star

The vertex group of the King has 50 tiles in its adjacent empire. One of the other four vertex groupings found in Exercise 9.39 has no tiles in its empire. In the connected empires of the others, one vertex grouping has two tiles, one has 11 tiles and the last has 22 tiles.

▶ **Exercise 9.40.** We have discussed the connected empires of the Star, the Sun, and the King. Construct the connected empires of the remaining four vertex groupings of Exercise 9.39.

◊ 9.4.3 Inflation and deflation

As we continue to try to construct Penrose tilings, two very natural questions arise. How do we know that we can tile the whole plane with the Penrose tiles and how do we know that such tilings are nonperiodic? The answers to these questions involve the concepts of composition, decomposition, inflation, and deflation. Note that the definitions of these concepts are not consistent in the literature. What we call composition in following Grünbaum and Shephard, is sometimes referred to as inflation in other sources.

Definition 9.9. A tiling T_1 is a *composition* of tiling T_2 if every tile in T_1 is a union of tiles from T_2.

In forming a composition, we take tiles and combine them to make larger tiles. For example, the tiling on the left below is a composition of the the tiling on the right:

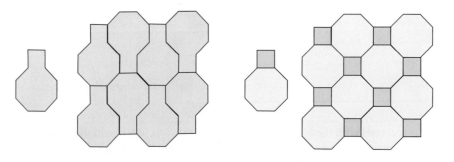

Similarly, a decomposition will take larger tiles and cut them into smaller tiles.

Definition 9.10. A tiling T_2 is a *decomposition* of tiling T_1 if every tile in T_1 is a union of tiles from T_2.

As another example, let T_2 be the Star and its empire. The Sun tiling T_1 shown below can be considered as a composition of the tiling T_2. Conversely, T_2 is a decomposition of T_1:

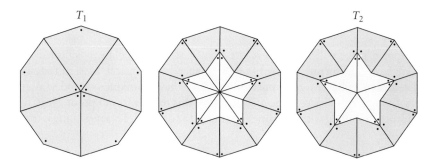

Definition 9.11. An *inflation* is a combination of two actions: First, decompose a tiling or section thereof into smaller tiles, then enlarge it so that the new tiles are the size of the original tiles. Similarly, a *deflation* is the reverse process: Compose a tiling into larger tiles and then shrink it so that the new tiles are the size of the original tiles.

Inflation is a valuable concept: If we understand how to inflate a section of a tiling (and if the tiles are closed topological discs, as they usually are), then we can use inflation and the Extension Theorem 9.6 to prove that the tiling can cover the entire plane. Thus our next task is to figure out how to inflate the Penrose tiles.

Assume that we have constructed a section of the tiling using the Penrose tiles and obeying both of the rules. First, cut each dart to form one kite and two half-darts. Then cut each kite to form two kites and two half-darts, as shown below:

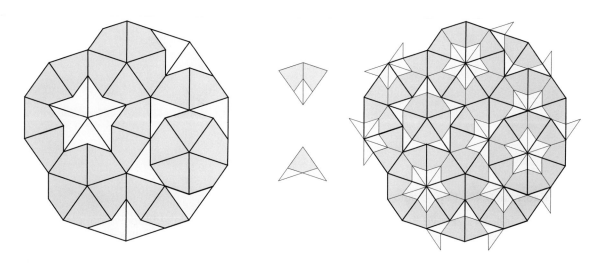

Now reassemble all of the half-darts to form whole darts. The Penrose rules will ensure that two half-darts will always be adjacent to each other, except along the edges of the section that has been completed. For the half-darts along the edges of the patch of tiles, complete these tiles to form whole darts. Note that while the new parts no longer have side lengths equal to one and ϕ, the ratio between the new lengths is still the golden ratio.

▶ **Exercise 9.41.** For a kite and a dart, compute the edge lengths of the pieces after the operation described above.

Each reassembled piece will be either a kite or a dart, and they are positioned according to the Penrose rules. We have thus formed a new tiling by smaller, but identical, tiles, in fact, a decomposition of the original section of tiling. More importantly, we now have a unique method of decomposition for the Penrose tiles, though we will accept the uniqueness of this result without proof. Next, rescale your new tiling (by ϕ) to form a new tiling with the same size tiles as the original. This new tiling is the inflation of the original.

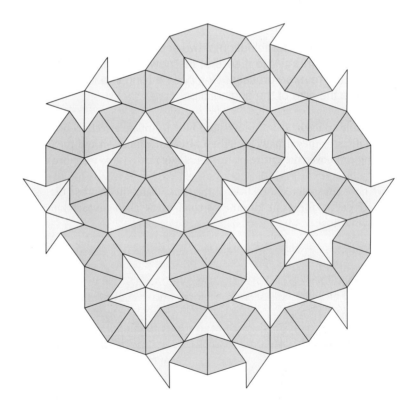

▶ **Exercise 9.42.** Figure out and describe how to deflate a tiling.

▶ **Exercise 9.43.** Prove that if a set of tiles (each of which is a closed topological disk) can tile a disc in the plane and can be inflated, then that set of tiles can tile the entire plane. In particular, the Penrose tiles tile the plane.

How do we know that a Penrose tiling is not periodic? Since these tilings are infinite, how do we know that we just have not found a big enough section to see that it repeats? Composition allows us to make the necessary argument.

Theorem 9.12. No tiling of the plane by Penrose tiles is periodic.

Proof: Suppose there is a Penrose tiling T_1 that is periodic. Note that in composing a tiling, the same thing is done to all parts of the tiling. Thus if the original tiling was periodic, the new tiling T_2 formed by the composition must also be. We accept without proof that the composition of a Penrose tiling is unique. From this, it follows that the composed tiling T_2 has the same period as the original tiling T_1. Let d denote the translation distance for the given original tiling, so that a translation of distance d moves the tiling to itself. Choose a point P in the tiling T_1 and consider one of the translations of this point by distance d. This translation will take point P to another point P'. By periodicity, if we slide the entire tiling T_1 from P to P', all the translated tiles will match up exactly with the original tiles. If we compose a Penrose tiling, the new larger tiling T_2 will have the same distance d between P and P'. If we slide the new tiling T_2 from P to P', the translated new tiles should exactly overlap the untranslated new tiles. We can repeat the composition process for the new tiling T_2 to get tiling T_3, etc. However, if we compose a sufficient number of times, the tiles in T_n will be so big that the both P and P' will lie in the same tile. This is the contradiction: If both P and P' lie in the same tile, when we translate the nth tiling from P to P', the translated tiles cannot overlap the original ones. Therefore, no Penrose tiling of kites and darts can be periodic. □

There are many more interesting phenomena regarding Penrose tilings. We mention some of them here (without proof) and encourage readers to pursue the ideas that interest them.

The Star and the Sun have fivefold rotational symmetry and they also have five lines of reflection. It is possible to maintain the rotational symmetry for the Star and the Sun when extending to a tiling, but there is only one way to do this for each vertex. Thus, there are only two Penrose tilings that have rotational symmetry.

The King and all of the vertex groupings of Exercise 9.39 have at least one line of reflectional symmetry. It is also possible to maintain reflectional symmetry if you start with one of these vertex groupings. Specifying reflectional symmetry does not yield a unique tiling for any vertex, but there are many more nonsymmetric tilings than there are symmetric ones.

▶ **Exercise 9.44.** Use the idea of inflation to prove that there exist tilings that have five-fold rotational symmetry and tilings that have a single line of reflectional symmetry.

▷ **Activity 9.22.** Compare the compositions and decompositions, inflations and deflations of the Sun and the Star and the infinite symmetric tilings that surround them.

◆ 9.5 Projects

Project 9.1. Explore Escher's tiling patterns.[6]

---

[6] Some useful references are Doris Schattschneider, *Visions of Symmetry: Notebooks, Periodic Drawings, and Related Work of M.C. Escher* or F.H. Bool, J.R. Kist, J.L. Locher, and F. Wierda, *M.C. Escher: His Life and Complete Graphic Work*.

Project 9.2. Read and report on Roger Penrose, "Pentaplexity: A Class of Non-Periodic Tilings of the Plane," from the *Mathematical Intelligencer* 2 (1979), pp. 32–37.

Project 9.3. Read and report on Robert Ammann's work on aperiodic tilings.

Project 9.4. Use geometric software to draw some hyperbolic tilings.

Project 9.5. Investigate the use of cartwheel tilings to show that every tiling by Penrose kites and darts is *metrically balanced*. In other words, in any infinite tiling of the plane by kites and darts, the number of kites used is the golden ratio times the number of darts. In any tiling of a finite section of the plane, the ratio of kites to darts will approximate the golden ratio. The approximation improves as the area tiled increases.[7]

---

[7] See Grünbaum and Shephard, *Tilings and Patterns*.

10. Geometry in three dimensions

In this chapter, we generalize many of our previous findings to three-dimensional space. Section 10.1 forms the foundation, mirroring and extending some of the issues we studied in Chapter 2.1–2.3. In the second section, we develop the regular and semiregular polyhera, in analogy with and by similar methods as our study of tilings in Chapter 9.1. In Section 10.3, we turn our attention to questions about the volumes of solids, generalizing the approach used for area in Chapter 2.4. In 10.4, we give a few examples of the little-known infinite polyhedra that are then used as examples for the study of three-dimensional symmetry in Section 10.5, which extends Chapter 8. In Section 10.6 we discuss the symmetries of polyhedra, paralleling our discussion of the rosette groups for finite figures in the plane in Section 8.2. We end with a brief mention of the possibilities that four-dimensional space would offer. You will find that in the chapter we open many doors to show you what is possible in higher dimensions, but space does not allow us to pursue these avenues in much depth. However, the foundations are in place for you to explore on your own.

◆ 10.1 Euclidean geometry in three dimensions

◇ 10.1.1 Foundations

In order to extend euclidean geometry to three dimensions, we must add some additional axioms and definitions about the behavior of basic primitive notions: planes, lines, and points in three-dimensional space. We assume all of the postulates we adopted in Chapter 2 and all of the results stemming from those. The only care we need to have in applying these postulates and theorems is that they assume the objects discussed are contained in a single plane. Thus, we must be careful to specify which plane we are in whenever we use one of the results from Chapter 2. We denote the standard euclidean three-dimensional space as $\mathbb{R}^3$. Note that there are also three-dimensional counterparts to the hyperbolic plane and elliptic space, but we will not study those.

Postulates 1–13 carry over to $\mathbb{R}^3$, though we will add clauses to Incidence Postulate 6 and Congruence Postulate 9. The following postulate extends Incidence Postulate 6 for use in three-dimensional space:

Incidence Postulate 6, extended for three dimensions:

1. There exists at least one plane.
2. Every plane contains at least three noncollinear points.
3. Every line contains at least two points.
4. For any three noncollinear points, there is exactly one plane containing these points.
5. If a plane contains two distinct points of a line, then every point on the line lies on the plane.
6. If two planes $\mathcal{P}_1$ and $\mathcal{P}_2$ have one point in common, then they have at least one other point in common.
7. There are at least four points that do not all lie on a single plane.

We use the following result so very often that we state it as a lemma:

Lemma 10.1. *If two distinct planes intersect, then they intersect in a line.*

▶ **Exercise 10.1.** Prove Lemma 10.1

▶ **Exercise 10.2.** Prove that any pair of distinct intersecting lines lies in a unique plane, and that if a third line intersects these lines as shown below, then the triangle formed lies in the same plane.

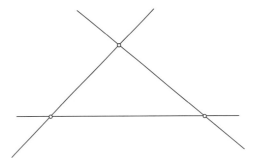

One can use the extended Incidence Postulate 6 and the Plane Separation Property Postulate 8 to prove the analogous property for planes.

Theorem 10.2. [**Space Separation Property**] Given a plane $\mathcal{P}$ in $\mathbb{R}^3$, $\mathcal{P}$ separates the points in $\mathbb{R}^3$ and not on $\mathcal{P}$ into two disjoint nonempty convex sets, S_1 and S_2, called the sides of the plane, so that whenever point A is in S_1 and point B is in S_2, the line segment $\overline{AB}$ must intersect $\mathcal{P}$.

Proof: We are given plane $\mathcal{P}$ in $\mathbb{R}^3$. By Incidence Postulate 6(2), there are two points C and D lying on $\mathcal{P}$, and by Incidence Postulate 6(7) there is another point A in $\mathbb{R}^3$ which does not lie on $\mathcal{P}$. Let S_1 be the set containing A and all points X so that the segment $\overline{AX}$ does not intersect the plane $\mathcal{P}$. Also, let S_2 be the set of points Y so that $\overline{AY}$ intersects $\mathcal{P}$. The sets S_1 and S_2 are clearly disjoint. We first show that the set S_2 is nonempty. By Incidence Postulate 6(4), there is a plane $\mathcal{P}'$ that contains the points A, C, and D. Since $\mathcal{P}$ and $\mathcal{P}'$ have points C and D in common, by Lemma 10.1, these two planes intersect in the line $\ell = CD$. By the Plane Separation Postulate 8, ℓ divides $\mathcal{P}'$ into two nonempty disjoint convex sets. Since A does not lie on the intersection ℓ of the two planes, A lies in one of these sets and there must exist a point B on the side of ℓ in $\mathcal{P}'$ opposite from A. Note that $\overline{AB}$ lies on plane $\mathcal{P}'$ by Incidence Postulate 6(5). Thus $\overline{AB}$ intersects ℓ at some point E and so $\overline{AB}$ intersects $\mathcal{P}$ at E. Therefore, B lies in the set S_2 and so S_1 and S_2 are nonempty.

Next we show that the plane $\mathcal{P}$ divides $\mathbb{R}^3 - \mathcal{P}$ into exactly two sets. Let $A \in S_1$ and $B \in S_2$ and let C denote the intersection of $\overline{AB}$ with $\mathcal{P}$. Let X be another point in $\mathbb{R}^3$ which does not lie on $\mathcal{P}$. If A, B, and X are collinear on some line ℓ, then by the Line Separation Theorem 2.10, X lies either on the A-side of C on ℓ or on the B-side. Thus, either $\overline{BX}$ or $\overline{AX}$ will contain the point C and so must intersect $\mathcal{P}$. In the first case X will lie in the set S_1, while in the second case X will lie in S_2.

If A, B, and X are not collinear, let $\mathcal{P}''$ be the plane determined by the points A, B, and X and again let C denote the intersection of $\overline{AB}$ and $\mathcal{P}$. Since $\overline{AB}$ lies on $\mathcal{P}''$ and also intersects $\mathcal{P}$, by Lemma 10.1, these planes must intersect in a line ℓ' through C. By the Plane Separation Property 8, the line ℓ' divides $\mathcal{P}''$ into exactly two sets. Thus X must lie on the same side of ℓ' as A or the same side as B. Without loss of generality, we assume that X lies on the plane $\mathcal{P}''$ and on the same side of ℓ' as A. Thus by Exercise 2.6, X lies on the side of ℓ' opposite B. By Incidence Postulate 6(5), $\overline{BX}$ lies on $\mathcal{P}''$. Since $\overline{BX}$ intersects ℓ', $\overline{BX}$ also intersects $\mathcal{P}$. Thus X is not an element of S_2. The line segment $\overline{AX}$, which also lies on $\mathcal{P}''$, does not intersect ℓ', so $\overline{AX}$ does not intersect $\mathcal{P}$. Therefore, X must be in S_1.

Finally, we show that S_1 is convex. Let $A \in S_1$ and $B \in S_2$ be points as above and let Y be another point in S_1. We want to show that $\overline{AY}$ is contained in S_1. As above, if A, B, and Y are collinear, then the result follows from the Line Separation Theorem 2.10. Assume that A, B and Y are not collinear and let $\mathcal{P}'''$ be the plane determined by these three points. By Incidence Postulate 6(5), $\overline{AB}$ and $\overline{BY}$ are on $\mathcal{P}'''$. Since A and B lie on opposite sides of $\mathcal{P}$, $\overline{AB}$ intersects $\mathcal{P}$. Thus $\mathcal{P}$ and $\mathcal{P}'''$ intersect and so by Lemma 10.1, these planes intersect in a line ℓ''. Thus $\overline{AB}$ intersects ℓ'' and similarly, $\overline{BY}$ also intersects ℓ''. Thus B and Y lie on opposite sides of ℓ'' on $\mathcal{P}'''$. By the Plane Separation Property 8, A and Y must lie on the same side of ℓ'' on plane $\mathcal{P}'''$. Since $\overline{AY}$ lies on plane $\mathcal{P}'''$, $\overline{AY}$ cannot intersect $\mathcal{P}$ or $\overline{AY}$ would also intersect ℓ''. Thus $\overline{AY}$ lies inside the set S_1 so S_1 is convex. Similarly, S_2 is also convex. □

As in Chapter 2, this theorem allows us to make sense of the idea of the sides of the plane:

Definition 10.3. Given a plane $\mathcal{P}$ in $\mathbb{R}^3$, let A and B be two points on that do not lie on $\mathcal{P}$. Then A and B are on the *same side* of the plane if the line segment $\overline{AB}$ does not intersect $\mathcal{P}$. A and B lie on *opposite sides* of $\mathcal{P}$ if $\overline{AB}$ intersects $\mathcal{P}$.

We can then define half-planes and half-spaces just as we defined half-lines or rays in Definition 2.2.

Definition 10.4. Given a plane $\mathcal{P}$ and a line ℓ in $\mathcal{P}$, a *half-plane* is the line ℓ together with all points lying on one side of ℓ in $\mathcal{P}$.

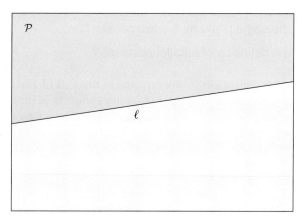

Definition 10.5. Given a plane $\mathcal{P}$ in $\mathbb{R}^3$, a *half-space* is the plane $\mathcal{P}$ together with all points lying on one side of $\mathcal{P}$ in $\mathbb{R}^3$.

We must also add two clauses to the Congruence Postulate, to extend those ideas to three-dimensional figures. We will discuss volume extensively in Chapter 10.3, after laying out the foundational definitions and exploring some three-dimensional objects to find the volumes of.

Congruence Postulate 9, extended for three dimensions:

1. Equality in length is an additive equivalence relation on line segments.
2. Equality in measure is an additive equivalence relation on angles.
3. Equality in area is an additive equivalence relation on polygons.

4. Congruent triangles have equal areas.
5. Equality in volume is an additive equivalence relation for polyhedra.
6. Congruent polyhedra have equal volumes.

One last note on our postulates: In the statement of the SAS Postulate 12, there is no implication that the two triangles being compared need to lie in the same plane. Of course, in Chapters 2–6, the triangles under consideration always inhabited the same plane, but now we are not so confined.

◇ 10.1.2 Parallel and perpendicular planes

By Definition 2.19 of parallel lines the lines must lie in the same plane. If two lines do not intersect but there is no plane containing both of them, then they are called *skew*.

▶ **Exercise 10.3.** Prove that given two lines ℓ and m in $\mathbb{R}^3$, then exactly one of the following must be true: ℓ and m are parallel, ℓ and m intersect at a single point, ℓ and m coincide, or ℓ and m are skew.

▶ **Exercise 10.4.** Define what it means for a line and a plane to be parallel.

▶ **Exercise 10.5.** Given a line ℓ and a plane $\mathcal{P}$ in $\mathbb{R}^3$, prove that exactly one of the following must be true: ℓ is parallel to $\mathcal{P}$, ℓ intersects $\mathcal{P}$ at a single point, or ℓ lies on $\mathcal{P}$.

▶ **Exercise 10.6.** Prove that if two lines are parallel in $\mathbb{R}^3$, then any transversal intersecting both of them lies in the same plane as the two parallel lines.

▶ **Exercise 10.7.** Give a definition of parallel planes.

Basic properties of lines and planes in $\mathbb{R}^3$ are explored in Book XI of *The Elements*, which we follow quite closely. Life in three-dimensional space is somewhat more complicated (or perhaps, more interesting), since there are more possibilities for relationships among objects, as you have seen in Exercises 10.3 and 10.5. In Definition 2.18, we defined perpendicular lines when those lines lay in a plane. In $\mathbb{R}^3$ we can have two lines that are perpendicular, or a line and a plane, or two planes, all of which need to be defined and have their basic properties laid out.

Definition 10.6. A line ℓ is *perpendicular* to a plane $\mathcal{P}$ in $\mathbb{R}^3$ if ℓ intersects $\mathcal{P}$ at a point X and whenever m is a line in $\mathcal{P}$ through X, then ℓ and m form a right angle. We then write $\ell \perp \mathcal{P}$.

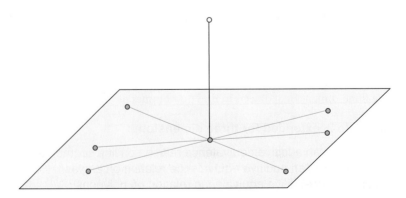

When proving theorems in three-dimensional euclidean geometry, we can use most of the propositions and theorems proved in Chapters 2 and 3, but only after we have shown that the relevant lines, angles, and triangles lie a particular plane.

Theorem 10.7. [Proposition XI.4] If a line is perpendicular to two intersecting lines at their point of intersection, then it is perpendicular to the plane containing those lines.

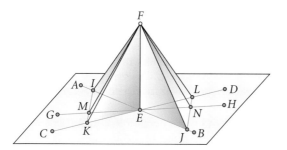

Proof: Let EF be perpendicular to lines AB and CD that intersect at E and let $\mathcal{P}$ denote the plane determined by AB and CD as in Exercise 10.2. Draw another line GH though E in the plane $\mathcal{P}$ and without loss of generality assume that G lies inside $\angle AEC$. We need to show that FE is perpendicular to GH at E. Cut AB and CD at points I, J, K, and L so that $\overline{EI} = \overline{EJ} = \overline{EK} = \overline{EL}$. Draw $\overline{IK}$ and $\overline{JL}$. Since $\angle IEK = \angle JEL$ by Proposition I.15, $\triangle IEK \cong \triangle JEL$ by SAS. Thus, $\overline{IK} = \overline{JL}$ and $\angle IKE = \angle JLE$. By the Crossbar Theorem 2.12, $\overrightarrow{EG}$ intersects $\overline{IK}$ at point M and $\overrightarrow{EH}$ intersects $\overline{JL}$ at N. Since $\angle KEM = \angle LEN$ by Proposition I.15 again, $\triangle KEM \cong \triangle LEN$ by ASA and so $\overline{ME} = \overline{NE}$ and $\overline{KM} = \overline{LN}$. Also, $\triangle FEI \cong \triangle FEJ \cong \triangle FEK \cong \triangle FEL$ by SAS, and so $\overline{FI} = \overline{FJ} = \overline{FK} = \overline{FL}$. Therefore, $\triangle FKI \cong \triangle FLJ$ by SSS and so $\angle FKM = \angle FLN$. Thus $\triangle FKM \cong \triangle FLN$ by SAS and therefore, $\overline{FM} = \overline{FN}$. Finally, $\triangle FME \cong \triangle FNE$ by SSS, and so in the plane determined by FE and GH, $\angle FEM = \angle FEN = 90°$. Thus $FE \perp GH$. Since the line GH was chosen to be an arbitrary line in the plane through E, it follows from Definition 10.6 that FE is perpendicular to the plane $\mathcal{P}$. □

We give Legendre's proof of the following theorem:

Theorem 10.8. [Proposition XI.8] If two lines are parallel and one of them is perpendicular to a given plane, then the other line is also perpendicular to the plane.

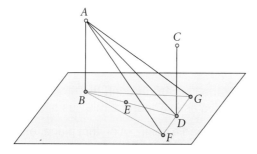

Proof: Let AB and CD be the given parallel lines and $\mathcal{P}$ the given plane. We assume that AB is perpendicular to the plane $\mathcal{P}$ at B. By Definition 2.19 of parallel lines, AB and CD lie on some plane $\mathcal{P}'$. By Lemma 10.1, the intersection of $\mathcal{P}$ and $\mathcal{P}'$ must be a line BE through B. If CD were parallel to $\mathcal{P}$, then CD will not intersect BE since it lies in $\mathcal{P}$. This contradicts Playfair's Postulate (and thus the Parallel Postulate 5) unless AB and BE coincide, but that contradicts our assumption that AB is perpendicular to $\mathcal{P}$. Therefore, CD intersects $\mathcal{P}$ and we can assume that this point of intersection is D, which must lie on the line BE.

Draw $\overline{BD}$ and construct a line FG on $\mathcal{P}$ so that $FG \perp BD$. Cut this line so that $\overline{FD} = \overline{GD}$. Then $\triangle BDF \cong \triangle BDG$ by SAS, so $\overline{BF} = \overline{BG}$. Since AB is perpendicular to the plane $\mathcal{P}$, by Definition 10.6, $AB \perp BF$ and $AB \perp BG$. Thus, $\triangle ABF \cong \triangle ABG$ by SAS and so $\overline{AF} = \overline{AG}$. Then $\triangle ADF \cong \triangle ADG$ by SSS and so $\angle ADF = \angle ADG$ and thus $AD \perp FG$. By Theorem 10.7, FG is perpendicular to the plane $\mathcal{P}''$ determined by A, B, and D. Note that CD also lies in $\mathcal{P}''$ by Definition 2.19. By Proposition I.29, $CD \perp BD$. Since FG is perpendicular to $\mathcal{P}''$, $FG \perp CD$ by Definition 10.6. Thus, CD is perpendicular to both BD and FG, and by Theorem 10.7, CD must be perpendicular to the plane $\mathcal{P}$ containing these lines. □

Euclid explains how to drop a perpendicular from a point to a plane, in analogy with Proposition I.12:

Theorem 10.9. [**Proposition XI.11**] Given a plane and a point that does not lie on the plane, one can drop a perpendicular from the point to the plane.

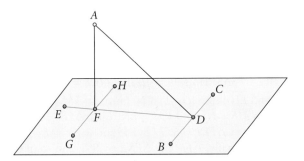

Proof: Let $\mathcal{P}$ be the given plane and A the point not on $\mathcal{P}$. Choose any two points B and C on $\mathcal{P}$ and draw the line BC, which will lie on the plane by Incidence Postulate 6(5). By Incidence Postulate 6(4), A, B, and C determine a unique plane $\mathcal{P}'$. In this plane $\mathcal{P}'$, we can use Proposition I.12 to drop a perpendicular $AD \perp BC$. If AD happens to be perpendicular to $\mathcal{P}$, then we have the desired perpendicular from A to the plane $\mathcal{P}$. Otherwise, use Proposition I.11 in the plane $\mathcal{P}$ to erect a perpendicular $DE \perp BC$. In the plane $\mathcal{P}''$ determined by A, D, and E, drop $AF \perp DE$ by Proposition I.12. In plane $\mathcal{P}$, use Proposition I.31 to construct GH through F and parallel to BC. Since $BC \perp DA$ and $BC \perp DE$, by Theorem 10.7, BC must be perpendicular to the plane $\mathcal{P}''$ determined by A, D, and E. By Theorem 10.8, since $GH \| BC$, GH is also perpendicular to $\mathcal{P}''$. Since F lies on line DE, F lies on $\mathcal{P}''$ and thus by Incidence Postulate 6(5), the line AF also lies on plane $\mathcal{P}''$. Thus, by Definition 10.6, $GH \perp AF$. By Theorem 10.7, since $AF \perp GH$ and $AF \perp DE$, AF must be perpendicular to the plane determined by D, E, and G, which is the original plane $\mathcal{P}$. □

▶ **Exercise 10.8.** Prove that the line AF, through the given point A and perpendicular to the given plane $\mathcal{P}$ and constructed as in Theorem 10.9, is unique.

▶ **Exercise 10.9.** Given a plane and a point not on that plane, explain how to construct a plane through the given point and parallel to the given plane.

▶ **Exercise 10.10.** Given two parallel planes $\mathcal{P}$ and $\mathcal{P}'$ and a line ℓ that is perpendicular to $\mathcal{P}$, prove that ℓ is perpendicular to $\mathcal{P}'$.

◇ 10.1.3 Angles in three dimensions

Now we can now use the procedure of Theorem 10.9 to define the angle formed between an arbitrary line and a plane:

Definition 10.10. Given a plane $\mathcal{P}$ and a line *AB* intersecting the plane at *B*, drop a perpendicular *AC* from the line to the plane. The *angle* between the line *AB* and the plane $\mathcal{P}$ is defined to be the angle $\angle ABC$.

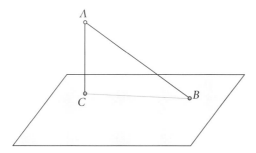

Next, we discuss the various relationships and angles that can be formed by planes in $\mathbb{R}^3$. To begin, consider parallel planes cut by a transversal plane, reducing the question to the relationship between their lines of intersection:

Theorem 10.11. [Proposition XI.16] If two parallel planes are cut by a third plane, then the lines of intersection of the parallel planes and the third plane are parallel.

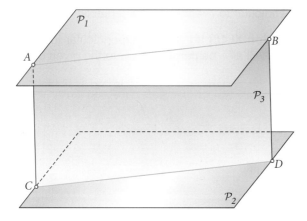

Proof: Let $\mathcal{P}_1$ and $\mathcal{P}_2$ be the parallel planes, and $\mathcal{P}_3$ a plane that cuts both of them. Let line *AB* be the line of intersection of $\mathcal{P}_1$ and $\mathcal{P}_3$ and *CD* the line of intersection of $\mathcal{P}_2$ and $\mathcal{P}_3$. We need to show that *AB* and *CD* are parallel. Note that by Incidence Postulate 6(5), both *AB* and *CD* lie on plane $\mathcal{P}_3$. If we assume that *AB* and *CD* are not parallel, then they must intersect at some point *E* on $\mathcal{P}_3$. Since point *E* lies on line *AB*, *E* must lie on plane $\mathcal{P}_1$. Similarly, *E* lies on $\mathcal{P}_2$. Therefore, $\mathcal{P}_1$ must intersect $\mathcal{P}_2$. This contradiction implies that *AB* must be parallel to *CD*. □

We go on to define the angle formed by two planes, called the *dihedral angle*. Picture two intersecting planes as if they are two adjacent pages in a book with the line of intersection formed by the spine of the book. If we draw a line on each page that is perpendicular to the spine, the dihedral angle is the angle formed by these lines, and so it is the angle formed by the pages.

Definition 10.12. Given planes $\mathcal{P}$ and $\mathcal{P}'$ that intersect in line AB, choose a point C on AB and draw $DC \perp AB$ in $\mathcal{P}$. Draw $CE \perp AB$ in $\mathcal{P}'$. Then the *dihedral angle* formed by planes $\mathcal{P}$ and $\mathcal{P}'$ is defined to be $\angle DCE$.

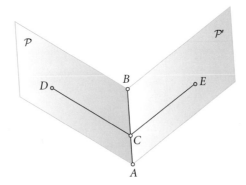

▶ **Exercise 10.11.** Show that the dihedral angle of two intersecting planes is well-defined: i.e., that it does not depend on the particular point C chosen.

▶ **Exercise 10.12.** Give a definition of perpendicular planes.

▶ **Exercise 10.13.** Given a plane $\mathcal{P}$ and a line ℓ lying on $\mathcal{P}$, prove that you can construct a plane $\mathcal{P}'$ that is perpendicular to $\mathcal{P}$ and whose intersection with $\mathcal{P}$ is the given line ℓ.

Theorem 10.13. [Proposition XI.18] If a line is perpendicular to a given plane, then all of the planes that contain this line will also be perpendicular to the given plane.

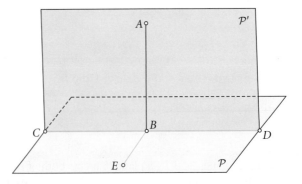

Proof: Let AB be the line perpendicular to plane $\mathcal{P}$, with B lying in $\mathcal{P}$. Consider a plane $\mathcal{P}'$ that contains AB. Since $\mathcal{P}$ and $\mathcal{P}'$ have point B in common, by Lemma 10.1, they must have a line CD in common, with B lying on CD. In $\mathcal{P}$, erect a perpendicular $BE \perp CD$. Since AB is perpendicular to $\mathcal{P}$, by Definition 10.6 $AB \perp BE$. Thus by Definition 10.12 and Exercise 10.12, $\mathcal{P}$ and $\mathcal{P}'$ are perpendicular. □

Corollary 10.14. Given a plane $\mathcal{P}$ and a line ℓ that does not lie on $\mathcal{P}$, if ℓ is not perpendicular to $\mathcal{P}$, then there is a unique plane $\mathcal{P}'$ that contains ℓ and is perpendicular to $\mathcal{P}$.

Proof:

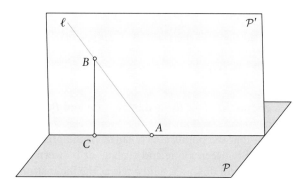

If ℓ intersects $\mathcal{P}$, let A be the point of intersection and, using Incidence Postulate 6(3), choose any other point B on ℓ. Use Theorem 10.9 to drop $BC \perp \mathcal{P}$. Since ℓ is not perpendicular to $\mathcal{P}$, A, B, and C are not collinear. Consider the plane $\mathcal{P}'$ determined by A, B, and C. Since $\mathcal{P}'$ contains B and C, by Incidence Postulate 6(5), the line BC lies on $\mathcal{P}'$. By Theorem 10.13, $\mathcal{P}'$ is perpendicular to $\mathcal{P}$.

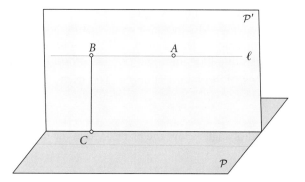

If ℓ is parallel to $\mathcal{P}$, choose any two points A and B on ℓ. Drop $BC \perp \mathcal{P}$, and consider the plane $\mathcal{P}'$ determined by A, B, and C. As above, Theorem 10.13 implies that $\mathcal{P}'$ is perpendicular to $\mathcal{P}$. In both cases, uniqueness follows from Incidence Postulate 6(4). □

Theorem 10.15. Given skew lines ℓ and m, there is a line n that is perpendicular to both ℓ and m.

Proof: Choose any point A that lies neither on ℓ nor on m. In the plane determined by A and the line ℓ, construct a line ℓ' through A and parallel to ℓ. In the plane determined by A and the line m, construct a line m' through A and parallel to m. The lines ℓ' and m' intersect at A, so there is a plane $\mathcal{P}$ containing both ℓ' and m'. By Corollary 10.14, construct another plane $\mathcal{P}'$ that contains ℓ and is perpendicular to $\mathcal{P}$. The intersection of $\mathcal{P}'$ and $\mathcal{P}$ is a line ℓ'' which is parallel to ℓ and to ℓ'. Construct a third plane $\mathcal{P}''$ that contains m and is also perpendicular to $\mathcal{P}$. The intersection of $\mathcal{P}''$ and $\mathcal{P}$ is a line m'' which is parallel to m and to m'. Since ℓ'' and m'' lie in $\mathcal{P}$ and are not parallel, they must intersect. Thus, the planes $\mathcal{P}'$ and $\mathcal{P}''$ must also intersect in a line n. Since ℓ is parallel to ℓ'' and ℓ'' is perpendicular to $\mathcal{P}''$, by Theorem 10.8, ℓ is perpendicular to $\mathcal{P}''$. Similarly, m is perpendicular to $\mathcal{P}'$. Thus the intersection of these planes, n, is perpendicular to both ℓ and m. □

▶ **Exercise 10.14.** Prove that given three planes $\mathcal{P}_1$, $\mathcal{P}_2$, and $\mathcal{P}_3$ so that $\mathcal{P}_1$ and $\mathcal{P}_2$ intersect in line ℓ and are both perpendicular to $\mathcal{P}_3$, then their line of intersection ℓ is perpendicular to $\mathcal{P}_3$.

As in Chapter 2, $\angle ABC$ denotes the angle formed by rays $\overrightarrow{BA}$ and $\overrightarrow{BC}$ at B. Implicit in this definition is the fact that $\angle ABC$ lies in the plane determined by the points A, B, and C. Such angles are sometimes called *plane angles*, to emphasize that the points are coplanar. In the preceding discussion, whenever we have mentioned the angle formed between a line and a plane or between two planes, we have done so in terms of certain plane angles. Euclid defines a *solid angle* as the inclination formed by more than two lines that do not lie on the same plane meeting one another at a point, i.e., the angle formed by several plane angles meeting at a point, such as the apex of a pyramid. There is, unfortunately, no convenient way to measure such a solid angle. However, it does make sense to talk of two solid angles being congruent. We can also give a more rigorous definition of such an angle, first in the case when it is formed by three planes.

Definition 10.16. Let $\mathcal{P}_1$, $\mathcal{P}_2$, and $\mathcal{P}_3$ be three distinct planes that intersect at A. Let line AB denote the intersection of $\mathcal{P}_1$ and $\mathcal{P}_2$, AC the intersection of $\mathcal{P}_2$ and $\mathcal{P}_3$, and AD the intersection of $\mathcal{P}_1$ and $\mathcal{P}_3$. Then three plane angles are formed: $\angle BAD$ in plane $\mathcal{P}_1$, $\angle BAC$ in $\mathcal{P}_2$, and $\angle CAD$ in $\mathcal{P}_3$. These three plane angles and their interiors form the *trihedral angle* $\angle A\text{-}BCD$ with *apex* A. The plane angles are the *faces* of the trihedral angle. The rays $\overrightarrow{AB}$, $\overrightarrow{AC}$, and $\overrightarrow{AD}$ are the *edges* of the trihedral angle.

In the illustration of a trihedral angle below, the plane $\mathcal{P}_1$ is at the back and is thus obscured in this viewpoint.

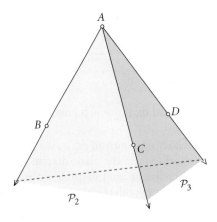

Thus, consider the three angles formed at the vertices of three triangles as shown below. These three plane angles can be lifted out of the plane where they were originally formed and attached to each other at their vertices and along their edges to form a trihedral angle at the vertex of a pyramid.

A *solid angle* is defined analogously by three or more plane angles that have a point in common. Any solid angle can be subdivided into trihedral angles as follows: Let A be the apex of the solid angle and choose $B_1, B_2, \ldots B_n$ on each of the edges of the solid angle, considered sequentially around the angle. Then $\angle A\text{-}B_1B_2B_3$, $\angle A\text{-}B_1B_3B_4$, $\angle A\text{-}B_1B_4B_5$, $\ldots$, $\angle A\text{-}B_1B_{n-1}B_n$ are trihedral angles and their union forms the solid angle. In the picture below three of the faces are obscured in the back.

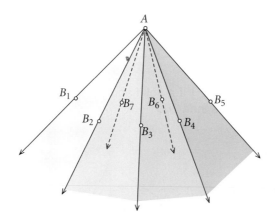

▷ **Activity 10.1.** Cut three plane angles from a piece of cardboard. Try to fit your angles together at their vertices to make a trihedral angle. Now cut several other angles, some obtuse and some acute. Choose three at a time and try to make trihedral angles. Conjecture what condition must be met for the creation of a trihedral angle.

▷ **Activity 10.2.** Repeat Activity 10.1 with four or more angles meeting at the vertex. When do you get a solid angle and what condition must be met for the plane angles in a solid angle?

The following proposition is a variation of the Triangle Inequality (Proposition I.20) for the plane angles forming a trihedral angle. Note that Proposition I.25, whose use is essential in this proof, applies even when the two triangles being compared lie in different planes.

Theorem 10.17. [Proposition XI.20] If a solid (trihedral) angle is formed by three plane angles, then the sum of the measures of any two of these plane angles is greater than the measure of the third plane angle.

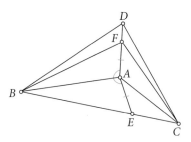

Proof: Let the solid angle at A be formed by the plane angles $\angle BAC$, $\angle CAD$, and $\angle DAB$ and consider the bird's-eye view above. If all three of these angles are equal, then the result follows easily. Otherwise, assume that $\angle BAC$ is greater than $\angle DAB$. In the plane $\mathcal{P}$ determined by A, B, and C, use Proposition I.23 to construct $\angle BAE = \angle DAB$. By the Crossbar Theorem 2.12 we may assume that E lies on $\overline{BC}$. On the ray $\overrightarrow{AD}$, cut $\overline{AF} = \overline{AE}$. Draw $\overline{BF}$ and $\overline{CF}$. Then $\triangle BAE \cong \triangle BAF$ by SAS, so $\overline{BE} = \overline{BF}$. By Proposition I.20 applied to $\triangle BFC$, $\overline{BF} + \overline{FC} > \overline{BC}$. Thus, $\overline{FC} > \overline{BC} - \overline{BF} = \overline{BC} - \overline{BE} = \overline{EC}$. Considering the triangles $\triangle AFC$ and $\triangle AEC$, note that $\overline{AE} = \overline{AF}$, $\overline{AC}$ is common to both triangles, and $\overline{FC} > \overline{EC}$. By Proposition I.25, $\angle DAC > \angle EAC$. Thus $\angle DAB + \angle DAC = \angle BAE + \angle DAC > \angle BAE + \angle EAC = \angle BAC$. □

Corollary 10.18. [Proposition XI.21] If a solid angle is formed by several plane angles, then the sum of the plane angles is less than 360°.

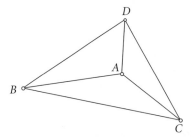

Proof: We only prove the case for a trihedral angle formed by three plane angles (as does Euclid in *The Elements*). The general convex case follows by dividing the solid angle into trihedral angles, each formed by three plane angles. One must imagine the illustration above as a bird's-eye view of a three-dimensional solid figure.

Let the solid angle at A be formed by plane angles $\angle BAC$, $\angle CAD$, and $\angle DAB$. Draw $\overline{BC}$, $\overline{CD}$, and $\overline{DB}$, forming an irregular solid with four faces and four solid angles. Applying Theorem 10.17 to the solid angle at B, $\angle ABC + \angle ABD > \angle CBD$. Similarly, considering the solid angle at C, $\angle BCA + \angle ACD > \angle BCD$ and finally, considering the angle at D, $\measuredangle CDA + \measuredangle ADB > \measuredangle CDB$. Thus,

$$\measuredangle ABC + \measuredangle ABD + \measuredangle BCA + \measuredangle ACD + \measuredangle CDA + \measuredangle ADB > \measuredangle CBD + \measuredangle BCD + \measuredangle CDB.$$

Applying Proposition I.32 to $\triangle BCD$, $\triangle ABC$, $\triangle ACD$, and $\triangle ABD$, we have

$$\measuredangle CBD + \measuredangle BCD + \measuredangle CDB = 180°,$$
$$\measuredangle ABC + \measuredangle BAC + \measuredangle ACB = 180°,$$
$$\measuredangle ACD + \measuredangle ADC + \measuredangle CAD = 180°,$$
$$\measuredangle ABD + \measuredangle ADB + \measuredangle BAD = 180°.$$

Combining these gives us:

$$\measuredangle ABC + \measuredangle BAC + \measuredangle ACB + \measuredangle ACD + \measuredangle ADC + \measuredangle CAD + \measuredangle ABD + \measuredangle ADB + \measuredangle BAD = 540°,$$

and so

$$\measuredangle ABC + \measuredangle ABD + \measuredangle BCA + \measuredangle ACD + \measuredangle CDA + \measuredangle ADB = 540° - \measuredangle BAC - \measuredangle CAD - \measuredangle BAD.$$

From this it follows that

$$540° - \measuredangle BAC - \measuredangle CAD - \measuredangle BAD > \measuredangle CBD + \measuredangle BCD + \measuredangle CDB = 180°.$$

Thus, we find that

$$\measuredangle BAC + \measuredangle CAD + \measuredangle BAD < 360°.$$

$\square$

The remainder of Euclid's Book XI includes propositions such as the three-dimensional analog of Proposition I.22, showing that it is possible to construct a solid angle from any three plane angles, as long as they satisfy the conditions of Corollary 10.18. He also proves the analog of Proposition I.23, that one can construct a solid angle congruent to any given solid angle, and then goes on to develop higher dimensional versions of Propositions I.33–46 for parallelepipeds, triangular prisms, and pyramids. We have followed Euclid only so far as to have the basic definitions and properties of planes and angles.

♦ 10.2 Polyhedra

Before we discuss the formal definition of a polyhedron, let us recall from Chapter 2 some basic definitions regarding polygons. To begin, consider a triangle. We adopted the following definition:

Definition 2.3. Given three noncollinear points *A*, *B*, and *C*, the *triangle* $\triangle ABC$ is the set of points *A*, *B*, and *C*, called the *vertices* of the triangle, and the line segments $\overline{AB}$, $\overline{AC}$, and $\overline{BC}$, called the *edges* or sides of the triangle.

There is an alternate definition we could have used, suggested by Exercise 2.11: A triangle can be defined as the intersection of three half-planes. That is, $\triangle ABC$ is the intersection of the half-plane consisting of the A-side of line BC, the half-plane consisting of the B-side of the line AC, and the half-plane consisting of the C-side of AB. Note that the second definition would include the interior as well as the edges and vertices of the triangle, while the first definition omits the interior. There is no other functional difference between the two definitions.

In Exercise 2.30 of Chapter 2, we asked you to define a polygon. There are several acceptable solutions to this request. One approach would be modeled on Definition 2.3, but would have to explicitly state that the edges and vertices must be disjoint, or this definition would allow self-intersecting figures, such as the star polygons of Section 4.5. Another model would be to generalize the idea of intersecting half-planes, but then care must be taken that the resulting figure be bounded, rather than infinite in extent. Also, a side effect of this definition is that such a figure would be forced to be convex, so one must first decide if this is acceptable. All of these are issues we wanted to encourage you to consider in assigning this problem.

For polyhedra, things are a bit more complicated. In fact, while most texts obviously have the same family of solids in mind, there is no standard definition. Branko Grünbaum, one of the foremost researchers in the area, said: "The Original Sin in the theory of polyhedra goes back to Euclid, and through Kepler, Poinsot, Cauchy, and many others continues to afflict all work on this topic . . . at each stage . . . the writers failed to define what are the 'polyhedra' . . ."[1]

In point of fact, Euclid is clever enough to avoid the issue, only defining the actual figures he wishes to study: the Platonic solids. We have chosen to adopt a definition modeled on the idea of intersecting half-spaces, which means all of our polyhedra are necessarily finite and convex. We will discuss the nonconvex and infinite analogs in Section 10.5.

Definition 10.19. A *polyhedron* is a bounded figure formed by the intersection of a finite number of half-spaces. The points of the polyhedron that lie on a single plane form a *face* of the polyhedron. The points of the polyhedron formed by the intersection of two of these half-spaces are the *edges* of the polyhedron. A point in the polyhedron where three or more of the half-spaces intersect is a *vertex* of the polyhedron.

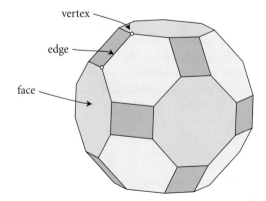

---

[1] B. Grünbaum, "Polyhedra with Hollow Faces," *Proc. NATO-ASI Conference on Polytopes*, 1994, quoted in Cromwell, *Polyhedra*, p. 286.

From this definition, it follows that the faces are convex polygons. We will refer to the number of faces in a polyhedron as f, e is the number of edges, and v is the number of vertices for the polyhedron.

◇ 10.2.1 Pyramids, prisms, and antiprisms

Definition 10.20. A *pyramid* is a polyhedron with one polygonal face designated as the *base* and one vertex, called the *apex*, that is not on the same plane as the base. The apex is connected to each vertex of the base by an edge and there is a triangular face connecting the apex to each edge of the base.

A pyramid may have any polygon, regular or irregular, for its base, though the sides will always be triangles. The Egyptian pyramids are examples of what is called a *right square pyramid*, which means that the base is a square and the apex is directly above the point at the center of the square. Note that this pyramid has five vertices, eight edges, and five faces, one square and four triangles. On the right is a diagram called a *net* for this pyramid: a two-dimensional drawing of the faces of the pyramid that can be cut out and taped together to build it. The net has the advantage that all the polygonal faces are presented without the distortion inherent to the perspective drawing on the left. However, one must remember which edges get glued together to assemble the polyhedron.

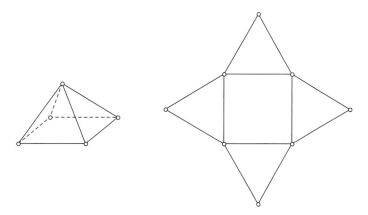

▶ **Exercise 10.15.** If the pyramid pictured above has a square base that is one inch on a side and the sides are equilateral triangles, how tall is it?

▶ **Exercise 10.16.** While clearly there is no limit for how tall the isosceles triangles forming the side faces of the right square pyramid may be, there is a limit to how short they can be. Express this with an inequality.

▶ **Exercise 10.17.** Find all the pyramids such that all of the faces are regular polygons (including the triangular side faces), and draw a net for each. Prove that you have listed all possible pyramids satisfying these conditions.

▶ **Exercise 10.18.** If a pyramid has a polygonal base with n sides, find f, the number of faces; e, the number of edges; and v, the number of vertices.

Another family of polyhedra is the class of *prisms*.

Definition 10.21. A *prism* is a polyhedron with congruent and parallel polygonal base and top faces and side faces joining corresponding parallel edges of the top and bottom.

A *regular n-sided prism* has regular polygonal base and top. A *right prism* has the side faces perpendicular to the base and the top.

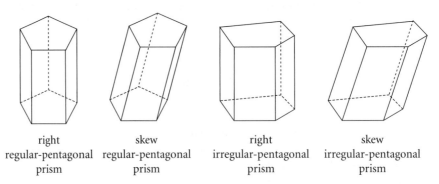

right regular-pentagonal prism   skew regular-pentagonal prism   right irregular-pentagonal prism   skew irregular-pentagonal prism

▶ **Exercise 10.19.** If a prism has a base with n sides, find f, the number of faces; e, the number of edges; and v, the number of vertices.

▶ **Exercise 10.20.** How many prisms are there such that all of the faces are regular polygons? Prove your answer.

Definition 10.22. A *regular antiprism* is a polyhedron with congruent parallel regular polygons for the base and top faces, with the top polygon rotated by the angle equal to half the central angle of the polygon forming the base, joined by a band of triangles around the sides.

▶ **Exercise 10.21.** If an antiprism has a base with n sides, find f, the number of faces; e, the number of edges; and v, the number of vertices.

◇ 10.2.2 Platonic Solids

We are interested in the regular and semiregular polyhedra, defined in analogy with the regular and semiregular tilings of Chapter 9.

Definition 10.23. A polyhedron is *regular* if all of the faces are congruent regular polygons and any two vertices are identical in the sense that there is a rotation of the polyhedron taking one vertex and all its neighboring faces to the other vertex and its faces.

These three conditions, regularity of the polygons forming the faces, congruence of these faces, and identical vertices, are all necessary. These conditions guarantee that the polyhedron will have identical arrangements of faces around each vertex. It can also be shown that for any of the regular polyhedra, all dihedral and solid angles are equal and the polyhedron can be circumscribed by a sphere, which Euclid proves. Note that except for the right square prism or cube, prisms fail the last condition of Definition 10.23. Similarly, only the right regular triangular pyramid and the regular triangular antiprism are regular.

The regular polyhedra are named by the number of faces they have, using the Greek numerical prefixes. Thus, a *tetrahedron* should have four faces. A cube, which has, of course, six faces, is sometimes called a *hexahedron*. Since a cube has three squares meeting at each vertex, we denote it by the symbol 4.4.4, in analogy with the notation we used to describe tilings.

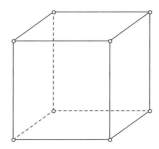

You will need to cut sets of congruent equilateral triangles, squares, regular pentagons, etc., out of cardboard to use in the following activities. Note that if you tape two identical polygons together, you get a sort of sandwich, which we do not consider truly three-dimensional. Therefore, any polyhedron must have at least three polygons meeting at each vertex. Also, by Corollary 10.18, we know that the sum of the plane angles at each vertex must be less than 360°.

▷ **Activity 10.3.** Build models of regular polyhedra using only equilateral triangles. You should be able to find three such polygons. Give their symbols and their names.

▷ **Activity 10.4.** What is the greatest number of regular pentagons that can be used at each vertex to build a polyhedron? There is only one regular polyhedron formed by regular pentagons. Build a model and name it.

▷ **Activity 10.5.** What happens if you tape three regular hexagons at each vertex?

At this point, you have found and named the five *Platonic solids*, the only regular polyhedra. The construction of these five regular polyhedra is the culmination of Euclid's *The Elements*, comprising the final chapter, Book XIII. They are named for their earliest mention by the Greek philosopher Plato. Following earlier writers such as Empedocles, Plato described nature as built of four basic elements: water, earth, air, and fire, in differing combinations and proportions.

Let us assign the cube to earth, for it is the most immobile of the four bodies and the most retentive of shape, the least mobile of the remaining figures (the icosahedron) to water, the most mobile (the tetrahedron) to fire, the intermediate in mobility (the octahedron) to air, the smallest (the tetrahedron) to fire, and the largest (the icosahedron) to water, the sharpest and most penetrating (the tetrahedron) to fire, the least sharp (the icosahedron) to water.[2]

There are, of course, five regular polyhedra, so the remaining one, the dodecahedron, was assigned to represent the universe or cosmos. These associations were revived by Johannes Kepler.

▷ **Activity 10.6.** Fill in the table below, giving the type of faces (triangles, squares, etc.), the symbol, and the numbers of faces, edges, and vertices.

Polyhedron	face type	symbol	f	e	v
Tetrahedron					
Cube (hexahedron)	squares	4.4.4			
Octahedron					
Dodecahedron					
Icosahedron					

▷ **Activity 10.7.** You will note some similarities in the numbers in the table of Activity 10.6. State a conjecture about these.

Dual polyhedra are defined much as dual tilings were in Section 9.1. For a polyhedron, place a vertex at the center of each face, and if the faces of the original polyhedron share an edge, then draw an edge in the dual polyhedron connecting the corresponding vertices. The dual polyhedron will then have one face for each vertex of the original polyhedron, with edges corresponding to the faces surrounding that vertex. For example, if this procedure is followed for a cube, we get an octahedron neatly embedded inside the the cube.

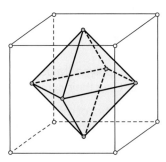

Similarly, if one places a new vertex in the center of each triangular face of an octahedron and connects these new vertices, one gets a cube embedded inside the octahedron.

▷ **Activity 10.8.** Find the dual of each of the Platonic solids and prove your answer.

The ideal image of a polyhedron is, of course, a carefully built three-dimensional model such as you built in Activities 10.3 and 10.4, but these are awkward to carry around. Nets have the advantage of

---

[2] Plato, *Timaeus*, paragraph 56.

representing the polygonal faces without distortion, but need to be assembled before you can see and study the vertex configurations. Below is a net for the cube:

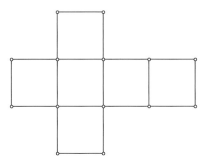

▷ **Activity 10.9.** Determine which of the following are nets for the cube:

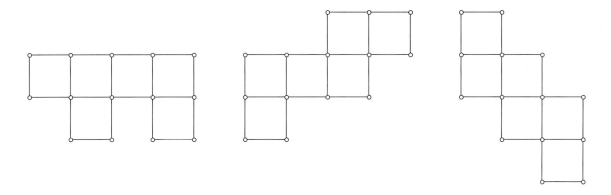

▷ **Activity 10.10.** Determine which of the following are nets for the tetrahedron:

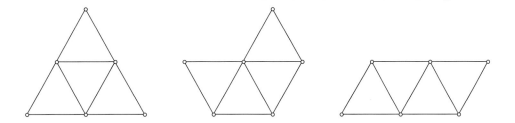

▷ **Activity 10.11.** Draw nets for the octahedron, icosahedron, and dodecahedron.

Irregular polyhedra can have either irregular faces or irregular vertex configurations. One class of polyhedra are the *convex deltahedra* which have only equilateral triangles as faces. For a model of any polyhedron with triangular faces, one starts with a pile of triangles. Each of these triangles has three edges.

If we let f denote the number of triangles, we have $3f$ edges before we start building the model. When we tape the triangles together, we match up the edges in pairs, so that the completed polyhedron must have $e = \frac{3f}{2}$. Since one cannot have a fractional number of edges, this means that f must be even. Thus, all deltahedra must have an even number of triangular faces. We must have at least three but no more than five triangles meeting at each vertex. Therefore, the tetrahedron is the smallest (in the sense of having the fewest faces) deltahedron with only four triangles, and the icosahedron the largest with 20 triangles so that five meet at each vertex. The octahedron is another regular deltahedron using eight triangles. Since the others must be irregular, they will have, for example, three triangles meeting at one vertex and four at another.

▷ **Activity 10.12.** There are five convex irregular deltahedra.
  a. Build a convex deltahedron (the *triangular dipyramid*) that has six equilateral triangular faces.
  b. Build a convex deltahedron (the *pentagonal dipyramid*) that has ten equilateral triangular faces.
  c. Build a convex deltahedron (the *Siamese dodecahedron* or *snub disphenoid*) that has twelve equilateral triangular faces.
  d. Build a convex deltahedron (the *tri-augmented triangular prism*) that has fourteen equilateral triangular faces.
  e. Build a convex deltahedron (the *gyro-elongated square dipyramid*) that has sixteen equilateral triangular faces.
  f. Although the argument above would lead one to think that there would be an irregular convex deltahedron with 18 triangles, it is impossible to assemble such a polyhedron. Try it!

◇ 10.2.3 Euler's formula

The great Swiss mathematician Leonhard Euler [1707–1783] introduced a new invariant for polyhedra.

Definition 10.24. Let v denote the number of vertices for a polyhedron P, e the number of edges, and f the number of faces, and define the *euler characteristic* of the polyhedron P to be

$$\chi(P) = v - e + f.$$

The surprising thing about the euler characteristic is that even though the regular and irregular polyhedra we have studied have many different types of faces and vertex configurations, the euler characteristic is the same for each of them.

Theorem 10.25. [Euler's formula] For any polyhedron P, $\chi(P) = 2$.

Proof: To see why Euler's formula is true for any of the convex polyhedra, consider an arbitrary polyhedron P. The proof is easier if we assume that each face is a triangle, so, if necessary, subdivide each polygonal face into triangles to obtain a new polyhedron P'. If a polygon has vertices $A_1, A_2, A_3, \ldots, A_n$ for $n > 3$, draw edges $\overline{A_1 A_3}, \overline{A_1 A_4}, \ldots, \overline{A_1 A_{n-1}}$. In doing this we have replaced the original face with $(n-2)$ faces and we have added $(n-2)$ edges and no new vertices. Thus, the net change to the euler characteristic is zero, and $\chi(P) = \chi(P')$.

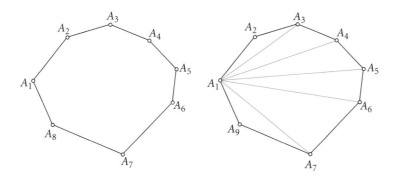

Once we have a triangulated polyhedron P', remove one of the faces, leaving the edges and vertices behind to get a new figure P'_1.

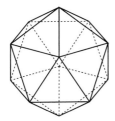 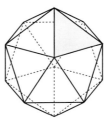

Since P'_1 has one face fewer than P' did, we have $\chi(P'_1) = \chi(P') - 1 = \chi(P) - 1$. Next remove the faces one by one. There are three cases to consider.

In the first case, suppose we wish to remove a face that is connected to the rest of the figure along two edges. Leave those edges and their endpoint vertices behind. Thus, after throwing away this triangle, we have left the same number of vertices, one less face, and one fewer edges as in the original. The new euler characteristic is thus $\chi = v - (e-1) + (f-1) = v - e + f$. Therefore, there is no net change in the euler characteristic.

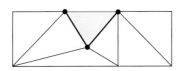 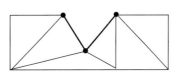

▶ **Exercise 10.22.** Suppose we wish to remove a face that is connected to the rest of the figure along one edge. Leave that edge and its endpoint vertices behind. What is the net change in the euler characteristic χ?

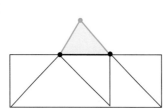 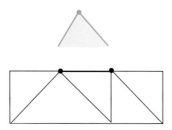

▶ **Exercise 10.23.** Suppose we wish to remove a face that is connected to the rest of the figure only at a vertex. Leave that vertex behind. What is the net change in the euler characteristic χ?

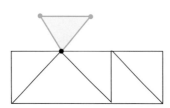 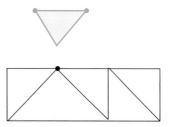

Finally after a finite number of steps, we are left with a single triangular face P'_k, with all of its edges and vertices intact. By Exercises 10.22 and 10.23, we know that $\chi(P'_k) = \chi(P) - 1$. This triangular face will have $f = 1$, $e = 3$, and $v = 3$. Thus, $\chi(P'_k) = 3 - 3 + 1 = 1$. Therefore, $\chi(P) = \chi(P'_k) + 1 = 2$, so for any polyhedron, Euler's formula is true. □

We can use the theorem above to show that there are exactly five regular polyhedra, confirming our constructions.

Theorem 10.26. *There are exactly five regular polyhedra.*

Proof: Let P be a regular polyhedron. If v, e, and f denote the vertices, edges, and faces of P, by Theorem 10.25, $\chi(P) = v - e + f = 2$. Since P is regular, it was constructed from a pile of f congruent polygons, each with n vertices and n edges before they were assembled to form P. Let V denote the number of vertices in the pile of polygons before assembly, E the number of edges, and F the number of faces. Note that obviously $F = f$. Each polygon has n edges and n vertices, so $E = V = nf$. Since the edges of the polygons are matched up in pairs to assemble P, we also know that $E = 2e = nf$. In assembling P, each vertex configuration is identical, formed by joining m polygon vertices, so $V = mv = nf = 2e$. Therefore,

$$2 = v - e + f$$
$$2 = \frac{2e}{m} - e + \frac{2e}{n}$$
$$2 = e\left(\frac{2}{m} - 1 + \frac{2}{n}\right)$$
$$\frac{1}{e} = \frac{1}{m} - \frac{1}{2} + \frac{1}{n}$$

Of course, e, n, and m must be integers greater than 2, and so $\frac{1}{e} < \frac{1}{2}$. We analyze each case separately:

Case 1: $n = 3$, so the polygons are triangles.

$$\frac{1}{2} > \frac{1}{e} = \frac{1}{m} - \frac{1}{2} + \frac{1}{3} = \frac{1}{m} - \frac{1}{6} > 0,$$

$$\frac{2}{3} > \frac{1}{m} > \frac{1}{6},$$

$$\frac{3}{2} < m < 6.$$

Thus, the only possibilities are $m = 3, 4,$ or 5.

$m = 3$: Then $\frac{1}{e} = \frac{1}{3} - \frac{1}{2} + \frac{1}{3} = \frac{1}{6}$, so $e = 6, f = \frac{2e}{n} = 4$, and $v = \frac{2e}{m} = 4$. This is the tetrahedron.

$m = 4$: Then $\frac{1}{e} = \frac{1}{4} - \frac{1}{2} + \frac{1}{3} = \frac{1}{12}$, so $e = 12, f = 8$, and $v = 6$. This is the octahedron.

$m = 5$: Then $\frac{1}{e} = \frac{1}{5} - \frac{1}{2} + \frac{1}{3} = \frac{1}{30}$, so $e = 30, f = 20$, and $v = 12$. This is the icosahedron.

Case 2: $n = 4$, so the faces are squares.

$$\frac{1}{2} > \frac{1}{e} = \frac{1}{m} - \frac{1}{2} + \frac{1}{4} = \frac{1}{m} - \frac{1}{4} > 0,$$

$$\frac{3}{4} > \frac{1}{m} > \frac{1}{4},$$

$$\frac{4}{3} < m < 4.$$

The only possibility is $m = 3$, and then $e = 12, f = 6$, and $v = 8$. This is the cube.

Case 3: $n = 5$, so the faces are pentagons.

$$\frac{1}{2} > \frac{1}{e} = \frac{1}{m} - \frac{1}{2} + \frac{1}{5} = \frac{1}{m} - \frac{3}{10} > 0,$$

$$\frac{4}{5} > \frac{1}{m} > \frac{3}{10},$$

$$\frac{5}{4} < m < \frac{10}{3}.$$

The only possibility is $m = 3$, and then $e = 30, f = 12$, and $v = 20$. This is the dodecahedron.

Case 4: $n \geq 6$.

$$0 < \frac{1}{e} = \frac{1}{m} - \frac{1}{2} + \frac{1}{n} \leq \frac{1}{m} - \frac{1}{2} + \frac{1}{6} = \frac{1}{m} - \frac{1}{3},$$

$$\frac{1}{m} > \frac{1}{3},$$

$$m < 3.$$

This contradicts our stipulation that $m > 2$ so there are no additional cases. □

◊ 10.2.4 Archimedean solids

In Section 9.1 we investigated the three regular tilings (by squares, equilateral triangles, and regular hexagons) and then went on to find the eight semiregular, or Archimedean, tilings. For polyhedra, the five Platonic solids are the analogs of the regular tilings. We now wish to find the three-dimensional analogs of the semiregular tilings.

Definition 10.27. A polyhedron is *semiregular* or *Archimedean* if all of the faces are regular polygons and if any two vertices are identical in the sense that there is a rotation of the polyhedron taking one vertex and all its neighboring faces to the other vertex and its faces.

While Archimedes apparently knew of all thirteen of the solids that bear his name, his book on the subject was lost many centuries ago. Several of the semiregular polyhedra appear in texts and pictures of the Renaissance, but the complete list was rediscovered by Kepler, who also defined the classes of prisms and antiprisms. We proceed as he did in finding them, as described in Cromwell's *Polyhedra*. The rules we devised in Section 9.1 for semiregular tilings carry over to our search for semiregular polyhedra, with only one necessary change. We modify Rule 1 to obey Corollary 10.18:

Rule 1: *For every polyhedron, the sum of the angles of the polygons meeting at any vertex must be less than 360°.*

Rule 2: *At each vertex of a regular or semiregular polyhedron, at least three but no more than five polygons meet.*

Rule 3: *There can be no more than three different types of polygons at each vertex of a regular or semiregular polyhedron.*

Rule 4: *No semiregular polyhedron can have vertex configuration k.n.m where k is odd and $n \neq m$.*

Rule 5: *No semiregular polyhedron can have vertex configuration 3.k.n.m unless $k = m$.*

We will investigate the possible semiregular polyhedra systematically by types of vertex configurations, beginning with combinations consisting of triangles and one other type of polygon. Ambitious readers should attempt to find the polyhedra on their own before reading further. Slightly less ambitious readers might want to read the first few cases before venturing out on their own. The goal is to find 13 Archimedean polyhedra.

Theorem 10.28. *There are thirteen semiregular polyhedra.*

Proof:

Case 1: The vertex configuration consists of triangles and one other type of n-sided polygon.

Case 1.a. $n = 4$, so the vertex configuration is some combination of triangles and squares. By Rule 2, we must consider the following: 3.3.3.3.4 (the snub cube), 3.3.3.4 (the square antiprism), and 3.3.4. This last possibility can be eliminated by Rule 4. If the vertex involves two squares, we have 3.3.4.4 which is eliminated by Rule 1, 3.3.4.4 which violates Rule 5, 3.4.3.4 (the cuboctahedron), and 3.4.4 (the triangular prism). If we use three squares, we have 3.3.4.4.4 which violates Rule 1 or 3.4.4.4 (the rhombicuboctahedron). Four or more squares would violate Rule 1. The prism and the antiprism fail the regularity condition that a rotation preserve the configuration of each vertex.

▷ **Activity 10.13.** Build models of the snub cube, the cuboctahedron, and the rhombicuboctahedron.

Case 1.b. $n = 5$, so the vertex configuration is some combination of triangles and pentagons.

▶ **Exercise 10.24.** Find all the valid vertex configurations for Case 1.b.

▷ **Activity 10.14.** Build models of the two semiregular polyhedra that result from Exercise 10.24, the snub dodecahedron and the icosidodecahedron. (Note that another possibility is the pentagonal antiprism.)

Case 1.c. $n = 6$, so the vertex configuration is some combination of triangles and hexagons.

▶ **Exercise 10.25.** Find all the valid vertex configurations for Case 1.c.

▷ **Activity 10.15.** Build a model of the one semiregular polyhedron that results from Exercise 10.25, the truncated tetrahedron. (Note that another possibility is the hexagonal antiprism.)

Case 1.d. $n \geq 7$, so the vertex configuration is some combination of triangles and n-gons. Some of the possibilities that can be eliminated are 3.3.3.3.n which violates Rule 1, 3.3.3.n which yields the n-gon antiprism, and 3.3.n which violates Rule 4.

▶ **Exercise 10.26.** Find all the valid vertex configurations for Case 1.d.

▷ **Activity 10.16.** Build models of the two semiregular polyhedra that result from Exercise 10.26, the truncated cube and the truncated dodecahedron. (Note that another possibility is the n-gon antiprism.)

Case 2: The vertex configuration consists of only two types of polygons and no triangles.

Case 2.a. The vertex configuration is some combination of squares and n-sided polygons, with $n \geq 5$. Note that 4.4.4.n violates Rule 1. Vertex configuration 4.4.n gives us the n-gon prism. The only other possibilities will be of the form 4.n.n and by Rule 4 we see that n must be even. Thus, we consider 4.6.6 (the truncated octahedron) and 4.8.8, which violates Rule 1. No other possibilities exist.

▷ **Activity 10.17.** Build a model of the truncated octahedron.

Case 2.b. The vertex configuration is some combination of pentagons and n-sided polygons, with $n \geq 6$.

▶ **Exercise 10.27.** Find all the valid vertex configurations for Case 2.b.

▷ **Activity 10.18.** Build a model of the one semiregular polyhedron that results from Exercise 10.27, the truncated icosahedron.

Case 2.c. The vertex configuration is some combination of hexagons and n-sided polygons, with $n \geq 7$.

▶ **Exercise 10.28.** Find all the valid vertex configurations for Case 2.c.

Case 3: The vertex configuration consists of three types of polygons.

Case 3.a. The vertex configuration is some combination of triangles, squares and n-sided polygons, with $n \geq 5$. Note that 3.3.3.4.n violates Rule 1, Rule 5 eliminates 3.3.4.n and 3.4.3.n, and Rule 4 eliminates 3.4.n. If only one triangle is involved, 3.4.4.6 and 3.4.6.4 violate Rule 1. Rule 5 eliminates 3.4.4.5, but rearranging gives us 3.4.5.4, the rhombicosidodecahedron.

▷ **Activity 10.19.** Build a model of the rhombicosidodecahedron.

Case 3.b. The vertex configuration involves three types of polygons, none of which are triangles.

▶ **Exercise 10.29.** Find all the valid vertex configurations for Case 3.b.

▷ **Activity 10.20.** Build models of the two semiregular polyhedra that result from Exercise 10.29, the truncated cuboctahedron (or great rhombicuboctahedron) and the truncated icosidodecahedron (or great rhombicosidodecahedron). □

You should now have models of all of the thirteen Archimedean solids. These will be helpful in our later study of the symmetries of polyhedra, so keep them on hand. The names we have used for the Archimedean solids were given to them by Kepler, and we want to explain these names briefly.

One way of getting new polyhedra from known ones is *truncation*. For example, take a cube and slice off each of the vertices. Cutting off the vertices introduces equilateral triangular faces where each vertex was, and cuts the corners off the squares to turn them into octagons.

Gradually enlarge the triangles until the octagons are regular and all of the edges are equal in length.

  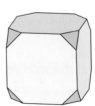 

At this point, one has a semiregular polyhedron, called a *truncated cube*. It has two regular octagons and an equilateral triangle meeting at each vertex and so is denoted by 3.8.8.

The cube had 6 faces, 12 edges, and 8 vertices. In truncation, the 8 vertices are replaced by 8 triangles, so the truncated cube has $f = 14$, with 6 octagonal faces and 8 triangular faces. The truncated cube has the 12 edges from the original cube, plus three new edges around each of the triangles, for a total of $e = 12 + 3 \cdot 8 = 36$. None of the vertices of the original cube survive, but there are three new vertices at each of the triangles, so $v = 3 \cdot 8 = 24$. As a check on these computations, note that $\chi = f - e + v = 14 - 36 + 24 = 2$, as it should by Theorem 10.25.

If we continue to slice away at the truncated cube until the slices meet at the midpoint of the edge of what was once the cube, we get another semiregular polyhedron, the cuboctahedron, with pattern 3.4.3.4:

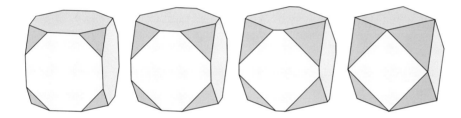

The cuboctahedron has 14 faces, 6 squares from the original cube and 8 triangles.

▶ **Exercise 10.30.** Find the number of edges and vertices in the cuboctahedron.

▷ **Activity 10.21.** Truncate an octahedron as demonstrated above for the cube. At a point halfway through the truncation process and again where the slices meet, semiregular polyhedra are formed. Which of the Archimedean solids are these?

▶ **Exercise 10.31.** Give the number of faces, edges, and vertices formed by the Archimedean solids of Activity 10.21.

▷ **Activity 10.22.** Truncate a tetrahedron as above for the cube. At a point halfway through the truncation process and again where the slices meet, semiregular polyhedra are formed. Which of the Archimedean solids are these?

▶ **Exercise 10.32.** Give the number of faces, edges, and vertices formed by the Archimedean solids of Activity 10.22.

▷ **Activity 10.23.** Truncate a dodecahedron as above for the cube. At a point halfway through the truncation process and again where the slices meet, semiregular polyhedra are formed. Which of the Archimedean solids are these?

▶ **Exercise 10.33.** Give the number of faces, edges, and vertices formed by the Archimedean solids of Activity 10.23.

▷ **Activity 10.24.** Truncate an icosahedron as above for the cube. At a point half way through the truncation process and again at the time when the slices meet, semiregular polyhedra are formed. Which of the Archimedean solids are these?

▶ **Exercise 10.34.** Give the number of faces, edges, and vertices formed by the Archimedean solids of Activity 10.24.

This process explains how the cuboctahedron, icosidodecahedron, and the five Archimedean solids whose names begin with "truncated" got their names.

If we return to the transformation of the cube, after truncating the cube to get the cuboctahedron, we could have decided to truncate the vertices of the cuboctahedron. If we had done this, we would have gotten a new figure, the *great rhombicuboctahedron* (which Kepler called the truncated cuboctahedron) shown below on the right, with vertex configuration 4.6.8.

 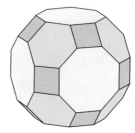

In point of fact, truncating the vertices of the cuboctahedron actually gives the figure shown below, whose faces are not all regular polygons. It is necessary to inflate the figure a bit to produce only regular polygonal faces and form the semiregular polyhedron that we want. Kepler explains this, "I call it a truncated cuboctahedron: not because it can be formed by truncation but because it is like a cuboctahedron that has been truncated."

▶ **Exercise 10.35.** Find the numbers of faces, edges, and vertices for the great rhombicuboctahedron.

▷ **Activity 10.25.** Continue the truncation series above until the squares of the great rhombicuboctahedron meet. Which of the Archimedean solids is formed?

▶ **Exercise 10.36.** Give the number of faces, edges, and vertices formed by the Archimedean solid of Activity 10.25.

▷ **Activity 10.26.** Describe the similar set of operations on an icosidodecahedron, by truncating its vertices. Which two Archimedean solids are formed?

▶ **Exercise 10.37.** Give the number of faces, edges, and vertices formed by the Archimedean solids of Activity 10.26.

Again, the semiregular polyhedra obtained in Activity 10.26 cannot be formed by true truncation, but need to be inflated to make the faces regular.

There are only two of the Archimedean solids left: the snub cube and the snub dodecahedron. To form the snub cube, in the center of each face of the cube place a smaller square, slightly rotated. Surround this smaller square by a belt of equilateral triangles. The original cube had six square faces, so the snub cube will have six square faces. Each square will share an edge with four triangles, as shown below. Six of these will fit together to form the snub cube, with the addition of one triangle where each of the original vertices of the cube was.

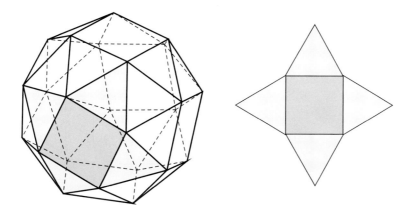

Thus, the snub cube will have $f = 6 + 6 \cdot 4 + 8 = 38$. The vertices are easier: Each of the central square faces has four vertices, and all the triangles share these vertices, so $v = 6 \cdot 4 = 24$. Since $\chi = f - e + v = 38 - e + 24 = 2, e = 60$.

▶ **Exercise 10.38.** Find the number of faces, edges, and vertices in a snub dodecahedron.

▷ **Activity 10.27.** What happens when you snub a tetrahedron?

◇ 10.2.5 Descartes' formula

Another interesting quantity associated with polyhedra is the *angular deficit*. The tetrahedron has three equilateral triangles meeting at each vertex, for a vertex angle sum of $3 \cdot 60° = 180°$. If you cut a small region surrounding a vertex of the tetrahedron, you will have a small triangular pyramid that will not lie flat. If you slit it open along one edge up to the vertex in question and lay it out flat, it forms a gap of $180°$.

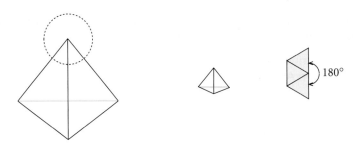

Definition 10.29. Consider a vertex *v* of a polyhedron and the plane angles formed by the polygons that meet at *v*. The quantity $\delta(v) = 360° -$ (vertex angle sum) is called the *angular deficit at the vertex v*.

The angular deficit at a vertex measures how far the vertex configuration is from being flat. Thus any tiling has a deficit of $\delta = 0°$ at each vertex, while the tetrahedron has a deficit of $\delta = 180°$ at each vertex. The icosahedron, on the other hand, has five equilateral triangles meeting at each vertex, giving a deficit of $\delta = 360° - 5(60°) = 60°$. Informally, the smaller the angular deficit is at a vertex, the flatter the figure is, while a larger angular deficit implies a pointier vertex.

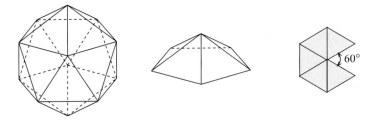

Definition 10.30. The *total angular deficit* for a polyhedron P is the sum of the vertex angle deficits for each vertex:

$$\Delta(P) = \sum_{v \text{ vertex in } P} \delta(v).$$

▶ **Exercise 10.39.** Fill in the table below:

Polyhedron	δ	# vertices	Δ
Tetrahedron	180°	4	720°
Cube			
Octahedron			
Dodecahedron			
Icosahedron			

▷ **Activity 10.28.** In irregular polyhedra, the angular deficit will vary from vertex to vertex.

  a. Find the total angular deficit Δ for a right square pyramid, with a square base and four equilateral triangles for sides.
  b. Find the total angular deficit Δ for a regular-pentagonal prism.
  c. Find the total angular deficit Δ for a pentagonal antiprism.
  d. Using the computations above, formulate a conjecture for the total angular deficit of any convex polyhedron.

The conjecture from Activity 10.28 seems surprising, considering how much the angular deficits of each vertex can vary. This seeming oddity is similar to that of Euler's formula, which proved that the euler characteristic is also an invariant for all convex polyhedra.

Theorem 10.31. [Descartes' formula] If P is a polyhedron, then $\Delta(P) = 720°$.

Proof: Let P be a polyhedron with v vertices, e edges, and f faces. Label the vertices $V_1, V_2, V_3, \ldots, V_v$, with vertex angular deficits $\delta(V_1), \delta(V_2), \delta(V_3), \ldots, \delta(V_v)$. Label the faces $F_1, F_2, \ldots, F_f$ and let $N(F_j)$ be the number of vertices (and edges) of face F_j. From Exercise 2.57, the sum of the vertex angles of a face with N vertices (and edges) is $180°(N-2)$.

$$\Delta = \delta(V_1) + \delta(V_2) + \cdots + \delta(V_v)$$
$$= [360° - (\text{sum of the angles meeting at } V_1)] + [360° - (\text{sum of the angles meeting at } V_2)] + \cdots$$
$$+ [360° - (\text{sum of the angles meeting at } V_v)]$$
$$= 360°v - [(\text{sum of the angles meeting at } V_1) + (\text{sum of the angles meeting at } V_2) + \cdots$$
$$+ (\text{sum of the angles meeting at } V_v)]$$
$$= 360°v - [\text{sum of all the angles of all the polygonal faces}]$$
$$= 360°v - [(\text{sum of the angles of the vertices of face } F_1) + (\text{sum of the angles of the vertices}$$
$$\text{of face } F_2) + \cdots + (\text{sum of the angles of the vertices of face } F_f)]$$
$$= 360°v - [180°(N(F_1) - 2) + 180°(N(F_2) - 2) + \cdots + 180°(N(F_f) - 2)]$$
$$= 360°v - 180°[N(F_1) + N(F_2) + \cdots + N(F_f)] + [180° \cdot 2 + 180° \cdot 2 + \cdots + 180° \cdot 2]$$
$$= 360°v - 180°[(\text{total number of edges for all the faces before assembling the polyhedron}] + 360°f$$

Note that the total number of edges for all the faces is $2e$, since these edges are glued together in pairs to form the polyhedron, so

$$\Delta = 360°v - 180° \cdot 2e + 360°f$$
$$= 360°(v - e + f)$$
$$\Delta = 360° \cdot 2 = 720°.$$

$\square$

◆ 10.3 Volume

Euclid's treatment of volume is based on Eudoxus' method of exhaustion, a forerunner of calculus. Euclid's first key result on area is an analog of Proposition I.36 that states that triangular pyramids with congruent bases and equal heights have equal volumes. He proves this using exhaustion, and then proceeds much as in Book I, discussing the volumes of n-sided pyramids, prisms, cones, cylinders, and finally spheres.

The theory of volume for solid figures acts much like area for planar figures, with one significant exception. In Section 10.1, we added two clauses to the Congruence Postulate 9 to address basic properties of volume:

Congruence Postulate 9, extended for three dimensions:

5. Equality in volume is an additive equivalence relation for polyhedra.
6. Congruent polyhedra have equal volumes.

We must also introduce another postulate for volume, analogous to Area Postulate 13. This postulate allows us to state that the volume of a rectangular box is given by the familiar formula: $V = lwh$, where l is the length of the box, w its width, and h its height. A *parallelepiped* is the three-dimensional equivalent of a parallelogram: a polyhedron with six faces so that the pairs of opposite faces are parallel.

Postulate 14. [Volume] The volume of a parallelepiped is the product of the area of its base and its height.

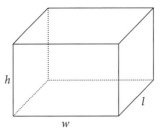

A more modern treatment of area, due to Janos Bolyai and David Hilbert, is to use the theory of decomposition, as we saw in Chapter 7.4, in place of exhaustion. In 1900, Hilbert gave an address to the International Congress of Mathematicians and suggested a list of problems, which guided the development of much of mathematics during the twentieth century. The third problem on the list was the problem of developing a similar theory for decomposition of solids and volume. The theory breaks down in trying to find an equivalent to the Bolyai-Gerwien Theorem 7.21, which proved that polygons are equivalent by decomposition if and only if they have the same area. That no such theorem can hold in three dimensions was shown by one of Hilbert's students, Max Dehn [1878–1952], a little later in 1900. Dehn showed that a regular tetrahedron cannot be decomposed into a finite number of pieces that can be reassembled into a cube of equal volume.

Thus, we find it necessary to add an additional postulate to make up for this deficiency. Bonaventura Cavalieri [1598–1647] was a disciple of Galileo. He studied and extended the method of exhaustion, using Kepler's work on indivisible geometric quantities and in the process taking a major step towards the invention of integration. The following axiom is distilled from his work.

Postulate 15. [Cavalieri's Principle] Given two solids P and Q and a fixed plane, if every plane parallel to the given plane intersects the solids to form regions with equal areas, then the volume of P is equal to the volume of Q.

To help understand this concept, one can cut a dozen identical shapes from cardboard and stack them up. If the stack is pushed sideways, like a deck of cards, then the resulting stack is the same height and has the same cross-sections, the individual cards. Both stacks clearly have the same volume.

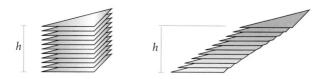

Cavalieri's Principle provides us with a powerful tool for studying volumes of polyhedra.

Theorem 10.32. The volume of a triangular prism is $V(prism) = hA(base)$.

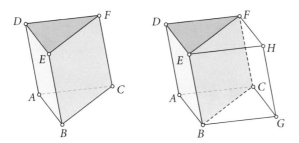

Proof: Let the figure on the left with vertices A, B, C, D, E, and F be the given triangular prism. Use Exercise 10.9 to construct a plane through BE parallel to the plane containing the parallelogram $ACFD$ and another plane through CF parallel to $ABED$. These planes form another triangular prism with base $\triangle BCG$ which is congruent to $\triangle ABC$. Thus by Cavalieri's Principle 15, the volume of these prisms are equal. The two prisms form a parallelepiped. By the Volume Postulate 14, the volume of this parallelepiped is $V = hA(ABGC) = 2V(prism)$. By Proposition I.35, the triangular base $\triangle ABC$ of the prism is half of the area of the parallelogram $ABGC$ that forms the base of the parallelepiped. Therefore, the volume of the original prism is $V = hA(\triangle ABC)$. □

▶ **Exercise 10.40.** Find and prove a formula for the volume of a prism with polygonal base by explaining how to cut up a prism with an *n*-sided base into triangular prisms, each with the same height.

To find the formula for the volume of a pyramid, let us start with a pyramid with a triangular base.

Theorem 10.33. The volume of a triangular pyramid is $V(pyramid) = \frac{1}{3}h\,A(base)$.

Proof: Label the pyramid by its vertices as $A\text{-}BCD$, where A is the apex and $\triangle BCD$ is the base. Attach at vertex A another triangle $\triangle AEF$ congruent and parallel to $\triangle BCD$, as shown below. A skew prism is formed with $\triangle BCD$ on the base and $\triangle AEF$ on the top. The pyramids $A\text{-}BCD$ and $D\text{-}AEF$ have congruent bases and equal heights, so they will have equal volumes by Cavalieri's Principle 15: $V(A\text{-}BCD) = V(D\text{-}AEF)$.

Pyramid *D-AEF* is the same as pyramid *A-DEF*, since this is just a change in which vertex is considered as the apex and which vertices form the base.

Now consider the parallelogram *CDEF*. The diagonal *DF* cuts this into two congruent triangles, △*CDF* and △*DEF*. The pyramids *A-DEF* and *A-CDF* thus have congruent bases and the same height, and so have equal volumes: $V(A\text{-}DEF) = V(A\text{-}CDF)$. We have cut the prism up into three pyramids, each of which has the same volume as the original pyramid.

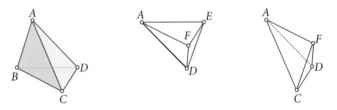

Therefore, the pyramid has volume equal to one-third of the volume of the prism with the same base and the same height, so $V(pyramid) = \frac{1}{3}hA(base)$. □

▶ **Exercise 10.41.** Explain how to extend the proof of Theorem 10.33 to any pyramid.

▶ **Exercise 10.42.** Find the volume of the tetrahedron, assuming an edge length of one unit.

▶ **Exercise 10.43.** Find the volume of the octahedron, assuming an edge length of one unit.

Next we discuss the volumes of circular figures: the cylinder, cone, and sphere. Volume Postulate 14 would lead us to believe that the volume of a cylinder with circular base of radius *r* and height *h* should be $\pi r^2 h$. While this is indeed true, we cannot apply Volume Postulate 14 as justification, since its use is restricted to parallelepipeds. However, if we could construct a prism with square base of area πr^2 and height *h*, then the cross-sections of this prism would have area equal to the cross-sections of the cylinder. Note that it is impossible to construct such a square using only straightedge and compass. Indeed, this is the famous problem of squaring the circle. But if such a square could be constructed by some other means, then Cavalieri's Principle 15 would apply and prove that the prism and the cylinder have equal volumes and so the volume of the cylinder must be $\pi r^2 h$. We assume that such a prism has been constructed by some means and so we have the general formula $V(cylinder) = hA(base)$.

▶ **Exercise 10.44.** Find and prove a formula for the volume of a cone that has a circle of radius *r* as its base and that has height *h*, by considering a pyramid that has a base of equal area and equal height.

To find the volume of a sphere $\mathcal{S}$ of radius r, we follow in the steps of Archimedes. First circumscribe the sphere by a cylinder $\mathcal{C}$. This is pictured below along with a vertical cross-section:

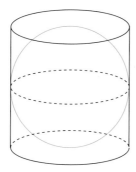
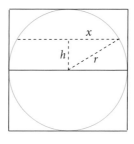

We proceed by considering the space between the sphere and the cylinder, $\mathcal{C} - \mathcal{S}$. At height h above the center of the sphere, a horizontal cross-section will consist of two circles, one of radius x formed by the intersection of the horizontal plane and the sphere and another of radius r formed by the intersection with the cylinder:

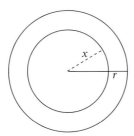

Referring to the vertical cross-section and applying the Pythagorean Theorem, we see that $x^2 + h^2 = r^2$. Thus, the area of the cross-section of $\mathcal{C} - \mathcal{S}$ is $A = \pi r^2 - \pi x^2 = \pi h^2$. Now, consider the seemingly irrelevant figure shown below, a double cone of height $2r$ with base a circle of radius r:

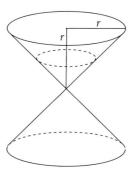
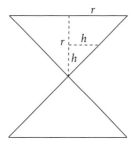

Similar triangles show that the horizontal cross-sections at height h from the center are circles of radius h. Thus, these cross-sections have area $A = \pi h^2$. Therefore, Cavalieri's Principle 15 applies to $\mathcal{C} - \mathcal{S}$ and this cone.

▶ **Exercise 10.45.** Finish the argument above to find a formula for the volume of the sphere.

Archimedes is said by some to have been one of the best and most versatile mathematicians to have ever lived. He was a native of the Greek colony of Syracuse in Sicily, though he studied for some years in Alexandria and continued to correspond with the mathematicians there, the successors of Euclid. Among other stories of his ingenuity are legends of mechanisms he invented to protect Syracuse during the siege by the Roman general Marcellus. After holding off the Romans for almost three years, the city's defenses were finally broken and Archimedes was killed during the sack of the city. The legend of how he was killed is first described by Plutarch several centuries later.

He was alone, examining a diagram closely; and having fixed both his mind and his eyes on the object of his inquiry, he perceived neither the inroad of the Romans nor the taking of the city. Suddenly a soldier came up to him and bade him follow to Marcellus, but he would not go until he had finished the problem... The soldier became enraged and dispatched him....

And though he made many elegant discoveries, he is said to have besought his friends and kinsmen to place on his grave... a cylinder enclosing a sphere, with an inscription giving the proportion by which the volume of the cylinder exceeds that of the sphere.[3]

Later writers improve on the legend, adding details that he scolded the soldier for stepping on the geometric drawing he had made in the sand. There is some evidence that the claim about his gravestone is true. In 75 B.C., Cicero [106–43 B.C.] visited Syracuse in an official capacity:

When I was quaestor I searched out his tomb, which was shut in on every side and covered with thorns and thickets. The Syracusans did not know of it: they denied it existed at all. I knew some lines of verse which I had been told were inscribed on his gravestone, which asserted that there was a sphere and cylinder on top of the tomb.

Now while I was taking a thorough look at everything—there is a great crowd of tombs at the Agrigentine Gate—I noticed a small column projecting a little way from the thickets, on which there was a representation of a sphere and cylinder. I immediately told the Syracusans—their leading men were with me—that I thought that was the very thing I was looking for. A number of men were sent in with sickles and cleared and opened up the place.

When it had been made accessible, we went up to the base facing us. There could be seen the epitaph with about half missing where the ends of the lines were worn away. So that distinguished Greek city, once also a centre of learning, would have been unaware of the tomb of its cleverest citizen, if it had not learned of it from a man from Arpinum....

---

[3] Plutarch, *Parallel Lives*.

> Who in the world is there, who has any dealings with ... culture and learning, who would not rather be this mathematician than that tyrant? [Dionysius, ruler of Syracuse for 38 years] ...The mind of the one was nourished by weighing and exploring theories, along with the pleasure of using one's wits, which is the sweetest food of souls, the other's in murder and unjust acts...[4]

♦ 10.4 Infinite polyhedra

In our discussion of tilings in Section 9.1 we noted that in every tiling the sum of the angles at each vertex is 360°, while by Corollary 10.18, every convex polyhedron must have angle sum strictly less than 360° at each vertex. In this section we will investigate some structures that occur when we allow the angle sum at a vertex to be greater than 360°.

▷ **Activity 10.29.** Build a structure from squares put together edge to edge so that only two edges meet at a time and five squares meet at each vertex. There are several ways to do this, but try to build as symmetrical a structure as possible.

Definition 10.19 of a polyhedron implies convexity, and the structure of Activity 10.29 does not satisfy this property. If, however, we defined a regular polyhedron as some books do, as an assembly of regular polygonal faces, meeting edge to edge, so that each vertex is identical, then this alternate definition would be satisfied by structures such as the one of Activity 10.29. We define such structures as infinite polyhedra. Some texts call such a figure a *sponge* or *honeycomb*.

There are, naturally enough, infinitely many infinite polyhedra, so we will not try to build all of them. Our intention is to develop certain examples and types of infinite polyhedra that will later serve as examples for the isometries in three dimensions. Thus, this discussion is by no means exhaustive. The interested reader is referred to *Infinite Polyhedra* by Wachman, Burt, and Kleinmann. There are only three completely regular infinite polyhedra, first discovered by J. F. Petrie and H. S. M. Coxeter:

> One day in 1926, J. F. Petrie told me with much excitement that he had discovered two new regular polyhedra; infinite, but without false vertices. When my incredulity had begun to subside he described them to me: one consisting of squares, six at each vertex and one consisting of hexagons, four at each vertex. It was useless to protest that there is no room for more than four squares round a vertex. The trick is to let the faces go up and down in a kind of zig-zag formation so that the faces that adjoin a given "horizontal" face lie alternately "above" and "below" it. When I understood this, I pointed out a third possibility: hexagons, six at each vertex.[5]

Below is illustrated a section of the infinite regular polyhedron 4.4.4.4.4.4 or 4^6 that Coxeter describes. It is understood that the structure should go on forever. The polyhedron divides the rest of space up into two pieces: the space inside the tunnels of the polyhedron, and the space outside of the polyhedron. Both of these spaces are exactly the same shape! Imagine walking around, first inside the tunnels and then outside to convince yourself of this.

---

[4] Quoted in Sherman Stein, *Archimedes*.
[5] H.S.M. Coxeter, "Regular Skew polyhedra in three and four dimensions, and their topological analogues," *Proceedings of the London Mathematical Society* 43 (1937), pp. 33–34.

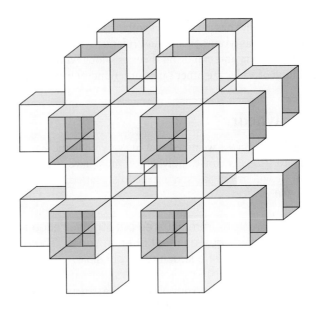

▷ **Activity 10.30.** The second regular infinite polyhedron described by Coxeter can be built by making a number of truncated octahedra, but leaving all of the squares off to make holes. Stack the truncated octahedra together so that the square holes match up. Build a section of this infinite polyhedron, using eight truncated octahedra stacked two polyhedra wide, two deep, and two tall. How would you denote this polyhedron? How do the spaces enclosed by the polyhedron and outside the polyhedron compare?

The first infinite regular polyhedron illustrated above is built from cubes, missing a pair of opposite faces, assembled in a symmetric manner. But another even simpler infinite structure could be built from similar cubes: Imagine taking a pile of cubes, removing their tops and bottoms, and stacking them to form an infinite square cylinder:

This square cylinder is an infinite polyhedron, but is not considered completely regular, since the fourfold symmetry of each square face is not carried over to the infinite polyhedron.

▶ **Exercise 10.46.** How would you denote this polyhedron? How do the spaces enclosed by the polyhedron and outside the polyhedron compare?

Cylindrical polyhedra can be built by stacking copies of any prism. In studying these, two things rapidly become clear: First, that there are infinitely many cylindrical prismatic polyhedra, and second, that the notation we have been using for tiling and finite polyhedra is inadequate to distinguish among the infinite polyhedra. Similar cylindrical structures can be built by stacking antiprisms, without their tops and bottoms, or by alternating prisms and antiprisms with the same base.

▷ **Activity 10.31.** Build a section of a cylindrical antiprismatic polyhedron by stacking four square antiprisms.

▷ **Activity 10.32.** Build a section of another cylindrical polyhedron by stacking alternately square prisms and square antiprisms.

So far we have three types of cylindrical polyhedra: prismatic, antiprismatic, and alternate prisms and antiprisms. There is another class of cylindrical polyhedra, and in some ways it is the most interesting. One example of this class is called the *tetrahelix* and will provide an excellent example of a certain type of symmetry we will discuss in the next section.

▷ **Activity 10.33.** Here is a way of building a section of the tetrahelix, due to Buckminster Fuller, as shown to Anthony Pugh. Make an enlarged copy of the picture below. Crease with a mountain fold (so the fold lies higher than its surroundings) along the lines marked ─────, and with a valley fold (so the fold lies lower than its surroundings) along the lines marked ─ ─ ─ ─ ─. Roll up into a cylinder, so that the shaded cells lie under the top row of triangles and then slide and twist so that the triangle marked *A* lies underneath triangle *B*.

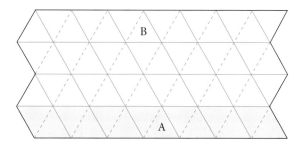

Another group of infinite polyhedra, the double-sided infinite polyhedra, are formed from the semiregular tilings we found in Section 9.1. To build a double-sided infinite polyhedron, take two copies of one of these tilings and place them parallel to one another. Then punch out copies of one of the tiles, and connect these holes with polygonal tunnels. For example, if we take the tiling 4.8.8 and replace all of the squares with tunnels made of squares, we get the following infinite polyhedron, which can be denoted (not uniquely) by 4.4.8.8 since at every vertex two squares and two octagons meet.

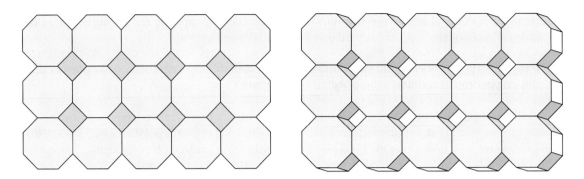

Similarly, we could remove some of the octagons and build octagonal tunnels connecting the two sheets of tilings. We cannot take all the octagons out or the tiling would be disconnected, so we take out every other one to get an infinite polyhedron denoted by 4.4.4.8.

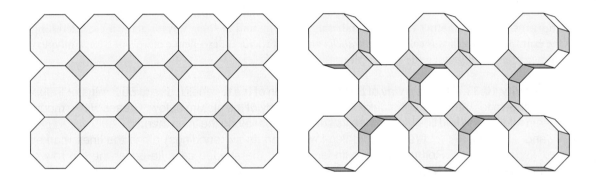

In building infinite polyhedra, we must make sure that the polygons meet edge to edge, and that the edges meet in pairs only. The following sort of intersection, where more than two edges meet, is not allowed:

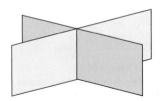

▷ **Activity 10.34.** Describe the three different double-layered infinite polyhedra that can be formed from the tiling 4.6.12.

There are 19 double-layered infinite polyhedra, all derived from the regular and semiregular tilings.

▷ **Activity 10.35.** Mark which squares you could replace by tunnels in the tiling 3.3.4.3.4 below to build a double-layered infinite polyhedron, being sure that the result will have only two polygons meeting at each edge, that you remove as many squares as possible, and that each vertex will have the same configuration.

We have given the slightest of introductions to this interesting family of three-dimensional objects. The interested reader is encouraged to pursue this study further in Projects 10.7 and 10.8. We have only gone far enough to generate examples for the study of isometries in $\mathbb{R}^3$.

◆ 10.5 Isometries in three dimensions

Definition 8.1 of an isometry on the plane extends easily to the higher dimensional $\mathbb{R}^n$. We found that there are five isometries in the euclidean plane: the identity, translation, rotation about a point, reflection across a line, and glide reflection along a line. All of these extend to $\mathbb{R}^3$ after a few minor modifications. There are two additional isometries, for a total of seven, in euclidean three-dimensional space. The results of Exercises 8.1 and 8.3, Theorems 8.2 and 8.15, and Corollaries 8.3 and 8.4 are simple to modify so that they apply to higher dimensions. Thus, we assume that we know that any isometry of $\mathbb{R}^3$ preserves lines, congruence, parallelism, and angle measure. We also assume that if an isometry fixes two points, then it must fix the line determined by these points.

Our objective in this section is to explore three-dimensional isometries, prove analogs of Theorem 8.16, 8.18, 8.19 and 8.20, and then use these to prove an equivalent statement to Theorem 8.21 on the classification of isometries.

◇ 10.5.1 The seven isometries

Our definition of a translation needs no change to adapt to $\mathbb{R}^3$. An example of an infinite polyhedron with translational symmetry in three separate directions is Coxeter's first infinite polyhedron:

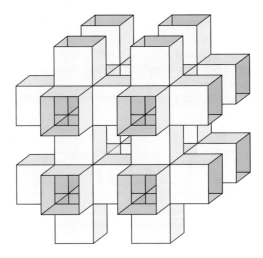

▷ **Activity 10.36.** Give an example of an infinite polyhedron that has translational symmetry in exactly two directions.

▷ **Activity 10.37.** Give an example of an infinite polyhedron that has translational symmetry in exactly one direction.

In three dimensions, we must reflect across a plane instead of a line as in Chapter 8.1, and we rotate about a line rather than about a point. Thus the definitions need some rewriting, which we ask you to do:

▶ **Exercise 10.47.** Rewrite Definition 8.9 for a reflection $F_\mathcal{P}$ across a plane $\mathcal{P}$ in $\mathbb{R}^3$.

▷ **Activity 10.38.** Give three examples of polyhedra that have reflectional symmetry.

▶ **Exercise 10.48.** Prove that if $\mathcal{P}'$ is a plane perpendicular to $\mathcal{P}$, then $F_\mathcal{P}$ takes $\mathcal{P}'$ to itself. Prove that, when restricted to points X on $\mathcal{P}'$, $F_\mathcal{P}(X) = F_\ell(X)$ where ℓ is the line of intersection of $\mathcal{P}$ and $\mathcal{P}'$.

▶ **Exercise 10.49.** Explain how to write the translation $T_{A,B}$ in $\mathbb{R}^3$ as the composition of reflections in two planes. What is the relationship between the two planes of reflection? What is the relationship between the planes and the line AB?

▶ **Exercise 10.50.** Rewrite Definition 8.7 for the rotation $R_{\ell,\theta}$ about a line ℓ by angle θ in $\mathbb{R}^3$.

▶ **Exercise 10.51.** Explain how to write the rotation $R_{\ell,\theta}$ as the composition of reflections in two planes. What is the relationship between the two planes of reflection? What is the relationship between the planes and the line ℓ? What is the relationship between the planes and the angle θ?

▷ **Activity 10.39.** Give three examples of polyhedra that have rotational symmetry.

▷ **Activity 10.40.** Give an example of an infinite polyhedron that has rotational symmetry.

▶ **Exercise 10.52.** Rewrite Definition 8.11 for the glide reflection $G_{\mathcal{P},A,B}$ along a plane $\mathcal{P}$ in the direction of AB for a distance of $\overline{AB}$ in $\mathbb{R}^3$, where either AB lies on P or is parallel to P.

▶ **Exercise 10.53.** Explain how to write the glide reflection $G_{\mathcal{P},A,B}$ as the composition of reflections in three planes.

▷ **Activity 10.41.** Give an example or build a section of an infinite polyhedron that has glide reflectional symmetry in only one direction.

▷ **Activity 10.42.** a. By Theorem 8.16, if an isometry fixes three noncollinear points, it must fix the plane determined by these points. What types of isometries have this property?

b. If an isometry fixes two points, then it fixes the line between them. What types of isometries have this property?

c. Try to find an isometry that fixes a single point.

In $\mathbb{R}^3$, there is a new type of isometry that fixes one point, the *rotary reflection* or *rotary inversion*. The fixed point is called the *center of symmetry*. Illustrated below is a typical rotary reflection, in which point 1 moves to point 2, point 2 to 3, 3 to 4, 4 to 5, 5 to 6, and point 6 back to point 1:

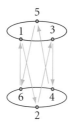

A rotary reflection is a rotation (in the example above by 60°) followed by a reflection in a plane perpendicular to the axis of rotation (or vice versa). An antiprism has rotary reflectional symmetry:

Definition 10.31. Given a plane $\mathcal{P}$, a point O on $\mathcal{P}$, and an angle θ, for any point $A \in \mathbb{R}^3$, define the *rotary reflection* $H_{O,\mathcal{P},\theta}(A)$ to be the unique point defined by reflecting A across $\mathcal{P}$, and then rotating the reflected point by θ about the line through O perpendicular to $\mathcal{P}$.

▷ **Activity 10.43.** For the pentagonal antiprism above, draw the axis of rotation and the plane of reflection. What is the angle of rotation?

▶ **Exercise 10.54.** Explain how to write a rotary reflection as the composition of reflections in three planes.

In the case when the angle of rotation for a rotary reflection is 180°, the isometry is called the *reflection across the point O*.

▶ **Exercise 10.55.** Prove that the reflection across the point O can be written as the composition of reflections across three mutually perpendicular planes.

▷ **Activity 10.44.** Give an example (other than the antiprism) of a polyhedron that has a reflection through a point.

The last of the seven isometries is called a *screw rotation* or *screw displacement*. This is a rotation followed by a translation in the direction of the rotation axis (or vice versa). A typical machine screw has, of course, screw rotational symmetry. Another figure that has screw rotational symmetry is the *tetrahelix*, made by gluing tetrahedra together face to face, or constructed as in Activity 10.33.

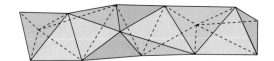

Definition 10.35. Given a line ℓ, an angle θ, and two points P and Q on ℓ in $\mathbb{R}^3$, for any point $A \in \mathbb{R}^3$, define the *screw rotation* $S_{\ell,\theta,P,Q}(A)$ to be the unique point defined by rotating A by θ about ℓ, and then translating the rotated point by the distance $\overline{PQ}$ in the direction from P to Q.

▶ **Exercise 10.56.** Explain how to write a screw rotation as the composition of reflections in four planes.

▶ **Exercise 10.57.** Describe the fixed points, lines, and planes for each type of isometry in $\mathbb{R}^3$.

We have found seven isometries of $\mathbb{R}^3$: the identity, translation, rotation about a line, reflection across a plane, glide reflection along a plane, rotary reflection, and screw rotation. We must show that this list is complete, as we did in Chapter 8.1 for the plane.

◊ 10.5.2 Classification of isometries of $\mathbb{R}^3$

To verify that we have described all possible isometries of $\mathbb{R}^3$, we need counterparts of the theorems of Section 8.1.7. Note that Theorem 8.15 carries over to $\mathbb{R}^3$ without change, so we know that if an isometry fixes two points, then it must fix the line given by those points. Most of the other proofs for euclidean isometries on the plane generalize in a straightforward manner to $\mathbb{R}^3$.

Theorem 10.36. If T is an isometry of $\mathbb{R}^3$ and T fixes four noncoplanar points, then T is the identity transformation.

▶ **Exercise 10.58.** Prove Theorem 10.36, generalizing Theorem 8.16.

Corollary 10.37. If T and S are isometries of $\mathbb{R}^3$ and there are four noncoplanar points A, B, C, and D such that $T(A) = S(A)$, $T(B) = S(B)$, $T(C) = S(C)$, and $T(D) = S(D)$, then $T = S$.

▶ **Exercise 10.59.** Prove Corollary 10.37.

Theorem 10.38. Any isometry of $\mathbb{R}^3$ can be written as the composition of at most four reflections.

Proof: Let S be an isometry of $\mathbb{R}^3$. Choose four noncoplanar points A, B, C, and D and construct $A' = S(A)$, $B' = S(B)$, $C' = S(C)$, and $D' = S(D)$.

Case 1: S fixes all four points: By Theorem 10.36, S is the identity.

Case 2: S fixes three points: If $A' = A$, $B' = B$ and $C' = C$, let $\mathcal{P}$ denote the plane determined by these three points. By Theorem 8.16, S fixes this plane. By Theorem 10.9, drop $DX \perp \mathcal{P}$. Choose a point Y on $\mathcal{P}$ and distinct from X and note that $S(X) = X$ and $S(Y) = Y$. Thus $\angle DXY = \angle S(D)S(X)S(Y) = \angle D'XY$ and $\overline{DX} = \overline{S(D)S(X)} = \overline{D'X}$. Therefore, $\mathcal{P}$ must be the perpendicular bisector of $\overline{DD'}$ and so $F_\mathcal{P}(D) = D'$. Since S and $F_\mathcal{P}$ agree on the points A, B, C, and D, by Corollary 10.37, $S = F_\mathcal{P}$.

Case 3: S fixes two points: If $A' = A$ and $B' = B$, construct the plane $\mathcal{P}$ that forms the perpendicular bisector of $\overline{CC'}$. Let $F_\mathcal{P}$ denote reflection across this plane and note that $F_\mathcal{P}(C) = C'$. Let X be the point of intersection of $\overline{CC'}$ and $\mathcal{P}$ and note that $\overline{C'X} = \overline{CX} = \overline{S(C)S(X)} = \overline{C'S(X)}$. Thus $S(X) = X$. Since $\overline{AC} = \overline{S(A)S(C)} = \overline{AC'}$ and $\angle CXA = \angle S(C)S(X)S(A) = \angle C'XA$, A must lie on the perpendicular bisector of $\overline{CC'}$ which is $\mathcal{P}$. Similarly, B also lies on $\mathcal{P}$. Then $F_\mathcal{P} \circ S(A) = A$, $F_\mathcal{P} \circ S(B) = B$, and $F_\mathcal{P} \circ S(C) = F_\mathcal{P}(C') = C$, so by Case 2, $F_\mathcal{P} \circ S = F_1$. Therefore, $S = F_\mathcal{P} \circ F_1$ and S is written as the composition of two reflections.

Case 4: S fixes one point: If $A' = A$, construct the plane $\mathcal{P}$ that forms the perpendicular bisector of $\overline{BB'}$. Let $F_\mathcal{P}$ denote reflection across this plane. As in Case 3, we can show that A lies on the plane $\mathcal{P}$. Then $F_\mathcal{P} \circ S(A) = A$ and $F_\mathcal{P} \circ S(B) = B$, so by Case 3, $F_\mathcal{P} \circ S(A) = F_1 \circ F_2$ and thus S can be written as the composition of three reflections.

Case 5: S fixes no points: Construct the plane $\mathcal{P}$ that forms the perpendicular bisector of $\overline{AA'}$. Let $F_\mathcal{P}$ denote reflection across the plane $\mathcal{P}$. Then $F_\mathcal{P} \circ S(A) = A$, so by Case 4, $F_\mathcal{P} \circ S(A) = F_1 \circ F_2 \circ F_3$ and thus S can be written as the composition of four reflections. □

▶ **Exercise 10.60.** Let S be an isometry of $\mathbb{R}^3$ such that S is the identity on a plane $\mathcal{P}$. Let X be a point that does not lie on this plane. Prove that if $S(X)$ and X lie on the same side of $\mathcal{P}$, then $S(X) = X$. Thus, S is the identity on $\mathbb{R}^3$.

The next three theorems are the three-dimensional counterparts of Theorems 8.19 and 8.20. The proofs of the first two are essentially the same as their two-dimensional analogs, though we must replace reflections across lines with reflections across planes and rotations about points with rotations about lines.

Theorem 10.39. An isometry of $\mathbb{R}^3$ is either a rotation or a translation if and only if the isometry can be written as the composition of reflections in two distinct planes $\mathcal{P}_1$ and $\mathcal{P}_2$.

Proof: If an isometry S of $\mathbb{R}^3$ is a rotation or a translation, then by Exercises 10.49 and 10.51, S can be written as the composition of two reflections in distinct planes. Conversely, if $S = F_2 \circ F_1$ where F_1 denotes reflection across $\mathcal{P}_1$ and F_2 reflection across $\mathcal{P}_2$, then either $\mathcal{P}_1$ intersects $\mathcal{P}_2$ or they are parallel. If the planes intersect in line ℓ, then as in Exercise 10.51, S is a rotation about ℓ of angle 2α, where α is the dihedral angle between the planes. If the planes are parallel, then by Theorem 10.9 and Exercise 10.10, there is a line ℓ perpendicular to both $\mathcal{P}_1$ and $\mathcal{P}_2$. Let d be the distance between these planes. By Exercise 10.49, $S = F_2 \circ F_1$ is a translation in the direction of line ℓ by a distance of $2d$. □

The next theorem is the $\mathbb{R}^3$ counterpart of Theorem 8.20 for the euclidean plane, in which we proved that an isometry on the plane is a glide reflection if and only if it can be written as the composition of three reflections. In $\mathbb{R}^3$, we also have the rotary reflections to deal with, but the proof is much like the earlier one, which you may want to review.

Theorem 10.40. An isometry of $\mathbb{R}^3$ is either a rotary reflection or a glide reflection if and only if the isometry can be written as the composition of reflections in three distinct planes $\mathcal{P}_1$, $\mathcal{P}_2$, and $\mathcal{P}_3$.

Proof: If an isometry S is a glide reflection or a rotary reflection, then by Exercises 10.53 and 10.54, we can write S as a composition of three reflections. Assume that $S = F_3 \circ F_2 \circ F_1$ for reflections in the distinct planes $\mathcal{P}_1$, $\mathcal{P}_2$, and $\mathcal{P}_3$.

Case 1: There is a plane $\mathcal{P}_4$ that is perpendicular to each of $\mathcal{P}_1$, $\mathcal{P}_2$, and $\mathcal{P}_3$:

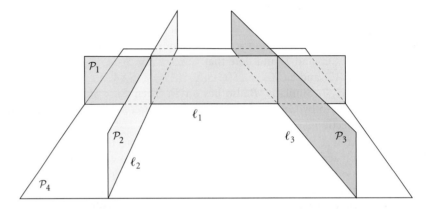

Let ℓ_i be the intersection of $\mathcal{P}_i$ with $\mathcal{P}_4$ for $i = 1, 2, 3$. By Exercise 10.48, F_i restricted to plane $\mathcal{P}_4$ is the planar isometry F_{ℓ_i}. Therefore, when restricted to $\mathcal{P}_4$, $S|_{\mathcal{P}_4} = F_{\ell_3} \circ F_{\ell_2} \circ F_{\ell_1}$. By Theorem 8.20, this restriction of S is a glide reflection defined on $\mathcal{P}_4$ which we denote by $g = S|_{\mathcal{P}_4}$. By Definition 8.11, we can write g as a composition of reflections in lines m_1, m_2 and m_3 where $m_2 \| m_1$ and $m_3 \perp m_2$ and so $g = F_{m_3} \circ F_{m_2} \circ F_{m_1}$. Use Exercise 10.13 to construct planes $\mathcal{P}'_1$, $\mathcal{P}'_2$, and $\mathcal{P}'_3$ all perpendicular to $\mathcal{P}_4$ so that the intersection of $\mathcal{P}'_i$ with $\mathcal{P}_4$ is line m_i.

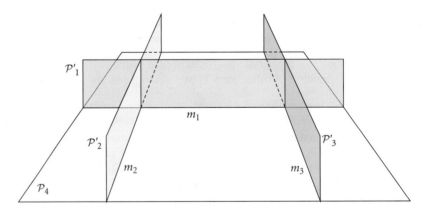

Let $F'_i = F_{\mathcal{P}'_i}$ for $i = 1, 2, 3$ and consider $F'_3 \circ F'_2 \circ F'_1$, which is a glide reflection on $\mathbb{R}^3$. Note that if we restrict to $\mathcal{P}_4$, $F'_3 \circ F'_2 \circ F'_1 = g = S|_{\mathcal{P}_4}$. Furthermore, it is clear that for any point X lying on one side of $\mathcal{P}_4$, $F'_3 \circ F'_2 \circ F'_1(X)$ will lie on the same side of $\mathcal{P}_4$. Similarly, X and $S(X)$ lie on the same side of $\mathcal{P}_4$, so $F'_3 \circ F'_2 \circ F'_1$ and S are identical if restricted to plane $\mathcal{P}_4$ and also take points on one side of this plane to points on the same side. Consider $F'_3 \circ F'_2 \circ F'_1 \circ S^{-1}$. This will be the identity transformation on $\mathcal{P}_4$ and takes points on one side of $\mathcal{P}_4$ to points on the same side. By Exercise 10.60, the only isometry that has this property is the identity, so $S = F'_3 \circ F'_2 \circ F'_1$. Thus, S is a glide reflection on $\mathbb{R}^3$.

Case 2: There is no plane $\mathcal{P}_4$ that is perpendicular to all three of $\mathcal{P}_1$, $\mathcal{P}_2$, and $\mathcal{P}_3$: In this case, $\mathcal{P}_1$, $\mathcal{P}_2$, and $\mathcal{P}_3$ cannot be parallel, so at least two of these planes intersect.

Case 2a: $\mathcal{P}_2$ and $\mathcal{P}_3$ intersect in line ℓ. In this case, $F_3 \circ F_2$ is a rotation about ℓ. We generalize the construction in Case 2a of Theorem 8.20. By Corollary 10.14, we can construct $\mathcal{P}_4$ containing ℓ so that $\mathcal{P}_4 \perp \mathcal{P}_1$. Let m be the line of intersection of $\mathcal{P}_1$ and $\mathcal{P}_4$. Construct $\mathcal{P}_5$ so that the dihedral angle formed by $\mathcal{P}_4$ and $\mathcal{P}_5$ at ℓ is the same as the dihedral angle formed by $\mathcal{P}_2$ and $\mathcal{P}_3$ at ℓ. Then $F_3 \circ F_2 = R_{\ell,\theta} = F_5 \circ F_4$. Construct a plane $\mathcal{P}_6$ containing m and perpendicular to $\mathcal{P}_5$ by Corollary 10.14. Then use Exercise 10.13 to construct another plane $\mathcal{P}_7$ through m and perpendicular to $\mathcal{P}_6$. Since $\mathcal{P}_5$ and $\mathcal{P}_7$ are both perpendicular to $\mathcal{P}_6$, they must intersect in a line we denote by n. Let $\frac{\alpha}{2}$ denote the dihedral angle formed between $\mathcal{P}_5$ and $\mathcal{P}_7$. Then $F_5 \circ F_7 = R_{n,\alpha}$. Since $\mathcal{P}_4 \perp \mathcal{P}_1$, $F_4 \circ F_1 = R_{m,180°} = F_7 \circ F_6$. Therefore, $S = F_3 \circ F_2 \circ F_1 = F_5 \circ F_4 \circ F_1 = F_5 \circ F_7 \circ F_6 = R_{n,\alpha} \circ F_6$. Note that $\mathcal{P}_6$ is perpendicular to n, the line of intersection of $\mathcal{P}_5$ and $\mathcal{P}_7$. Thus, $S = F_5 \circ F_7 \circ F_6 = R_{n,\alpha} \circ F_6$ is a rotary reflection with axis n.

Case 2b: $\mathcal{P}_1$ and $\mathcal{P}_2$ intersect in line ℓ. The proof of this case is similar to Case 2a above.

Case 2c: $\mathcal{P}_1$ and $\mathcal{P}_3$ intersect in line ℓ. The proof of this case is similar to Case 2c of Theorem 8.20. □

Recall Definition 8.22, that an isometry is called *even* if it can be expressed as the composition of an even number of reflections, and an isometry is called *odd* if it can be written as the composition of an odd number of reflections. Theorem 10.40 implies that an odd isometry in $\mathbb{R}^3$ must be a reflection, a glide reflection, or a rotary reflection. Theorem 10.39 shows that the identity, rotation and translation are even isometries. There remains one case to be considered.

Theorem 10.41. *An isometry of $\mathbb{R}^3$ is a screw rotation if and only if the isometry can be written as the composition of reflections in four distinct planes $\mathcal{P}_1$, $\mathcal{P}_2$, $\mathcal{P}_3$, and $\mathcal{P}_4$.*

Proof: If S is a screw rotation, then S can be written as the composition of four reflections by Exercise 10.56. Conversely, if S is the composition of four reflections, there must be some point A so that $S(A) = A' \neq A$ or by Theorem 10.36, S would be the identity. Let F denote reflection across the plane $\mathcal{P}$ that is the perpendicular bisector of $\overline{AA'}$. Let T denote the isometry $T = F \circ S$. Then T is an odd isometry and so must be a reflection, a glide reflection, or a rotary reflection. Since $T(A) = F \circ S(A) = F(A') = A$, T has a fixed point and so cannot be a glide reflection. Thus T must be a rotary reflection, though perhaps with a rotation of angle 0 if it is a simple reflection. By Theorem 10.40, T can be written as a composition $T = F_3 \circ F_2 \circ F_1$ for reflections across planes $\mathcal{P}_1$, $\mathcal{P}_2$, and $\mathcal{P}_3$ where $\mathcal{P}_3$ and $\mathcal{P}_2$ intersect along line ℓ with dihedral angle $\frac{\theta}{2}$ and $\mathcal{P}_1$ is perpendicular to both $\mathcal{P}_2$ and $\mathcal{P}_3$.

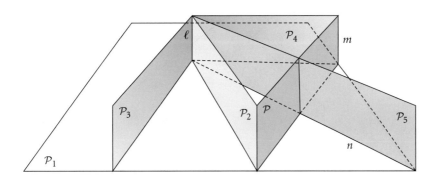

Construct plane $\mathcal{P}_4$ through ℓ and perpendicular to $\mathcal{P}$. Let m be the line of intersection of $\mathcal{P}$ and $\mathcal{P}_4$. Since $\mathcal{P}_4 \perp \mathcal{P}$, $F \circ F_4 = R_{m,180°}$. Construct another plane $\mathcal{P}_5$ through ℓ so that the dihedral angle formed by $\mathcal{P}_4$ and $\mathcal{P}_5$ at ℓ is $\frac{\theta}{2}$. Thus, $F_4 \circ F_5 = R_{\ell,\theta} = F_3 \circ F_2$. Since $\ell \perp \mathcal{P}_1$, by Theorem 10.13, $\mathcal{P}_5 \perp \mathcal{P}_1$. Let n denote the line of intersection of $\mathcal{P}_5$ and $\mathcal{P}_1$, so $F_5 \circ F_1 = R_{n,180°}$. We then have

$$T = F \circ S = F_3 \circ F_2 \circ F_1 = F_4 \circ F_5 \circ F_1,$$

and thus

$$S = F \circ F_4 \circ F_5 \circ F_1.$$

By Exercise 10.3, exactly one of the following must be true: m and n are parallel, intersect, or are skew.

Case 1: m and n are parallel:

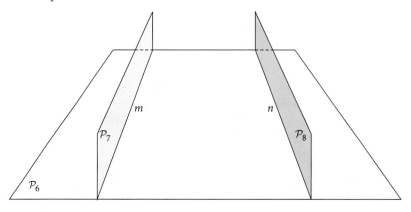

If m and n are parallel, then there is a plane $\mathcal{P}_6$ containing m and n. By Exercise 10.13, we can construct plane $\mathcal{P}_7$ through m perpendicular to $\mathcal{P}_6$ and plane $\mathcal{P}_8$ through n perpendicular to $\mathcal{P}_6$. Then $F \circ F_4 = R_{m,180°} = F_7 \circ F_6$ and $F_5 \circ F_1 = R_{n,180°} = F_6 \circ F_8$. Thus

$$S = F \circ F_4 \circ F_5 \circ F_1 = F_7 \circ F_6 \circ F_6 \circ F_8 = F_7 \circ F_8.$$

Since $\mathcal{P}_7$ is parallel to $\mathcal{P}_8$, S is a translation (a screw rotation with angle 0).

Case 2: m and n intersect:

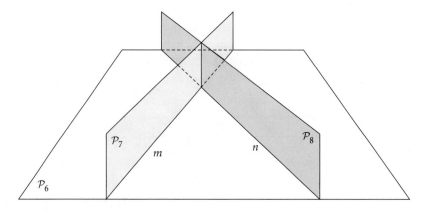

Let $\mathcal{P}_6$ be the plane containing m and n. Use Exercise 10.13 again to construct plane $\mathcal{P}_7$ through m perpendicular to $\mathcal{P}_6$ and plane $\mathcal{P}_8$ through n perpendicular to $\mathcal{P}_6$. Then $F \circ F_4 = R_{m,180°} = F_7 \circ F_6$ and $F_5 \circ F_1 = R_{n,180°} = F_6 \circ F_8$. Thus

$$S = F \circ F_4 \circ F_5 \circ F_1 = F_7 \circ F_6 \circ F_6 \circ F_8 = F_7 \circ F_8.$$

Since $\mathcal{P}_7$ intersects $\mathcal{P}_8$, S is a rotation (a screw rotation with translation distance 0).

Case 3: m and n are skew:

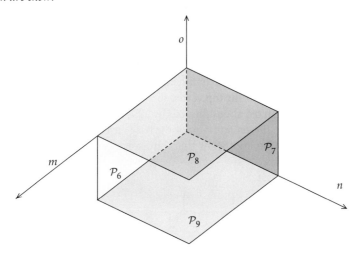

By Theorem 10.15, there is a line o perpendicular to both m and n. Let $\mathcal{P}_6$ be the plane that contains lines m and o and $\mathcal{P}_7$ the plane containing lines n and o. By Exercise 10.13, construct planes $\mathcal{P}_8 \perp \mathcal{P}_6$ through m and $\mathcal{P}_9 \perp \mathcal{P}_7$ through n. Then $F \circ F_4 = R_{m,180°} = F_6 \circ F_8$ and $F_5 \circ F_1 = R_{n,180°} = F_7 \circ F_9$. Thus

$$S = F \circ F_4 \circ F_5 \circ F_1 = F_6 \circ F_8 \circ F_7 \circ F_9.$$

Since $\mathcal{P}_7$ contains line o and o is perpendicular to $\mathcal{P}_8$, by Theorem 10.13, $\mathcal{P}_7 \perp \mathcal{P}_8$ and so $F_7 \circ F_8 = F_8 \circ F_7$. Thus

$$S = F_6 \circ F_8 \circ F_7 \circ F_9 = F_6 \circ F_7 \circ F_8 \circ F_9.$$

The plane $\mathcal{P}_6$ intersects $\mathcal{P}_7$ along line o, so $F_6 \circ F_7$ is a rotation about o. Since $\mathcal{P}_8$ is parallel to $\mathcal{P}_9$ and both are perpendicular to o, $F_8 \circ F_9$ is a translation along o. Thus, S is a screw rotation in the o-direction. □

Theorem 10.42. **[Classification of Isometries on $\mathbb{R}^3$]** Every isometry of $\mathbb{R}^3$ is either the identity, a reflection, a translation, a rotation, a glide reflection, a rotary reflection, or a screw rotation.

▶ **Exercise 10.61.** Prove Theorem 10.42.

▶ **Exercise 10.62.** Show that every isometry with at least one fixed point is the composition of at most three reflections.

The isometries of the sphere are precisely the $\mathbb{R}^3$ isometries that leave the sphere in place. Thus spherical isometries must leave at least one point, the center of the sphere, fixed.

Corollary 10.43. [**Classification of Isometries on the Sphere**] Every isometry of the sphere is either the identity, a rotation about a diameter, a reflection across a plane passing through the center of the sphere, or a rotary reflection across a plane passing through the center of the sphere.

▶ **Exercise 10.63.** Prove Corollary 10.43.

◆ 10.6 Symmetries of polyhedra

In this section we will investigate the symmetries of finite polyhedra, extending our work on rosette groups from Chapter 8.2. This is far easier to visualize with models in hand, as constructed in Section 10.2.

First we'll consider the simplest solids, the tetrahedron and the cube. If you set the tetrahedron on its base, and consider the line passing from the top vertex through the center of the base, you have an *axis of rotation*. The tetrahedron can be rotated about this axis by 120° to land back in the same position, though all points except the top vertex and the center point of the base have moved. Below is an illustration of the tetrahedron with this axis of rotation and below that, bird's-eye views of the side faces, shaded to show the effect of the three 120° rotations.

We say that the tetrahedron has a 3-fold axis of rotation, with a rotation angle of $120° = \frac{360°}{3}$.

▷ **Activity 10.45.** How many other 3-fold axes of rotation does the tetrahedron have?

The tetrahedron has another type of rotational symmetry. Below is a tetrahedron with an axis passing through the midpoint of one edge and coming out at the midpoint of the opposite edge. This is a 2-fold axis of rotation. Sighting down the axis, you will see only two of the triangular faces, and a 180° rotation will interchange these.

▷ **Activity 10.46.** How many other 2-fold axes of rotation does the tetrahedron have?

As we did for rosette groups in Section 8.2, we can define the symmetry groups for polyhedra. We can also consider the subset (actually, the subgroup) that consists of just the rotations of the figure. From Activities 10.45 and 10.46, we know that the rotation group for the tetrahedron has 12 elements: the identity, rotations of 120° and 240° for each of the four vertices, and rotations of 180° for each of three pairs of opposite sides. We say that the rotation group for the tetrahedron has *order* 12.

In addition to rotational symmetry, three-dimensional figures can have reflectional symmetry. For example, the tetrahedron is cut into two identical pieces by a plane through one edge and passing through the middles of the two faces opposing that edge:

▷ **Activity 10.47.** How many other reflection planes does the tetrahedron have?

The full symmetry group of the tetrahedron has order 24, combining the 12 rotations with this reflection. Any other reflection can be written as some combination of these. We cannot pursue these computations without more knowledge of group theory that we are willing to assume. The interested reader is referred to Project 10.11, Hartshorne's *Geometry: Euclid and Beyond*, or George Martin's *Transformation Geometry*.

Next, consider the cube. Below is a cube with a 4-fold axis of rotation passing through the center points of two opposite faces. We make use of a representation called a *Schlegel diagram*, which is a sort of squashed bird's-eye view. More precisely, the Schlegel diagram is formed by imagining the wire framework of the polyhedron and the shadow it would cast if a light bulb were suspended over the center of the figure. The Schlegel diagrams below the cube show the effect of 90° rotations around this axis. After four rotations through 90°, the cube returns to its original configuration:

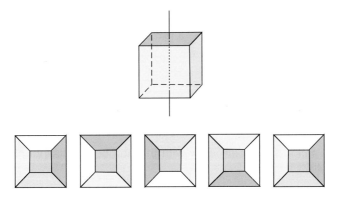

▷ **Activity 10.48.** How many other 4-fold axes of rotation does the cube have?

▶ **Exercise 10.64.** An axis of rotation for the cube is shown below, passing from the front left top vertex to the rear right bottom vertex. What is the order of the axis? What is the angle of rotation? How many other axes of rotation of the same type are there?

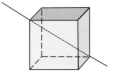

▶ **Exercise 10.65.** An axis of rotation for the cube is shown below, passing from the midpoint of one edge to the midpoint of the edge diametrically opposite. What is the order of the axis? What is the angle of rotation? How many other axes of rotation of the same type are there?

The cube can be cut by a plane of reflection in two ways, by a plane parallel to the faces or diagonally:

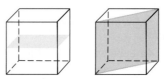

▷ **Activity 10.49.** How many reflection planes does a cube have?

▶ **Exercise 10.66.** Fill in the following table with the number of distinct axes of rotation for each Platonic solid:

Polyhedron	2-fold axes	3-fold axes	4-fold axes	5-fold axes
Tetrahedron		4	0	0
Cube				
Octahedron				
Dodecahedron				
Icosahedron				

▶ **Exercise 10.67.** Fill in the table below with the number of distinct planes of reflection for each Platonic solid:

Polyhedron	reflection planes
Tetrahedron	
Cube	
Octahedron	
Dodecahedron	
Icosahedron	

▷ **Activity 10.50.** You will notice certain coincidences in the information you have collected in Exercises 10.66 and 10.67. Formulate a conjecture on when two polyhedra will have the same numbers of lines of rotation and planes of reflection.

▶ **Exercise 10.68.** What are the symmetries of the truncated forms of the Platonic solids?

▶ **Exercise 10.69.** What are the symmetries of the cuboctahedron and the icosidodecahedron?

▶ **Exercise 10.70.** What are the symmetries of the snub cube and the snub dodecahedron?

◊ **10.6.1 Orthoschemes**

In this section we take advantage of the fact that any isometry of a Platonic solid must be a composition of at most three reflections to build the three-dimensional analog of a kaleidoscope.

To start, take three mirrors and place one flat on the table and the other two perpendicular to the table and meeting at an angle of 90°. Thus, a right angle is formed between each pair of mirrors:

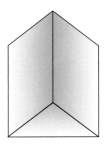

Place a small cube (about an inch on each side) in the three-mirror assembly where the three mirrors meet. You should see a bigger cube in the reflection, with each side twice the length of the original cube.

▷ **Activity 10.51.** Cut an equilateral triangle (about an inch on each side) out of cardboard and place it symmetrically in your three-mirror assembly where the three mirrors meet. Describe the solid that appears in the mirrors.

▷ **Activity 10.52.** Copy the figure below and bend along the dotted lines. Place it symmetrically in your 3-mirror assembly where the three mirrors meet, and adjust the three 45°–45°–90° isosceles triangles so that each meets the join of two mirrors at a right angle. Describe the solid that appears in the mirrors.

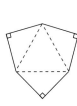
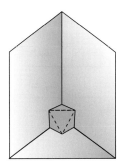

Recall from Exercise 10.67 that the cube has nine planes of reflection. Illustrated below in the first picture are the three mirror planes that form the 90°–90°–90° assembly we just constructed. The other six mirror planes are illustrated in pairs:

The cube, sliced by all nine mirror planes, is illustrated below:

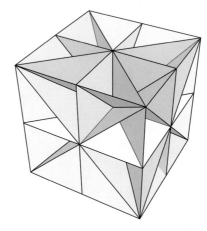

Note that the cube is sliced by these mirror planes into 48 congruent irregular tetrahedra. We would like to build a mirror assembly in the form of one of these wedges, called an *orthoscheme*.

The first three planes cut the big cube into eight smaller cubes. If the original cube were 2 inches on a side, then the little cube would be 1 inch on each side. Throw everything out except the upper left front little cube:

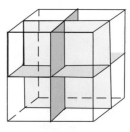

Only one of the next two planes cuts through the upper left front cube, and it slices it diagonally in half:

10.6. SYMMETRIES OF POLYHEDRA • 411

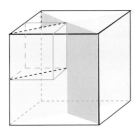

▶ **Exercise 10.71.** If the small cube measures 1 inch on a side, find the length of each edge of the wedge-shaped region pictured above.

Only one of the next two planes cuts through the wedge shaped region, and it slices it diagonally into two pieces:

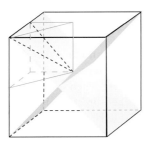

▶ **Exercise 10.72.** Find the length of each edge of the triangular wedge pictured above.

Neither of the last two planes cut through the triangular wedge, so this wedge is the final orthoscheme.

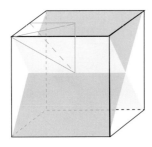

▶ **Exercise 10.73.** Below is a net for the wedge, or orthoscheme. Using your results from Exercise 10.72, label the lengths of each unmarked edge:

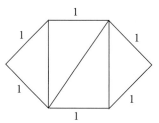

▷ **Activity 10.53.** Build the orthoscheme out of cardboard, using 1 inch as your basic unit of measure and the net from Exercise 10.73. Next, build the orthoscheme out of mirrors (with the mirrored surfaces facing in), leaving off the face that formed part of the original cube and using 6 inches for the basic measure instead of 1 inch. Place the small cardboard orthoscheme so that it nests inside the mirrored orthoscheme. Describe the results you see in the mirrors.

◆ 10.7 Four-dimensional figures

With some imagination, we can extend many of the results we have found to higher dimensions. To try to visualize four dimensions, consider the progression that leads us to the three we are familiar with: start with a single point, which we consider as the origin, or center of the universe. Then walk back and forth, forming an axis called the x-axis. Turn at right angles and walk in the other direction, forming the y-axis and a plane spanned between the two axes. The direction perpendicular to both the x- and y-axes is the z-axis, forming a three-dimensional space. To find a fourth dimension all we need is to find another direction perpendicular to all of the previous directions. If x denotes back and forth, y left and right, and z up and down, we need words to describe movement in the fourth dimension. C. H. Hinton coined the words *ana* and *kata* for these movements (from the Greek for "up towards" and "down from"). For lack of better, we will use these. Four-dimensional space is called *hyperspace*.

The one-dimensional analog of the unit cube is a line segment 1 inch long lying on the x-axis. Stacking many copies of this line segment on top of each other in the y-direction builds a square. Stacking squares together in the z-direction, like stacking sheets of paper, builds up a cube. In four dimensions, stacking cubes together in the *ana* direction would build a *hypercube*.

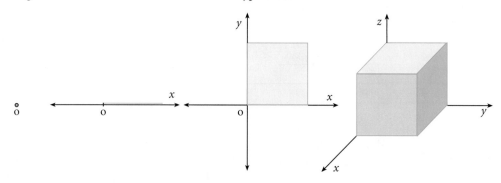

▷ **Activity 10.54.** In the plane, note that any two nonparallel lines intersect in a point.

   a. Generalize this idea to the intersection of two nonparallel planes in the third dimension.
   b. Generalize this idea to the intersection of two nonparallel three-dimensional spaces in hyperspace.

▷ **Activity 10.55.** In the illustration below, you see that in general, a point and a line on the plane will not intersect at all. The probability that the point will land on the line is, in fact, zero, given that there are so many other points from which to choose. In the second illustration, we see that the most general, or most likely, intersection of a line and a plane in three-dimensional space is a point. Of course, this does not always occur, since the line could be parallel to the plane or the line could lie on the plane, but these are considered special cases. In each of those cases, jiggling the line a little bit would result in the general case where the line intersects the plane at a single point.

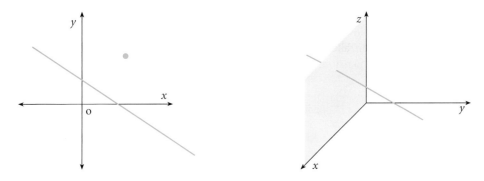

   a. What is the most general intersection of two lines in the plane?
   b. What is the most general intersection of two lines in $\mathbb{R}^3$?
   c. Using the results of this activity and the previous one, conjecture a formula for the dimension of the intersection of a k-dimensional line or plane with an ℓ-dimensional line or plane in n-dimensional space.
   d. What is the most general intersection of a line and a plane in hyperspace?
   e. What is the most general intersection of two planes in hyperspace?
   f. What is the most general intersection of a plane and a copy of $\mathbb{R}^3$ in hyperspace?

To see what is possible in hyperspace, we reason by analogy. First, imagine two points on a line, and let the first point hop up off of the line and back onto the line on the other side of the fixed second point. As a resident of the line, the fixed point sees the first point disappear and then reappear on the other side and is completely mystified. To the observer in the plane, it is quite clear what has happened: The first point has taken advantage of the second dimension to perform a maneuver impossible if all action is confined to the line.

Next, imagine a point inside a circle in the plane. The point can jump up into the third dimension and land back on the plane outside the circle. As a resident of the plane, the circle sees the point inside the circle disappear and then reappear outside. To the observer in the third dimension, it is obvious how the point escaped, but the circle is clueless.

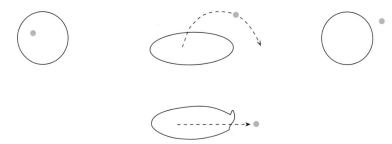

Alternatively, one could imagine that the circle lifted a small section of itself up into the third dimension. To the point in the plane, this would look as though a gap appeared in the circle, and then the point could make a dash through the gap to escape from the circle.

This sort of maneuver allows us to unlink two circles in $\mathbb{R}^4$: One circle could lift a bit of itself *ana* or *kata* into the fourth dimension, allowing the other to slide through and unlink. Thus, the following figures are the same in the fourth dimension:

▷ **Activity 10.56.** Explain, with a series of pictures, how to untie the knot below, without cutting the string. Note exactly where you need to slip into hyperspace.

▷ **Activity 10.57.** Explain informally how an object could be removed from a sealed three-dimensional box in hyperspace.

▷ **Activity 10.58.** Explain informally how a left glove could be changed into a right glove in hyperspace.

From the exercises and examples, we see that if a fourth dimension exists, then beings could walk through walls, disappear and reappear, remove things from sealed containers, untie knots, etc. What sort of beings have these capabilities? Ghosts!

The *hypercube* or *tesseract* is the four-dimensional analog of the familiar cube. Start with a single point. Moving this point one unit right traces out a line segment. Moving the line segment one unit up or north traces out a square. Moving the square one unit forwards or backwards traces out a cube. Moving the cube one unit *ana* or *kata* traces out a hypercube.

▶ **Exercise 10.74.**

a. Illustrate the process described above, showing a point, the line segment formed by moving the point, and the square formed by moving the line segment. How many vertices, edges, and two-dimensional faces does the square have?
b. Now think about moving the square parallel to itself. What figure is formed? How many vertices, edges, two-dimensional faces, and three-dimensional solids are there?
c. Now think about moving the figure of (b) parallel to itself. The result is called a four-dimensional *hypercube*. How many vertices, edges, two-dimensional faces, three-dimensional solids, and four-dimensional regions are there?
d. Now think about moving the figure of (c) parallel to itself to form a five-dimensional *hyper-hypercube*. How many vertices, edges, two-dimensional faces, three-dimensional solids, four-dimensional regions, and five-dimensional regions are there?

One representation of the hypercube is shown below:

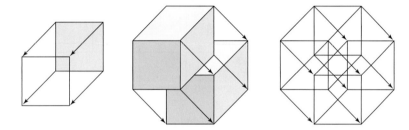

▷ **Activity 10.59.** From Exercise 10.74, a hypercube contains eight cubes. Two are illustrated above. Find the other six in the projection below.

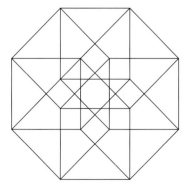

Another way of viewing the hypercube is the Schlegel diagram: Consider the cube as a wire framework that can be stretched and laid flat as below on the left. One can think of the center square as the bottom face of the cube, and the outer square the upper face. The side faces of the cube are stretched to form the trapezoids surrounding the inner square in this projection. The four-dimensional analog of this projection, distorting the hypercube to fit it into three-dimensional space, is shown on the right:

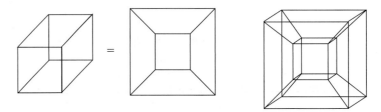

▷ **Activity 10.60.** Describe the positions of the eight cubes that make up the hypercube in the projection above on the right.

Yet another way of envisioning the hypercube is by analogy with the formation of a cube from a net or pattern that can be cut up and folded to form a cube. One net for the cube is shown below on the right.

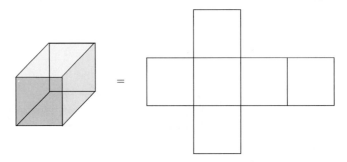

Similarly, a hypercube could be assembled from eight cubes with their faces or sides glued together in pairs correctly. A net for the hypercube is shown below:

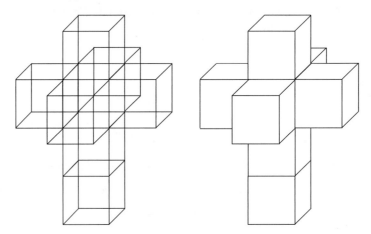

▷ **Activity 10.61.** Build the net above (sugar cubes work well). Figure out how the sides of the cubes should be glued together to build the hypercube and label or color-code them.

Another sequence of figures can be built up in analogy with the triangle. Start with a single point. This is called a 0-*simplex*. In the next generation, add another point above the first and connect the two, getting a line segment or 1-simplex. Add another point above the line segment and connect this point with each of the points on the line segment, obtaining a triangle or 2-simplex. Place the triangle flat on the floor and add another point above it; connecting the new point to each of the points in the triangle gives a three-dimensional solid called the tetrahedron or 3-simplex.

Adding another point in the fourth dimension, *ana* or *kata* the figure, and connecting this point to each point in the tetrahedron gives a regular four-dimensional figure called the *hypertetrahedron* or *pentahedroid* or *4-simplex*.

▶ **Exercise 10.75.** Complete the table below, where *v* indicates the number of vertices, *e* the number of edges or line segments, *f* the faces or polygons, *s* the solids or polyhedra, and *t* the four-dimensional regions:

dimension	figure	v	e	f	s	t
1	line segment	2	1			
2	triangle					
3	tetrahedron					
4	hypertetrahedron					

The higher-dimensional analogs of the Platonic solids are called *regular polytopes*. The interested reader is referred to H. S. M. Coxeter's *Regular Polytopes*.

◆ 10.8 Projects

Project 10.1. Compare and contrast several definitions of polyhedra, explaining the figures that each includes and excludes and the advantages and disadvantages of each potential definition.

Project 10.2. Investigate the history of the Platonic and/or Archimedean solids.[6]

Project 10.3. Build origami models of some polyhedra.[7]

Project 10.4. Read A. R. Pargeter, "Plaited polyhedra," and build several models.

Project 10.5. Read Peter Hilton and Jean Pedersen, *Build your own Polyhedra*, and build several models.

Project 10.6. Alexander G. Bell's solution to the problem of flight was large manned kites. Read and report on Bell, "The Tetrahedral Principle in Kite Structure" (published just months before the Wright brothers' first flight of a powered airplane).

---

[6] A good place to start is Cromwell's *Polyhedra*.
[7] Possible references are Tomoko Fusè, *Unit Origami*, Kunihiko Kasahara, *Origami Omnibus*, or Kunihiko Kasahara and Toshie Takahama, *Origami for the Connoisseur*.

Project 10.7. Investigate other types of infinite polyhedra other than those discussed in Section 10.4.

Project 10.8. A *close packing*, or *space filling*, of polyhedra is the three-dimensional analog of a tiling on the plane: a way of fitting together polyhedra face to face so that they fill up all of three-dimensional space. Investigate this phenomenon and build several suitable models.

Project 10.9. Periodicity is a naturally occurring phenomenon in crystal structures and as such is interesting to chemists and physicists, among others. A crystal is formed when identical molecules align themselves to fit a lattice. These form the three-dimensional analog of the wallpaper patterns of Chapter 8. Investigate the classification of such periodic crystallographic lattices.

Project 10.10. Explore quasi-crystals, the three-dimensional analogs of the Penrose tilings of Chapter 8.

Project 10.11. For those who have studied group theory: Derive the full symmetry group of the cube, including both rotations and reflections. Derive the symmetry groups of the other Platonic solids.

Project 10.12. Build a tetrahedral orthoscheme.

Project 10.13. Investigate the three-dimensional analog of the hyperbolic plane.

Project 10.14. Investigate the three-dimensional analog of the elliptic plane.

Project 10.15. Read and report on fictional representations of life in a two-dimensional universe, such as E. A. Abbott's *Flatland* and A. K. Dewdney's *The Planiverse*.

Appendix A. Logic and proofs

This brief optional appendix is a crash course in logic and the writing of proofs. We have included only those laws of logic that will be needed in the main text. Many students will not need to read this material, but we have included it for those who have less exposure or comfort in devising proofs.

A.1 Mathematical systems

A *mathematical system* consists of the following:

1. A set of undefined or primitive terms,
2. Definitions of objects, relations, and operations,
3. A list of axioms or postulates,
4. Theorems with proofs,
5. An underlying system of logic and set theory.

In a modern deductive system, undefinable terms are allowed. These are called the *primitive terms*. It is impossible to define everything. Take any dictionary and choose a word. Look up that word and choose the primary defining term and then look that up. One quickly finds that doing this almost always results in a circular web of words defining words defining words that define the first word. To avoid this type of circular reasoning, one allows primitive undefined terms. Note that we cannot then use properties of these undefined terms in arguments, except for those properties that are clearly specified in the axioms that follow. Examples of such primitive terms in geometry are points and lines. While we all have a clear image of a point, it is difficult to precisely define what one is. On the other hand, it is easy to describe the properties that we want a point to have, and these are then given in the axioms for geometry.

The *objects* referred to in part (2) of our mathematical system above are things like triangles and quadrilaterals, which are defined in terms of primitives such as lines and points. The geometric *relations* we are interested in are properties such as parallelism, which is defined as a relation between two straight lines. Another example of a relation in geometry is similarity, which is defined as a relation between two triangles. *Operations* are actions such as bisection of angles or line segments, reflection of an object across a line, or rotation of an object about a point.

The list of *axioms* for a mathematical system is a set of statements that one is willing to accept without proof. Since these are not proven, but based on faith alone, the classical view was that this list should be comprised of statements that everyone is willing to accept as true. A more modern view is to consider these merely as assumptions one chooses to make. They are thus true by definition, and need not even be plausible, as long as they do not lead to logical contradictions. Furthermore, in an ideal mathematical system, we would want to keep this list of unproven statements as minimal as possible. One rule of thumb in logic is known as Occam's Razor: *Entia non sunt multiplicanda praeter necessitatem*: "No more things should be presumed to exist than are absolutely necessary." William of Ockham [ca. 1285–1349] was an English logician and Franciscan friar who formulated this general principle, saying that in trying to explain any phenomenon, one should try to make as few assumptions as possible, eliminating (with the razor) any that are unnecessarily complicated. In other words, given two explanations for something, one should adopt the simplest. In mathematics, adding redundant axioms can make a system unwieldy and can accidentally introduce contradictions.

The *theorems* of mathematics are proved using the definitions, axioms, any previously proven theorems, and the rules of logical deduction. Euclid, and some other authors, use the term *proposition* instead of theorem. Theorems form the main substance and the challenge of mathematics. A *conjecture* is a potential theorem that has not yet been proved or disproved. Other terms you will see are *lemma* (plural, *lemmata*) and *corollary*. A lemma is usually a subtheorem necessary for the proof of a major theorem. It is often advantageous to separate some portion of a long proof as a separate subproof, and this is called a lemma. Thus, the statement of the lemma is usually less interesting than the proof. In contrast, a corollary is a theorem that follows easily from another theorem. The proof of a corollary is typically quite short, and depends heavily on the previous theorem. Corollaries are usually interesting and useful results but have easy proofs.

Another definition that we will need in this text is that of an *equivalence relation* among the objects of study. Equivalence relations are modeled on the properties of equality. We write $A \equiv B$ if A and B are equivalent. A relation among the objects of study is an equivalence relation if it is:

1. Reflexive: For all A, $A \equiv A$,
2. Symmetric: If $A \equiv B$, then $B \equiv A$,
3. Transitive: If $A \equiv B$ and $B \equiv C$, then $A \equiv C$.

In every field of mathematics, particular equivalence relations are used, depending on the objects of study and the properties of interest. Some examples of geometric equivalence relations we will study are congruence and similarity.

A.2 Logic

The system of logic underlying any mathematical system is itself a mathematical system. We will not go into the axioms behind standard logic and set theory, but assume that you have some acquaintance with these. We will, however, summarize the bits of logic we will need for this course. Let P and Q denote statements or propositions. These are sentences which have a well-defined truth value: i.e., sentences that are either true or false but not both. We are interested in five connectives:

$P \wedge Q$ is read "P and Q." This is true only if both P and Q are true.
$P \vee Q$ is read "P or Q." This is true if either P is true or Q is true or both.
$\neg P$ is read "not P." This statement is true if and only if P is false.
$P \implies Q$ is read "P implies Q" or "if P, then Q."
$P \iff Q$ is read "P if and only if Q."

These are descriptions or interpretations of the connectives. We will formally define these in terms of truth tables:

Logical Connectives						
P	Q	$\neg P$	$P \wedge Q$	$P \vee Q$	$P \implies Q$	$P \iff Q$
T	T	F	T	T	T	T
T	F	F	F	T	F	F
F	T	T	F	T	T	F
F	F	T	F	F	T	T

A warning: Note that if P is false then the statement $P \implies Q$ is *vacuously true*. This may not make a lot of sense at first but the alternative makes even less.

One can find the truth values for a compound statement by building up a truth table. For example, if one wants to find the truth values for $(\neg P) \wedge \neg(Q \vee R)$, make a table with all possible combinations of T and F for P, Q, and R on the left, and then build up, using the values from the table above, first the truth values for $Q \vee R$, then $\neg(Q \vee R)$ and $\neg P$, and finally the values for $(\neg P) \wedge \neg(Q \vee R)$. Note that the latter columns could have been done in a different order.

P	Q	R	$Q \vee R$	$\neg(Q \vee R)$	$\neg P$	$(\neg P) \wedge \neg(Q \vee R)$
T	T	T	T	F	F	F
T	T	F	T	F	F	F
T	F	T	T	F	F	F
T	F	F	F	T	F	F
F	T	T	T	F	T	F
F	T	F	T	F	T	F
F	F	T	T	F	T	F
F	F	F	F	T	T	T

▶ **Exercise A.1.** Give the truth tables for the following expressions:

a. $\neg(P \wedge Q)$
b. $\neg P \vee \neg Q$
c. $\neg Q \implies \neg P$
d. $\neg(P \implies Q)$
e. $P \wedge \neg Q$
f. $P \vee (Q \wedge R)$
g. $P \implies (Q \vee R)$

Most theorems are of the form $P \implies Q$, or can be rewritten in that form. P is called the *hypothesis* of the statement and Q the *conclusion*. Other ways of restating such a proposition are:

> P implies Q.
> If P, then Q.
> P only if Q.
> Whenever P is true, then Q is also true.
> Q is true if P is true.
> P is sufficient for Q.
> Q is necessary for P.

For example, the following are different ways of saying the same thing:

> Being square implies being rectangular.
> If a figure is a square, then it is a rectangle.
> A figure is a square only if it is a rectangle.
> Whenever a figure is a square, it is also a rectangle.
> A figure is a rectangle if it is a square.

Being square is sufficient to be a rectangle.
Being rectangular is necessary for a figure to be a square.

When you first meet a statement to be proved, rewrite it in the form $P \implies Q$, which will help clearly identify the hypotheses given and the conclusion to be deduced.

The statement $P \iff Q$ is shorthand for the compound statement $(P \implies Q) \wedge (Q \implies P)$. Other ways of stating this are:

P if and only if Q.
Q if and only if P.
P is necessary and sufficient for Q.
P is logically equivalent to Q.

Proofs of theorems of this form are often done in two parts: First prove the implication $P \implies Q$ and then separately prove $Q \implies P$. We occasionally make use of the symbol $\equiv$ to denote logical equivalence. We say two statements are logically equivalent if they always have the same truth values.

Next, we give a list of some of the rules of logic. These are actually theorems (also called *tautologies*) in logic, but we omit the proofs. Note that this list is hardly inclusive. We have omitted some very obvious rules (such as $P \implies P$, "If P is true, then P is true"), since you almost certainly know those already, and some other more complex ones that we won't need for this text. The common names are given for these laws. Many come in two forms: one for $\wedge$ (and) and another for $\vee$ (or).

Modus ponens: $P \wedge (P \implies Q) \implies Q$
Syllogism: $(P \implies Q) \wedge (Q \implies R) \implies (P \implies R)$
Commutativity of *and*: $P \wedge Q \equiv Q \wedge P$
Commutativity of *or*: $P \vee Q \equiv Q \vee P$
Associativity of *and*: $(P \wedge Q) \wedge R \equiv P \wedge (Q \wedge R)$
Associativity of *or*: $(P \vee Q) \vee R \equiv P \vee (Q \vee R)$
Distributive Law for *and*: $P \wedge (Q \vee R) \equiv (P \wedge Q) \vee (P \wedge R)$
Distributive Law for *or*: $P \vee (Q \wedge R) \equiv (P \vee Q) \wedge (P \vee R)$
DeMorgan's Law for *and*: $\neg(P \wedge Q) \equiv (\neg P) \vee (\neg Q)$
DeMorgan's Law for *or*: $\neg(P \vee Q) \equiv (\neg P) \wedge (\neg Q)$
Contraposition: $(P \implies Q) \equiv (\neg Q \implies \neg P)$
Negation of Implication: $\neg(P \implies Q) \equiv P \wedge \neg Q$
Law of Contradiction: $(P \implies Q) \equiv [(P \wedge \neg Q) \implies (R \wedge \neg R)]$

The last three of these laws of logic are important in structuring proofs and will be discussed in detail in the next section. To show that two logical statements are logically equivalent, we compare their truth values. For example, to prove the first of DeMorgan's Laws, we prepare the truth tables for each of the statements and see that the results have the same truth values:

P	Q	$P \wedge Q$	$\neg(P \wedge Q)$	$\neg P$	$\neg Q$	$(\neg P) \vee (\neg Q)$
T	T	T	F	F	F	F
T	F	F	T	F	T	T
F	T	F	T	T	F	T
F	F	F	T	T	T	T

▶ **Exercise A.2.** Prove that the following are equivalent:
   a. $\neg(P \vee Q) \equiv (\neg P) \wedge (\neg Q)$
   b. $(P \implies Q) \equiv (\neg Q \implies \neg P)$
   c. $\neg(P \implies Q) \equiv P \wedge \neg Q$
   d. $[(P \wedge \neg Q) \implies (R \wedge \neg R)] \equiv (P \implies Q)$
   e. $P \wedge (Q \vee R) \equiv (P \wedge Q) \vee (P \wedge R)$
   f. $P \vee (Q \wedge R) \equiv (P \vee Q) \wedge (P \vee R)$

Most theorems can be restated in the form $P \implies Q$. If we call such a statement the original proposition, then we can name the other combinations of P and Q:

> Proposition: $P \implies Q$,
> Contrapositive: $\neg Q \implies \neg P$,
> Converse: $Q \implies P$,
> Inverse: $\neg P \implies \neg Q$,
> Negation: $P \wedge \neg Q$.

Compare the truth values for these:

hypothesis P	conclusion Q	proposition $P \implies Q$	contrapositive $\neg Q \implies \neg P$	converse $Q \implies P$	inverse $\neg P \implies \neg Q$	negation $P \wedge \neg Q$
T	T	T	T	T	T	F
T	F	F	F	T	T	T
F	T	T	T	F	F	F
F	F	T	T	T	T	F

From this table (or from the laws of logic), we can determine the relationship between the truth of the original statement and its derivative statements. We see that the original proposition and its contrapositive always have the same truth values and thus are logically equivalent. Note that the inverse of the proposition is equivalent to the contrapositive of the converse. Therefore, the converse and the inverse are logically equivalent. However, the proposition and its converse may have different truth values: when the proposition is true, the converse may or may not be true. When the proposition is true, its negation must be false and vice versa.

Some theorems use *quantifiers*. These are of two forms: "for all" or "for every," abbreviated with the symbol $\forall$, and "there exists" or "there is at least one," for which we use the symbol $\exists$. For example, some theorems we will prove may state that all triangles have a certain property, and other theorems may state that one triangle exists satisfying certain precise conditions.

Statements we will most commonly use that involve these quantifiers can be rewritten in the following forms:

> $\forall x \in S, P(x)$
>> For all x in some set S, the statement P is true for x.
>
> $\exists x \in S, P(x)$
>> There exists (at least one) x in the set S so that the statement P is true for x.

The set S is the appropriate group of objects under consideration, such as the set of all triangles, or the set of all right triangles, etc. The rules of logic that govern the negation of these statements are variants of De Morgan's Laws:

$$\neg(\forall x \in S, P(x)) \equiv \exists x \in S, \neg P(x)$$
There is at least one x such that P is false for this x.
$$\neg(\exists x \in S, P(x)) \equiv \forall x \in S, \neg P(x)$$
For all x, P is false.

▶ **Exercise A.3.** State the negative of the following statements in plain English.

a. All dogs are retrievers.
b. Some cats are retrievers.
c. All retrievers are dogs.
d. Some cats and all retrievers are pink.

One further symbol we occasionally make use of is →←, used to designate a contradiction, such as $R \wedge \neg R$. We also use the symbol □ to indicate the end of a proof.

A.3 Structuring proofs

The basic logical structure of the proofs we will do in this course can be classified as one of three types. If the statement of the theorem is rewritten in the form $P \implies Q$, then a *direct proof* starts with assumption P and proceeds to conclusion Q. Thus, if the proof were written in two-column or outline format, it would look like this:

1. P 1. Hypothesis
2. ...   2. ...
...
n. Q n. Steps 1 through $(n-1)$

A direct proof has the simplest logical structure, though it may not be the simplest proof. Usually, one first tries to find a direct proof.

However, if you ever have trouble proving $P \implies Q$ directly, you can use the Law of Contraposition and try to prove $\neg Q \implies \neg P$ instead. The statement $P \implies Q$ says that if P is true, then Q must be true. Therefore if Q is false, then P cannot be true. The initial hypothesis is thus $\neg Q$ and the final conclusion is $\neg P$, and so we have we have a proof of the form:

1. $\neg Q$ 1. Hypothesis
2. ...   2. ...
...
n. $\neg P$ n. Steps 1 through $(n-1)$
$n+1$. $P \implies Q$ $n+1$. Step n, Law of Contraposition

If you suspect that $P \implies Q$ is not true, try to think of an example in which P is true and Q is false. This makes use of the Law of Negation. Such an example would be called a *counterexample* to the proposition. The existence of such a counterexample shows that the theorem cannot be true.

The Law of Contradiction says, $(P \Longrightarrow Q) \equiv [(P \wedge \neg Q) \Longrightarrow (R \wedge \neg R)]$. If you fail to find either a direct proof, a counterexample, or a proof by contraposition, this is the next thing to try. In practice, this rule of logic is used as follows: Assume that P is true and Q is false. Argue logically from these hypotheses until you get a contradiction of some sort (this is denoted above by $R \wedge \neg R$). It then follows that something is wrong: either with your hypotheses or with the logical flow of your argument. Assuming that your logic is sound, it must be that Q is true rather than false, so $P \Longrightarrow Q$. Such a proof is called a *proof by contradiction*. Such proofs have the advantage of allowing you two hypotheses instead of only one: the original hypothesis P and the *contradiction hypothesis* $\neg Q$. This frequently makes proving the theorem easier. Such proofs also have the advantage of ending with *any* contradiction, rather than a fixed objective. In other words, after starting the proof, you just need to play around until you find a contradiction. Once such a contradiction is found, you can conclude, by the Law of Contradiction, that $\neg Q$ must be false, and therefore Q must be true. A proof by contradiction is also called a proof by *reductio ad absurdum*. In two-column format, such a proof will look like the following:

1. P 1. Hypothesis
2. $\neg Q$ 2. Contradiction hypothesis
...
n. $\rightarrow \leftarrow$ n. Step $(n-1)$ and some previous step or theorem
$n+1$. Q $n+1$. Steps 2, n and Law of Contradiction
$n+2$. $P \Longrightarrow Q$ $n+2$. Law of Contradiction

The contradiction is indicated in Step n. The statement of Step $(n-1)$ contradicts some previous step of the proof or some previously proven theorem.

In the following examples we prove some elementary facts about even and odd numbers. We will assume the usual laws of arithmetic and also the fact that any *even integer p* can be written as $p = 2k$ for some integer k and any *odd integer q* can be written in the form $q = 2n + 1$ for some integer n. Every integer is either even or odd but cannot be both.

Example 1: Direct Proof: If p is even, then p^2 is even.

Proof:

1. p is even   1. Hypothesis
2. $p = 2k$ 2. Definition of even integer
3. $p^2 = (2k)^2 = 4k^2 = 2(2k^2)$ 3. Arithmetic
4. p^2 is even   4. Step 3, Definition of even integer   □

Example 2: Proof by Contraposition: If p is an integer so that p^2 is even, then p is even.

Proof by contraposition: If p is an integer and p is not even, then p^2 is not even.

1. p is an integer   1. Hypothesis
2. p is not even   2. Contraposition hypothesis
3. p is odd   3. Every integer is either even or odd but not both.
4. $p = 2k + 1$ 4. Definition of odd integer
5. $p^2 = (2k+1)^2 = 4k^2 + 4k + 1 = 2(2k^2 + 2k) + 1$ 5. Arithmetic

6. p^2 is odd	6. Step 5, Definition of odd integer
7. p^2 is not even	7. Step 6, Every integer is either even or odd but not both.
8. If p is an integer and p^2 is even, then p is even.	8. Step 1–7, Law of Contraposition □

Example 3: Proof by Contradiction: If p is an integer and p^2 is odd, then p is odd.

Proof by contradiction:

1. p is an integer	1. Hypothesis
2. p^2 is odd	2. Hypothesis
3. p is not odd	3. Contradiction hypothesis
4. p is even	4. Every integer is either even or odd but not both.
5. $p = 2k$	5. Definition of even integer
6. $p^2 = (2k)^2 = 4k^2 = 2(2k^2)$	6. Arithmetic
7. p^2 is even	7. Step 6, Definition of even integer
8. $\rightarrow \leftarrow$	8. Steps 2 and 7
9. If p is an integer and p^2 is odd, then p is odd.	9. Step 8, Law of Contradiction □

For a theorem that involves quantifiers, we usually use the following procedures. To prove a theorem that is written in the form $\exists x \in S, P(x)$, one must first find an appropriate x, and then show that the statement P is true for your chosen x. To prove a theorem of the form $\forall x \in S, P(x)$, choose an arbitrary x in the set S. Then show that the statement P is true for this x. It is important that no restrictions or conditions be placed on x, so don't assume that it is special in any way. It then follows that if a completely arbitrary, randomly chosen, member of S has the desired property, then all members of S must have this property.

▶ **Exercise A.4.** Prove that the sum of two even integers is even.

▶ **Exercise A.5.** Prove that the sum of two odd integers is even.

▶ **Exercise A.6.** Prove that the sum of an even integer and an odd integer is odd.

▶ **Exercise A.7.** Prove that the product of two odd integers is odd.

Lastly, we must make some mention of *mathematical induction*. There are only a few theorems in this geometry text that will use this technique, but it is in general very important and powerful. It is used to prove some statements of the form $\forall n \in \mathbb{N}, P(n)$, where $\mathbb{N}$ denotes the set of natural (counting) numbers, $\mathbb{N} = \{1, 2, 3, 4, \ldots\}$. In other words, for any natural number n, the statement P is true for that value of n. The procedure goes as follows:

Basis Step: Prove $P(1)$: i.e., show that P is true for $n = 1$.
Induction Step: Prove that $P(n) \implies P(n+1)$: i.e., show that if P is true for n, then P is true for $n+1$.
Conclusion: P is true for all $n \in \mathbb{N}$.

This time, we choose to write our proof in paragraph form rather than the two-column format used in the previous examples.

Example 4: Proof by Induction: If p is odd and n is any positive integer, then p^n is odd.

Proof by induction: Basis step: For $n = 1$, note that $p^n = p^1 = p$ is odd by the hypothesis.

Induction step: We assume that p^n is odd. Therefore, $p^n = 2m + 1$ for some integer m. We also know that p itself is odd, so $p = 2k + 1$ for some integer k. Then we compute

$$p^{n+1} = p \cdot p^n = (2k+1)(2m+1) = 4km + 2k + 2m + 1 = 2(2km + k + m) + 1.$$

Therefore, p^{n+1} is odd. We conclude that p^n is odd for every positive integer n. □

A.4 Inventing proofs

Coming up with a proof is the true art of mathematics and involves both logic and imagination. Sometimes the whole proof will spring spontaneously into your mind, especially after you've practiced enough to get good at writing proofs, but other times you won't even know how to start. In this section, we offer some very general suggestions on how to go about inventing a proof for a theorem. J. V. Armitage says,

> Mathematics is an attractive subject because it is difficult. Obstinacy and the ability to concentrate are a necessary part of the equipment of a mathematician and, in other contexts, they are desirable qualities for life in general. Geometry is a good discipline in which to encounter difficulties for the first time, because its problems lend themselves to pictorial exploration and investigation and that combination of intuition and logic which is the essence of mathematics.[1]

Let us assume that we are working to prove a statement of the form $P \implies Q$. You should always start with a rough draft. In the next section we will discuss turning this draft into a neat written proof suitable to be handed in and graded. For now we only consider how to come up with the idea or ideas behind the proof and its logical structure.

- To start, write down precisely the hypothesis P of the theorem and what you know about this. Then write 'we want to show' and state the conclusion Q of the theorem. Try to find a direct proof first, a straight line of reasoning from the hypothesis P to the conclusion Q. Try to think of ideas or previously proven theorems that connect the ideas of P with the ideas of Q. Then try to arrange these to mark out a path of reasoning starting at P and ending at Q.
- Draw a picture and mark it up. Geometers use a system of slashes to mark equal line segments and angles, so that two equal line segments or angles will have the same number of slashes. If you are using geometric software, construct a drawing according to the hypothesis and use the drag capability to study the objects involved in the conclusion.
- Above all, convince yourself that the theorem is true. If you don't believe in a theorem, there is very little hope of proving it. If you come to believe that the theorem is not true, try to construct a counterexample: a special case where P is true and Q is false. This disproves the theorem and you're done.
- Try rewriting the definitions of the objects. For example, if you know a triangle is isosceles, you know that it has two equal sides and perhaps thinking about these equal line segments may help you derive the desired conclusion. Write down everything you think might help, though some of the things you may write down will not be used. You can always cross them out later.
- Sometimes, it is appropriate to divide a theorem up into separate cases. For example, if you are trying to show that all angles have some property, you might want to consider acute, right, and obtuse

---
[1] From "The Place of Geometry in a Mathematical Education" in *The Changing Shape of Geometry*.

angles separately. You will then have to do three separate proofs, but each case will have an additional hypothesis, called the *case hypothesis*. You are substituting three easier proofs for one general proof.
- If you cannot find a path from P to Q, you can try reverse reasoning. Start with the conclusion Q and see if you can trace back a connection to the hypothesis P. This is like working out a path through a maze from the objective back to the entrance, and sometimes this is easier. Once you have this path marked out, check and make sure you can convert it to a straight path of implications from P to Q. The final proof will have to be written in the forward direction.
- If both direct and reverse reasoning fail to help you figure out a direct proof, consider either a proof by contraposition or a proof by contradiction. Both of these begin with hypothesis $\neg Q$. See if you can think of a path connecting $\neg Q$ to $\neg P$. If so, then you have a proof by contraposition.
- If you can't find a proof by contraposition starting with $\neg Q$ as a hypothesis, then add the hypothesis P and try to find a path from $P \wedge \neg Q$ into the wilderness and look for a contradiction. Remember that it doesn't matter what form the contradiction takes, just that you find some statement that contradicts a previous step of the proof or some known fact.

Sometimes, you will have to try a variety of techniques before you come up with a proof. And then you may end up with a technique you tried earlier and rejected. With practice, you will gain confidence and ability. The suggestions above are very general and elementary. In the course of this text you will meet some theorems whose proofs are somewhat more sophisticated, such as a proof by cases in which each case is proven by contradiction. But the basic principles of proof-writing remain the same.

A.5 Writing proofs

Once you have the road map for the proof developed as in the previous section, you must write it up neatly and formally to hand into your teacher or to preserve for posterity. Usually your rough draft is somewhat sketchy and the discipline of formalizing it also acts as a check of your work and reasoning. Go back over your draft and cross out any dead-ends that you may have explored and rejected since they didn't go anywhere. Think about whether your proof can be simplified. If you broke the proof into cases, consider if they can be combined into a single proof.

Aside from the logical structure of the proof, there are two main methods for the presentation: paragraph style and two-column style. Each has advantages and disadvantages. The two-column style reinforces the idea that all steps must be justified (as they must in the paragraph style, but then some of the explanations are omitted). However, these proofs quickly become too lengthy and repetitive. Justifying each step in the two-column presentation is good training for students who are still gaining comfort with proofs. For each step, give one of the following reasons:

- Hypothesis
- Contradiction hypothesis (for a proof by contradiction only)
- Definition # (give term defined or definition number)
- Postulate # (give postulate number)
- Proposition # (give the proposition number, but this must precede current proposition)
- Step (or Steps) # (preceding step(s) of current proof)
- Logic (cite rule of logic)

However, two-column proofs for long theorems are very unwieldy and this often obscures the underlying structure of the proof.

Paragraph style is generally considered more elegant, but requires judgment on which steps to give justification for and how much detail to put in. This style allows for commentary on the structure of the proof, and also one can break up a long proof into paragraphs, each devoted to some significant portion. In writing up any proof, write clearly and use complete sentences. Try to organize longer proofs so that they are easy to follow by breaking them up into paragraphs and putting in comments on exactly what you are trying to prove in each paragraph. Clearly specify when you are using contraposition or proof by contradiction.

Finally, end your proof by clearly stating what you have proved. It also helps the reader to put some sort of marker to show the end of the proof. Classical translations of Euclid end most of his proofs with the Latin phrase "quod erat demonstrandum" or Q.E.D., which means "that which was to be demonstrated." Some of his propositions, specifically those in which a geometric object is constructed, end with the phrase "quod erat faciendum" or Q.E.F. instead, which means "that which was to be made." Modern mathematicians use a slug □ for the same purpose, to indicate that the proof ends here.

Once you are finished with a proof, take a moment and consider it. Try to think if the result of the theorem can be generalized or expanded to a larger class of objects. Think if your theorem opens up any new avenues for investigation. Your theorem may suggest a relationship between two unexplored objects or ideas. Take some time to think about these ideas.

A.6   Geometric diagrams

The mathematician Henri Poincaré once said that geometry is the art of good reasoning from bad diagrams. In Euclid, as in most elementary geometry books, the drawings are essential companions to the proofs. Often this is too true, since sometimes the drawings can lead one to make assumptions about the situation which are not explicitly given by the hypotheses and can thus lead to errors. David Hilbert tried to eliminate this hazard by giving a complete axiom set for geometry, saying, "Instead of points, straight lines, and planes, it would be perfectly possible to use the words tables, chairs, and tankards."[2] In other words, he meant that theorems in geometry should be true regardless of what the points and lines look like, as long as one obeys the axioms assumed. In his *Foundations of Geometry*, he carried out a program for geometry in which many of the hidden assumptions that lurk in Euclid's drawings were made explicit. This approach is discussed in detail in Chapter 2, especially in Section 2.6.

However, drawings have a role to play in any introductory geometry text, helping one to visualize the relationships between geometric objects. As you develop a conjecture or proof, simultaneously create the corresponding diagram, either by hand, with straightedge and compass, or with geometric software such as *The Geometer's Sketchpad*, *Cabri*, *Cinderella*, *GeoGebra* or *Geometry Playground*. It is standard practice to mark equal line segments with slashes and equal angles with some similar markings, as shown in the isosceles triangle below:

---

[2]Quoted in Gray, *Worlds out of Nothing*, p. 254.

In computer-generated drawings, line segments can be color-coded to indicate equality.

One error that drawings can lead you into is to assume more than is given. For example, consider the triangle △ABC shown below, in which $\overrightarrow{BD}$ bisects ∠ABC:

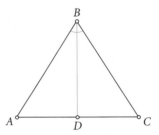

In the illustration, it looks like BD is perpendicular to AC and also that D is the midpoint of $\overline{AC}$. But this cannot be declared as true on the strength of the diagram alone. If we consider other configurations for the triangle, note that the angle bisector $\overrightarrow{BD}$ no longer looks like it bisects $\overline{AC}$ nor does it seem perpendicular:

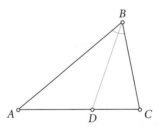

However, note that in an isosceles triangle, and only in an isosceles triangle, the angle bisector $\overrightarrow{BD}$ can be proven to be perpendicular to AC and to bisect segment $\overline{AC}$. This must be proved. (It is easy enough to prove this, since △BAD ≅ △BCD by SAS, and both results follow from corresponding parts of congruent triangles.) In general, one can construct a line either to bisect ∠ABC, to be the perpendicular bisector of $\overline{AC}$, or to connect vertex B with midpoint D of $\overline{AC}$. One line can only be assumed to do one thing. Any other properties must be proved.

In this, as in most textbooks, only the final diagram is given though it is more natural to consider such diagrams as works in progress. Often only one case is pictured, though you should consider all possible cases: acute, right, and obtuse angles; equilateral, isosceles, and scalene triangles; etc. Sometimes, more than one diagram is necessary. For example, in a proof by cases, you will need to create a separate diagram for each case. For more complicated theorems, again several drawings may be helpful, since a single diagram may get too confusing. Instead, separate drawings may show particular features under study.

In a proof by contradiction, one assumes a hypothesis that will later be shown to be false. Therefore, the illustration ought to look somewhat wrong. In fact, the diagram can help you figure out precisely what property leads to the desired contradiction.

Any diagram must be viewed with some suspicion; it is entirely too easy to let a drawing lead you into unjustified assumptions. For example, consider the "proof" below, which assumes some familiarity with geometric facts and notation.

Theorem: Every obtuse angle is equal to a right angle.

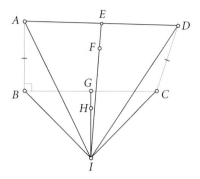

Proof: Construct a right angle $\angle ABC$ and an obtuse angle $\angle BCD$ so that $\overline{AB} = \overline{CD}$. Draw the line segment $\overline{AD}$. Construct the perpendicular bisectors EF to $\overline{AD}$ and GH to $\overline{BC}$. Since AD is not parallel to BC, it follows that EF and GH are not parallel and so EF intersects GH at some point called I. Draw $\overline{IA}$, $\overline{IB}$, $\overline{IC}$, and $\overline{ID}$. Then $\triangle BGI \cong \triangle CGI$ by SAS (Side-Angle-Side congruence) and similarly, $\triangle AEI \cong \triangle DEI$. Therefore, $\overline{BI} = \overline{CI}$, $\overline{AI} = \overline{DI}$, and $\angle GBI = \angle GCI$. Then $\triangle ABI \cong \triangle DCI$ by SSS (Side-Side-Side congruence), so $\angle ABI = \angle DCI$. Therefore $90° = \angle ABC = \angle ABI - \angle GBI = \angle DCI - \angle GCI = \angle DCB$. Thus the angle $\angle DCB$, which was constructed to be obtuse, is equal to a right angle. □

▶ **Exercise A.8.** Find the error in the proof above. A large and careful drawing may help, perhaps drawn with the aid of geometric software.

A.7 Using geometric software

Geometric software is quite useful in exploring a situation. Construct a drawing according to the hypotheses of the theorem and use the measurement capabilities and the dynamic ability to drag points around to explore the situation. This will naturally lead one to form conjectures, though a formal proof will still be necessary.

It is very easy to construct something that looks like a square using only the **Segment** tool of *The Geometer's Sketchpad*®, or other software such as *Cabri*, *Cinderella*, *GeoGebra*, or *Geometry Playground*. However, this kind of drawing will not pass what the software creators call "the drag test": If you click and hold on one of the points or segments of your sketch and then move it, your figure may no longer be a square. Using more involved techniques and a knowledge of geometry, you can create a square that will remain a square no matter how you move the points or change the segments.

A.8 Van Hiele levels of geometric thought

In the 1950's two Dutch high school teachers began to speculate on the causes of the disappointing performance of their students. Pierre van Hiele and Dina van Hiele-Geldof together formulated a theory for these results and some suggestions for overcoming their students' difficulties. In essence, Pierre van Hiele suggested that all students must progress through certain levels or stages of learning in geometry. Furthermore, if the teacher assumes knowledge or vocabulary at a higher level than the student currently possesses, then very little knowledge is imparted. Dina van Hiele-Geldof focused her studies on methods to help students progress from their current level to the next.

Briefly, the van Hiele levels of geometric thought can be summarized:

- **Level 0: Recognition or Visualization:** At this level students recognize figures as global objects, such as triangles or rectangles, but do not explicitly identify these by their properties. For example, a student at Level 0 may not recognize a long skinny triangle as being a triangle, or might not recognize a square if it is rotated out of the more normal configuration. Thus, if one rotates a square by 45°, these students would call this a diamond and would not perceive that it is also a square. Level 0 thought is typical of young children and early elementary students.

- **Level 1: Analysis:** Properties of geometric figures are identified, but relationships between these properties are not perceived. At Level 1, students will categorize both of the figures above correctly. At this level, while a student might notice that a rectangle has four right angles and that the opposite sides are equal, he or she would not assume any implication from these facts. Students will recognize and describe properties of figures, such that parallelograms have parallel opposite sides but will not make a connection between this fact that the fact that opposite angles are equal. Students will not yet see relationships such as all rectangles are parallelograms or all squares are rectangles. After discovering a property of some class of geometric figures by experimentation, such as the angle sum of a triangle, students at this level do not understand the necessity for proof of geometric properties. Students usually attain this level of thinking during the later elementary or early middle school grades.
- **Level 2: Informal deduction:** Students at this level will see connections between properties, such as that all squares are rectangles or the sum of the angles in any triangle is 180°. They can identify a figure by its characteristics and are comfortable with arguments such as all rectangles are parallelograms. Thus, there is some idea of implication, but most students cannot organize a sequence of statements logically. Students will be able to follow an informal proof but are unable to understand how to develop a different proof. At this level, for example, if one gives two angles in a triangle, the student should be able to quickly calculate the measure of the third angle. Also, these students should be able to deduce and give an informal explanation of the sum of the angles for a quadrilateral by dividing it up into two triangles. However, at this level, students may not distinguish between a statement and its converse and will not understand the necessity for an axiomatic system. Level 2 is usually reached during middle school or early high school.
- **Level 3: Deduction:** At this level, students can understand and construct proofs and understand the role of definitions and axioms. These students will be aware of the difference between properties given in the definition of an object and those later proven. For example, the definition of a square is a quadrilateral with four right angles and four equal sides. From this, one can prove that the diagonals are equal and perpendicular. Students at this level should be capable of making statements such as "all squares

are rectangles, but not all rectangles are squares" and provide appropriate examples and arguments. Students at this level will understand the necessity for axioms, primitive terms, and formal definitions. They will be able to state the converse and contrapositive of a statement. They can compare different proofs of the same theorem and develop chains of related theorems. Typically, students attain Level 3 during the standard high school geometry course.
- **Level 4: Rigor:** Students at this level can understand the logical relationship between different axiom systems. They can understand the concepts of consistency and completeness of an axiom system. At this level, students should be comfortable with different geometries. Students should be able to generalize theorems and their proofs to broader classes of objects.

Typically, college students studying geometry are at Level 3, though some may lag behind at Level 2 due to inadequate high school instruction in geometry. Our goal in this course is to help them attain Level 4.

It is important to remember that one cannot skip a level and that if instruction is directed at a higher level than the student is in, then very little learning will take place. Dina van Hiele-Geldof's work on transitions between the levels of thought are summarized below. Using these as guides, activities can be devised to help a student progress from their current level to the next.

- **Phase 1: Inquiry:** The student acquires familiarity with the objects of study, considering examples and counterexamples. The teacher engages the student in a dialogue about the objects of study, introducing vocabulary, soliciting observations, and clarifying students' conceptions and interpretations of the relevant definition.
- **Phase 2: Directed orientation:** The students perform tasks directed at discovering key properties. Carefully sequenced explorations familiarize students with the objects of study and the definitions. Most of these activities involve only one step or one implication.
- **Phase 3: Explication:** Students, with only minimal input from the teacher, begin to refine their use of the vocabulary and to develop conjectures on the relationships between properties of the objects. Students gain comfort with the technical vocabulary and learn to express geometric properties with precision.
- **Phase 4: Free orientation:** The teacher now presents multi-step tasks and tasks that can be done in several ways. The students gain familiarity and confidence. Students begin to draw implications on the relationships between geometric figures.
- **Phase 5: Integration:** Students have an overview of the objects, their properties, and the interrelationships among these. They can summarize their findings clearly, form conjectures, and begin to prove their theories.

Appendix B. Postulates and theorems

The primitive terms are *point, line, plane, on, between* (and *separate*), *congruent* (or *equal*). Postulates and theorems are denoted by [E] if true in euclidean geometry, [N] if true in neutral geometry (and thus true in both euclidean and hyperbolic geometry), [H] if true in hyperbolic geometry, or [S] if true in single elliptic geometry. If there is no such notation, then that postulate or theorem is true in all three of our primary geometries: euclidean, hyperbolic, and elliptic.

♦ B.1 Postulates

Postulate 1. Given two distinct points, there exists exactly one line through them.

Postulate 2. [N] One can extend a given segment from either end to form a ray, or from both ends to form a line.

Elliptic Postulate 2. [S] One can extend a given segment to form a line.

Postulate 3. Given a point and a length, there exists exactly one circle with the given point as center and the given length as radius.

Elliptic Postulate 3. [S] Given a point and a length less than $\frac{\pi r}{2}$, there exists exactly one circle with the given point as center and radius equal to the given length.

Postulate 4. All right angles are congruent.

Euclidean Postulate 5. [Parallel Postulate 5] [E] If two lines are cut by a transversal so that the sum of the interior angles on one side of the transversal is less than 180°, then the two lines meet on that side of the transversal.

Hyperbolic Postulate 5. [H] Given a line AB and a point C not on the line, there are at least two lines through C that are parallel to AB.

Elliptic Postulate 5. [S] Any two distinct lines in a plane meet in exactly one point.

Postulate 6. [Incidence]

1. There exists at least one plane.
2. Every plane contains at least three noncollinear points.
3. Every line contains at least two points.
4. For any three noncollinear points, there is exactly one plane containing these points.
5. If a plane contains two distinct points of a line, then every point on the line lies on the plane.
6. If two planes P_1 and P_2 have one point in common, then they have at least one other point in common.
7. There are at least four points that do not all lie on a single plane.

Postulate 7. [N] [Betweenness]

1. If B is between A and C, then A, B, and C lie on a line and B is also between C and A. I.e., if $A - B - C$, then $C - B - A$.
2. Given any two distinct points A and C, there exists at least one other point B on the line that lies between them.
3. Given any two distinct points A and B, there exists at least one other point C so that B lies between A and C.
4. Given three distinct points on a line, only one of them is between the other two.

Elliptic Postulate 7. [S] [Separation]

1. If A and B separate C and D, then A, B, C, and D are distinct points and lie on a line.
2. If A and B separate C and D, then C and D separate A and B, and B and A separate C and D.
3. If A and B separate C and D, then A and C do not separate B and D.
4. Given any four distinct points A, B, C and D on a line, then A and B separate C and D, or A and C separate B and D, or A and D separate B and C.
5. Given three distinct points A, B, and C on a line, then there is a point D so that A and B separate C and D.
6. Given any five distinct points A, B, C, D, and E on a line, if A and B separate D and E, then either A and B separate C and D or A and B separate C and E.

Postulate 8. [N] [Plane Separation Property] Given a plane and a line ℓ on that plane, ℓ divides the points on the plane and not on ℓ into two disjoint convex sets, S_1 and S_2, called the sides of the line, so that whenever point A is in S_1 and point B is in S_2, the line segment $\overline{AB}$ must intersect ℓ.

Elliptic Postulate 8. [S] [Pasch's Postulate] If A, B, and C are noncollinear points and a line ℓ intersects $\widehat{AB}$ at a single point D between A and B, then ℓ must intersect exactly one of the following: $\widehat{BC}$ between B and C, $\widehat{AC}$ between A and C, or the vertex C.

Postulate 9. [Congruence]

1. Congruence or equality in length is an additive equivalence relation on line segments. (**[S]**: *on arcs.*)
2. Congruence or equality in measure is an additive equivalence relation on angles.
3. Equality in area is an additive equivalence relation on polygons.
4. Congruent triangles have equal areas.
5. Equality in volume is an additive equivalence relation for polyhedra.
6. Congruent polyhedra have equal volumes.

Postulate 10. [N] [Archimedes' Axiom] If a and b are positive real numbers with $a < b$, then there exists $n \in \mathbb{N}$ so that $na > b$. In particular,

1. Given line segments $\overline{AB}$ and $\overline{CD}$, there exists $n \in \mathbb{N}$ and a point X on $\overrightarrow{CD}$ so that $\overline{CX} = n \cdot \overline{AB} > \overline{CD}$ and thus $C - D - X$.
2. Given angles $\angle ABC$ and $\angle DEF$, there exists $n \in \mathbb{N}$ and a ray $\overrightarrow{EX}$ so that $\angle XEF = n \cdot \angle ABC$ and $\angle XEF > \angle DEF$.

Elliptic Postulate 10. [S] [Archimedes' Axiom] If a and b are positive real numbers with $a < b$, then there exists $n \in \mathbb{N}$ so that $na > b$. In particular,

1. Given arcs $\widehat{AB}$ and $\widehat{CD}$ with $\widehat{CD}$ longer than $\widehat{AB}$, there exists $n \in \mathbb{N}$ and a point X on $\overrightarrow{CD}$ so that $\widehat{CDX} = n \cdot \overline{AB}$ or $\widehat{CDX}$ exceeds πr.
2. Given angles $\angle ABC$ and $\angle DEF$, there exists $n \in \mathbb{N}$ and a ray $\overrightarrow{EX}$ so that $\angle XEF = n \cdot \angle ABC$ and $\angle XEF > \angle DEF$.

Postulate 11. [N] [Circular Continuity Principle]

1. A line segment with one endpoint outside a given circle and the other endpoint inside the circle will intersect the circle exactly once.
2. A circle passing through a point inside a given circle and a point outside that circle will intersect the given circle twice.

Elliptic Postulate 11. [S] [Circular Continuity Principle]

1. An arc with one endpoint outside a given circle and the other endpoint inside the circle will intersect the circle exactly once.
2. A circle passing through a point inside a given circle and a point outside that circle will intersect the given circle twice.

Postulate 12. [SAS] If two triangles have two sides equal to two sides respectively, and the angles contained by the equal sides are also equal, then the triangles will be congruent.

Postulate 13. [E] [Area] The area of a rectangle is the product of the lengths of its base and height.

Postulate 14. [E] [Volume] The volume of a parallelepiped is the product of the area of its base and its height.

Postulate 15. [E] [Cavalieri's Principle] Given two solids P and Q and a fixed plane, if every plane parallel to the given plane intersects the solids to form regions with equal areas, then the volume of P is equal to the volume of Q.

◆ B.2 Book I of Euclid's *The Elements*

◇ B.2.1 Propositions that do not use Postulate 5

Theorem 2.7. [N] Given a line ℓ, let A be a point on line ℓ and B another point that does not lie on ℓ.

1. All of the points on $\overline{AB}$, except A itself, lie on the same side of ℓ as B.
2. All of the points on the ray $\overrightarrow{AB}$, except A itself, lie on the same side of ℓ as B.

Theorem 2.10. [N] [Line Separation] A point P on a line ℓ divides the points on the line distinct from P into two nonempty disjoint sets, called the sides of the point P on line ℓ.

Theorem 2.11. [N] [Pasch's Theorem] If A, B, and C are noncollinear points and a line ℓ intersects $\overline{AB}$ at a single point D between A and B, then ℓ must intersect exactly one of the following: $\overline{AC}$ between A and C, $\overline{BC}$ between B and C, or the vertex C.

Theorem 2.12. [N] [Cross-bar Theorem] If $\overrightarrow{AD}$ is between $\overrightarrow{AB}$ and $\overrightarrow{AC}$, then $\overrightarrow{AD}$ intersects the segment $\overline{BC}$ at a point between B and C.

Corollary 2.13. [N] If a line segment connects a point outside $\triangle ABC$ to a point inside the triangle, then it must intersect the triangle.

Theorem 2.16. [N] Given a circle with center O, any ray $\overrightarrow{OA}$ emanating from O must intersect the circle.

Theorem 2.25. [N] Given a circle and a point A inside the circle, any ray $\overrightarrow{AB}$ emanating from A must intersect the circle.

Lemma 2.26. The following are equivalent (i.e., either all true or all false):

   a. $\angle 3 = \angle 6$
   b. $\angle 4 = \angle 5$
   c. $\angle 1 = \angle 5$
   d. $\angle 2 = \angle 6$
   e. $\angle 3 = \angle 7$
   f. $\angle 4 = \angle 8$
   g. $\angle 3 + \angle 5 = 180°$
   h. $\angle 4 + \angle 6 = 180°$

Proposition I.1. On a given line segment one can construct an equilateral triangle.

Proposition I.3. Given two unequal line segments, one can cut off from the longer given segment from either endpoint a line segment equal in length to the shorter given segment.

Proposition I.5. In isosceles triangles, the angles at the base are equal to one another, and, if the equal sides are extended, the angles under the base will be equal to one another.

Proposition I.6. If a triangle has two equal angles, then the sides which subtend the equal angles will be equal in length.

Proposition I.7. Given $\triangle ABC$ with apex C, we cannot construct another $\triangle ABD$ with D lying on the same side of AB as C and so that $\overline{AD} = \overline{AC}$ and $\overline{BD} = \overline{BC}$.

Proposition I.8. [SSS] If two triangles have three pairs of equal sides, then the triangles are congruent.

Proposition I.9. Given an angle, one can construct the angle bisector.

Proposition I.10. Given a line segment, one can construct the midpoint.

Proposition I.11. Given a line and a point on the line, one can construct a line through the given point and perpendicular to the given line.

Proposition I.12. Given a line and a point not on the line, one can construct a line through the given point and perpendicular to the given line.

Proposition I.13. The sum of the angle measures of two supplementary angles is 180°.

Proposition I.14. If the angle measures of two adjacent angles sum to 180°, then the sides of these angles that are not shared form a straight line.

Proposition I.15. [Vertical Angle Theorem] If two lines intersect, then the vertical angles are equal.

Proposition I.16. [N] [Weak Exterior Angle Theorem] In any triangle, the exterior angle is greater than either of the nonadjacent interior angles.

Proposition I.17. [N] In any triangle, the sum of the angle measures of any two angles is less than 180°.

Proposition I.18. [N] In any triangle, the greater side subtends the greater angle.

Proposition I.19. [N] In any triangle, the greater angle is subtended by the greater side.

Proposition I.20. [Triangle Inequality] In any triangle, the sum of the lengths of any two sides is greater than the length of remaining side.

Proposition I.21. Given $\triangle ABC$, let D be an arbitrary point inside this triangle. Then $\overline{DB} + \overline{DC} < \overline{AB} + \overline{AC}$ and $\angle BDC > \angle BAC$.

Proposition I.22. Given three lengths, it is possible to construct a triangle with sides of these lengths if the sum of any two of the given lengths is greater than the third length.

Proposition I.23. Given an angle and a line segment, one can construct an angle equal to a given angle with the given line segment as one of the sides and one of the endpoints of the segment as the vertex of the angle.

Proposition I.24. [Hinge Theorem] Given $\triangle ABC$ and $\triangle DEF$ with $\overline{AB} = \overline{DE}$ and $\overline{AC} = \overline{DF}$. If $\angle BAC > \angle EDF$, then $\overline{BC} > \overline{EF}$.

Proposition I.25. Given $\triangle ABC$ and $\triangle DEF$ with $\overline{AB} = \overline{DE}$ and $\overline{AC} = \overline{DF}$. If $\overline{BC} > \overline{EF}$, then $\angle BAC > \angle EDF$.

Proposition I.26. [N] [ASA and AAS] If two triangles have two angles of the first triangle equal to two angles of the second triangle respectively, and one side of the first triangle equal to one side of the second, either the side adjoining the equal angles, or the side subtending one of the equal angles, then the triangles are congruent.

Proposition I.27. [N] If two lines are cut by a transversal so that the alternate interior angles are equal, then the given lines are parallel.

Proposition I.28. [N] If two lines are cut by a transversal so that an exterior angle is equal to the opposite interior angle on the same side of the transversal, or so that the sum of the interior angles on one side of the transversal is 180°, then the given lines are parallel.

Proposition I.31. [N] Given a line and a point not on that line, one can construct a line through the given point and parallel to the given line.

◊ **B.2.2 Propositions that rely on Postulate 5**

Proposition I.29. [E] If two parallel lines are cut by a transversal, then the alternate interior angles are equal, and consequently, each exterior angle is equal to the opposite interior angle and the sum of the interior angles on the same side of the transversal is 180°.

Proposition I.30. [E] Lines parallel to the same line are also parallel to each other.

Proposition I.32. [E] [Exterior Angle Theorem] In any triangle, the exterior angle is equal to the sum of the two opposite interior angles. Furthermore, the sum of the three interior angles of the triangle is 180°.

Proposition I.33. [E] The figure formed by two parallel line segments of equal lengths, joined by line segments connecting the corresponding endpoints at each side, is a parallelogram. Furthermore, the joining line segments are themselves equal in length.

Proposition I.34. [E] In a parallelogram, the opposite sides and angles are equal to one another and the diagonal bisects the area.

Proposition I.35. [E] Parallelograms with the same base and in the same parallel are equal in area.

Proposition I.36. [E] Parallelograms with equal bases and in the same parallel are equal in area.

Proposition I.37. [E Triangles with the same base and in the same parallel are equal in area.

Proposition I.38. [E] Triangles with equal bases and in the same parallel are equal in area.

Proposition I.41. [E] If a parallelogram has the same base as a given triangle and lies in the same parallel, then the area of the parallelogram is double the area of the triangle.

Proposition I.46. [E] Given a line segment, one can construct a square with the given line segment as one of the sides.

Proposition I.47. [E] [Pythagorean Theorem] In a right triangle, the square on the side subtending the right angle is equal to the sum of the squares on the sides containing the right angle.

Proposition I.48. [E] If in a triangle the square on one of the sides is equal to the sum of squares on the remaining two sides, then the angle contained by the two sides of the triangle is right.

◆ B.3   More euclidean geometry

Theorem 3.1. [E] [Thales' Theorem] An angle inscribed in a semicircle is a right angle.

Theorem 3.2. [E] If triangle $\triangle ABC$ is a right triangle with $\angle ACB = 90°$, then there is a circle that passes through C and with diameter $\overline{AB}$.

Theorem 3.3. [E] Given a circle with center O and an arc $\overset{\frown}{BC}$ and a point A on the circle but not on the arc, the measure of the central angle is twice the measure of the inscribed angle.

Corollary 3.4. [E] The angle formed by the endpoints of a fixed arc or chord of a circle and an arbitrary third point on the circle exterior to the chord is always the same.

Theorem 3.7. [E] Given a circle and a point outside the circle, one can construct a line through the point tangent to the circle.

Theorem 3.9. [E] From a point outside a given circle, draw a line tangent to the circle and another line through the point intersecting the circle at two points. Then the square of the distance from the given point to the point of tangency is equal to the product of the distances from the point to the endpoints of the chord formed by the second line.

Theorem 3.10. [E] From a point outside a given circle, draw a line intersecting the circle at two points and another line touching the circle. If the square of distance from the given point to the point of intersection of the second line with the circle is equal to the product of the distances from the given point to the endpoints of the chord formed by the first line, then the second line is tangent to the circle.

Theorem 3.13. [E] If a triangle is cut by a line parallel to one of its sides, then this line cuts the sides of the triangle proportionally.

Corollary 3.14. [E] If a triangle is cut by a line parallel to one of its sides, then the triangle formed is similar to the original triangle.

Theorem 3.15. [E] If a triangle is cut by a line so that this line cuts the sides of the triangle proportionally, then the line is parallel to the base.

Theorem 3.16. [E] [AAA similarity] If two triangles have equal angles, then they are similar.

Corollary 3.17. [E] [AA similarity] If two triangles have two equal angles, then they are similar.

Theorem 3.18. [E] [SSS similarity] Given two triangles such that the sides are proportional, the triangles are similar.

Theorem 3.19. [E] [SAS similarity] Given two triangles such that an angle in the first triangle is equal to an angle of the second triangle and the corresponding sides enclosing these angles are proportional, then the triangles are similar.

Theorem 3.20. [N] [Hypotenuse-Side] If two right triangles have their hypotenuses equal and one other pair of sides equal, then the triangles are congruent.

Theorem 3.21. [N] [SSA+] If two triangles have two pairs of equal sides and if the angles subtended by one of these pairs are equal and the angles subtended by the other pair of equal sides are known to be both acute or both obtuse, then the triangles are congruent.

Lemma 3.22. [E] Given three distinct parallel lines cut by two transversals so that the parallel lines cut the first transversal into equal segments, they also cut the second transversal into equal segments.

Theorem 3.23. [E] The medians of a triangle intersect at a single point, called the centroid, and cut one another so that the distance from a vertex to the centroid is twice the distance from the centroid to the midpoint of the side opposite that vertex.

Theorem 3.24. [N] The angle bisectors of a triangle meet at a single point, the incenter of the triangle.

Corollary 3.25. [N] A circle can inscribed in any triangle.

Theorem 3.26. [E] The perpendicular bisectors of the sides of a triangle meet at a single point, the circumcenter of the triangle.

Corollary 3.27. [E] Any triangle can be circumscribed by a circle.

Theorem 3.28. [E] The altitudes of a triangle meet at a single point, the orthocenter of the triangle.

Theorem 3.29. [E] For any $\triangle ABC$, the circumcenter P, centroid Q, and orthocenter O are collinear, lying on a line called the Euler line.

Theorem 3.30. [E] Given a triangle, the midpoints of the three sides, the feet of the three altitudes, and the three midpoints of the line segments connecting the orthocenter to the vertices lie on a circle, called the nine-point circle.

Lemma 3.32. [E] Given a circle $\mathcal{C}$ and P and R two arbitrary points inside this circle, invert to get points P' and R'. Then $\triangle OPR$ is similar to $\triangle OR'P'$.

Theorem 3.33. [E] Inversion in a circle centered at point O will take any line not passing through O onto a circle through O, excluding the point O itself. Conversely, inversion takes any circle through O, excluding O itself, to a line that does not pass through O.

Theorem 3.34. [E] Inversion in a circle centered at O takes any circle that does not pass through O to another circle.

Theorem 3.35. [E] If a circle $\mathcal{C}'$ is perpendicular to the circle $\mathcal{C}$ of inversion, then inversion takes $\mathcal{C}'$ to itself.

Theorem 3.36. [E] If a circle $\mathcal{C}'$ contains a point P and its inverse P' where $P \neq P'$, then $\mathcal{C}'$ is perpendicular to the circle of inversion.

Theorem 3.37. [E] Inversion in a circle preserves angle measure.

Theorem 3.39. [E] Inversion in a circle preserves the cross-ratio.

◆ B.4 Constructions

Theorem 4.2. [E] Given two constructible numbers a and b, their product ab is constructible.

Corollary 4.3. [E] Given two constructible numbers a and b, their quotient $\frac{a}{b}$ is constructible.

Theorem 4.4. [E] Given a constructible number a, then $\sqrt{a}$ is also constructable.

Theorem 4.5. [E] Given a line segment of unit length, let $\mathcal{F}$ denote the set of lengths constructible by straightedge and compass. Then $\mathcal{F}$ is a field.

Theorem 4.7. [E] Two lengths x and y with $x > y$ are related by the golden ratio if $\frac{x}{y} = \phi = \frac{1+\sqrt{5}}{2}$.

Theorem 4.8. [E] Given a line segment, there is a point dividing it into two parts, so that the whole is to the larger piece as the larger piece is to the smaller.

Theorem 4.9. [E] [Gauss's Theorem] A regular n-sided polygon can be constructed with straightedge and compass alone if and only if all the odd prime factors of n are distinct Fermat primes; i.e., if all the odd prime factors are different and if each is of the form $F_k = 2^{2^k} + 1$.

Theorem 4.10. [E] [Mohr-Mascheroni Theorem] Anything that is constructible with a straightedge and a compass can be constructed with a compass alone.

Theorem 4.13. [E] Anything that is constructible with a straightedge and a compass can be constructed with origami.

Theorem 4.14. [E] Given a line segment, we can fold so as to divide this line segment by the golden ratio, so that the whole is to the larger piece as the larger piece is to the smaller.

Theorem 4.15. [E] It is possible to duplicate the cube using origami.

Theorem 4.16. [E] Knotting a strip of paper with parallel edges with an overhand knot forms a regular pentagon.

Theorem 4.17. [E] [Kempe's Universality Theorem] If $f(x,y)$ is any algebraic curve, then there is a linkage that will trace out a section of this curve.

◆ B.5   Neutral geometry

Theorem 5.1. [N] [Saccheri-Legendre Theorem] The angles in any triangle sum to less than or equal to 180°.

Corollary 5.3. [N] The sum of the angles in any convex quadrilateral is less than or equal to 360°.

Theorem 5.4. [N] [Omar Khayyam's Theorem] If $ABCD$ is a quadrilateral with $\angle B = \angle C = 90°$, then

$$\overline{AB} = \overline{DC} \iff \angle A = \angle D,$$
$$\overline{AB} > \overline{DC} \iff \angle A < \angle D,$$
$$\overline{AB} < \overline{DC} \iff \angle A > \angle D.$$

Corollary 5.6. [N] The summit angles of a saccheri quadrilateral are equal and the midline connecting the midpoints of base and summit is perpendicular to both the base and the summit.

Corollary 5.7. [N] The summit and base of a saccheri quadrilateral are parallel.

Corollary 5.8. [N] If $ABCD$ is a quadrilateral with $\angle ABC = \angle DCB = 90°$ and $\angle BAD = \angle CDA$, then $ABCD$ is a saccheri quadrilateral.

Theorem 5.9. [N] Given $\triangle ABC$, let D and E be the midpoints of $\overline{AB}$ and $\overline{AC}$. Drop perpendiculars $BF \perp DE$ and $CG \perp DE$. Then $BFGC$ is the saccheri quadrilateral associated with $\triangle ABC$ with base $\overline{FG}$. Furthermore, $\overline{FG} = 2\overline{DE}$, and $\angle ABC + \angle ACB + \angle BAC = \angle FBC + \angle GCB$.

Corollary 5.10. [N] If the angles in a triangle sum to 180°, then the associated saccheri quadrilateral is a rectangle.

Corollary 5.11. [N] The summit of a saccheri quadrilateral is greater than or equal to the base, and the midline is less than or equal to the legs.

Theorem 5.12. [N] [Base-Leg Congruence] If the base of one saccheri quadrilateral is equal to the base of a second saccheri quadrilateral, and if the legs of the first are equal to the legs of the second, then the saccheri quadrilaterals are congruent.

Theorem 5.13. [N] [Base-Midline Congruence] If the base of one saccheri quadrilateral is equal to the base of a second saccheri quadrilateral, and if the midline of the first is equal to the midline of the second, then the saccheri quadrilaterals are congruent.

Theorem 5.16. [N] If there exists one saccheri quadrilateral with acute summit angles, then all saccheri quadrilaterals have acute summit angles.

Theorem 5.17. [N] If one rectangle exists, then the sum of the angles in any triangle is 180°.

Corollary 5.18. [N] If there exists a triangle whose angles sum to 180°, then all triangles have angle sum 180°.

Corollary 5.19. [N] If there exists a triangle whose angles sum to less than 180°, then all triangles have angle sum less than 180°.

Theorem 5.20. [N] Playfair's's Postulate [Given a line and a point not on the line, exactly one line can be drawn through the given point and parallel to the given line.] is logically equivalent to the Parallel Postulate 5.

Theorem 5.21. [N] Poseidonios's Postulate [Parallel lines are equidistant.] is logically equivalent to the Parallel Postulate 5.

Theorem 5.22. [N] Proclus' Postulate [If a straight line intersects one of two parallel lines, it will also intersect the other.] is logically equivalent to the Parallel Postulate 5.

Theorem 5.23. [N] Saccheri's Postulates [1: In any saccheri quadrilateral, the summit angles are right, and thus all saccheri quadrilaterals are rectangles. 2: At least one rectangle exists. 3: The angle sum of any triangle is 180°. 4: At least one triangle with angle sum 180° exists.] are logically equivalent to one another.

Theorem 5.24. [N] Saccheri's Postulates are logically equivalent to the Parallel Postulate 5.

Theorem 5.25. [N] The Pythagorean Theorem is logically equivalent to the Parallel Postulate 5.

Theorem 5.26. [N] Wallis's Postulate [Given a line segment, one can construct on it a triangle similar to a given triangle.] is logically equivalent to the Parallel Postulate 5.

♦ B.6   Hyperbolic geometry

Theorem 6.1. [N] The negation of the Parallel Postulate 5 is logically equivalent to the Hyperbolic Postulate 5.

Theorem 6.2. [H] There are parallel lines that are not equidistant.

Theorem 6.3. [H] If two triangles are similar, then they must be congruent.

Theorem 6.4. [H] The angles in any triangle sum to less than 180°.

Corollary 6.5. [H] The angles in any convex quadrilateral sum to less than 360°.

Theorem 6.6. [H] Rectangles do not exist.

Theorem 6.7. [H] The summit angles of any saccheri quadrilateral are acute.

Theorem 6.8. [H] There is a pair of parallel lines so that the distance between them is unbounded.

Theorem 6.9. [H] There is an upper bound for the possible area of a triangle.

Theorem 6.12. [H] Given a line AB and a point C not on the line, there are two lines XCW and YCZ through C that are parallel to AB so that any line through C entering angle ∠WCZ will intersect AB and any line through C entering angle ∠YCW will be parallel to AB.

Theorem 6.14. [H] Given a line AB and a point C not on the line and asymptotic parallels WCX and YCZ, drop CD ⊥ AB. Then the angle of parallelism $\Pi(\overline{CD}) = \angle WCD = \angle ZCD < 90°$.

Theorem 6.15. [H] $\Pi(\overline{CD})$ depends only on the length of $\overline{CD}$ and not on the point C or on the line AB.

Theorem 6.16. [H] If YZ is asymptotic parallel to AB at C in the direction of B and E is another point on YZ, then YZ is asymptotic parallel to AB at E in the direction of B.

Theorem 6.17. [H] If AB is asymptotic parallel to CD in the direction of B and D and CD is asymptotic parallel to EF in the direction of D and F, then AB is asymptotic parallel to EF in the direction of B and F.

Theorem 6.18. [H] Asymptotic parallels get arbitrarily close together in the direction of parallelism and arbitrarily far apart in the other direction.

Theorem 6.21. [H] **[Angle-Base Congruence]** Given biangles ⌐XABY and ⌐WCDZ with ∠BAX = ∠DCW and $\overline{AB} = \overline{CD}$, then ⌐XABY ≅ ⌐WCDZ.

Theorem 6.22. [H] **[Exterior Angle Theorem for Biangles]** Given biangle ⌐XABY, extend AB to C. The exterior angle ∠CBY is greater than the opposite interior angle ∠CAX.

Theorem 6.23. [H] **[Angle-Angle Congruence]** Given biangles ⌐XABY and ⌐WCDZ with ∠BAX = ∠DCW and ∠ABY = ∠CDZ, then ⌐XABY ≅ ⌐WCDZ.

Theorem 6.24. [H] If ⌐XABY is a biangle and WCDZ is a figure formed so that $\overline{AB} = \overline{CD}$, ∠BAX = ∠DCW, and ∠ABY = ∠CDZ, then ⌐WCDZ is a biangle.

Corollary 6.25. [H] The angle of parallelism decreases as the distance of C from the line AB increases.

Theorem 6.26. [H] $\Pi(d) = 90° - 2\arctan\left(\frac{e^d - 1}{e^d + 1}\right)$.

Corollary 6.27. [H] $\lim_{d \to \infty} \Pi(d) = 0$ and $\lim_{d \to 0^+} \Pi(d) = 90°$.

Theorem 6.28. [H] If a line is divergent parallel to a given line at one point, then it is divergent parallel to that line everywhere.

Theorem 6.29. [H] If two lines are cut by a transversal so that the alternate interior angles are equal (or if the exterior angle is equal to the opposite interior angle or if the interior angles on one side of the transversal sum to 180°), then the lines are divergent parallel.

Theorem 6.30. [H] If two lines are divergent parallel, then they have a common perpendicular.

Corollary 6.31. [H] If two lines are divergent parallel, then the parallel angle relationships hold at one point.

Theorem 6.32. [H] Divergent parallels get farther apart on either side of the common perpendicular.

Theorem 6.33. [H] [AAA Congruence] If △ABC and △DEF have ∠ABC = ∠DEF, ∠BAC = ∠EDF, and ∠ACB = ∠DFE, then the triangles are congruent.

Theorem 6.34. [H] There are triangles that cannot be circumscribed.

Theorem 6.35. [H] Given any angle ∠XOY there is a line that is asymptotic parallel to both OX and OY.

Theorem 6.36 [H] Given two asymptotic parallel lines, there is a third line asymptotic parallel to both of the given lines at opposite ends.

Theorem 6.37. [H] [Summit-Summit Angle] If two saccheri quadrilaterals are such that the summit of one is equal to the summit of the other and the summit angles of the two saccheri quadrilaterals are equal, then the saccheri quadrilaterals are congruent.

Theorem 6.38. [H] [Base-Summit Angle] If two saccheri quadrilaterals are such that the base of one is equal to the base of the other and the summit angles of the two saccheri quadrilaterals are equal, then the saccheri quadrilaterals are congruent.

♦ B.7 Other geometries

Theorem 7.2. [S] The following statements are true in elliptic geometry:

1. Given a point on a line and a length less than or equal to $\frac{\pi r}{2}$, one can cut off a segment of the given length.
2. The base angles of an isosceles triangle are equal.
3. If the base angles of a triangle are equal, then the triangle is isosceles.
4. Vertical angles are equal.
5. Two adjacent angles form a line if and only if their sum is 180°.
6. Any arc can be bisected.
7. Any angle can be bisected.
8. Given a line and a point on the line, a unique perpendicular can be constructed to the given line through the given point.
9. Given a line and a point not on that line, at least one perpendicular can be constructed to the given line through the given point.
10. Given an angle and a segment, one can construct on that segment at an endpoint an angle equal to the given angle.
11. If two triangles satisfy the SSS criterion, then they are congruent.
12. If two triangles satisfy the ASA criterion, then they are congruent.
13. The summit angles of any saccheri quadrilateral are equal and the midline is perpendicular to both the base and the summit.

Theorem 7.3. [S] Let C be a line. There is a point P, called the pole for C, such that every straight line connecting P to a point on C is perpendicular to C. The distance from P to any point X on C is called the polar distance and is the same for any X on C.

Corollary 7.4. [S] All lines have the same polar distance, defined to be $k = \frac{\pi r}{2}$.

Corollary 7.5. [S] All lines have the same length, πr.

Theorem 7.6. [S] Consider a right triangle $\triangle ABC$ with $\angle ABC = 90°$. Then $\angle ACB < 90°$ if and only if $\widehat{AB} < \frac{\pi r}{2}$, $\angle ACB = 90°$ if and only if $\widehat{AB} = \frac{\pi r}{2}$, and $\angle ACB > 90°$ if and only if $\widehat{AB} > \frac{\pi r}{2}$.

Theorem 7.7. [S] In elliptic geometry, the summit angles of a saccheri quadrilateral are obtuse.

Theorem 7.9. [S] In elliptic geometry, the sum of the angles in a triangle is greater than 180°.

Theorem 7.10. [S] [AAA congruence] If $\triangle ABC$ and $\triangle DEF$ have $\angle ABC = \angle DEF$, $\angle BAC = \angle EDF$, and $\angle ACB = \angle DFE$, then the triangles are congruent.

Theorem 7.11. [E] [Heron's formula] $A(\triangle ABC) = \sqrt{s(s-a)(s-b)(s-c)}$ where $s = \frac{1}{2}(a+b+c)$.

Theorem 7.15. Equivalence by decomposition is an equivalence relation.

Theorem 7.16. If triangles $\triangle ABC$ and $\triangle DEF$ are equivalent by decomposition, then $A(\triangle ABC) = A(\triangle DEF)$.

Theorem 7.17. [E] Any parallelogram is equivalent by decomposition to a rectangle.

Theorem 7.19. [E] If $ABCD$ is a rectangle and a line segment $\overline{EF}$ is given, then $ABCD$ is equivalent by decomposition to a rectangle with $\overline{EF}$ as one side.

Corollary 7.20. [E] Given a line segment $\overline{EF}$, any polygon is equivalent by decomposition to some rectangle with side $\overline{EF}$.

Theorem 7.21. [E] [Bolyai-Gerwien] Two polygons P and Q are equivalent by decomposition if and only if they have the same area.

Theorem 7.23. The defect is additive. I.e., if $\triangle ABC$ is a triangle, and ABC is divided into polygons P_i, for $i = 1, 2, \ldots, n$, then

$$def(\triangle ABC) = \sum_{i=1}^{n} def(P_i).$$

Theorem 7.24. If triangles $\triangle ABC$ and $\triangle DEF$ are equivalent by decomposition, then $def(\triangle ABC) = def(\triangle DEF)$.

Theorem 7.26. [H] If $\triangle ABC$ and $\triangle DEF$ are two triangles such that $def(\triangle ABC) = def(\triangle DEF)$, then $\triangle ABC$ has the same area as $\triangle DEF$.

Theorem 7.27. [Gauss-Bonnet Theorem] If $\triangle ABC$ is a triangle, then $def(\triangle ABC) = \kappa A(\triangle ABC)$ for some constant κ. Note that the constant κ depends only on the space and not on a particular triangle.

Theorem 7.28. The SAS Postulate 12 is not true in the taxicab geometry.

♦ B.8 Isometries

Theorem 8.2. If T is an isometry, then T is one-to-one and T takes lines to lines.

Corollary 8.3. If T is an isometry, then T preserves angle measure.

Corollary 8.4. Let ℓ_1 and ℓ_2 be parallel lines on the plane and T an isometry of the plane. Then $T(\ell_1)$ is parallel to $T(\ell_2)$.

Theorem 8.6. [N] Translation is an isometry.

Theorem 8.8. [N] Rotation is an isometry.

Theorem 8.10. [N] Reflection across a line is an isometry.

Theorem 8.12. [N] Glide reflection is an isometry.

Theorem 8.15. [N] If an isometry T fixes two distinct points A and B, then T fixes all points on the line AB.

Theorem 8.16. [N] If an isometry T of the plane has three noncollinear fixed points, then T is the identity transformation.

Corollary 8.17. [N] If T and S are isometries of the plane, and there are three noncollinear points A, B, and C so that $T(A) = S(A)$, $T(B) = S(B)$, and $T(C) = S(C)$, then $T = S$.

Theorem 8.18. [N] Any isometry of the plane can be expressed as the composition of at most three reflections.

Theorem 8.19. [E] A euclidean isometry S is either a rotation or a translation if and only if S can be written as the composition of two reflections in distinct lines ℓ_1 and ℓ_2.

Theorem 8.20. [E] A euclidean isometry S is a glide reflection if and only if S can be written as the composition of three reflections in distinct lines ℓ_1, ℓ_2, and ℓ_3.

Theorem 8.21. [E] [Classification of Euclidean Isometries] Every isometry of the euclidean plane is either the identity, a reflection, a translation, a rotation, or a glide reflection.

Isometry Postulate: For every line ℓ, there is a function F_ℓ defined on the plane, called the reflection across ℓ. This function takes lines to lines, preserves distance and angle measure, has the property that for any point A on ℓ, $F_\ell(A) = A$, and also has the property that for points B not on ℓ, B and $F_\ell(B)$ lie on opposite sides of ℓ.

Theorem 8.23. [N] In the context of neutral geometry, the Isometry Postulate is logically equivalent to the SAS Postulate 12.

Theorem 8.26. [N] [Leonardo's Theorem] The rosette group of a finite figure is either a cyclic group or a dihedral group.

Theorem 8.27. [E] [Classification of Frieze Patterns] There are exactly seven different symmetry groups for frieze patterns.

Theorem 8.28. [E] Let ℓ be a line of reflection for a wallpaper pattern, and let $T_{P,Q}$ be a translation for the pattern which is not parallel to ℓ, so $T_{P,Q}(\ell)$ is another line of reflection. Let m be the line parallel to ℓ and $T_{P,Q}(\ell)$ and lying midway between the two. Then

1. If PQ is perpendicular to ℓ, then m is a line of reflection for the pattern.
2. If PQ is not perpendicular to ℓ, then m is a line of glide reflection.

Theorem 8.30. [E] [The Crystallographic Restriction] The only possible angles of rotation for a wallpaper pattern on the plane are 180°, 120°, 90°, and 60°.

Theorem 8.31. [E] [Classification of Wallpaper Patterns] There are exactly 17 different symmetry groups for wallpaper patterns.

Theorem 8.34. [H] A hyperbolic isometry S is either a rotation, a translation, or a horolation if and only if S can be written as the composition of two reflections in distinct lines ℓ_1 and ℓ_2.

Theorem 8.35. [H] A hyperbolic isometry S is glide reflection if and only if S can be written as the composition of three reflections in distinct lines ℓ_1, ℓ_2, and ℓ_3.

Theorem 8.36. [H] [Classification of Hyperbolic Isometries] Every isometry of the hyperbolic plane is either the identity, a reflection, a translation, a rotation, a horolation, or a glide reflection.

♦ B.9 Tilings

Theorem 9.4. [E] [Classification of Euclidean Tilings] There are three regular tilings and eight semiregular tilings of the euclidean plane.

Theorem 9.5. [E] [Conway Criteria] A simple region will tile the plane if the boundary can be divided into six sections by six points labeled A, B, C, D, E, and F in order as one travels around the boundary, and

1. The curve AB from A to B is the translation of the curve ED.
2. The curves BC, CD, EF, and FA have rotational symmetry about their midpoints.

Theorem 9.6. [E] [The Extension Theorem] Let S be any finite set of tiles, each of which is a closed topological disc. If S tiles arbitrarily large circular discs, then S admits a tiling of the plane.

Theorem 9.12. [E] No tiling of the plane by Penrose tiles is periodic.

♦ B.10 Geometry in three dimensions

Lemma 10.1. [N] If two distinct planes intersect, then they intersect in a line.

Theorem 10.2. [E] [Space Separation Property] Given a plane $\mathcal{P}$ in $\mathbb{R}^3$, $\mathcal{P}$ separates the points in $\mathbb{R}^3$ and not on $\mathcal{P}$ into two disjoint nonempty convex sets, S_1 and S_2, called the sides of the plane, so that whenever point A is in S_1 and point B is in S_2, the line segment $\overline{AB}$ must intersect $\mathcal{P}$.

Theorem 10.7. [N] If a line is perpendicular to two intersecting lines at their point of intersection, then it is perpendicular to the plane containing those lines.

Theorem 10.8. [E] If two lines are parallel and one of them is perpendicular to a given plane, then the other line is also perpendicular to the plane.

Theorem 10.9. [N] Given a plane and a point that does not lie on the plane, one can drop a perpendicular from the point to the plane.

Theorem 10.11. [E] If two parallel planes are cut by a third plane, then the lines of intersections of the parallel planes and the third plane are parallel.

Theorem 10.13. [E] If a line is perpendicular to a given plane, then all of the planes that contain this line will also be perpendicular to the given plane.

Corollary 10.14. [E] Given a plane $\mathcal{P}$ and a line ℓ that does not lie on $\mathcal{P}$, if ℓ is not perpendicular to $\mathcal{P}$, then there is a unique plane $\mathcal{P}'$ that contains ℓ and is perpendicular to $\mathcal{P}$.

Theorem 10.15. [E] Given skew lines ℓ and m, there is a line n that is perpendicular to both ℓ and m.

Theorem 10.17. [E] If a solid (trihedral) angle is formed by three plane angles, then the sum of the measures of any two of these plane angles is greater than the measure of the third plane angle.

Corollary 10.18. [E] If a solid angle is formed by several plane angles, then the sum of the plane angles is less than $360°$.

Theorem 10.25. [E] [Euler's formula] For any polyhedron P, $\chi(P) = 2$.

Theorem 10.26. [E] [Classification of Regular Polyhedra] There are exactly five regular polyhedra.

Theorem 10.28. [E] [Classification of Semiregular Polyhedra] There are 13 semiregular polyhedra.

Theorem 10.31. [E] [Descartes' formula] If P is a polyhedron, then $\Delta(P) = 720°$.

Theorem 10.32. [E] The volume of a triangular prism is $V(\textit{prism}) = h\,A(\textit{base})$.

Theorem 10.33. [E] The volume of a triangular pyramid is $V(\textit{pyramid}) = \frac{1}{3} h\,A(\textit{base})$.

Theorem 10.36. [E] If T is an isometry of $\mathbb{R}^3$ and T fixes four noncoplanar points, then T is the identity transformation.

Corollary 10.37. [E] If T and S are isometries of $\mathbb{R}^3$ and there are four noncoplanar points A, B, C, and D such that $T(A) = S(A)$, $T(B) = S(B)$, $T(C) = S(C)$, and $T(D) = S(D)$, then $T = S$.

Theorem 10.38. [E] Any isometry of $\mathbb{R}^3$ can be written as the composition of at most four reflections.

Theorem 10.39. [E] An isometry of $\mathbb{R}^3$ is either a rotation or a translation if and only if the isometry can be written as the composition of reflections in two distinct planes $\mathcal{P}_1$ and $\mathcal{P}_2$.

Theorem 10.40. [E] An isometry of $\mathbb{R}^3$ is either a rotary reflection or a glide reflection if and only if the isometry can be written as the composition of reflections in three distinct planes $\mathcal{P}_1$, $\mathcal{P}_2$, and $\mathcal{P}_3$.

Theorem 10.41. [E] An isometry of $\mathbb{R}^3$ is screw rotation if and only if the isometry can be written as the composition of reflections in four distinct planes $\mathcal{P}_1$, $\mathcal{P}_2$, $\mathcal{P}_3$, and $\mathcal{P}_4$.

Theorem 10.42. [E] [Classification of Isometries on $\mathbb{R}^3$] Every isometry of $\mathbb{R}^3$ is either the identity, a reflection, a translation, a rotation, a glide reflection, a rotary reflection, or a screw rotation.

Corollary 10.43. [S] [Classification of Isometries on the Sphere] Every isometry of the sphere is either the identity, a rotation about a diameter, a reflection across a plane passing through the center of the sphere, or a rotary reflection across a plane passing through the center of the sphere.

Bibliography

Edwin A. Abbott, *Flatland*, Princeton University Press, Princeton, NJ, 1991.

Roger C. Alperin, "A Mathematical Theory of Origami Constructions and Numbers," *New York Journal of Mathematics* 6 (2000), pp. 119–133.

Apollonios of Perga, *Conics*, translated by R. C. Taliaferro, Green Lion Press, Santa Fe NM, 2000.

J.W. Armitage, "The Place of Geometry in a Mathematical Education," *Mathematical Gazette* 57 (1973), pp. 267–278.

Benno Artmann, *Euclid—The Creation of Mathematics*, Springer, New York, 1999.

I.I. Artobolevski, *Mechanisms for the Generation of Plane Curves*, Macmillan, New York, 1964.

Clifford W. Ashley, *The Ashley Book of Knots*, Doubleday, New York, 1944.

David Auckly and John Cleveland, "Totally Real Origami and Impossible Paper Folding," *American Mathematical Monthly* 102 (1995), pp. 215–226.

Anatole Beck and Michael N. Bleicher, and Donald W. Crowe, *Excursions into Mathematics*, Worth Publishers, New York, 1969.

sarah-marie belcastro and Thomas C. Hull, "Classifying Frieze patterns Without Using Groups," *College Mathematics Journal* 33 (2002), pp. 93–98.

Jessica Benashaski, John Meier, Kevin O'Brien, Paige Reinheimer, Margaret Skarbek, "Introducing Hyperbolicity via Piecewise Euclidean Complexes," *College Mathematics Journal* 31 (2000), pp. 213–217.

Allan Berele and Jerry Goldman, *Geometry: Theorems and Constructions*, Prentice Hall, Upper Saddle River NJ, 2001.

J. L. Berggren and R. S. D. Thomas, *Euclid's Phaenomena*, AMS, Providence RI, 1996.

Jinny Beyer, *Designing Tessellations*, Contemporary Books, Lincolnwood IL, 1999.

George David Birkhoff and Ralph Beatley, *Basic Geometry*, Chelsea Publishing Company, New York, 1999 (reissue of Scott, Foresman and Company, 1940).

Robert Bix, *Topics in Geometry*, Academic Press, New York, 1994.

Alexander Bogomolny, "Geometrical Exploits," at www.cut-the-knot.org/proof.shtml, 2007.

Benjamin Bold, *Famous Problems of Geometry and How to Solve Them*, Dover, New York, 1969.

Brian Bolt, *Mathematics meets Technology*, Cambridge University Press, New York, 1991.

Roberto Bonola, *Non-Euclidean Geometry*, translated by H.S. Carswell, Dover, New York, 1955.

F. H. Bool, J. R. Kist, J. L. Locher, and F. Wierda, *M. C. Escher: His Life and Complete Graphic Work*, Harry N. Abrams, New York, 1992.

K. Borsuk and W. Szmielew, *Foundations of Geometry*, North Holland, Amsterdam, 1960.

O. Bottema, *Topics in Elementary Geometry*, translated by Reinie Erné, Springer, New York, 2008.

S. Allen Broughton, "Constructing Kaleidoscopic Tiling Polygons in the Hyperbolic Plane," *American Mathematical Monthly* 107 (2000), pp. 689–710.

W. K. Bühler, *Gauss, a Biographical Study*, Springer, New York, 1981.

Oliver Byrne, *The First Six Books of The Elements of Euclid in which Coloured Diagrams and Symbols are used instead of Letters for the Greater Ease of Learners*, from http://www.sunsite.ubc.ca/DigitalMathArchive/Euclid/byrne.html, 2009.

Oliver Byrne, *The Geometry of Compasses*, Crosby, Lockwood, and Co., London, 1877.

Paul A. Calter, *Squaring the Circle: Geometry in Art and Architecture*, Key College Publishing, Emeryville CA, 2008.

James Casey, *Exploring Curvature*, Friedrich Vieweg & Sohn, Braunschweig, 1996.

James Casey, "Using a Surface Triangle to Explore Curvature," *The Mathematics Teacher* 87 (1994), pp. 69–76.

John H. Conway, "The Orbifold Notation for Surface Groups," in *Groups, Combinatorics, and Geometry*, ed. Liebeck and Saxl, Cambridge University Press, New York, 1992.

John H. Conway, Heidi Burgiel, Chaim Goodman-Strauss, *The Symmetries of Things*, A. K. Peters, Wellesley MA, 2008.

Richard Courant and Herbert Robbins, *What is Mathematics?*, Oxford University Press, London, 1941.

H. S. M. Coxeter, *Introduction to Geometry*, Wiley & Sons, New York, 1969.

H. S. M. Coxeter, *Non-Euclidean Geometry*, Mathematical Association of America, Washington DC, 1998.
H. S. M. Coxeter, *Regular Polytopes*, Dover, New York, 1973.
H. S. M. Coxeter, "Regular Skew polyhedra in three and four dimensions, and their topological analogues," *Proceedings of the London Mathematical Society* 43 (1937), pp. 33–62.
Peter R. Cromwell, *Polyhedra*, Cambridge University Press, New York, 1997.
Donald W. Crowe and Dorothy K. Washburn, *Symmetries of Culture*, University of Washington Press, Seattle WA, 1988.
H.M. Cundy and A.P. Rollett, *Mathematical Models*, Oxford University Press, New York, 1961.
Erik Demaine and Joseph O'Rourke, *Geometric Folding Algorithms*, Cambridge University Press, New York, 2007.
René Descartes, *Geometry*, trans. David E. Smith and Marcia L. Latham, Dover, New York, 1954.
Frank De Sua, "Consistency and Completeness—A Resume," American Mathematical Monthly 63(1956), pp. 295–305.
A. K. Dewdney, "Computer Recreations: Imagination meets geometry in the crystalline realm of latticeworks," *Scientific American*, June 1988, pp. 120–123.
A. K. Dewdney, *The Planiverse*, Poseidon Press, New York, 1984.
Charles L. Dodgson, "What the Tortoise Said to Achilles," *Mind* 4, pp. 278–280.
Underwood Dudley, *A Budget of Trisections*, Springer-Verlag, New York, 1987.
Underwood Dudley, *Mathematical Cranks*, Mathematical Association of America, Washington CD, 1992.
Underwood Dudley, *Numerology, or what Pythagoras wrought*, Mathematical Association of America, Washington DC, 1997.
Richard A. Dunlap, *The Golden Ratio and Fibonacci Numbers*, World Scientific, Singapore, 1997.
G. Waldo Dunnington, *Carl Friedrich Gauss—Titan of Science*, Hafner Publishing, New York, 1955.
B. Carter Edwards and Jerry Shurman, "Folding quartic roots," *Mathematics Magazine* 74 (2001), pp. 19–25.
Euclid, *The Thirteen Books of The Elements*, translated by Sir Thomas L. Heath, Dover, New York, 1956.
Howard Eves, *Foundations and Fundamental Concepts of Mathematics*, PWS-Kent Publishing, Boston MA, 1990.
Howard Eves, *Fundamentals of Modern Elementary Geometry*, Jones and Bartlett, Boston MA, 1992.
Howard Eves, *A Survey of Geometry*, Allyn and Bacon, Boston MA, 1972.
David W. Farmer, *Groups and Symmetry*, American Mathematical Society, Providence RI, 1991.
L. Fejes Tóth, *Regular Figures*, Pergamon Press, New York, 1964.
D. H. Fowler, *The Mathematics of Plato's Academy*, Clarendon Press, Oxford, 1987.
Tomoko Fusè, *Unit Origami*, Japan Publications, Tokyo, 1990.
David Fuys, Dorothy Geddes, and Rosamund Tischler, "The Van Hiele Model of Thinking in Geometry among Adolescents," *Journal for Research in Mathematics Education*, Monograph #3, National Council of Teachers of Mathematics, Reston VA, 1988.
David Gale, "Egyptian Rope, Japanese Paper, and High School Math," *Math Horizons*, September 1998, pp. 5–7.
David Gans, *An Introduction to Non-Euclidean Geometry*, Academic Press, New York, 1973.
Martin Gardner, *Penrose Tiles to Trapdoor Ciphers*, W. H. Freeman, New York, 1989.
Martin Gardner, *Time Travel and Other Mathematical Bewilderments*, W. H. Freeman, New York, 1988.
Martin Gardner, *The Unexpected Hanging and Other Mathematical Diversions*, Simon & Schuster, New York, 1969.
Robert Geretschläger, "Euclidean Constructions and the Geometry of Origami," *Mathematics Magazine* 68 (1995), pp. 357–371.
Livia Giacardi, "Scientific Research and Teaching Problems in Beltrami's Letters to Hoüel," in *Using History to Teach Mathematics*, ed. Victor Katz, Mathematical Association of America, Washington DC, 2000.
Andrew M. Gleason, "Angle trisection, the heptagon, and the triskaidecagon," *American Mathematical Monthly* 95 (1988), pp. 185–194.
Roe Goodman, "Alice through Looking Glass after Looking Glass: The Mathematics of Mirrors and Kaleidoscopes," *American Mathematical Monthly* 111 (2004), pp. 281–298.
Chaim Goodman-Strauss, "Compass and Straightedge in the Poincaré Disk," *American Mathematical Monthly* 108 (2001), pp. 38–49.
Jeremy Gray, *Ideas of Space*, Clarendon Press, Oxford, 1979.
Jeremy Gray, *Janos Bolyai, Non-Euclidean Geometry, and the Nature of Space*, Burndy Library Publications, Cambridge MA, 2004.
Jeremy Gray, *Worlds out of Nothing*, Springer, New York, 2007.

Marvin J. Greenberg, *Euclidean and Non-Euclidean Geometries*, W. H. Freeman, New York, 1993.

Branko Grünbaum and G.C. Shephard, "Ceva, Menelaus, and the Area Principle," *Mathematics Magazine* 68 (1995), pp. 254–268.

Branko Grünbaum and G.C. Shephard, "Interlace Patterns in Islamic and Moorish Art," *Leonardo* 25 (1992), pp. 331–339.

Branko Grünbaum and G.C. Shephard, *Tilings and Patterns*, W. H. Freeman, New York, 1987.

Kazuo Haga, *Origamics*, World Scientific, Hackensack NJ, 2008.

Robin Hartshorne, *Geometry: Euclid and Beyond*, Springer Verlag, New York, 2000.

Robin Hartshorne, "Teaching geometry according to Euclid," *Notices of the American Mathematical Society* 47 (2000), pp. 460–465.

Thomas Heath, *A History of Greek Mathematics*, Clarendon Press, Oxford, 1921.

J. L. Heilbron, *Geometry Civilized*, Clarendon Press, Oxford, 1998.

David W. Henderson and Diana Taimina, "Crocheting the Hyperbolic Plane," *The Mathematical Intelligencer* 23 (2001), pp. 17–28.

David W. Henderson and Daina Taimina, *Experiencing Geometry on Plane and Sphere*, Prentice Hall, Upper Saddle River NJ, 2005.

Pierre M. van Hiele, *Structure and Insight: A Theory of Mathematics Education*, Academic Press, New York, 1986.

David Hilbert, *Foundations of Geometry*, translated by Leo Unger, Open Court, LaSalle IL 1971.

David Hilbert and S. Cohn-Vossen, *Geometry and the Imagination*, Chelsea Publishing Co., New York, 1990.

Peter Hilton and Jean Pedersen, *Build your own Polyhedra*, Addison Wesley, Menlo Park CA, 1994.

Charles Howard Hinton, *Scientific Romances*, Merchant Books, 1886, digitized by Watchmaker Publishing, Seaside OR, 2008.

E. W. Hobson, *Squaring the Circle*, Chelsea Publishing Company, New York, 1953.

Alan Holden, *Shapes, Space, and Symmetry*, Dover, New York, 1971.

Thomas Hull, "A Note on 'Impossible' Paper Folding," *American Mathematical Monthly* 103 (1996), pp. 240–241.

Thomas Hull (ed.) *Origami*[3], A. K. Peters, Natick MA, 2002.

Thomas Hull, *Project Origami*, A. K. Peters, Wellesley MA, 2006.

H. E. Huntley, *The Divine Proportion*, Dover, New York, 1970.

Humiaki Huzita, "Drawing the Regular Heptagon and the Regular Nonagon by Origami," *Symmetry: Culture and Science* 5(1994), pp. 69–83.

Roger V. Jean, *Phyllotaxis*, Cambridge University Press, Cambridge, 1994.

Donovan Johnson, *Paper Folding for the Mathematics Classroom*, National Council of Teachers of Mathematics, Washington DC, 1957.

Sasho Kalajdzievski, *Math and Art: An Introduction to Visual Mathematics*, CRC Press, Boca Raton FL, 2008.

Michael Kapovich and John J. Millson, "Universality theorems for configuration spaces of planar linkages," *Topology* 41 (2002), pp. 1051–1107.

Jay Kappraff, *Connections: The Geometrical Bridge between Art and Science*, McGraw-Hill, New York, 1991.

Kunihiko Kasahara, *Origami Omnibus*, Japan Publications, Tokyo, 1988.

Kunihiko Kasahara and Toshie Takahama, *Origami for the Connoisseur*, Japan Publications, Tokyo, 1987.

David Kay, *College Geometry*, Harper Collins, New York, 1994.

A. B. Kempe, *How to Draw a Straight Line*, National Council of Teachers of Mathematics, Reston VA, 1977.

A. B. Kempe, "On a general method of describing plane curves of the nth degree by linkwork," *Proceedings of the London Mathematical Society* 7 (1875), pp. 213–216.

Henry C. King, "Configuration spaces of linkages in $\mathbb{R}^n$," from arxiv.org/abs/math/9811138, 2009.

L. Christine Kinsey, *Topology of Surfaces*, Springer-Verlag, New York, 1993.

L. Christine Kinsey and Teresa E. Moore, *Symmetry, Shape, and Space*, Key College Publishing, Emeryville CA, 2002.

Aleksandr N. Kostovskii, *Geometrical Constructions using Compasses Only*, Blaisdell Publishing Company, New York, 1961.

Thomas S. Kuhn, *The Structure of Scientific Revolutions*, University of Chicago Press, Chicago IL, 1962.

Robert Lang (ed.), *Origami*[4], A. K. Peters, Natick MA, 2009.

Robert Lang, "Origami and Geometric Constructions," www.langorigami.com/science/ hha/hha.php4, 2003.

Reinhard Laubenbacher and David Pengelley, *Mathematical Expeditions*, Springer, New York, 1999.

Richard Lesh and Marsha Landau (eds.), *Acquisition of Mathematical Concepts and Processes*, Academic Press, New York, 1983.

Elisha S. Loomis, *The Pythagorean Proposition*, National Council of Teachers of Mathematics, Washington DC, 1968.
Stephen Luecking, "Introducing Geometry with a Neolithic Tool Kit," http://facweb.cs.depaul.edu/sluecking/pullers%201.htm, 2007.
George Markowsky, "Misconceptions about the Golden Ratio," *College Mathematics Journal* 23 (1992), pp. 2–19.
George E. Martin, *The Foundations of Geometry and the Non-Euclidean Plane*, Springer, New York, 1975.
George E. Martin, *Geometric Constructions*, Springer-Verlag, New York, 1998.
George E. Martin, *Transformation Geometry: An Introduction to Symmetry*, Springer-Verlag, New York, 1982.
Joanne Mayberry, "The Van Hiele Levels of Geometric Thought in Undergraduate Preservice Teachers," *Journal for Research in Mathematics Education* 14 (1983), pp. 58–69.
Peter Messer, "Problem 1054," *Crux Mathematicorum* 12 (1986), pp. 284–285.
Walter Meyer, *Geometry and Its Applications*. Harcourt Brace, San Diego CA, 1999.
John Milnor, "Hyperbolic geometry: the first 150 years," *Bulletin of the American Mathematical Society* 6 (1982), pp. 9–24.
David Mitchell, *Mathematical Origami*, Tarquin Publications, Stradbroke, 1997.
Edwin E. Moise, *Elementary Geometry from an Advanced Standpoint*, Addison-Wesley, New York, 1990.
Patricia F. O'Grady, *Thales of Miletus: the beginnings of Western science and philosophy*, Ashgate Publishing Company, Burlington VT, 2002.
Alton T. Olson, *Mathematics through Paper Folding*, National Council of Teachers of Mathematics, Reston VA, 1975.
A. R. Pargeter, "Plaited polyhedra," *Mathematical Gazette* 43 (1959), pp. 88–101.
Roger Penrose, "Pentaplexity: A Class of Non-Periodic Tilings of the Plane," *Mathematical Intelligencer* 2 (1979), pp. 32–37.
John Playfair, *Elements of Geometry*, J. Howe, Philadelphia PA, 1832.
Henri Poincaré, *Science and Hypothesis*, Dover, New York, 1952.
G. Polya, *How to Solve It*, Princeton University Press, Princeton, 1945.
Alfred Posamentier, *Advanced Euclidean Geometry*, Key College Publishing, Emeryville CA, 2002.
Chris Pritchard (ed.), *The Changing Shape of Geometry*, Cambridge University Press, Cambridge MA, 2003.
Proclus, *A Commentary on the First Books of Euclid's Elements*, translated by Glenn R. Morrow, Princeton University Press, Princeton NJ, 1970.
Anthony Pugh, *Polyhedra: A Visual Approach*, University Of California Press, Berkeley CA, 1976.
Raymond Queneau, Italo Calvino, Paul Fournel, Claude Berge, Jacques Jouet, and Harry Mathews, *Oulipo Laboratory*, Atlas Press, London, 1995.
Barbara Reynolds and William Fenton, *College Geometry using the Geometer's Sketchpad*, Key College Publishing, Emeryville CA, 2006.
John Robertson, *A Treatise of Mathematical Instruments*, The Invisible College Press, Woodbridge VA, 2002 (a reprint of the 1775 third edition)
B. A. Rosenfeld, *A History of Non-Euclidean Geometry*, Springer-Verlag, New York, 1988.
Sundara Row, *Geometric Exercises in Paper Folding*, Open Court Publishing Company, La Salle IL, 1958.
C. A. Rupp, "On a transformation by paper-folding," *American Mathematical Monthly* 31 (1924), pp. 432–435.
Lucio Russon, *The Forgotten Revolution*, Springer-Verlag, New York, 2004.
G. Girolamo Saccheri, *Euclides Vindicatus*, translated by G. B. Halstead, Chelsea Publishing Company, New York, 1986.
Doris Schattschneider, "The Plane Symmetry Groups: Their Recognition and Notation," *American Mathematical Monthly* 85 (1978), pp. 439–450.
Doris Schattschneider, *Visions of Symmetry: Notebooks, Periodic Drawings, and Related Work of M. C. Escher*, W. H. Freeman, New York, 1990.
Doris Schattschneider, "Will It Tile? Try the Conway Criterion!," *Mathematics Magazine* 53 (1980), pp. 224–233.
Caspar Schwabe, "Perfect Polyhedral Kaleidoscopes, " *Forma* 21 (2006), pp. 29–35.
Brigitte Servatius, "The Geometry of Folding Paper Dolls," *Mathematical Gazette* 81 (1997), pp. 29–36.
Dale Seymour and Jill Britton, *Introduction to Tessellations*, Dale Seymour Publications, Palo Alto CA, 1989.
A. V. Shubnikov and V. A. Koptsik, *Symmetry in Science and Art*, translated by G. D. Archard, Plenum Press, New York, 1974.
James T. Smith, *Methods of Geometry*, John Wiley and Sons, New York, 2000.
William George Spencer, *Inventional Geometry*, American Book Company, New York, 1876.
Sherman Stein, *Archimedes, What Did He Do Besides Cry Eureka?*, Mathematical Association of America, Washington DC, 1999.

Peter S. Stevens, *Handbook of Regular Patterns*, MIT Press, Cambridge, 1980.
Ian Stewart, "Daisy, Daisy, Give Me Your Answer, Do," *Scientific American*, January 1995, pp. 96–99.
John Stillwell, *Sources of Hyperbolic Geometry*, American Mathematical Society, Providence RI, 1996.
David Sutton, *Islamic Design: a Genius for Geometry*, Walker & Company, New York, 2007.
Daina Taimina, *Crochet Adventures with Hyperbolic Planes*, A. K. Peters, Wellesley MA, 2009.
Ivor Thomas (trans.) *Greek Mathematical Works*, Harvard University Press. Cambridge MA, 1939.
Richard J. Trudeau, *The Non-Euclidean Revolution*, Birkhäuser, Boston MA, 1987.
Gerard A. Venema, *Foundations of Geometry*, Pearson-Prentice Hall, Upper Saddle River NJ, 2002.
Avraham Wachman, Michael Burt, Menachem Kleinman, *Infinite Polyhedra*, Technion, Haifa Israel, 1974.
Bartel Leendert van der Waerden, *Science Awakening*, Oxford University Press, New York, 1961.
Edward C. Wallace and Stephen F. West, *Roads to Geometry*, Prentice Hall, Upper Saddle River NJ, 1998.
Hans Walser, *The Golden Section*, trans. Peter Hilton, Mathematical Association of America, Washington DC, 2001.
Pierre Laurent Wantzel, "Recherches sur les moyens de reconnaître si un Problème de Géométrie peut se résoudre avec la règle et le compas," *Journal de Mathématiques Pures et Appliquées*, 1837, pp. 366–372, translated by Brian Hayes as "Research on the means of knowing whether a problem in geometry can be solved with ruler and compass" at sigmaxi.org/amscionline/Wantzel1837english.pdf, 2006.
Jeffrey Weeks, *The Shape of Space*, Marcel Dekker, New York, 1985.
Magnus Wenninger, *Polyhedron Models*, Cambridge University Press, New York, 1971.
Hermann Weyl, *Symmetry*, Princeton University Press, Princeton NJ, 1952.
Robert C. Yates, *Geometrical Tools*, Educational Publishers, St. Louis MO, 1949.
Robert C. Yates, *The Trisection Problem*, National Council of Teachers of Mathematics, Washington DC, 1971.

Index

Abe, Hisashi, 122
Abbott, Edwin A., 418
additivity, 25, 246, 250
d'Alembert, Jean, 159
altitude, 98
ana, 412
angle, acute, 35
   adjacent, 35
   alternate interior, 37, 54, 208
   between curves, 86
   bisector, 26, 43, 97, 109, 130, 150, 232
   central, 81
   congruence, 25
   constructible, 114
   copy, 52, 232
   definition, 17, 231, 363
   dihedral, 364
   exterior, 35, 46, 58, 204, 225
   exterior of, 19, 231
   inscribed, 81
   interior of, 19, 231
   measure, 25, 105, 222, 272
   obtuse, 35
   of parallelism, 184, 199, 205
   plane, 366
   right, 34
   solid, 366
   supplementary, 34, 45
   trihedral, 366
   trisection of, 115, 121
   vertical, 35, 46
angle sum, polygon, 59
   quadrilateral, 58, 161, 182
   triangle, 47, 58, 160, 168, 182, 195, 225, 235, 240, 249
angular deficit, 385
anthropology, 295, 319
antipodal points, 80, 221
antiprism, 372, 399
Apollonios of Perga, 14
arc, 80, 223, 229
Archimedean solid, 380, 417
Archimedes, 9, 26, 121, 244, 264, 391
Archimedes' axiom, 26, 73, 79, 160, 173, 231, 245, 435
area, 25, 60, 182, 225, 242, 246
area postulate, 63, 243, 436
Aristophanes, 5, 14, 153
Aristotle, 5, 8, 14
Armitage, J. V., 427

asymptotic parallel, 185, 199
axiom, see postulate

Babylon, 4, 67
Bell, Alexander, 417
Beltrami, Eugenio, 186, 197
Beltrami-Klein model, 188
betweenness, 16, 20, 72, 228, 435
Berkeley, George, 157
Bhaskara, 68
biangle, 203, 211
Birkhoff, George, 75
Bogomolny, Alexander, 140
Bold, Benjamin, 116
Bolyai, Farkas, 176, 177, 179, 211
Bolyai, Janos, 176, 179, 218, 246, 388
Bolyai-Gerwien theorem, 249, 264, 388
Byrne, Oliver, 29, 67

cartesian coordinates, 76, 254
Casey, James, 241
Cavalieri, Bonaventure, 388
Cavalieri's principle, 388, 436
centroid, 95, 99
Ceva, Giovanni, 108, 124
chickens, 349
chord, 80
Cicero, 392
circle, 27, 49, 80, 131, 224, 256, 272
   definition, 27
   exterior of, 28
   great, 220
   interior of, 28
   invariant, 104
   inversion in, 101, 146, 192, 238
   nine-point, 100
   squaring, 121
circular continuity principle, 31, 48, 224, 231, 436
circumcenter, 97, 99
circumscribe, 98, 116, 211
classification, of euclidean planar isometries, 288
   of frieze patterns, 297
   of hyperbolic isometries, 318
   of regular polyhedra, 378
   of semiregular polyhedra, 380
   of spherical isometries, 406
   of three-dimensional isometries, 405
   of tilings, 325

classification, (*continued*)
   of wallpaper patterns, 312
Clavius, Christopher, 176, 177
commensurability, 7, 11
common notion, 13, 25, 26
compass, 27, 109, 124, 230
completeness, of line, 73, 184
completeness, logical, 197
composition, of isometries, 279
   of tiling, 352
concurrence, 94
congruence, 25, 72, 79
   of angles, 25
   of biangles, 203
   of saccheri quadrilaterals, 165, 218
   of segments, 25
   of triangles, 25, 36, 52, 53, 92, 210, 236, 272
   of triangles, angle-angle-angle (AAA), 210, 236
   of triangles, angle-angle-side (AAS), 53, 225, 257
   of triangles, angle-side-angle (ASA), 6, 53, 232, 257
   of triangles, hypotenuse-side, 93
   of triangles, side-angle-side (SAS), 36, 231, 257, 267, 288, 436
   of triangles, side-side-angle-plus (SSA+), 93
   of triangles, side-side-side (SSS), 42, 52, 232, 257
congruence postulate, 25, 60, 230, 243, 359, 388, 435
conjecture, 420
consistency, 197
construction, angle bisector, 43, 109, 130
   copy angle, 52, 110
   drop perpendicular, 49, 109, 130, 362
   erect perpendicular, 44, 109, 130
   midpoint, 43, 109, 129
   parallel, 56, 110
   perpendicular bisector, 43, 109, 129
   straightedge and compass, 8, 109, 264
   tangent, 84
constructability, with compass alone, 124
   with linkage, 123, 153
   with origami, 122, 129, 134, 137
   with straightedge, 129
   with straightedge and compass, 14, 110, 114, 116, 120, 153

continuity axiom, 18, 31, 73
contradiction, 422
contraposition, 422, 424
converse, 423
convexity, 18, 24, 49, 370
Conway criterion, 339
Conway, John H., 297, 319, 339, 348
corollary, 420
cosines, law of, 108
Coxeter, H. S. M., 343, 393, 417
cross-bar theorem, 22
cross-ratio, 106, 193
crystallography, 294, 297, 308, 418
crystallographic restriction, 307
curvature, 187, 241, 253
cycle, 141
cyclic group, 293
cyclic quadrilateral, 83, 98
cylinder, 390

da Vinci, Leonardo, 69, 293
decomposition, of polygon, 246
   of tiling, 352
Dedekind's axiom, 79
defect, 249
deflation, 353
degree, 34
Dehn, Max, 388
deltahedron, 375
DeMorgan's laws, 422
Descartes, René, 76, 144, 157, 254
Descartes' formula, 387
De Sua, Frank, 198
Dewdney, A. K., 418
diameter, 28, 80
dihedral angle, 364
dihedral group, 292
distance, 76, 254
divergent parallel, 185, 208
Dodgson, Charles, 14, 78
duality, 325, 374
Dudley, Underwood, 14, 124, 153
duplication of the cube, 113, 138, 153

Egypt, 3, 67
Einstein, Albert, 11, 158, 199
elliptic geometry, 226, 252, 418
   projective plane model, 226
   Riemann disc model, 236
elliptic postulate, 231, 260, 434
empire, 351
enveloping line, 213
equivalence relation, 19, 25, 89, 247, 263, 420
Eratosthenes, 113
Erlanger Program, 242, 268
Escher, M. C., 195, 320, 329, 332, 334, 337, 343, 355
Euclid, 9, 198, 223

Euclid, *The Elements*, 6, 9, 10, 15, 28, 34, 39, 64, 67, 80, 119, 157, 169, 177, 178, 244, 267, 360, 366, 370, 373
Eudoxus, 8, 244, 387
Euler, Leonhard, 67, 99, 376
euler characteristic, 376
Euler line, 99
Euler's formula, 376
exhaustion, method of, 8, 244, 387
extension theorem, 346
exterior angle theorem, 58, 160, 204, 225

Fano, Gino, 258, 261
Fedorov, Yevgraf, 294
Fermat, Pierre, 121
Fermat prime, 120
Feuerbach, Karl, 100
Fibonacci number, 153
field, 113
finite geometry, 258
   affine, 262
   Fano's 261
   model, 259, 264
   projective, 263
   Young's, 262
fixed point, 282
frieze pattern, 295, 297
Fuller, Buckminster, 395

Galilei, Galileo, 157, 388
Gardner, Martin, 348
Garfield, James, 70
Gauss, Carl F., 114, 120, 176, 177, 178, 197, 218, 240
Gauss' theorem, 120, 153
Gauss-Bonnet theorem, 252
geodesic, 241
geometric algebra, 24
geometry, analytic, 76, 254
   caliper, 75
   differential, 241
   elliptic, 226
   euclidean, 15, 56, 80, 99
   Fano's, 261
   finite, 258
   four-dimensional, 412
   hyperbolic, 181
   neutral, 46, 57, 99, 159, 219
   noneuclidean, 46, 57, 99, 181, 223
   projective, 242
   spherical, 220
   taxicab, 255
   three-dimensional, 357
   transformation, 267
   Young's, 262
Gerbert, 159
Geretschläger, Robert, 129, 139
ghosts, 414
Girard, Albert, 253
Gleason, Andrew, 139

glide reflection, 269, 276, 281, 285, 295, 302, 315, 333, 398, 401
Godel's Incompleteness theorem, 198
golden ratio, 118, 135, 153, 348
great circle, 220
group, 291, 319, 407, 418
   cyclic, 293
   dihedral, 292
   rosette, 292, 406
Grünbaum, Branko, 345, 348, 370

half-plane, 359
half-space, 359
Hartshorne, Robin, 107, 407
Hatori, Koshiro, 129, 136
Henderson, David, 186, 221
Hermes, Johann, 121
Heron of Alexandria, 243
hexagon, 66, 116, 135, 139, 244
Hilbert, David, 15, 60, 71, 188, 247, 251, 264, 388, 429
Hilton, Peter, 417
hinge theorem, 53
Hinton, C. H., 412
Hippocrates of Chios, 7
Hobbes, Thomas, 10, 157
horolation, 315
Hume, David, 157
Huzita, Humiaki, 129, 136, 153
Huzita-Hatori axioms, 129, 136
hyperbolic plane, 181, 418
   Beltrami-Klein model, 188
   isometry, 314, 318
   physical models, 183, 186, 197, 219
   Poincare disc model, 190
   tiling of, 342
   upper half-plane model, 219
hyperbolic postulate, 181, 260, 434
hypercube, 412, 415
hyperspace, 412
hypertetrahedron, 417
hypotenuse, 64, 93

ideal point, 191
identity, 282
incenter, 97
incidence postulate, 16, 72, 230, 259, 357, 434
induction, 426
inflation, 353
inscribe, 81, 97
invariant, 278
inverse, 423
inversion in circle, 101, 146, 192, 238
irrationality, 6, 11
Islamic lattice pattern, 319
isometry, 268, 271, 314, 397
   classification, 288, 318, 405, 406
   composition, 279
   direct, 279
   even, 288

INDEX • 457

isometry, (continued)
  fixed point, 282
  glide reflection, 269, 276, 281, 285, 295, 302, 315, 333, 398, 401
  horolation, 315
  hyperbolic, 314, 318
  identity, 282
  indirect, 279
  odd, 288
  orientation, 278
  postulate, 288
  reflection, 269, 275, 283, 295, 302, 314, 398, 407
  rotary reflection, 399, 401
  rotation, 268, 274, 281, 285, 295, 302, 316, 334, 398, 401
  screw rotation, 400, 403
  spherical, 405
  symmetry, 289, 406
  translation, 269, 272, 280, 285, 295, 302, 314, 330, 397, 401

Justin, Jacques, 137

kaleidoscope, 319, 409
Kant, Immanuel, 7, 157, 178, 197
Kapovich, Michael, 153
kata, 412
Kay, David, 105
Kempe, Alfred, B., 147, 150, 153
Kempe's Universality theorem, 153
Kepler, Johannes, 8, 370, 374, 382, 388
Khayyam, Omar, 159, 161, 177
Khayyam's theorem, 161
King, Henry, 153
Klein, Felix, 188, 226, 242, 268
knot, 139, 143

Lambert, Johann, 159, 168, 177
Lang, Robert, 137
Legendre, Adrien, 70, 159, 160, 177
Leibniz, Gottfried, 84, 157, 159
lemma, 420
Leonardo's theorem, 293
length, 25, 222, 234
  constructible, 110
line, 15, 221
  completeness of, 73, 184
  enveloping, 213
  Euler, 99
  of reflection, 275
  separation, 20, 222
  skew, 360
linkage, 123, 144
  pantograph, 148
  Peaucellier's, 146
  translator, 150
  variable-based triangle, 147
  Watt's, 145
Lipkin, Lippman, 147
Lobachevskii, Nikolai, 180, 218

Locke, John, 157
locus, 144
logic, 419
Loomis, Elisha, 68
Lorenz, Johann, 172
Luecking, Stephen, 3
lune, 225, 239, 252

Martin, George, 114, 116, 129, 153, 407
Mascheroni, Lorenzo, 124
median, 95
Menelaus of Alexandria, 107, 223
Menelaus' theorem, 107
Mercator projection, 189
Messer, Peter, 137
metric, 78, 241, 254
  euclidean, 76
  hyperbolic, 193, 219
  taxicab, 255
midline, 163, 212
midpoint, 26, 43, 76, 129
Millson, John, 153
mirror, 275, 409
Möbius transformations, 108
model, Beltrami-Klein, 188
  finite geometry, 259, 264
  hyperbolic plane, 183, 186, 197
  Poincare disc, 190, 206, 219
  projective plane, 226
  Riemann disc, 236
  upper half-plane, 219
modus ponens, 422
Mohr, Georg, 125
Mohr-Mascheroni theorem, 125
Moise, Edwin, 114
Morley's theorem, 108

negation, 422
net, 375
Newton, Isaac, 84, 157, 177
Niggli, Paul, 295
nine-point circle, 100

Occam's razor, 419
octagon, 66, 117, 135, 139
orientation, 278
origami, 122, 129, 417
orthocenter, 98, 99
orthoscheme, 409, 418

pantograph, 148
parallel, 19, 37, 54, 61, 73, 77, 89, 95, 181, 272, 360
  asymptotic, 185, 199
  construction, 56, 110
  definition, 37
  divergent, 185, 208
parallel postulate, 38, 56, 158, 169, 197, 260, 434
parallelepiped, 388
parallelogram, 38, 59, 77, 91
  area, 64

Pargeter, A. R., 417
Pasch, Moritz, 21, 71
Pasch's postulate, 79, 230
Pasch's theorem, 21, 79, 212
Peaucellier, Charles-Nicholas, 146
Peaucellier's linkage, 146, 153
Penrose, Roger, 348, 356
pentagon, 117, 136, 139
pentagram, 143
pentahedroid, 417
periodicity, 346
perpendicular bisector, 43, 97, 109, 129
perpendicularity, 34, 44, 49, 77, 130, 209, 360, 362
Petrie, J. F., 393
pi (π), 4, 121, 153, 244
Pieri, Mario, 71
Pierpont prime, 139
plane, 15, 357
  parallel, 360
plane separation, 18, 79, 221, 230, 358, 435
Plato, 5, 7, 113, 144, 157, 373
Platonic solids, 8, 373, 408, 417
Playfair, John, 57, 169
Playfair's postulate, 57, 73, 169, 260
Plutarch, 144, 392
Poincaré, Henri, 158, 190, 198, 219, 429
Poincaré disc model, 190, 206
point, 11, 15, 357
  ideal, 191
polar, 233
polar distance, 232
pole, 222, 232
Polya, George, 295, 329
polygon, 38, 320, 369
  angle sum, 59
  constructible, 116, 120, 139
  decomposition, 246
  star, 142
polyhedron, 8, 342, 370, 417
  deltahedron, 375
  dual, 374
  infinite, 393, 418
  regular, 372, 378, 408
  semiregular, 380
  symmetry of, 406
  tetrahelix, 395, 400
  truncated, 382
Poncelet, Jean-Victor, 129
Poncelet-Steiner theorem, 129
Poseidonios, 158, 170
postulate, 12, 15, 419, 434
  Archimedes' axiom, 26, 73, 79, 160, 173, 231, 245, 435
  area, 63, 243, 436
  betweenness, 16, 20, 72, 228, 435
  Cavalieri's principle, 388, 436
  circular continuity, 31, 48, 224, 231, 436
  compass, 27, 109, 230

postulate (*continued*)
  congruence, 25, 60, 230, 243, 359, 388, 435
  elliptic, 231, 260, 434
  hyperbolic, 181, 260, 434
  incidence, 16, 72, 230, 259, 357, 434
  isometry, 288
  parallel, 38, 56, 158, 169, 197, 260, 434
  Pasch's, 79, 230
  plane separation, 18, 79, 221, 230, 358, 435
  Playfair's, 57, 73, 169, 260
  right angle, 34, 36, 73, 230, 434
  separation, 228, 435
  side-angle-side (SAS), 36, 231, 257, 288, 436
  straightedge, 15, 18, 109, 221, 228, 434
  volume, 388, 436
primitive term, 12, 15, 16, 25, 72, 223, 258, 419
prism, 372, 389
proof, 5, 424
Proclus, 9, 29, 158, 171, 176, 177
projective plane, 226
proportions, 8, 88
pseudosphere, 186
Ptolemy's theorem, 107
pyramid, 371, 389
Pythagoras of Samos, 6, 108
Pythagorean spiral, 112
Pythagorean Theorem, 4, 10, 64, 78, 174, 197
Pythagorean triple, 66

quadrilateral, 38, 42
  angle sum, 58, 161, 182
  cyclic, 83, 98
  saccheri, see saccheri quadrilateral
  similarity of, 92
quantifier, 423
quasicrystal, 418
Q.E.D., 29, 429
Q.E.F., 29, 429
Queneau, Raymond, 74

Rabinowitz, Stanley, 137
radius, 27, 80
ray, 17
rectangle, 39, 61, 168, 182, 247
reflection, 269, 275, 283, 295, 302, 314, 398, 407
regularity, 320, 372
reptile, 328, 346
revolution, 142
Rhind papyrus, 4
rhombus, 39, 61, 243
Richelot, Friedrich, 121
Riemann, Georg, 240
Riemann disc model, 236
right angle postulate, 34, 36, 73, 230, 434
rigidity, 42, 150, 271, 282

rosette group, 292, 406
rotary reflection, 399, 401
rotation, 268, 274, 281, 285, 295, 302, 316, 334, 398, 401
Rousseau, Jean-Jacques, 157
Row, Sundara, 129
Russell, Bertrand, 11
Ryan, Matthew, 197

Saccheri, Fr. Girolamo, 159, 172, 176, 246
Saccheri-Legendre theorem, 160
saccheri quadrilateral, 162, 182, 194, 211, 225, 247
  congruence of, 165, 218
  definition, 162
  midline, 163
  summit, 162
  summit angle, 162, 182, 232, 235
Schattschneider, Doris, 330
Schlegel diagram, 407, 415
Schweikart, Karl, 178
screw rotation, 400, 403
secant line, 84
segment, 17, 223, 229
  bisector, 26, 43, 232
  congruence of, 25
  perpendicular bisector, 43, 232
separation, 223, 228, 435
  line, 20, 221
  plane, 18, 79, 221, 435
  space, 358
Servatius, Brigitte, 319
Shephard, G. C., 346, 348
similarity, 88, 102, 175, 178, 182, 236
skew, 360, 365
Socrates, 5
software, geometric, 28, 195, 221, 239, 278, 314, 342, 344, 356, 429, 431
space separation, 358
sphere, 220
  area, 253
  tiling of, 341
  volume, 391
square, 39, 45, 63, 134
squaring the circle, 121
star polygon, 142
Steiner, Jacob, 129
stereographic projection, 190, 237, 264
Stobaeus, 10
straightedge, 15, 18, 109, 129, 147
straightedge postulate, 15, 18, 109, 221, 228, 434
subtend, 35, 47, 80
superposition, 36, 42, 267
surveying, 3
syllogism, 422
Sylvester, James J., 147
symmetry, 289, 406, 418

Taimina, Daina, 186, 221
tangency, 32, 84

tangent, construction of, 84, 85
taxicab geometry, 255
tetrahelix, 395, 400
tesselation, see tiling
tesseract, 414
Thales, 6, 81, 268
Thales' theorem, 81, 237
Theon of Smyrna, 113
Theudius, 8
Thomson, William (Lord Kelvin), 147
Thurston, William, 186
tiling, 320, 396
  aperiodic, 348
  Archimedean, 321
  classification of, 325
  composition, 352
  decomposition, 352
  deflation, 353
  dual, 325
  hyperbolic, 342
  inflation, 353
  irregular, 327
  nonperiodic, 348
  Penrose, 348, 356, 418
  periodic, 346
  regular, 320
  semiregular, 321, 396
  spherical, 341
translation, 269, 272, 280, 285, 295, 302, 314, 330, 397, 401
transversal, 37, 54
transitivity, 57, 201, 420
trapezoid, 39, 64
triangle, acute, 38
  altitude of, 98
  angle-angle similarity (AA), 92
  angle-angle-angle congruence (AAA), 210, 236
  angle-angle-angle similarity (AAA), 91
  angle-angle-side congruence (AAS), 53, 225, 257
  angle-side-angle congruence (ASA), 53, 232, 257
  angle sum, 47, 58, 160, 168, 182, 195, 225, 235, 240, 249
  area, 64
  associated saccheri quadrilateral, 163
  centroid of, 95
  circumcenter of, 97
  congruence of, 6, 25, 92, 272
  definition, 17, 369
  doubly asymptotic, 213
  equilateral, 28, 66, 134
  exterior of, 20
  golden, 117
  hypotenuse-side congruence, 93
  ideal, 217, 345
  incenter of, 97
  interior of, 20
  isosceles, 6, 39, 232, 257

triangle (*continued*)
  median of, 95
  obtuse, 38
  orthocenter of, 98
  right, 38, 64, 81, 93, 235
  scalene, 38
  side-angle-side congruence (SAS), 36, 231, 257, 267, 288, 436
  side-angle-side similarity (SAS), 92
  side-side-angle-plus congruence (SSA+), 93
  side-side-side congruence (SSS), 42, 52, 232, 257
  side-side-side similarity (SSS), 92
  similarity of, 88, 102, 175, 178, 182, 236

triangle (*continued*)
  singly asymptotic, 211
  triply asymptotic, 217
triangle inequality, 48, 78, 219, 220, 254, 368
trigonometry, 108
trisection, 115, 121
Trudeau, Richard, 11, 170
truncation, 382

van Hiele, Pierre, 431
van Hiele-Geldof, Dina, 431
Varignon's theorem, 107
vertical angle theorem, 46, 232
Vitale, Giordano, 174

volume, 387
volume postulate, 388, 436

Wachter, Friedrich, 178
Wallis, John, 159, 175
wallpaper, 295, 301, 312
Wantzel, Pierre, 114, 120
Watt, James, 145
weak exterior angle theorem, 46, 160, 225
Weeks, Jeffrey, 186, 342

Yates, Robert, 153
Young, John, 262

Zeno, 9, 12